Healthier Meat Products

Ashok Kumar Pathera • Harsh Kumar
Sanjay Yadav
Editors

Healthier Meat Products

Springer

Editors
Ashok Kumar Pathera
Amity Institute of Food Technology
Amity University
Noida, Uttar Pradesh, India

Harsh Kumar
Centre of Advanced Technologies
Faculty of Science, University of Hradec Králové
Hradec Králové, Czech Republic

Sanjay Yadav
Department of Livestock Products Technology
Lala Lajpat Rai University of Veterinary and Animal Sciences
Hisar, Haryana, India

ISBN 978-3-031-78214-5 ISBN 978-3-031-78215-2 (eBook)
https://doi.org/10.1007/978-3-031-78215-2

This Springer imprint is published by the registered company Springer Nature Switzerland AG
The registered company address is: Gewerbestrasse 11, 6330 Cham, Switzerland

Preface

The increasing demand for healthier food choices has significantly influenced the global food industry, necessitating the development of innovative products for healthier eating. Meat, as a rich source of high-quality proteins, essential fatty acids, vitamins, and minerals, plays an integral role in human nutrition. However, processed meat products are often associated with concerns related to high levels of saturated fat, cholesterol, sodium, and certain chemical additives, which can contribute to health risks. In addressing these issues, the development of "healthier meat products" has emerged as a key area of research and innovation.

This book, "Healthier Meat Products," intends to provide an in-depth review of strategies, technologies, and practices aimed at improving the quality of meat products while maintaining nutritional profile and consumer acceptance.

The book begins by discussing consumer attitudes and consumption patterns to understand the status and the growing preference for healthier meat products. This book comprehensively addresses the development of meat products enriched with bioactive compounds, dietary fibers, antioxidants, and functional ingredients, alongside efforts to reduce undesirable components like saturated fats, cholesterol, salt, and nitrites. The development of meat products with specific functional ingredients, such as those enriched with micronutrients, encapsulated bioactive, soy, and modified fatty acid profiles, as well as fermented, marinated, and ethnic meat products, is covered in several chapters of this book.

This book aims to serve as a valuable resource for researchers, industry professionals, and students engaged in food science, nutrition, and product development. We extend our heartfelt gratitude to the contributors whose expertise and dedication have enriched this book, ensuring comprehensive coverage of the subject matter. We hope this book will inspire future research and innovation in the field of healthier meat products, contributing to the creation of health-conscious food systems.

Noida, Uttar Pradesh, India — Ashok Kumar Pathera
Hradec Králové, Czech Republic — Harsh Kumar
Hisar, Haryana, India — Sanjay Yadav

Contents

Consumer's Attitudes and Consumption Patterns of Meat Products

Sao-Mai Dam, Hoang-Duy Truong, Duyen-Anh Le, Tuan-Anh D. Le, and Duc-Vuong Nguyen

1 Introduction

In the continuously evolving landscape of the global food industry, there is an increasing recognition of the need to thoroughly investigate consumer attitudes and consumption patterns related to meat products. The subtleties within these preferences result from many factors, forming an intricate web that influences individuals' decisions regarding meat consumption. These considerations span beyond the fundamental aspects of culture, including health, ethics, the environment, lifestyle choices, and demographic characteristics.

Cultural dimensions are crucial in shaping how individuals perceive and interact with meat. The impact of cultural norms, intricately woven into culinary practices, significantly shapes the dynamics of meat consumption. Boer et al. (2014), Garnett (2011), Oude Ophuis and Van Trijp (1995), and Rozin et al. (2012) highlight that consumer attitudes toward meat products encapsulate a diverse array of emotions. Importantly, these attitudes towards are not fixed. Instead, they undergo dynamic evolution, influenced by the intricate interplay of psychological, social, and cultural factors [1–4].

Moreover, the meat industry is confronted with a range of challenges and opportunities stemming from the escalating interest in plant-based alternatives and the growing prevalence of vegetarian diets. While these shifts challenge traditional meat producers, they simultaneously open avenues for ground-breaking innovation. This innovation extends beyond mere product diversification, incorporating the development of sustainably sourced and ethically produced meat products-a testament to the industry's responsiveness to evolving consumer values and preferences [5].

S.-M. Dam (✉) · H.-D. Truong · D.-A. Le · T.-A. D. Le · D.-V. Nguyen
Industrial University of Ho Chi Minh City, Ho Chi Minh City, Vietnam
e-mail: damsaomai@iuh.edu.vn

A. Kumar Pathera et al. (eds.), *Healthier Meat Products*,
https://doi.org/10.1007/978-3-031-78215-2_1

A profound understanding of the consumption patterns of meat products is imperative for effectively navigating market trends. It furnishes businesses with valuable insights into current preferences and equips them to tailor their products in harmony with the ever-changing demands of consumers. The comprehensive research conducted by Verbeke et al. (2010) underscores the pivotal role of understanding these patterns for businesses seeking strategic alignment with market dynamics [6].

As we delve into the scrutiny of consumer attitudes and consumption patterns related to meat products, it becomes evident that individuals' choices transcend mere nutritional considerations. Instead, these choices are intricately entwined with complex threads of cultural heritage, ethical considerations, health consciousness, and environmental awareness. Deciphering this elaborate tapestry of factors is essential for businesses to endure and flourish in an industry where consumer preferences stand as pivotal drivers of change.

2 Consumer Attitudes Towards Meat

2.1 Definition and Importance of Consumer Attitudes

Perception, a multifaceted cognitive process encompassing the selection, organization, and interpretation of external information, is a fundamental mechanism crucial for comprehending the intricacies of our surroundings [7, 8]. This intricate process is not a standalone function but rather an amalgamation of cognitive elements influenced by affective and conative dimensions, collectively playing a pivotal role in shaping decisions related to products or services [9]. When consumers embark on the cognitive processing of information concerning food, they traverse three distinct phases: exposure, attention, and interpretation. The exposure phase marks the reception of sensory stimuli, primarily through vision or audition, initiating the intricate cognitive mechanism. Subsequently, attention comes into play, selectively filtering stimuli based on internal states such as goals or expectations. Finally, the interpretation phase represents the culmination of higher mental processes involving semantic processing that tap into personal knowledge and experiences [10]. In essence, this cyclical journey through exposure, attention, and interpretation elucidates the dynamic nature of how individuals engage with and make sense of information about the food they encounter. This nuanced process of perception is indispensable, as it not only forms the bedrock of cognitive understanding but also profoundly influences consumers' subsequent choices and behaviors.

Perception, as a psychological factor mediated by attitude, strongly influences behavior [6, 8, 10]. Perceptions of meat product characteristics, production methods, and distribution trends influence consumer attitude formation. Factors like eliminating unhealthy ingredients, animal welfare considerations, and processing technology impact perception and attitude formation. Although eliminating

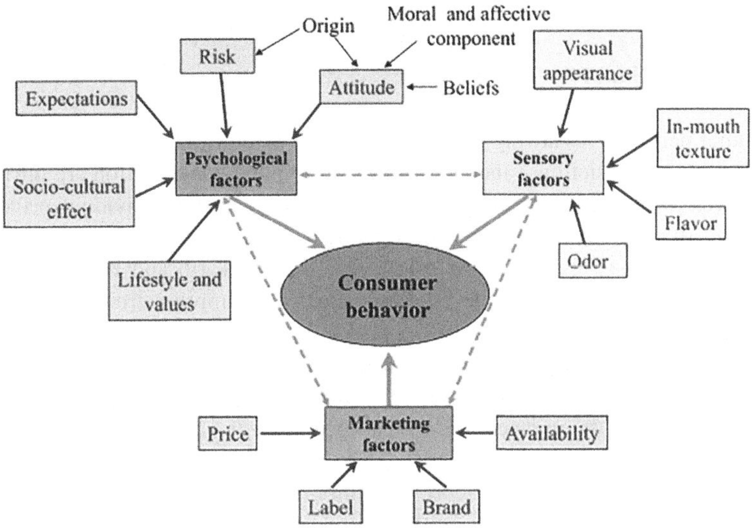

Fig. 1 Multidisciplinary model of the main factors affecting consumer behavior in the food domain [11]

unhealthy ingredients may improve general product perception, it can complicate attitude formation by affecting sensory properties. Animal welfare considerations, while valued, may sometimes be overshadowed by sensory attributes [6, 8, 9].

Maria Font-i-Furnols and Guerrero (2014) highlight that consumers, the final link in the production chain, play a crucial role in determining satisfaction and influencing shopping habits. Meeting their expectations is thus an essential aspect of consumer satisfaction. Examining the various factors at play is critical to understanding better what affects consumer behavior. The factors that influence consumer behavior are psychological (about individual factors), sensory (related to specific product attributes), and marketing (associated with environmental factors), as depicted in Fig. 1 [11].

Consumer attitudes towards meat are influenced by both inherent and external factors throughout various stages, encompassing the receipt of information and the manipulation of products [12]. Recognizing the profound impact of consumer attitudes on the demand for meat products, as illustrated by Verbeke et al. (2010), is imperative for the meat industry. The study emphasizes the interconnectedness of consumer preferences and market dynamics. Furthermore, Garnett (2011), Grunert et al. (2015), Grunert and Aachmann (2016), and Oude Ophuis and Van Trijp (1995) stress the pivotal role of perceived quality in consumer decision-making, emphasizing that attitudes towards meat products significantly shape their perceived value. In a market where consumer choices drive trends, businesses must intricately understand attitudes towards meat to align with evolving market dynamics strategically [3, 4, 13, 14].

2.2 *Factors Effect on Consumer Attitudes*

2.2.1 Cultural and Societal Factors

Consumer perspectives are significantly impacted by cultural and societal dynamics [6, 15]. Views, convictions, and actions related to meat consumption are intricately interwoven with cultural practices, societal standards, and individual principles. Unveiling these elements is essential for uncovering the core motivators steering consumer decisions about meat consumption. Beardsworth and Bryman's (2004) study provides insights into the evolving meat consumption trends, particularly among younger demographics. Their research delineates the positive sway of societal elements on attitudes and behaviors regarding meat, showcasing the dynamic influence of communal factors on dietary choices [15]. Furthermore, Verbeke et al. (2010) delved into the perspectives of European citizens and consumers, shedding light on the complex interplay between cultural backgrounds and preferences for beef and pork. The study underscores the importance of cultural diversity in influencing decisions related to meat, with cultural aspects serving as pivotal drivers of consumer attitudes [6].

This understanding is vital for businesses, policymakers, and researchers as they navigate the ever-evolving food landscape within diverse cultural and social milieus. Cultural and social impacts are pivotal in moulding consumer stances towards meat and influencing preferences and decisions nuancedly. The complex dynamics of meat consumption trends are intricately linked to cultural customs, societal norms, and individual values, underscoring the necessity of a comprehensive understanding of these factors when consumers make intricate decisions about their dietary habits.

2.2.2 Health and Nutrition

The dynamic connection between health and meat consumption is influenced by various factors, including individual health awareness, dietary patterns, and cultural transformations [16, 17]. The trend towards health consciousness, driven by increased awareness of health implications and improved access to nutritional information, guides consumers in making informed dietary decisions. Research highlights a shift towards leaner meat cuts and alternative protein sources [18]. Emerging dietary trends, particularly plant-based diets, emphasize the intersection of health and meat consumption, motivated by ethical, environmental, and perceived health benefits [19]. Cultural and social factors play a crucial role in shaping health-oriented consumer attitudes, with movements like "clean eating" prompting individuals to reconsider meat consumption for overall well-being [20]. These evolving attitudes present challenges and opportunities for the meat industry. As the demand for healthier options grows, the industry must adapt to meet consumer expectations [21]. Innovations, including plant-based and lab-grown products, signify an industry responding to the changing landscape [22]. Initiatives focused on sustainability and ethical sourcing resonate with consumers prioritizing health and considering broader environmental and social impacts [23, 24]. The modern consumer, increasingly mindful of health, is reassessing dietary habits, including meat consumption. Vartanian et al. (2007) research underscores a growing awareness of health implications related to excessive red and processed meat intake [20]. Health-conscious individuals now tend to make dietary choices aligned with long-term well-being, often preferring leaner meat cuts and exploring alternative protein sources. Access to nutritional information has emerged as a crucial factor shaping meat consumer attitudes. Studies indicate that individuals are more likely to consider the nutritional content of meat products when making purchasing decisions [25]. Transparent and comprehensive labelling on packaging and in restaurants empowers consumers to make informed choices based on their specific health objectives, whether centred on protein intake, fat content, or other nutritional considerations. Consumer perceptions of meat consumption are closely tied to health and nutritional considerations, representing an evolving recognition of the significant influence dietary choices can exert on overall well-being.

2.2.3 Ethical and Animal welfare Concerns

Growing consumer ethical awareness influences meat consumption trends [22, 26]. Consumers now prioritize ethical considerations in their food choices, weighing the impact on animals, the environment, and meat industry worker welfare. De Boer et al. (2007) highlight the importance of ethics in shaping preferences for sustainably and ethically sourced meat. Increasing concern for animal treatment during meat production suggests that humane farming and slaughter practices affect purchasing decisions, aligning with broader societal trends favoring ethical consumerism [26, 27].

Producers adapt by prioritizing animal welfare through spacious living conditions, avoiding antibiotics, and humane slaughter methods [28, 29]. The rise of alternative protein sources like plant-based and lab-grown meat addresses consumer worries about ethical issues tied to conventional meat production [22].

Ethical considerations and animal welfare concerns significantly shape consumer attitudes towards meat [30, 31]. The preference for ethically sourced meats mirrors a broader societal awareness of how dietary choices impact animals and the environment. The meat industry is pressured to adopt more sustainable and compassionate practices, driven by increasing transparency in production and a rising demand for ethically produced goods.

2.2.4 Environmental Awareness

Environmental impacts significantly shape consumer attitudes towards meat consumption. Consumers are re-evaluating their meat consumption patterns in the face of increasing global environmental worries, considering the associated ecological implications [32].

The shifting landscape of environmental awareness is influencing consumer perspectives on meat consumption. The substantial contribution of the meat industry to carbon emissions and deforestation urges consumers to scrutinize the carbon footprint tied to their meat preferences. Research by Tilman and Clark (2014) underscores the environmental benefits of plant-based diets in mitigating greenhouse gas emissions and land use. This trend towards sustainability reflects consumers' inclination to opt for eco-friendly choices, considering the broader ecological implications of their dietary selections [32].

Certifications and labels emphasizing sustainability, such as organic, grass-fed, or ethically sourced, play a pivotal role in influencing consumer decisions by providing insights into environmentally conscious practices in meat production [33]. Consumers prioritizing environmental sustainability are more likely to choose products with these certifications. Media coverage and environmental advocacy contribute to heightened awareness regarding the environmental impact of meat consumption, fostering a transition towards sustainable alternatives. This increased consciousness compels the meat industry to reevaluate production methods and adopt sustainable farming practices, including regenerative agriculture. The emergence of plant-based and lab-grown alternatives represents a strategic response to the environmental challenges associated with conventional meat production.

The burgeoning environmental consciousness among consumers significantly shapes attitudes towards meat consumption. Acknowledging the ecological repercussions of large-scale meat production has spurred a shift in consumer preferences towards sustainable and environmentally friendly alternatives. As consumers strive to align their dietary choices with environmental values, the meat industry faces pressure to embrace more eco-friendly practices, contributing to a more sustainable future for food production [34].

2.3 Measurement of Consumer Attitudes Towards Meat

Assessing consumer attitudes towards meat is an intricate endeavour that requires an examination of diverse psychological, social, and cultural factors impacting individual preferences and choices concerning meat consumption. Grasping these attitudes is imperative for businesses, policymakers, and researchers to address evolving consumer behaviors and market demands adeptly. Some standard methods used to measure consumers' attitudes towards meat include surveys and questionnaires, focus groups and qualitative research, experimental designs, social media and online platform analysis, and longitudinal studies.

Surveys and questionnaires are prevalent for understanding consumer attitudes towards meat, offering valuable insights into various dimensions of meat consumption. These surveys cover topics ranging from frequency of consumption to reasons for choosing specific meat types, exploring attitudes towards health, environmental impact, and animal welfare. They provide quantitative data for analyzing consumer sentiments and identifying patterns and trends in behavior, such as the influence of health or environmental concerns on meat choices [35]. In addition, focus groups and qualitative research provide a comprehensive understanding of consumer attitudes, delving into perceptions, emotions, and motivations. These methods uncover nuanced aspects of attitudes that surveys might miss, particularly revealing social influences on meat attitudes. Experimental designs, such as taste tests and variable manipulation, offer insights into immediate reactions and preferences, addressing sensory aspects and varying impacts on consumer attitudes [36–38].

In conclusion, measuring consumer attitudes towards meat involves a multifaceted approach combining quantitative and qualitative methods. Surveys, focus groups, experimental designs, neuroscientific approaches, and online analysis contributed to a comprehensive understanding of the diverse factors influencing consumer attitudes. Combining these methods can offer a nuanced and holistic perspective, enabling businesses and policymakers to respond effectively to the meat industry's dynamic landscape of consumer preferences.

3 Meat Consumption Patterns

3.1 Historical Trends in Meat Consumption

In the early stages of history, hunting played a crucial role in shaping meat consumption. Subsequently, the transition from nomadic lifestyles to settled agricultural communities marked a significant shift, accompanied by the domestication of animals, providing a primary source of protein and increasing meat consumption [39]. Throughout history, the motivations behind human meat consumption have undergone substantial changes influenced by cultural, economic, and technological factors [40–42].

Advanced preservation techniques such as refrigeration, coupled with developments in transportation in the nineteenth century, brought about substantial transformations. These technological advancements allowed for mass production and distribution of meat, and canning and preservation methods expanded the accessibility of meat to urban populations, resulting in increased consumption, particularly in industrialized nations [43–45].

Post-World War II, changes in dietary habits and economic development contributed to a surge in meat consumption, solidifying its status as a primary food in Western cultures. Additionally, the proliferation of fast food in the latter half of the twentieth century further intensified the demand for meat-based dishes [46].

However, by the late twentieth century, scientific studies linking excessive meat consumption, especially red and processed meat, to health issues such as cardiovascular diseases and certain cancers prompted changes in dietary recommendations. Calls for dietary diversity and reduced meat consumption garnered significant attention [47].

Furthermore, the twenty-first century has witnessed increased awareness of the ethical and environmental dimensions of meat production. Concerns regarding deforestation, greenhouse gas emissions, and animal welfare have prompted individuals to reconsider their meat consumption. The rise of vegetarianism and veganism reflects a societal shift towards more sustainable and ethically conscious dietary choices [33]. Alongside this, there is a growing interest in alternative protein sources, leading to the introduction of plant-based meat substitutes, lab-grown meat, and insect protein products as alternatives to traditional livestock farming [48].

The paradigm of meat consumption has evolved from early agricultural societies to the contemporary era, adapting to societal changes. Understanding these transformations is crucial for addressing challenges related to health, sustainability, and ethical considerations in meat consumption.

3.2 Regional Variations in Meat Consumption

Cultural, economic, and environmental factors contribute to diverse dietary habits and patterns of dietary restrictions in various regions worldwide. In countries with a longstanding tradition of livestock farming, such as those in Central Asia and the Middle East, dishes traditionally feature predominantly sheep or goat meat [49, 50]. Conversely, regions with a history of vegetarianism, such as parts of India, may exhibit lower meat consumption rates. This underscores the significant role that cultural traditions and dietary preferences play in shaping regional differences in meat consumption.

Additionally, economic factors significantly contribute to disparities in meat consumption between regions. Wealthier regions consume more meat due to increased purchasing power and affordability. For example, in Western countries, higher income correlates with higher average per capita meat consumption [51]. In contrast, regions with lower income may face limitations in accessing meat. Figures 2 and 3 present data on aggregate meat consumption (in million metric tons) across various regions and worldwide. Nations characterized by lower economic status or influenced by cultural preferences exhibit reduced levels of meat consumption in comparison to their counterparts.

Furthermore, the availability of natural resources influences meat consumption patterns in regions. Areas with abundant resources for livestock farming may exhibit higher meat consumption rates. Conversely, regions facing resource constraints may rely more on plant-based diets. The impact of available resources on meat

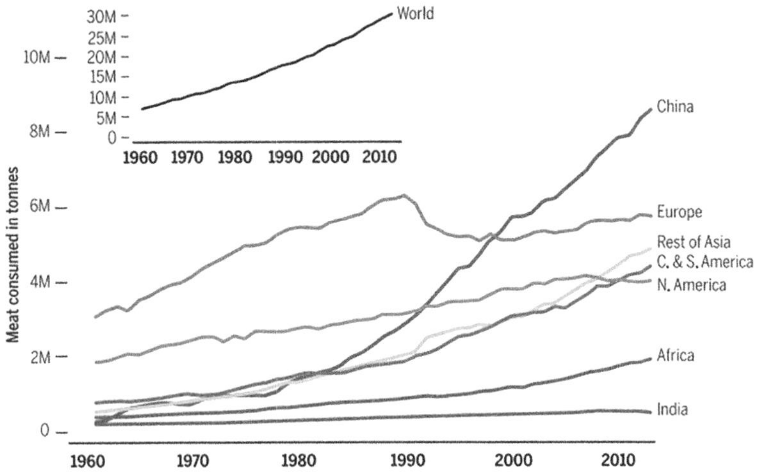

Fig. 2 Total consumption of meat (in million metric tons) in different regions and (inset) globally [52]

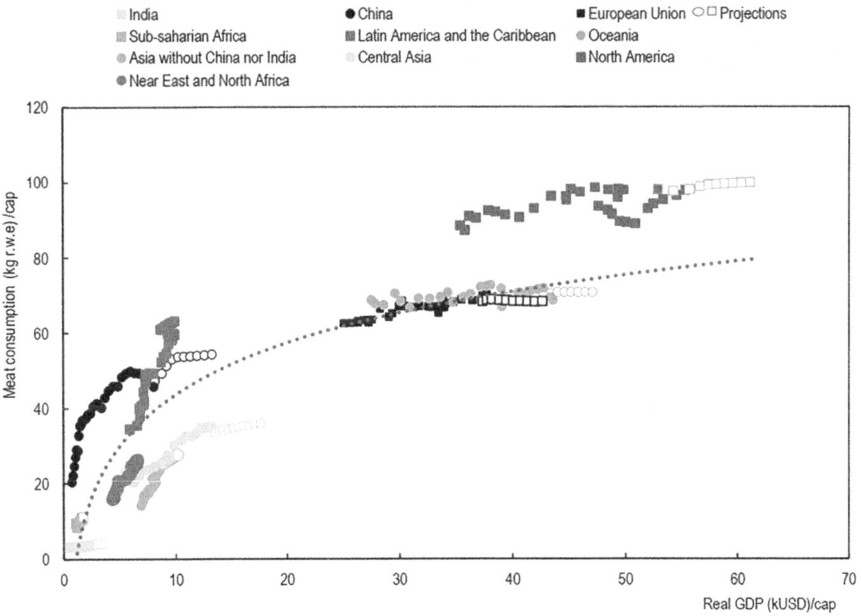

Fig. 3 Income impact on per capita meat consumption per region, 1999–2030 [53]

consumption is evident in countries like Argentina, where vast grasslands support a strong tradition of beef consumption [53, 54] (Fig. 4).

Environmental concerns also contribute to differences in meat consumption by region. This change is noteworthy in Western countries, where awareness of

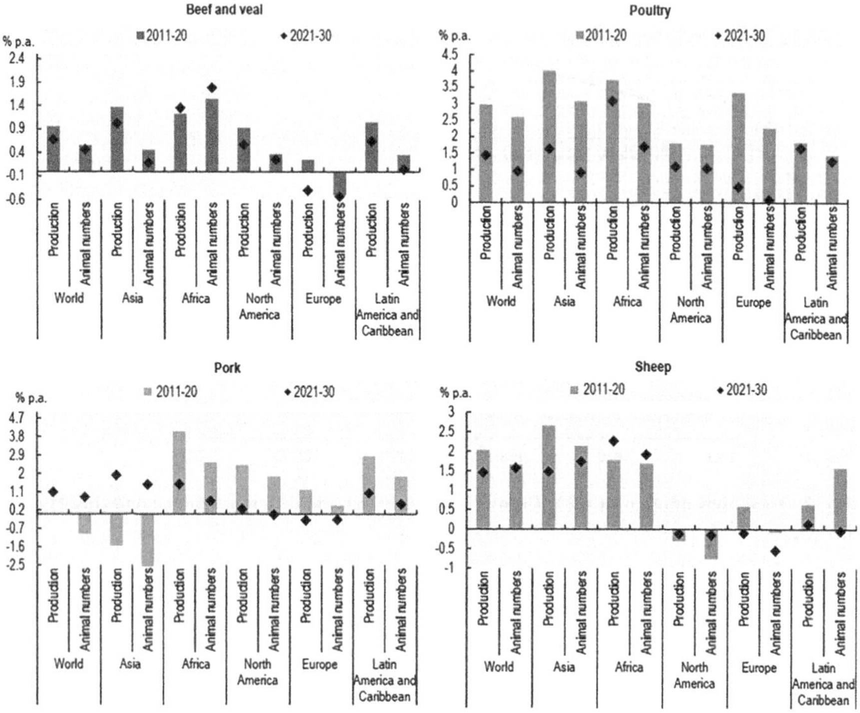

Fig. 4 Production of meat and animal inventories by type [53]

environmental issues has increased in plant-based and alternative protein diets [55–57]. Associated with ecological concerns is health awareness. Regions with higher awareness of health risks related to the consumption of excessive red and processed meat may witness a decrease in meat consumption [58].

3.3 *Types of Meat Consumed*

The consumption of various types of meat is influenced by economic, cultural, and environmental factors. Red meat, poultry, seafood, and processed meat are distinct types with diverse nutritional compositions (Table 1).

Figure 5 illustrates data on meat consumption in several countries. Prominent types of meat consumed include beef, veal, pigmeat, poultry, and sheep.

In general, "red" meat encompasses beef, pork, horse, veal, deer, and lamb [61]. They provide essential protein and necessary nutrients [62]. Preferences, culture, and traditions influence meat choices. Figure 5 illustrates the global prevalence of red meat consumption. Another type of meat is poultry, called "white" meat. Poultry is a widely chosen meat option globally. It is often considered a leaner and healthier

Table 1 Types of meat and processed meat [59]

Type	Description	Food examples
Red meat	Meat from mammals that possesses a higher myoglobin content compared to white meat.	Lamb, mutton, beef, pork, veal, goat, horse
White meat	White meat maintains its pale, light color both before and after cooking. While it contains fewer calories and less protein and iron compared to red meat, white meat's lower saturated fat content is associated with fewer adverse health effects compared to red meat.	Poultry—including chicken, turkey, or duck
Processed meats	Processed meat products undergo modifications to alter taste or prolong shelf life by undergoing curing processes (involving the addition of salt enriched with nitrates and nitrites), smoking, salting, or including preservatives. While most include beef or pork, they may also incorporate poultry, offal, red meats, or meat by-products.	Ham, sausages, salami, bacon, hot dogs, corned beef, beef jerky, ham, canned meat and meat-based sauces

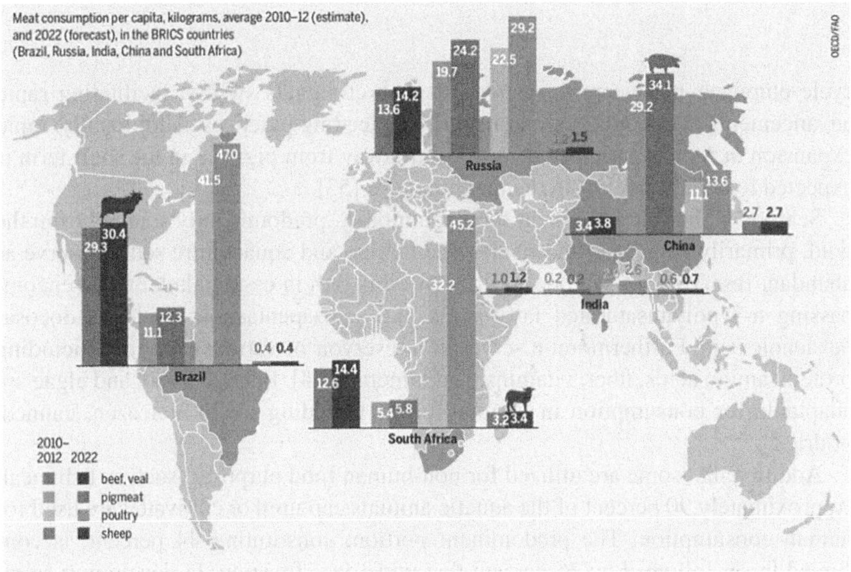

Fig. 5 Meat consumption over time, different types of meat, developing countries [60]

alternative to red meat due to its lower fat content. Its widespread availability contributes to its popularity. Figure 4 shows that poultry meat is expected to remain the primary growth driver in meat production. However, the rate of increase in the projected period is anticipated to be slower than in the past decade. The favorable meat-to-feed ratios, especially compared to other ruminants, and the short production

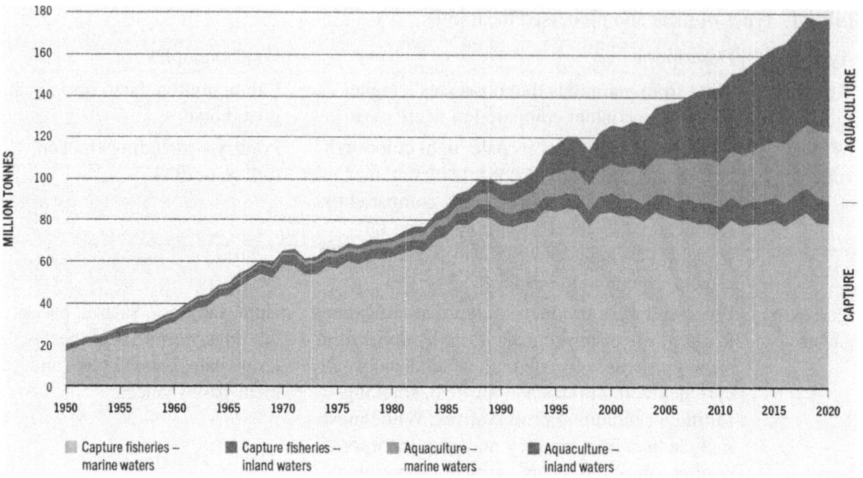

Fig. 6 Word capture fisheries and aquaculture production [65]

cycle empower producers to respond to market signals swiftly, facilitating rapid advancements in genetics, animal health, and feeding practices. Additionally, rapid expansion in Asia is anticipated, as the shift away from pigmeat in the short term is expected to benefit poultry in the medium term [53].

Seafood is the sole prominent food commodity, predominantly sourced from the wild, primarily from seas and rivers. The marine and aquaculture settings serve as abundant reservoirs of seafood [63]. Seafood is rich in essential elements, encompassing n-3 polyunsaturated fatty acids like eicosapentaenoic acid and docosahexaenoic acid. Furthermore, it serves as a reservoir of various nutrients, including protein, amino acids, fiber, vitamins, and minerals [64]. Fish, seafood, and algae are adaptable for consumption in various forms, including fresh, or frozen, canned, or dried.

Additionally, some are utilized for non-human food purposes, such as fish meal. Approximately 90 percent of the aquatic animals captured or cultivated are used for human consumption. The predominant portion, constituting 44 percent, is consumed fresh, followed by 35 percent that undergoes freezing. In developing countries, there is a higher prevalence of salting and drying fish [65]. Figure 6 shows the data for word-capture fisheries and aquaculture production.

Processed meats, which undergo various preservation and flavour enhancement methods, have become a significant component of modern diets. Figure 7 illustrates data on consumption levels of foods, mainly processed meat. Processed meat is widely consumed in some European and Asian countries.

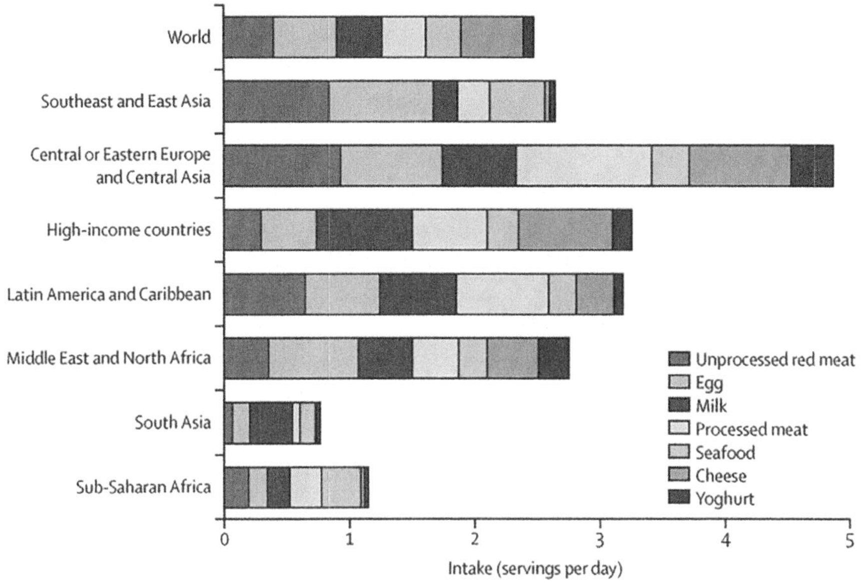

Fig. 7 Mean global and regional consumption of animal-source foods in 2018 [66]

4 Factors Influencing Meat Consumption Patterns

4.1 Socioeconomic Factors

4.1.1 Income Levels

Throughout history, the rise in meat consumption has been linked to an improvement in the nutritional content of diets. Kearney (2010) demonstrates how global increases in incomes and reduced food prices have driven the consumption of animal-based foods. This phenomenon is particularly evident in Europe and North America, explaining the growth in meat consumption over the past century [67]. Newly industrialized nations like China and Brazil have experienced this trend more recently. Additionally, as meat prices decrease, they become more accessible in developing countries, especially for inexpensive, highly processed, and lower-quality meat [68].

The interplay between financial means and patterns of meat consumption is nuanced and displays variation across socioeconomic brackets and regional divides. Whitton et al. (2021) showed a positive correlation between income and meat consumption to a certain extent. After reaching this apex, a further increase in per capita income is met with a downtrend in overall meat intake. At higher incomes, there tends to be a pivot towards less processed meat and an increased intake of fish, reflecting a potential move towards diet choices that are health-conscious or environmentally friendly. Furthermore, the research indicates that those from lower to middle-income backgrounds are more likely to include processed meats in their diet

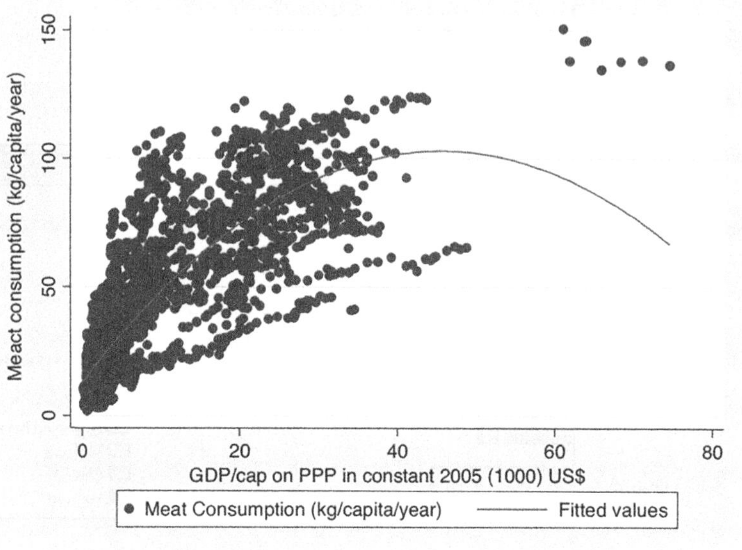

Fig. 8 Relationship between meat consumption and income per capita [71]

[69]. Figure 8 presents the correlation between meat consumption and income, with income measured in GDP/cap in constant 2005 international dollars. However, the relationship between meat consumption and income does not follow a linear pattern.

On a global scale, meat consumption exhibits diverse trends in established economies versus those still developing. In the former, there is an anticipated modest annual growth in meat consumption of about 0.24%, a marked reduction from past years. Conversely, the expected growth rate in developing nations is about 0.8% annually, suggesting a robust increase in meat consumption concurrent with rising prosperity [70].

4.1.2 Education

The impact of education on meat consumption patterns is an emerging field of research that correlates the level of education with dietary choices, particularly the consumption of meat and processed meat products [72]. Klink et al. (2022) underscore the environmental ramifications of excessive meat consumption, particularly its contribution to escalating greenhouse gas emissions. They advocate for an in-depth exploration of how educational attainment, as part of a broader socioeconomic context, shapes meat consumption habits, especially amidst the intensifying climate emergency [73].

The research conducted by Petroman et al. (2015) establishes a more pronounced correlation between educational levels and meat consumption. Their findings indicate that families with educated members tend to consume meat quantities below the national average. This trend may reflect the educational impact of elevating awareness about meat consumption's health impacts and ecological consequences, potentially driving dietary practices towards moderation [74]. Educational attainment significantly influences a person's understanding of nutrition, health, and the repercussions of their food choices. A more advanced education typically correlates with an enhanced consciousness of the potential health hazards and advantages linked to specific meat consumption patterns. Individuals with a robust educational background tend to favor diets that are diverse in protein sources, reflecting a reasonable approach to eating. Conversely, a lesser degree of academic exposure may lead to a narrower understanding and, as a result, dietary choices that may need to be better informed [75].

The role of education in dictating meat consumption patterns is pivotal. Individuals with higher educational backgrounds are often more aware of the health dangers associated with diets rich in processed meats and the extensive environmental effects of meat production. This awareness could lead to a decreased desire for meat consumption. The shift in consumption habits among the educated populace is not merely an isolated incident but also a significant element in the discourse on public health and ecosystem sustainability.

4.1.3 Occupation

The type of job a person has can affect what they eat, including how much meat they choose. Those in sedentary jobs may prefer different foods than those in labor-intensive tasks. Work schedules that vary or jobs with high-stress levels can influence eating habits, including meat choices. Additionally, specific jobs have cultural or workplace norms that can alter usual eating patterns. For instance, individuals in physically demanding jobs are often observed to consume more meat, notably beef, as such roles may heighten the need for protein-rich diets. Financial standing also comes into play; with increased earnings, people may opt for more beef, recognizing it as a pricier option [76].

According to the research of Tonsor and Lusk (2022), in 1960, the typical U.S. individual's annual meat intake comprised 150 pounds, split across beef, pork, and chicken. A surge in beef and pork consumption was noted in the 1970s, but a gradual preference for chicken emerged over time. This shift could reflect changing occupational demands and consequent dietary needs [77]. However, the findings of Klink et al. (2022) suggest that one's job needs to correlate with meat consumption levels more straightforwardly. This indicates that personal preferences and broader socioeconomic factors may exert a more pronounced influence on dietary choices [73].

In conclusion, while the nature of employment can steer meat consumption tendencies, with physically strenuous jobs possibly leading to higher meat intake, it's clear that a host of other factors are at play. These factors collectively shape dietary habits, suggesting that occupation is just one of many influences on an individual's choice of meat.

4.2 Cultural and Dietary Factors

Meat consumption and dietary preferences are deeply intertwined with cultural practices and religious beliefs. It is crucial to recognize the diversity of dietary practices.

Cultural factors have a significant influence on meat consumption. In a study by Choi and Lee (2023), it was observed that cultural differences among various ethnic groups are substantial and need to be considered when designing appropriate dietary interventions. This finding is particularly relevant in nutrition education and community health, where one-size-fits-all approaches often prove less effective than culturally tailored strategies [78].

Religion plays a vital role in shaping dietary habits. Islam (Halal) and Judaism (Kosher) dietary laws include detailed regulations regarding permissible meats and their preparation methods. These rules affect individual dietary choices and impact the global food supply chain, creating demand for products that adhere to these religious guidelines. Furthermore, beliefs and taboos within cultures, often propagated by families and communities, play a significant role in determining which foods are acceptable or prohibited. Religious or spiritual foundations strongly influence these beliefs and affect meat consumption patterns. For example, in Hinduism, cows are revered and commonly not consumed, while many Christian traditions do not have such restrictions. However, meat is also prominent in cultural festivals and religious traditions. In many cultures, meat is an integral part of celebrations and rituals. Therefore, abstaining from meat is not just a dietary choice but also a departure from traditional norms and culture, which can have social consequences for individuals within their communities [79–83].

It's essential to note that changes in meat consumption patterns are occurring in response to increasing awareness of health, environmental concerns, and animal welfare. More people are adopting reduced or meatless diets, reflecting cultural shifts that may diverge from traditional customs. This development in dietary preferences underscores the complex interplay between cultural preservation and modern values.

Dietary preferences, as well as religious and cultural practices, are factors influencing meat consumption patterns. Understanding these factors is essential for community health, nutrition education, and the food industry. The relationship between diet, culture, and religion highlights the need for adaptable and culturally sensitive approaches to dietary recommendations and interventions.

4.3 Health Considerations

With a growing consciousness about wellness, humans know more about the perceived dangers of eating meat. This is having a profound impact on global eating habits. A wealth of research indicates a link between the intake of red and processed

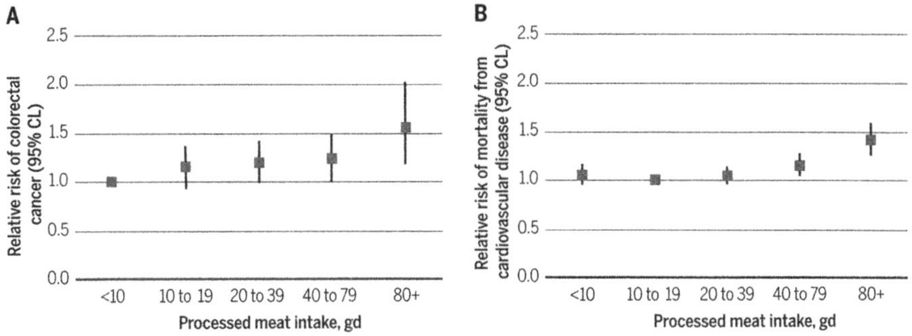

Fig. 9 Meat and health [52]. (**a**) The relative risk of colorectal cancer as a function of average processed meat intake. (**b**) The relative risk of cardiovascular death as a function of average processed meat intake

meats and a heightened probability of critical health issues, encompassing cardiac conditions, a spectrum of cancers, overall mortality, and diabetes mellitus type 2. Figure 9 illustrates the relationship between meat and health. The International Agency for Research on Cancer (IARC) posits that high processed meat consumption could be responsible for 34,000 global cancer deaths a year. If red meat's association with cancer is causal, this could rise to 50,000. In Western Europe, a daily average of 26.4 g of processed meat may increase colorectal cancer risk by 9%. Processed meats may also heighten the risk of heart disease and diabetes and contribute to weight gain, but the evidence isn't conclusive. The health implications of meat consumption are less evident in non-Western countries, where lower meat intake often reflects economic status rather than health choices. Studies from Asia and Iran report lower meat consumption than in the U.S., without a corresponding rise in mortality from heart disease or cancer [52].

Furthermore, the examination of global risks associated with meat consumption also sheds light on its environmental impacts, including the rise in greenhouse gas emissions and the potential transmission of viruses from animals to humans. These issues contribute to the theoretical basis for reducing meat intake as part of a strategy to mitigate the risks of chronic diseases and environmental harm [84–86]. Papier et al. (2021) conducted a study on nearly half a million middle-aged adults to investigate the relationship between meat consumption and the risk of 25 common conditions other than cancer. Such large-scale studies serve as valuable tools to paint a clearer picture of the health impacts of meat consumption and its influence on public awareness and dietary behavior [87].

The impact of health trends and risk perception on meat consumption patterns is evident. Health concerns, ethical considerations, and environmental awareness influence a noticeable shift towards reduced meat consumption. These changes are supported by substantial scientific evidence linking high meat consumption to various health risks.

4.4 Environmental Concerns

Environmental concerns regarding the meat consumption model are increasingly being focused on, reflecting a growing awareness of the relationship between the environment and meat consumption.

According to Eshel et al. (2014), livestock production significantly affects air and water quality, ocean health, and emissions of greenhouse gases (GHG) on both regional and global levels, and it also represents the predominant use of land around the world. Compared to other livestock types, beef production demands substantially more resources, such as 28 times more land, 11 times more irrigation water, five times more GHG emissions, and six times more nitrogen (Nr). An initial analysis of three fundamental plant-based foods indicates they require significantly less land, GHG emissions, and Nr—between two to six times lower—compared to the resources needed for producing non-beef animal-based calories. In contrast, their need for irrigation is roughly equivalent. The principal finding is that beef production requires approximately ten times more resources than other livestock categories, a conclusion that remains strong despite existing uncertainties [88].

In this context, beef production accounts for approximately 88% of all land in the US dedicated to generating calories from animals. This land use is divided into about 79% for pasture, around 7% for processed roughage, and about 2% for concentrated feed. The land requirements for dairy production are also presented using this categorization method (Fig. 10).

A recent study from the University of Bonn in Germany found that affluent nations need to significantly cut down on their meat consumption to maintain global climate targets and secure food availability for the future. This study is the latest among several that highlight the considerable environmental impact of extensive animal farming. By May 2022, the annual worldwide consumption of meat had escalated to around 350 million tons, twice the quantity consumed back in 1988. This increasing trend raises concerns due to the meat and dairy sectors' notable role in climate change. The Food and Agriculture Organization of the United Nations noted in 2021 that over one-third of the world's greenhouse gas emissions are produced by the global food system, with livestock farming being responsible for nearly half of these emissions. Consequently, it is crucial for wealthier nations to significantly cut down on their meat consumption to achieve global climate objectives and maintain food security for the future. It is suggested that countries with high meat consumption per person, such as the USA and European nations, need to aim for a reduction of at least 75% in their meat consumption [89].

Figure 11 illustrates the relationship between meat and climate change. Figure 11a shows that foods derived from animals are the primary contributors to food-system GHGs, and their significance is expected to rise. Figure 11b illustrates the climate warming effects of each gas under a scenario where current livestock-produced emissions are released at Year 0 and then maintained indefinitely. Methane causes substantial and rapid warming, but due to its shorter atmospheric lifespan, its warming effect stabilizes after about 20 years. In contrast, the warming effect of

Fig. 10 The share of the total environmental impact caused by different animal categories on a national level is calculated by combining the annual national calorie consumption data [88]

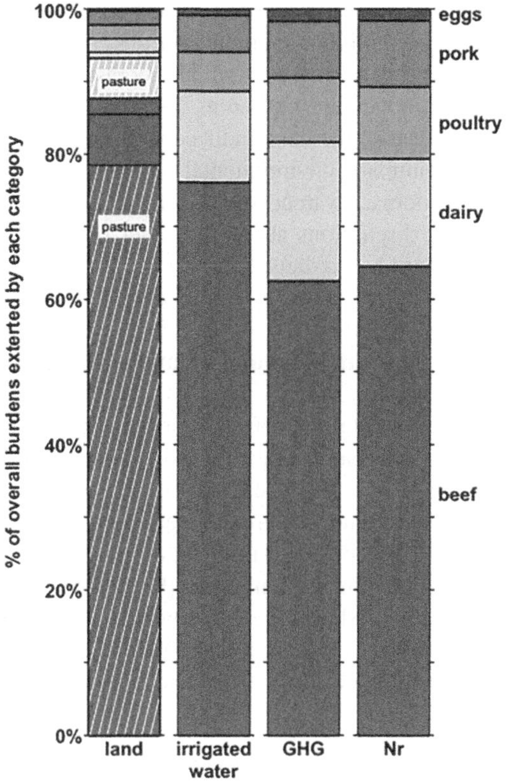

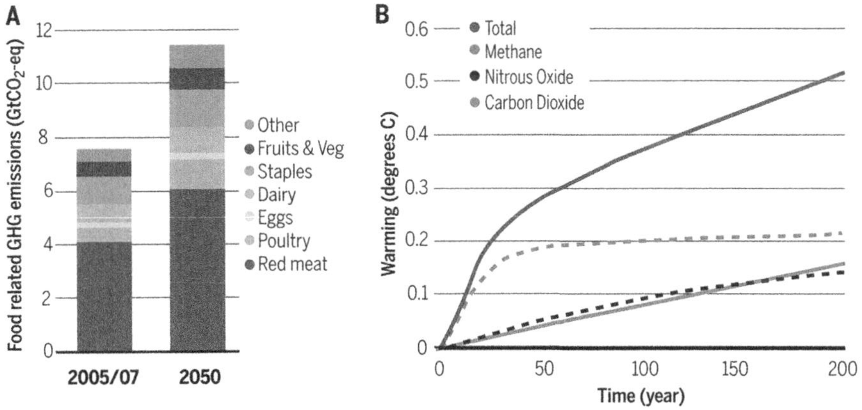

Fig. 11 Meat and climate change [52]. (**a**) Greenhouse gas (GHG) emissions from various food types during 2005–2007 and their projected impact in 2050. (**b**) The three main GHGs-carbon dioxide, methane, and nitrous oxide—differ significantly in their climatic impact

carbon dioxide has continued to increase over the past two centuries and will persist as long as emissions remain. Nitrous oxide's warming effect starts to plateau after two centuries, with a minimal increase. However, it's important to note that any growth in methane emissions due to rising livestock production would correspondingly escalate methane-induced warming.

In summary, environmental concerns are significantly impacting meat consumption patterns. Awareness of the ecological impact of the meat processing industry is driving discussions about consumer behavior and shifting towards more sustainable, plant-based diets.

4.4.1 Sustainability and Meat Consumption

The detrimental impact of meat on sustainable development is a critical factor in why an increasing number of individuals are adopting vegetarian or flexitarian diets, aiming to minimize their meat intake as much as possible. While alternatives like plant-based and lab-grown meats are gaining popularity as more eco-friendly choices, their potential implications on nutrition, health, and environmental aspects require additional examination [90].

Figure 12 shows the data about the Life cycle assessments (LCA) study from Beyond Meat® and Impossible™ Foods, revealing that their plant-based beef products have much lower eutrophication potential and land use than factory-farmed

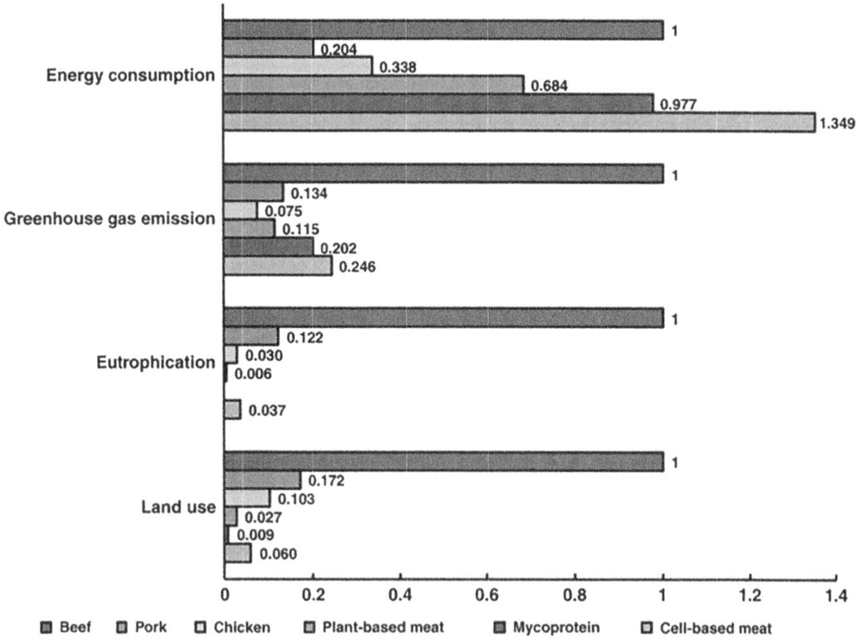

Fig. 12 Comparison of the environmental impact of meat and meat analogues [90]

beef, pork, and chicken. Their greenhouse gas emissions are higher than pork and chicken but lower than beef, while energy consumption exceeds pork and chicken. Mycoproteins (such as Quorn™) have a higher energy and emissions impact but lower land use requirements.

Overall, vegetarian and vegan diets are becoming increasingly popular as sustainable alternatives to meat-based diets, offering lower environmental impacts and a rich source of plant-based proteins and essential nutrients.

5 Changing Trends in Meat Consumption

5.1 Shift towards Plant-Based Diets

Plant-based diets, like vegetarian eating patterns, link to a reduced incidence of certain chronic illnesses and changes in the gut microbiome composition. This dietary choice may enhance the growth of beneficial bacteria while diminishing harmful ones. The abundance of dietary fiber, vegetable proteins, polyunsaturated fatty acids, and phytochemicals in vegetarianism likely contributes to this microbiome modulation [19, 90–92].

The benefits of plant-based diets validate the efficacy of vegan or vegetarian diets, facilitating weight loss and reducing the risks of heart disease, high blood pressure, diabetes, and obesity. In terms of diabetes prevention, plant-based diets establish that vegetarians face a significantly lower diabetes risk compared to nonvegetarians. A 2009 study with over 60,000 participants further illustrates a 2.9% diabetes prevalence in vegans versus 7.6% in nonvegetarians, emphasizing the potential of plant-based diets, especially low-fat vegan diets, in diabetes management [93].

Concerning cardiovascular health, Ornish's Lifestyle Heart Trial demonstrates an 82% regression of atherosclerosis. In comparison, de Lorgeril's Lyon Diet Heart Study indicates a 73% reduction in coronary events and a 70% decrease in all-cause mortality [94]. Exploring blood pressure, the Dietary Guidelines Advisory Committee (2010) and a randomized crossover trial underline the positive correlation between vegetarian diets and lower systolic and diastolic blood pressure. On the topic of mortality, plant-based diets, as suggested by the Dietary Guidelines Advisory Committee (2010), exhibit a diminished risk of cardiovascular disease and total mortality. This evidence underscores the diverse health benefits of plant-based dietary choices [95] (Fig. 13).

Plant-based eating remains highly popular today, as evidenced by a significant 600% increase in Americans adhering to a vegan diet from 2014 to 2018. This growing interest in plant-based diets stems from various motivations, such as health consciousness, ethical concerns for animal welfare, and the desire to reduce one's environmental impact. Celebrity endorsements, media coverage, and influential documentaries have also propelled this dietary shift. Regardless of this change,

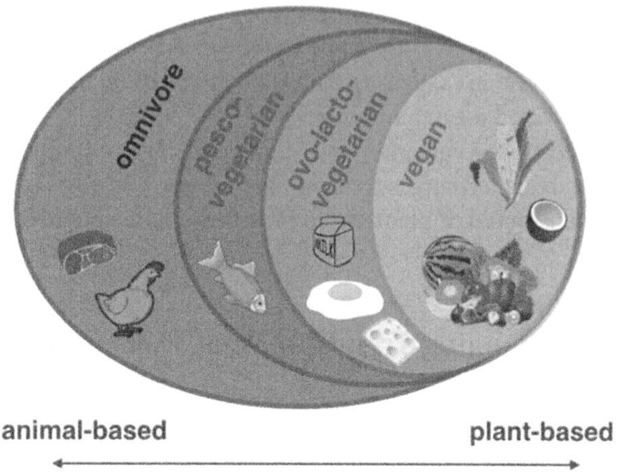

Fig. 13 From left to right: including all food items (omnivore), including all except for meat (pes-co-vegetarian) or meat and fish (ovo-lacto-vegetarian), to including only plant-based items (vegan) [96]

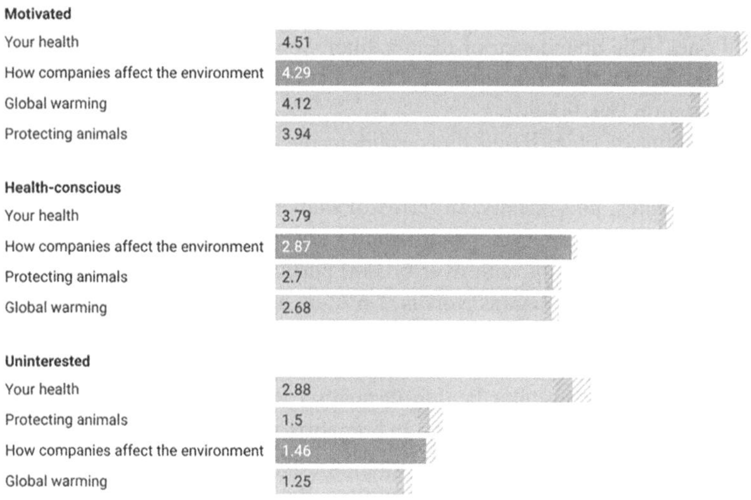

Fig. 14 Perceived motivators for eating plant-based foods by audience segment [99]

many Americans actively seek to incorporate more plant-based foods into their diets [97, 98].

Figure 14 presents the results of a survey of 940 Americans asked about "How important are each of the following to you personally, if at all, when you choose to purchase or eat plant-based foods (fruit, vegetables, meat or dairy alternative)? With "1" not important, "5″ is extremely important.

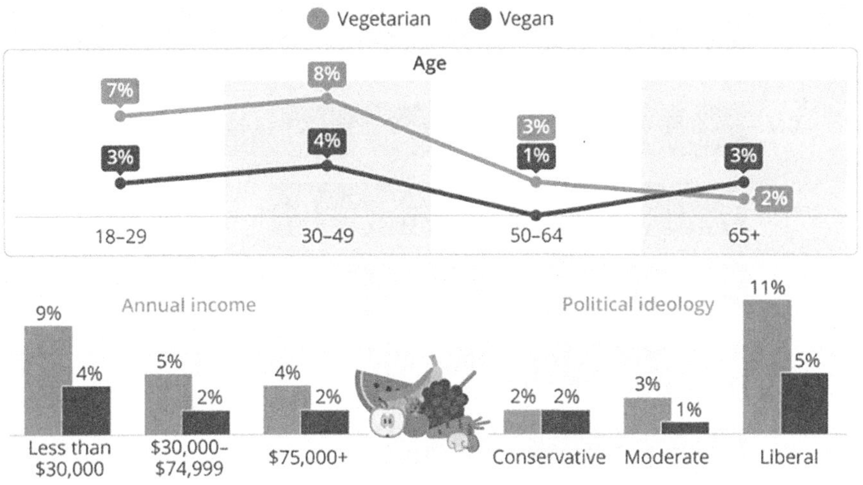

Fig. 15 Vegetarianism and Veganism in the U.S [101].

5.2 *Impact of Veganism and Vegetarianism*

The impact of vegetarian and vegan diets on meat consumption patterns and related environmental effects has been extensively studied, providing a nuanced and complex understanding. From 2018 to 2022, global retail sales of plant-based meat and seafood alternatives more than doubled to $6 billion, reflecting a significant shift in consumer habits towards plant-based diets [100].

For example, in recent years, a growing number of vegan and vegetarian restaurants have opened in the U.S., and supermarket shelves are consistently stocked with meat and dairy product alternatives. A recent Gallup poll reveals that fewer than 10% of Americans follow vegetarian or vegan diets, with 5% identifying as vegetarian and 3% as vegan [101].

In Fig. 15, Americans with an annual income below $30,000 are more likely to identify as vegetarians, with 9% of this group following a vegetarian diet compared to those in higher income brackets. The differences in the prevalence of veganism among these income groups are not statistically significant. In terms of age, vegetarianism is slightly more common among Americans under 50, with 8% of those aged 30 to 49 and 7% of those aged 18–29 identifying as vegetarians. In contrast, older Americans are less likely to be vegetarian, with only 3% of those aged 50–64 and 2% of those 65 or older following a vegetarian diet. However, there is a relatively slight variation by age regarding the percentage of people identifying as vegans.

Regarding the environment, vegetarians have a 25.1% lower impact on greenhouse gas emissions and land use than those who consume a lot of meat. Adopting a reduced meat diet (50 g or less per day) can halve the amount of carbon emissions, water pollution, and land use compared to a high meat diet (100 g or more per day) [102].

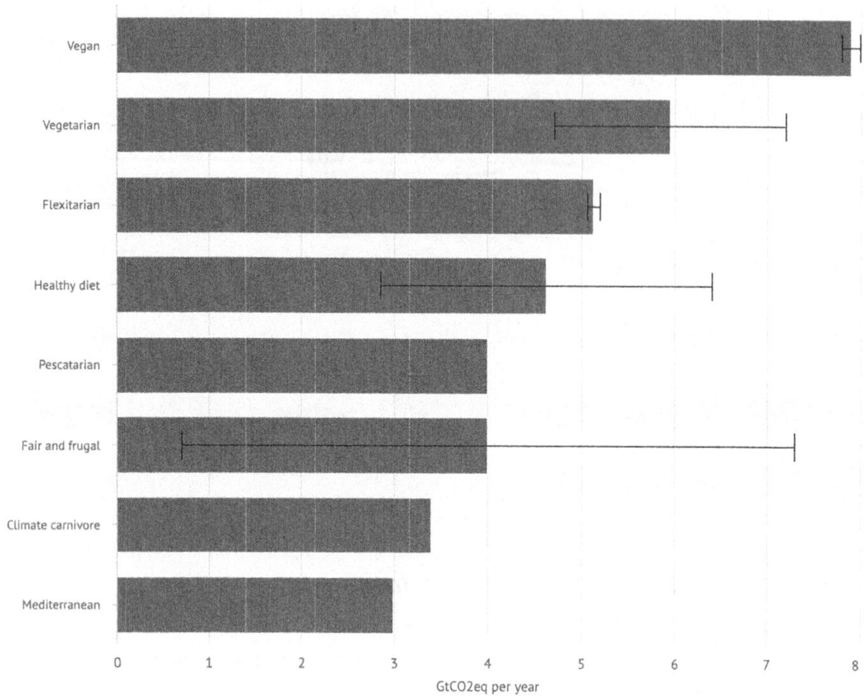

Fig. 16 The potential for greenhouse gas emissions reduction through the worldwide adoption of different dietary patterns [105]

However, when considering another aspect, the importation of soybean products and palm oil, which are often critical components of a vegetarian diet, is often associated with deforestation, habitat loss, and human rights violations in countries like India, Indonesia, and Malaysia. These issues highlight that reducing meat consumption moderately, rather than eliminating it, can bring significant environmental benefits [100, 103, 104].

Figure 16 illustrates the potential reduction in greenhouse gas emissions achievable through the global adoption of different dietary patterns.

In conclusion, the impact of vegetarianism and plant-based diets on meat consumption and the environment is substantial, but it also has many facets. The critical point is the importance of a balanced approach, incorporating plant-based foods and sustainably sourced animal products, to achieve optimal outcomes for the environment and ethics.

5.3 *Sustainable Meat Consumption Practices*

Practices that focus on sustainable meat consumption are crucial in tackling the various environmental, ethical, and health-related issues that arise from both the production and consumption of meat. Based on the continuous increase in the world's population, the demand for meat products has significantly increased, leading to concerns about the environmental consequences of this trend. Consumer perceptions of sustainability can be influenced by various factors, including income, age, and geographical location [106–108]. The research conducted in Europe and Asia has indicated a higher willingness to pay for sustainable food products than in North America and Oceania [109]. In countries like Japan and India, where meat consumption is minimal due to cultural factors, there may be little need to decrease meat consumption for environmental reasons [110].

Furthermore, consumers' willingness to invest in sustainable food may vary depending on the product category. For instance, research has demonstrated that consumers are more inclined to pay higher prices for meat products than seafood and beverages [111].

Additionally, eco-labels and certification mechanisms have proven effective in enhancing consumers' willingness to pay for sustainable products, underscoring the importance of effective communication and rigorous labelling practices in promoting sustainable consumption.

There are several recommended approaches to promoting environmentally friendly and mindful meat consumption. A key strategy is to promote predominantly plant-based diets, potentially reducing the environmental toll of food production. A different approach involves embracing flexitarian eating habits. This means people largely eat plant-based meals but sometimes include meat that has been responsibly sourced [19]. Making informed decisions about the meat one consumes is also crucial, as is opting for products from sustainable and ethical sources like organic, grass-fed, or pasture-raised meat. Moreover, educating the public about the environmental and health effects of eating meat via educational initiatives and public health drives is essential for encouraging sustainable eating habits. When these strategies are adopted together, they can lead to a more balanced and sustainable meat consumption pattern [17, 112].

6 Conclusion

Insights about consumers' attitudes and consumption patterns towards meat products illuminate the intricate factors influencing dietary choices and preferences. This deep dive reveals the significant impact of health awareness, ecological mindfulness, ethical considerations, and cultural norms on meat consumption habits. There was a discernible shift towards leaner meat or vegetarian alternatives, primarily motivated by health concerns. Simultaneously, environmental factors, especially the global impact of meat production, are prompting consumers to reconsider their options. The moral aspects, particularly animal welfare, are increasingly central to consumer choices.

Meanwhile, the enduring influence of cultural heritage and personal taste preferences remains significant. This changing landscape places the meat industry in a position where adopting sustainable and ethical practices is becoming imperative. This shift reflects a broader trend of consumers becoming more thoughtful and aware of the consequences of their food choices. This growing consciousness is steering a notable move towards different protein sources and a rise in vegetarian diet preferences, signalling a possible shift away from traditional meat-eating patterns. For the food industry, staying attuned to these trends is vital to evolving and meeting new consumer demands, thereby playing a role in fostering a more sustainable and ethically responsible food system in the years ahead.

References

1. Rozin P, Hormes JM, Faith MS, Wansink B. Is meat male? A quantitative multimethod framework to establish metaphoric relationships. J Consum Res. 2012;39:629–43. https://doi.org/10.1086/664970.
2. de Boer J, Schösler H, Aiking H. "Meatless days" or "less but better"? Exploring strategies to adapt Western meat consumption to health and sustainability challenges. Appetite. 2014;76:120–8. https://doi.org/10.1016/j.appet.2014.02.002.
3. Garnett T. Where are the best opportunities for reducing greenhouse gas emissions in the food system (including the food chain)? Food Policy. 2011;36:S23–32. https://doi.org/10.1016/j.foodpol.2010.10.010.
4. Oude Ophuis PAM, Van Trijp HCM. Perceived quality: a market driven and consumer oriented approach. Food Qual Prefer. 1995;6:177–83. https://doi.org/10.1016/0950-3293(94)00028-T.
5. Ruby MB. Vegetarianism. A blossoming field of study. Appetite. 2012;58:141–50. https://doi.org/10.1016/j.appet.2011.09.019.
6. Verbeke W, Pérez-Cueto FJA, de Barcellos MD, Krystallis A, Grunert KG. European citizen and consumer attitudes and preferences regarding beef and pork. Meat Sci. 2010;84:284–92. https://doi.org/10.1016/j.meatsci.2009.05.001.
7. de Araújo PD, Araújo WMC, Patarata L, Fraqueza MJ. Understanding the main factors that influence consumer quality perception and attitude towards meat and processed meat products. Meat Sci. 2022;193:108952. https://doi.org/10.1016/j.meatsci.2022.108952.
8. Šarčević D, Slobodan L, Đorđević V, Milićević D, Danijela V, Lakicevic B, et al. The role of consumer perception and attitude in purchasing of meat and meat products. Meat Technol. 2011;52:283–90.
9. Font-i-Furnols M, Skrlep M, Aluwé M. Attitudes and beliefs of consumers towards pig welfare and pork quality. IOP Conf Ser Earth Environ Sci. 2019;333:12002. https://doi.org/10.1088/1755-1315/333/1/012002.
10. Hati SRH, Zulianti I, Achyar A, Safira A. Perceptions of nutritional value, sensory appeal, and price influencing customer intention to purchase frozen beef: evidence from Indonesia. Meat Sci. 2021;172:108306. https://doi.org/10.1016/j.meatsci.2020.108306.
11. Font-i-Furnols M, Guerrero L. Consumer preference, behavior and perception about meat and meat products: an overview. Meat Sci. 2014;98:361–71. https://doi.org/10.1016/j.meatsci.2014.06.025.

12. Sbardelotto PRR, Balbinot-Alfaro E, da Rocha M, Alfaro AT. Natural alternatives for processed meat: legislation, markets, consumers, opportunities and challenges. Crit Rev Food Sci Nutr. 2022:1–16. https://doi.org/10.1080/10408398.2022.2081664.
13. Grunert KG, Aachmann K. Consumer reactions to the use of EU quality labels on food products: a review of the literature. Food Control. 2016;59:178–87. https://doi.org/10.1016/j.foodcont.2015.05.021.
14. Grunert KG, Loose SM, Zhou Y, Tinggaard S. Extrinsic and intrinsic quality cues in Chinese consumers' purchase of pork ribs. Food Qual Prefer. 2015;42:37–47. https://doi.org/10.1016/j.foodqual.2015.01.001.
15. Beardsworth A, Bryman A. Meat consumption and meat avoidance among young people: an 11-year longitudinal study. Br Food J. 2004;106:313–27. https://doi.org/10.1108/00070700410529573.
16. Clonan A, Wilson P, Swift J, Leibovici D, Holdsworth M. Red and processed meat consumption and purchasing behaviours and attitudes: impacts for human health, animal welfare and environmental sustainability. Public Health Nutr. 2015;18:1–11. https://doi.org/10.1017/S1368980015000567.
17. Willett W, Rockström J, Loken B, Springmann M, Lang T, Vermeulen S, et al. Food in the Anthropocene: the EAT – Lancet Commission on healthy diets from sustainable food systems. Lancet. 2019;393:447–92. https://doi.org/10.1016/S0140-6736(18)31788-4.
18. CAST. Scientific, ethical, and social aspects of farm animal welfare. Council for Agricultural Science and Technology; 2017.
19. Dinu M, Abbate R, Gensini GF, Casini A, Sofi F. Vegetarian, vegan diets and multiple health outcomes: a systematic review with meta-analysis of observational studies. Crit Rev Food Sci Nutr. 2017;57:3640–9. https://doi.org/10.1080/10408398.2016.1138447.
20. Vartanian LR, Herman CP, Polivy J. Consumption stereotypes and impression management: how you are what you eat. Appetite. 2007;48:265–77. https://doi.org/10.1016/j.appet.2006.10.008.
21. Laestadius LI, Neff RA, Barry CL, Frattaroli S. "We don't tell people what to do": an examination of the factors influencing NGO decisions to campaign for reduced meat consumption in light of climate change. Glob Environ Chang. 2014;29:32–40. https://doi.org/10.1016/j.gloenvcha.2014.08.001.
22. Bryant C, Dillard C. The impact of framing on acceptance of cultured Meat. Front Nutr. 2019;6:103. https://doi.org/10.3389/fnut.2019.00103.
23. Galiè A, Mancino L, Lerro M. The role of environmental, health, and social contexts in the frequency of plant-based versus animal-based food choices: a preliminary study. Ital J Food Sci. 2018;30:145–61.
24. Hopwood CJ, Bleidorn W, Schwaba T, Chen S. Health, environmental, and animal rights motives for vegetarian eating. PLoS One. 2020;15:e0230609. https://doi.org/10.1371/journal.pone.0230609.
25. Hoek AC, Luning PA, Weijzen P, Engels W, Kok FJ, de Graaf C. Replacement of meat by meat substitutes. A survey on person- and product-related factors in consumer acceptance. Appetite. 2011;56:662–73. https://doi.org/10.1016/j.appet.2011.02.001.
26. Norwood FB, Lusk JL. Compassion, by the pound: the economics of farm animal welfare. Oxford: Oxford University Press; 2011. https://doi.org/10.1093/acprof:osobl/9780199551163.001.0001.
27. de Boer J, Hoogland CT, Boersema JJ. Towards more sustainable food choices: value priorities and motivational orientations. Food Qual Prefer. 2007;18:985–96. https://doi.org/10.1016/j.foodqual.2007.04.002.
28. Whiting T. Understanding animal welfare: the science in its cultural context. Can Vet J. 2011;52:662.
29. Reimert I, Webb LE, van Marwijk MA, Bolhuis JE. Review: towards an integrated concept of animal welfare. Animal. 2023;17:100838. https://doi.org/10.1016/j.animal.2023.100838.
30. Hocquette J-F. Is in vitro meat the solution for the future? Meat Sci. 2016;120:167–76. https://doi.org/10.1016/j.meatsci.2016.04.036.

31. Jiang J, Xiong YL. Natural antioxidants as food and feed additives to promote health benefits and quality of meat products: a review. Meat Sci. 2016;120:107–17. https://doi.org/10.1016/j.meatsci.2016.04.005.
32. Tilman D, Clark M. Global diets link environmental sustainability and human health. Nature. 2014:515. https://doi.org/10.1038/nature13959.
33. Schösler H, de Boer J, Boersema JJ. Can we cut out the meat of the dish? Constructing consumer-oriented pathways towards meat substitution. Appetite. 2012;58:39–47. https://doi.org/10.1016/j.appet.2011.09.009.
34. Clonan A, Holdsworth M, Swift JA, Leibovici D, Wilson P. The dilemma of healthy eating and environmental sustainability: the case of fish. Public Health Nutr. 2012;15:277–84. https://doi.org/10.1017/S1368980011000930.
35. Verbeke W. Consumer acceptance of functional foods: socio-demographic, cognitive and attitudinal determinants. Food Qual Prefer. 2005;16:45–57. https://doi.org/10.1016/j.foodqual.2004.01.001.
36. Grunert KG, Wills JM. A review of European research on consumer response to nutrition information on food labels. J Public Health (Bangkok). 2007;15:385–99. https://doi.org/10.1007/s10389-007-0101-9.
37. Chen M, Yin S, Xu Y, Wang Z. Consumers' willingness to pay for tomatoes carrying different organic labels: evidence from auction experiments. Br Food J. 2015:117. https://doi.org/10.1108/BFJ-12-2014-0415.
38. Linder NS, Uhl G, Fliessbach K, Trautner P, Elger CE, Weber B. Organic labeling influences food valuation and choice. NeuroImage. 2010;53:215–20. https://doi.org/10.1016/j.neuroimage.2010.05.077.
39. Bocquet-Appel J-P. When the world's population took off: the springboard of the neolithic demographic transition. Science. 2011;333:560–1. https://doi.org/10.1126/science.1208880.
40. Chiles RM, Fitzgerald AJ. Why is meat so important in Western history and culture? A genealogical critique of biophysical and political-economic explanations. Agric Human Values. 2018;35:1–17. https://doi.org/10.1007/s10460-017-9787-7.
41. Mann NJ. A brief history of meat in the human diet and current health implications. Meat Sci. 2018;144:169–79. https://doi.org/10.1016/j.meatsci.2018.06.008.
42. Hoogland CT, de Boer J, Boersema JJ. Transparency of the meat chain in the light of food culture and history. Appetite. 2005;45:15–23. https://doi.org/10.1016/j.appet.2005.01.010.
43. Welch R, Mitchell P. Food processing: a century of change. Br Med Bull. 2000;56:1–17. https://doi.org/10.1258/0007142001902923.
44. Preserve L, Steven Pearson G, Steven G. The democratization of food: tin cans and the growth of the American food processing industry, 1810–1940. Lehigh University; 2016.
45. Smith DI. 19th century development of refrigeration in the American meat packing industry. Tenor Our Times. 2019;8:14.
46. Nestle M. Food politics: how the food industry influences nutrition, and health. University of California Press. 2002;
47. Hu FB, Manson JE, Willett WC. Types of dietary fat and risk of coronary heart disease: a critical review. J Am Coll Nutr. 2001;20:5–19. https://doi.org/10.1080/07315724.2001.10719008.
48. Bryant C, Barnett J. Consumer acceptance of cultured meat: a systematic review. Meat Sci. 2018;143:8–17. https://doi.org/10.1016/j.meatsci.2018.04.008.
49. Baltenweck I, Enahoro D, Frija A, Tarawali S. Why is production of animal source foods important for economic development in Africa and Asia? Anim Front. 2020;10:22–9. https://doi.org/10.1093/af/vfaa036.
50. Boyazoglu J, Hatziminaoglou I, Morand-Fehr P. The role of the goat in society: past, present and perspectives for the future. Small Rumin Res. 2005;60:13–23. https://doi.org/10.1016/j.smallrumres.2005.06.003.
51. Ritchie H, Rosado P, Roser M. Meat and dairy production. Our World Data 2019. https://ourworldindata.org/meat-production. Accessed 1 Nov 2023.

52. Godfray HCJ, Aveyard P, Garnett T, Hall JW, Key TJ, Lorimer J, et al. Meat consumption, health, and the environment. Science. 2018:361. https://doi.org/10.1126/science.aam5324.
53. OECD-FAO. Chapter 6: Meat. OECD-FAO Agric Outlook 2021–2030 2021:163–77.
54. FAO. Meat market review—overview of market and policy developments in 2022. FAO 2022.
55. Springmann M, Wiebe K, Mason-D'Croz D, Sulser TB, Rayner M, Scarborough P. Health and nutritional aspects of sustainable diet strategies and their association with environmental impacts: a global modelling analysis with country-level detail. Lancet Planet Heal. 2018;2:e451–61. https://doi.org/10.1016/S2542-5196(18)30206-7.
56. Biesbroek S, Kok FJ, Tufford AR, Bloem MW, Darmon N, Drewnowski A, et al. Toward healthy and sustainable diets for the 21st century: importance of sociocultural and economic considerations. Proc Natl Acad Sci USA. 2023;120:e2219272120. https://doi.org/10.1073/pnas.2219272120.
57. Alexandropoulou I, Goulis DG, Merou T, Vassilakou T, Bogdanos DP, Grammatikopoulou MG. Basics of sustainable diets and tools for assessing dietary sustainability: a primer for researchers and policy actors. Healthcare. 2022:10. https://doi.org/10.3390/healthcare10091668.
58. Swinburn BA, Kraak VI, Allender S, Atkins VJ, Baker PI, Bogard JR, et al. The global Syndemic of obesity, undernutrition, and climate change: the lancet commission report. Lancet. 2019;393:791–846. https://doi.org/10.1016/S0140-6736(18)32822-8.
59. Clonan A, Roberts KE, Holdsworth M. Socioeconomic and demographic drivers of red and processed meat consumption: implications for health and environmental sustainability. Proc Nutr Soc. 2016;75:367–73. https://doi.org/10.1017/S0029665116000100.
60. Lindberg JED, Lindberg G, Teräs J, Poulsen G, Solberg SO, Tybirk K, et al. Nordic alternative protein potentials – mapping of regional bioeconomy opportunities. 2016.
61. Guo H, Ding J, Liang J, Zhang Y. Association of red meat and poultry consumption with the risk of metabolic syndrome: a meta-analysis of prospective cohort studies. Front Nutr. 2021;8:691848. https://doi.org/10.3389/fnut.2021.691848.
62. Micha R, Peñalvo J, Cudhea F, Imamura F, Rehm C, Mozaffarian D. Association between dietary factors and mortality from heart disease, stroke, and type 2 diabetes in the United States. JAMA. 2017;317:912. https://doi.org/10.1001/jama.2017.0947.
63. Toyofuku H. Seafood. In: Smithers GWBT-E of FS. 2nd ed. Oxford: Academic; 2024. p. 127–36. https://doi.org/10.1016/B978-0-12-822521-9.00120-9.
64. Hosomi R, Yoshida M, Fukunaga K. Seafood consumption and components for health. Glob J Health Sci. 2012;4:72–86. https://doi.org/10.5539/gjhs.v4n3p72.
65. FAO. In brief to the state of world fisheries and aquaculture 2022. FAO; 2022. https://doi.org/10.4060/cc0463en.
66. Miller V, Reedy J, Cudhea F, Zhang J, Shi P, Erndt-Marino J, et al. Global, regional, and national consumption of animal-source foods between 1990 and 2018: findings from the Global Dietary Database. Lancet Planet Heal. 2022;6:e243–56. https://doi.org/10.1016/S2542-5196(21)00352-1.
67. Kearney J. Food consumption trends and drivers. Philos Trans R Soc B Biol Sci. 2010;365:2793–807. https://doi.org/10.1098/rstb.2010.0149.
68. Sheng TY, Shamsudin MN, Mohamed Z, Abdullah AM, Radam A. Demand analysis of Meat in Malaysia. J Food Prod Mark. 2010;16:199–211. https://doi.org/10.1080/10454440903415105.
69. Whitton C, Bogueva D, Marinova D, Phillips CJC. Are we approaching peak meat consumption? Analysis of meat consumption from 2000 to 2019 in 35 countries and its relationship to gross domestic product. Animals. 2021;11:3466. https://doi.org/10.3390/ani11123466.
70. Whitnall T, Pitts N. Meat consumption. Dep Agric Fish For – Aust Gov 2019. https://www.agriculture.gov.au/abares/research-topics/agricultural-outlook/meat-consumption. Accessed 1 Nov 2023.
71. Vranken L, Avermaete T, Petalios D, Mathijs E. Curbing global meat consumption: emerging evidence of a second nutrition transition. Environ Sci Pol. 2014;39:95–106. https://doi.org/10.1016/j.envsci.2014.02.009.

72. Mata J, Kadel P, Frank R, Schüz B. Education- and income-related differences in processed meat consumption across Europe: the role of food-related attitudes. Appetite. 2023;182:106417. https://doi.org/10.1016/j.appet.2022.106417.
73. Klink U, Mata J, Frank R, Schüz B. Socioeconomic differences in animal food consumption: education rather than income makes a difference. Front Nutr. 2022:9. https://doi.org/10.3389/fnut.2022.993379.
74. Petroman C, Bidireac IC, Petroman I, Sucan M, Marin D, Turc B, et al. The impact of education on the behaviour of the consumer of animal origin food products. Proc Soc Behav Sci. 2015;190:429–33. https://doi.org/10.1016/j.sbspro.2015.05.021.
75. Darmon N, Drewnowski A. Does social class predict diet quality? Am J Clin Nutr. 2008;87:1107–17. https://doi.org/10.1093/ajcn/87.5.1107.
76. Lachat C, Otchere S, Roberfroid D, Abdulai A, Seret FMA, Milesevic J, et al. Diet and physical activity for the prevention of noncommunicable diseases in low- and middle-income countries: a systematic policy review. PLoS Med. 2013;10:e1001465.
77. Tonsor GT, Lusk JL. U.S. perspective: meat demand outdoes meat avoidance. Meat Sci. 2022;190:108843. https://doi.org/10.1016/j.meatsci.2022.108843.
78. Choi SE, Lee KJ. Ethnic differences in attitudes, beliefs, and patterns of meat consumption among American young women meat eaters. Nutr Res Pract. 2023;17:73–90. https://doi.org/10.4162/nrp.2023.17.1.73.
79. Heiman A, Gordon B, Zilberman D. Food beliefs and food supply chains: the impact of religion and religiosity in Israel. Food Policy. 2019;83:363–9. https://doi.org/10.1016/j.foodpol.2017.07.007.
80. Ansari S, Soltero EG, Lorenzo E, Lee RE. The impact of religiosity on dietary habits and physical activity in minority women participating in the Health is Power (HIP) study. Prev Med Rep. 2017;5:210–3. https://doi.org/10.1016/j.pmedr.2016.12.012.
81. Sabaté J. Religion, diet and research. Br J Nutr. 2004;92:199–201. https://doi.org/10.1079/BJN20041229.
82. Filippini M, Srinivasan S. Impact of religious participation, social interactions and globalization on meat consumption: evidence from India. Energy Econ. 2019;84:104550. https://doi.org/10.1016/j.eneco.2019.104550.
83. Leitzmann C. Vegetarian nutrition: past, present, future. Am J Clin Nutr. 2014;100(Suppl):496S–502S. https://doi.org/10.3945/ajcn.113.071365.
84. Geiker NRW, Bertram HC, Mejborn H, Dragsted LO, Kristensen L, Carrascal JR, et al. Meat and human health-current knowledge and research gaps. Foods (Basel, Switzerland). 2021:10. https://doi.org/10.3390/foods10071556.
85. González N, Marquès M, Nadal M, Domingo JL. Meat consumption: which are the current global risks? A review of recent (2010-2020) evidences. Food Res Int. 2020;137:109341. https://doi.org/10.1016/j.foodres.2020.109341.
86. Grummon AH, Goodman D, Jaacks LM, Taillie LS, Chauvenet CA, Salvia MG, et al. Awareness of and reactions to health and environmental harms of red meat among parents in the United States. Public Health Nutr. 2022;25:893–903. https://doi.org/10.1017/S1368980021003098.
87. Papier K, Fensom GK, Knuppel A, Appleby PN, Tong TYN, Schmidt JA, et al. Meat consumption and risk of 25 common conditions: outcome-wide analyses in 475,000 men and women in the UK Biobank study. BMC Med. 2021;19:53. https://doi.org/10.1186/s12916-021-01922-9.
88. Eshel G, Shepon A, Makov T, Milo R. Land, irrigation water, greenhouse gas, and reactive nitrogen burdens of meat, eggs, and dairy production in the United States. Proc Natl Acad Sci USA. 2014;111:11996–2001. https://doi.org/10.1073/pnas.1402183111.
89. Adams V. Meat consumption and climate change: western countries must cut by 75%. EarthOrg 2022. https://earth.org/meat-consumption-and-climate-change/. Accessed 1 Nov 2023.
90. Rubio NR, Xiang N, Kaplan DL. Plant-based and cell-based approaches to meat production. Nat Commun. 2020;11:6276. https://doi.org/10.1038/s41467-020-20061-y.

91. Iikura M. 27 - Plant-based diets and asthma. In: F M, editor. Vegetarian and plant-based diets in health and disease prevention. Academic; 2017. p. 483–91. https://doi.org/10.1016/B978-0-12-803968-7.00027-7.
92. Yeh M-C, Glick-Bauer M. Chapter 8 – Vegetarian diets and disease outcomes. In: Watson RR, Preedy VR, editors. Fruits, vegetables, and herbs: bioactive foods in health promotion. Academic; 2016. p. 149–64. https://doi.org/10.1016/B978-0-12-802972-5.00008-1.
93. Tonstad S, Butler T, Yan R, Fraser GE. Type of vegetarian diet, body weight, and prevalence of type 2 diabetes. Diabetes Care. 2009;32:791–6. https://doi.org/10.2337/dc08-1886.
94. Ornish D, Brown SE, Billings JH, Scherwitz LW, Armstrong WT, Ports TA, et al. Can lifestyle changes reverse coronary heart disease?: the lifestyle heart trial. Lancet. 1990;336:129–33. https://doi.org/10.1016/0140-6736(90)91656-U.
95. Tuso PJ, Ismail MH, Ha BP, Bartolotto C. Nutritional update for physicians: plant-based diets. Perm J. 2013;17:61–6. https://doi.org/10.7812/TPP/12-085.
96. Medawar E, Huhn S, Villringer A, Veronica WA. The effects of plant-based diets on the body and the brain: a systematic review. Transl Psychiatry. 2019;9:226. https://doi.org/10.1038/s41398-019-0552-0.
97. The Good Food Institute. U.S. plant-based market overview: new SPINS retail sales data 2018. https://www.gfi.org/marketresearch. Accessed 1 Nov 2023.
98. Forgrieve J. The growing acceptance of veganism Forbes. Forbes 2018. https://www.forbes.com/sites/janetforgrieve/2018/11/02/picturing-a-kindler-gentler-world-vegan-month/?sh=70b4b4142f2b. Accessed 1 Nov 2023.
99. Lacroix K, Goldberg M, Marlon J, Rosenthal S, Ballew M, Wang X, et al. Understanding differences in Americans' motivations for eating plant-rich foods. Yale Progr Clim Chang Commun 2022. https://climatecommunication.yale.edu/publications/understanding-differences-in-americans-motivations-for-eating-plant-rich-foods/. Accessed 1 Nov 2023.
100. Durbin D, Mchugh D. Plant-based meat is a simple solution to climate woes — if more people would eat it. Columbian Logo 2023. https://www.columbian.com/news/2023/nov/18/meat-substitutes-solve-climate-woes/#:~:text=Between2018and2022%2Cglobal,that'sdwarfedbyglobal. Accessed 1 Nov 2023.
101. McCarthy N. Who are America's vegans & vegetarians? Statista 2019. https://www.statista.com/chart/14989/who-are-americas-vegans-and-vegetarians/. Accessed 1 Nov 2023.
102. The Sustainable Restaurant Association. Vegan or bust' Isn't helping: why simply reducing Meat consumption matters too. Sustain Restaur Assoc 2023. https://thesra.org/news-insights/insights/vegan-or-bust-isnt-helping-why-simply-reducing-meat-consumption-matters-too/. Accessed 1 Nov 2023.
103. University of Georgia. Veganism may not save the planet: study suggests limited meat consumption better for environment, animals. PhysOrg 2023. https://phys.org/news/2023-02-veganism-planet-limited-meat-consumption.html. Accessed 1 Nov 2023.
104. Trauger A. The vegan industrial complex: the political ecology of not eating animals. J Polit Ecol. 2022;29:639–55. https://doi.org/10.2458/JPE.3052.
105. Dunne D, Prater T, Goodman J. Interactive: what is the climate impact of eating meat and dairy? CarbonBrief 2018. https://interactive.carbonbrief.org/what-is-the-climate-impact-of-eating-meat-and-dairy/. Accessed 1 Nov 2023.
106. Siegrist M, Visschers VHM, Hartmann C. Factors influencing changes in sustainability perception of various food behaviors: results of a longitudinal study. Food Qual Prefer. 2015;46:33–9. https://doi.org/10.1016/j.foodqual.2015.07.006.
107. Camilleri MA, Cricelli L, Mauriello R, Strazzullo S. Consumer perceptions of sustainable products: a systematic literature review. Sustain For. 2023:15. https://doi.org/10.3390/su15118923.
108. Annunziata A, Mariani A. Consumer perception of sustainability attributes in organic and local food. Recent Pat Food Nutr Agric. 2018;9:87–96. https://doi.org/10.2174/2212798410666171215112058.

109. Li S, Kallas Z. Meta-analysis of consumers' willingness to pay for sustainable food products. Appetite. 2021;163:105239. https://doi.org/10.1016/j.appet.2021.105239.
110. Sasaki K, Motoyama M, Watanabe G, Nakajima I. Meat consumption and consumer attitudes in Japan: an overview. Meat Sci. 2022;192:108879. https://doi.org/10.1016/j.meatsci.2022.108879.
111. Bastounis A, Buckell J, Hartmann-Boyce J, Cook B, King S, Potter C, et al. The impact of environmental sustainability labels on willingness-to-pay for foods: a systematic review and meta-analysis of discrete choice experiments. Nutrients. 2021:13. https://doi.org/10.3390/nu13082677.
112. Kaygisiz F, Acar Bolat B, Bulut D. Determining factors affecting Consumer's decision to purchase organic chicken meat. Braz J Poult Sci. 2019:21. https://doi.org/10.1590/1806-9061-2019-1060.

Practices for Healthier Meat and Meat Products

Pradeep Kumar Singh, Namrata Agrawal, and Shreya Suman

1 Introduction

The focus in meat product development should prioritize assessing the impact of adding functional ingredients on their physical, chemical and sensory properties to enhance their nutritional value and health benefits. While direct use of additives and functional nutrients may affect taste, utilizing active packaging films with technologies like coatings, microencapsulation and nanoencapsulation can effectively introduce antioxidants, antimicrobial and other functionalities, improving safety, health and sensory qualities. Consumer perception studies are crucial to guide the meat industry for production healthier products. Additionally, there's a growing trend towards ready-to-eat meat products, requiring efforts to enhance their appeal and healthiness, possibly through technological innovations that are both feasible and perceived as healthy by consumers. Thus, ensuring the healthiness of meat products remains an ongoing challenge and a key focus in both research and future consumption trends [1].

Lean meat usually contains low levels of fat, contrary to common belief that it is free from fat. While excessive fat consumption has been linked to certain diseases, fat in meat provides flavour, aroma and texture and is essential for carrying fat-soluble vitamins and essential fatty acids. Recommendations in Germany and the UK suggest reducing saturated and trans fats while increasing unsaturated fats in the diet to promote health. Research shows that a lower ratio of n-6 to n-3 fatty acids may reduce the risk of chronic diseases [2]. Consumers are increasingly interested in the nutritional aspects of animal products, such as the fatty acid profile, which can be altered by changes in animal diets. Conjugated linoleic acids (CLA), found in ruminant products, are being studied for their potential health benefits, including

P. K. Singh (✉) · N. Agrawal · S. Suman
Department of Livestock Products Technology, College of Veterinary Science and Animal Husbandry, NDVSU, Jabalpur, Madhya Pradesh, India

A. Kumar Pathera et al. (eds.), *Healthier Meat Products*,
https://doi.org/10.1007/978-3-031-78215-2_2

anticarcinogenic effects. The diet of cattle can influence the fatty acid composition of meat, with forage leading to higher levels of beneficial fatty acids. Altering the diet may impact sensory quality and lipid stability of meat [3].

In addition to the formulation approach, there are production practices, that promise the desired changes in the meat produced after following standard protocols. Specially the feeding practices, treatment regimen with reduced antibiotic use, regenerative agriculture, animal welfare practices, resource management and supply chain management strengthened with a traceability system ensure safe and healthy meat production.

2 Producing Healthy Meat Products

The nutritional attributes of meat and poultry products encompass both positive and negative aspects. While they are significant sources of essential nutrients like iron, zinc, B vitamins and conjugated linoleic acid, they also contain high levels of saturated fats, cholesterol, sodium and calories. However, strategies such as selecting lean cuts, removing fat, altering fatty acid composition and controlling portion sizes can mitigate some of these negative aspects [4]. Functional foods, which exceed the nutritional profiles of conventional products, offer an opportunity to enhance the nutritional value of meat products further. In determining suitable nutrients for functional foods, factors like current dietary intake levels, biological efficacy, stability of food products and impact on quality parameters must be considered. In the United States, adults often consume inadequate amounts of calcium, potassium, magnesium, fiber, vitamins A, C and E, as well as omega-3 fatty acids [5]. Emerging health-promoting bioactive compounds like conjugated linoleic acid and bioactive peptides hold promise as potential additions to functional foods [6].

Over the past 50–60 years, research has linked high-fat and saturated fatty acid diets to various diseases like heart disease and colon cancer. However, the Greenlandic Inuit's low incidence of heart disease despite a fat-rich diet due to consuming n-3 fatty acids from fish spurred interest in these fats. Recommendations from the Food and Agriculture Organization suggest fat intake at 15–35% of energy, with limited saturated fats and an emphasis on polyunsaturated fats. Meat products can have high fat content, impacting their quality. Meat analogues, or plant-based meats, offer cholesterol-free, low saturated fat alternatives rich in essential nutrients, resembling meat in taste and texture. There's growing consumer demand for "clean-label" products free from synthetic additives, leading to a shift towards natural additives sourced from plants, animals, or microbes. These additives, particularly those from plants, offer antimicrobial and antioxidant properties, aiding in food preservation in addition to nutritional variations. The food industry is reformulating processed foods to improve nutrition by reducing salt, fat and replacing synthetic additives with natural alternatives [7].

3 Feeding Practices

Consumer perceptions of animal-source products involve multiple factors such as sensory appeal, nutritional value, image, convenience, safety, technological aspects and commercial considerations [8]. Beef primarily consists of water, protein and fat, with protein being crucial for essential amino acids. Red meat, in particular, is rich in high-quality protein. While protein content doesn't significantly differ between organic and conventionally produced beef, fat content varies widely depending on factors like the animal's breed and the production methods used. Fat contributes not only to energy but also provides essential vitamins and fatty acids while enhancing flavour [9]. Marbling, the intramuscular fat in beef, is influenced by various factors such as breed, diet, age at slaughter and muscle type. Diets high in concentrate feed and feedlot finishing tend to result in higher marbling [10]. The effect of feed system on the fatty acid profile of meat is a complex; has remarkable implications for human nutrition and health. While ruminant meat contributes significantly to saturated fatty acid (SFA) intake, the differences in SFA content between organic, grass-fed and conventional meat are inconsistent [11]. Organic and grass-fed beef tend to have lower SFA content compared to conventional beef. Grass-fed beef, in particular, has been shown to have a more favourable SFA profile, with lower concentrations of myristic and palmitic acids, which may have adverse effects on serum cholesterol [12].

Conventional beef generally contains higher levels of monounsaturated fatty acids (MUFA) compared to organic and grass-fed alternatives. However, organic and grass-fed meat may contain higher levels of vaccenic acid (VA), which can be converted into beneficial conjugated linoleic acid (CLA9) in the body, potentially contributing to positive health outcomes [13]. Total polyunsaturated fatty acid (PUFA) content is consistently higher in grass-fed and organic meat compared to conventional meat. While this increase in PUFA may have positive health implications, it is important to note that it often comes at the expense of MUFA, which are recommended as replacements for SFA in the diet [11]. Grass-fed and organic beef are also richer sources of omega-3 fatty acids, particularly very-long-chain (VLC) omega-3s, which have been associated with numerous health benefits; including reduced cardiovascular risk and improved cognitive function. The omega-6 to omega-3 ratio is lower in beef from grazing systems, reflecting the higher omega-3 content and potentially offering additional health benefits [14].

Conjugated linoleic acid (CLA) concentrations in beef increase with higher forage content in the animal's diet, similar to omega-3 fatty acids. However, the overall impact of increased CLA consumption on health outcomes is difficult to assess given the mixed evidence regarding the health effects of red meat consumption in general [15]. Meat colour may not be strongly linked to its nutritional or organoleptic qualities, although its importance lies in its ability to impact consumer acceptance [16]. Differences in colour can be ascribed to various factors. Typically, meat from animals grazing on grass appears darker in hue compared to meat sourced from animals fed concentrates [17]. These animals could have a greater concentration of haemic pigments in their muscles, as a result of physical activity. A well-established relationship exists between muscle ultimate pH and the lightness of meat, showing an inverse correlation [18]. Animals raised on grass have wider ultimate pH variability and lower glycolytic potential than animals given concentrates [19].

The flavour of meat derived from animals raised on pasture differs from that of animals fed concentrates [20]. The presence of distinct volatile compounds in meat is influenced by both the diet and the species of the animal. For instance, the flavour of sheep and beef meat varies. Branched-chain fatty acids (BCFAs) are recognized as playing a role in contributing to the species-specific flavour [21]. Elevated levels of branched-chain fatty acids are frequently linked to the intense flavour characteristic of sheep meat. Sheep and goats tend to accumulate branched-chain fatty acids (BCFA) when their diet is high in grains or low in fibre [22]. Skatole is a byproduct of tryptophan degradation and its concentration tends to be higher in animals raised on pasture [16]. Grass-based diets generally contain a higher proportion of protein compared to non-fibrous diets [19]. Consequently, in the rumen of animals consuming grass-based diets, there is typically a higher rate of amino acid deamination by microorganisms compared to animals fed concentrate diets [19]. Linolenic acid has a very important role in determining meat flavour. Unsaturated linolenic acid (C18:3) is a fatty acid characteristic of forage lipid. Flavour intensity was positively correlated with milky-oily aroma and flavour and with sour flavour. Terpenoids are additional compounds that contribute to meat flavour. These compounds are directly transmitted from pasture to animal tissue. The abundance of terpenoids in animal tissues is heavily reliant on the botanical composition of the pasture [23].

4 Antibiotics Residue in Meat: Its Control and Reduction

Antibiotics are defined as a category of natural, semi-synthetic or chemical compounds with anti-microbial activity that are extensively used to control and prevent infectious diseases in animals and humans and may be added to feed as growth promoters to promote the growth and development of animals. Feeding antibiotics at sub-therapeutic levels in animal feed can result in the presence of antibiotic residues in food products such as milk, eggs and meat [24]. These antibiotic residues can potentially lead to a range of toxic effects in humans, including the transfer of antibiotic-resistant bacteria, allergic reactions, immunopathological effects, nephropathy (associated with gentamicin), bone marrow toxicity (linked to chloramphenicol), carcinogenicity (sulphamethazine, oxytetracycline and furazolidone), mutagenicity, hepatotoxicity, reproductive disorders and in severe cases, anaphylactic shock [25]. The utilization of antibiotics in poultry and livestock production is perceived as beneficial to farmers and the economy, as it has consistently enhanced poultry performance in an efficient and cost-effective manner. However, there are concerns regarding the potential dissemination of antibiotic-resistant strains of both pathogenic and non-pathogenic organisms into the environment. These resistant strains could subsequently be transmitted to humans through the food chain, posing significant risks to public health. Therefore, while antibiotics offer advantages in terms of agricultural productivity, their overuse and the development of antibiotic resistance raise serious concerns that must be carefully managed to safeguard public health problems associated with antibiotic use in animals [24].

The European Union (EU) defines antibiotic residues as "pharmacologically active substances (including active ingredients, excipients or degradation products) and their metabolites that persist in food products derived from animals to which the relevant veterinary medicinal products have been administered". Allergic reactions, particularly to antibiotics like penicillin, are among the most significant adverse effects associated with antibiotic residues in food [26]. Idiosyncratic reactions, such as skin rashes, allergies and phototoxic dermatitis, have been reported in association with the use of tetracycline antibiotics [27]. Various pathological conditions have been reported in association with antibiotic residues in food. These include immunopathological effects, carcinogenicity (linked to substances like sulphamethazine, oxytetracycline and furazolidone), mutagenicity, teratogenicity, disruptions in the normal intestinal environment, nephropathy (related to Gentamicin), hepatotoxicity, bone marrow toxicity (associated with chloramphenicol) and various reproductive disorders [28]. Heat treatment of animal foods has the potential to deactivate antibiotics [25]. Numerous studies have indicated that the degradation of antibiotics such as quinolones, β-lactams, macrolides, aminoglycosides, sulphonamides and tetracyclines is influenced by temperature. Prolonged heating times contribute to increased degradation of these antibiotics [29].

Oxytetracycline (OTC) residues were detected in 10% of imported beef samples and 37.00% of local beef samples. The levels of OTC residues ranged from 12.7 to 56.0 μg/kg for imported beef and 17.5 to 601.8 μg/kg for local beef, with highly significant differences between the two. According to regulatory standards, the maximum residue limit (MRL) of OTC in meat is 200 μg/kg. It was observed that about 17.00% of local beef samples exceeded this limit [30]. The presence of OTC residues in meat raises concerns about potential health risks to consumers, including gastrointestinal disturbances and the development of resistant pathogens [31]. Penicillin residues were detected in around 27.00% of imported beef samples and 23.00% of local beef samples. The levels of penicillin residues ranged from 3.6 to 24.1 μg/kg for imported beef and 5.9 to 71.3 μg/kg for local beef, with highly significant differences between the two. The MRL of penicillin residues in meat is 50 μg/kg and it was observed that all imported beef samples complied with this limit, while 10% of local beef samples exceeded it [32].

A study evaluated the effect of different cooking methods on the reduction of antibiotic residues in beef. It was found that processing techniques like grilling, followed by microwave cooking, was the most effective method for reducing residues, where as boiling was comparatively less effective. Proper use of antibiotics and strict adherence to withdrawal periods, along with different heat treatments are recommended to minimize consumer exposure to residues. Additionally, discarding meat broth during boiling can help reduce exposure to residues [33].

5 Regenerative Agriculture in Meat Production

Regenerative agriculture involves a holistic approach to farming and grazing, where the aim is to maintain a symbiotic relationship between food production and the environment. This involves employing techniques like cover cropping, mulching, no-till farming, integrating livestock and rotational grazing to preserve or enhance ecosystems. These methods help in improving the soil health, managing water resources and carbon sequestration, and finally countering the ecological imbalances [34].

The rebranding of cattle within the framework of regenerative agriculture encompasses two key elements: the re-evaluation of ruminant methane emissions and the enhancement of soil carbon sequestration through ruminant grazing. Firstly, the discourse challenges the conventional understanding of ruminant methane emissions, arguing that standard metrics overstate their contribution to global warming [35]. This is supported by a new metric, GWP (Global Warming Potential), which emphasizes the temporal dynamics of emissions, suggesting that stable methane emissions do not significantly exacerbate warming. Advocates claim that historical baselines justify current methane emissions from livestock, particularly in pasture-based systems and argue for the adoption of GWP in greenhouse gas accounting [36]. Secondly, proponents highlight the role of ruminant grazing in soil health and biodiversity conservation. They advocate for holistic grazing practices that mimic natural ecosystems, promote soil carbon sequestration and enhance agricultural productivity [37]. By integrating livestock into arable systems and emphasizing the historical precedent of mixed farming, they position pasture-fed livestock as a sustainable alternative to intensive crop production. This approach challenges the environmental credentials of veganism based on monoculture crop systems and underscores the ecological benefits of traditional livestock management practices [38].

Researchers observed that White Oak Pastures achieved higher productivity when considering all animal products combined compared to other grass-finished operations focusing solely on beef production. However, when focusing solely on beef, the productivity was slightly lower. Over a 20-year period, several soil health indicators showed substantial improvements. Regarding greenhouse gas (GHG) emissions, White Oak Pastures had a carbon footprint 44% higher than conventional production methods for all livestock combined, mainly due to grass-finished cattle emitting more methane than grain-finished ones. However, accounting for soil carbon sequestration significantly reduced net emissions by five-fold, resulting in a level lower than conventional production. In fact, the beef produced at White Oak Pastures acted as a carbon sink, offsetting 100% of the emissions from raising the animals and more. Additionally, cattle at White Oak Pastures reached harvestable weight earlier, at 20–22 months compared to 30 in conventional production, resulting in 43% fewer GHG emissions over their lifetime. Overall, the GHG footprint at this ranch was 66% lower than comparable conventional production. While the benefits to soil health, GHG emissions and carbon sequestration are evident, there is a trade-off in terms of land use. White Oak Pastures requires 2.5 times more land for the same amount of meat produced compared to typical conventional operations. This difference can be attributed to the rotational grazing practices of White Oak Pastures, which allow for land regeneration and rest, contrasting with the confined space and continuous use of land in conventional systems [39].

6 Animal Welfare Practices

The quality of meat begins at the farm, where the care provided by producers significantly influences it. Implementing optimal practices in animal handling, healthcare, nutrition and transportation all contribute to the production of meat; are greatly appreciated [40]. Animal welfare pertains to the overall physical and emotional condition of animals, influenced by human attitudes and actions, the quantity and quality of resources accessible to them and the environment they inhabit [41]. Current research is centred on enhancing animal welfare by promoting positive states rather than solely focusing on eliminating negative aspects. A five-domain model, akin to the five freedoms concept, has been developed to advance animal welfare. These domains encompass nutrition, environment, health, and behaviour, along with an overarching mental component [42]. The OIE (Office International des Epizooties) has formulated guidelines on animal welfare aimed at facilitating international trade among World Trade Organization members. While not obligatory, these standards facilitate and streamline trade among OIE member states [43].

Society's perceptions of animal welfare are shaped by cultural and economic factors, leading to two main viewpoints: animal welfare and animal rights. The animal welfare perspective supports the responsible use of animals by humans as long as their welfare is maintained and unnecessary suffering is avoided. Conversely, animal rights advocates argue that animals should have rights similar to those of humans and should not be exploited in any way. This stance goes beyond basic welfare considerations to grant animals rights comparable to those of humans [44]. While these perspectives often differ, they both aim to enhance animal welfare. Some of their views are summarized in Table 1.

Recent outbreaks of zoonotic diseases, exemplified by the COVID-19 pandemic, have heightened public awareness regarding the increased risk of infectious diseases associated with animals [46]. A related concern is the welfare of animals in industrialized livestock production, transportation, and slaughter methods. Some suggest that lessons from efforts to enhance the welfare of laboratory animals through the ethical principles of the Three Rs—Reduction, Replacement and Refinement—could inform similar improvements in agricultural and aquacultural settings. These principles aim to reduce the number of animals used, replace animals with alternatives where possible and refine procedures to minimize animal pain and distress. This includes both Absolute Replacement (substituting animals with inanimate objects) and Relative Replacement (replacing more sentient animals with those with lower potential for pain perception). The concept is that applying

Table 1 A comparison of different views relating to animal welfare

Animal welfare advocates	Animal rights advocates
Judicious use of animals for human benefit is morally right	Using animals for human benefit is morally wrong
Human interests always come first before animal interests.	Human interests should not overrule animal interests
Humans should not cause animals unnecessary pain or death	Humans should not cause pain or death at all in animals
Treat animals as humanely as conveniently possible	Treat animals humanely always and eliminate human-made causes of animal suffering
Humane euthanasia/killing of animals	No killing of animals
Objects to cruel practices such as dog fighting, confinement of veal calves and pregnant sows	Abolish the use of animals in any form, it is exploitation.
	Animals have the same rights as humans

Adapted from [45]

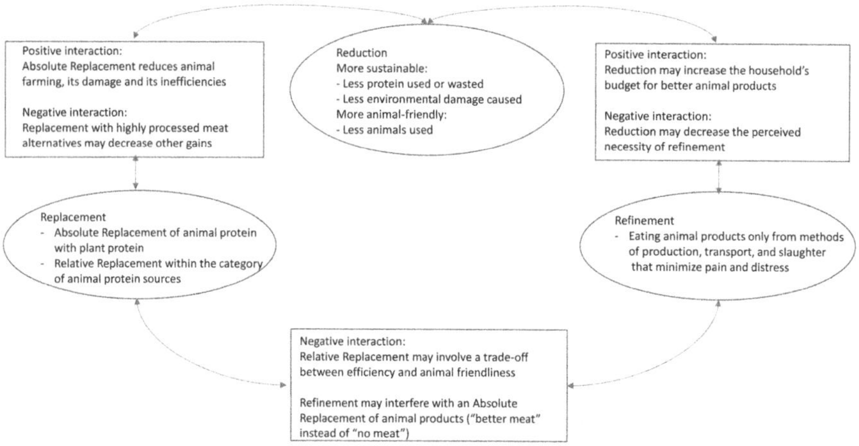

Fig. 1 Overview of the three Rs and their potential interactions. (Adapted from [49])

these principles effectively, as seen in animal testing, could be adapted to promote a shared ethic of animal use in food production [47].

Consumer responses to animal welfare concerns may not align with ethical principles, as many individuals tend to have modest moral aspirations, aiming to match the moral standards of their peers without striving to exceed or fall below them explicitly. This phenomenon is exemplified by "do-gooder derogation," where people react unfavourably to those perceived as advocating exceptionally moral conduct. This reaction can significantly impede communication between omnivores and vegetarians, as demonstrated by research [48].

The Fig. 1 illustrates three principles and their potential interactions. For instance, Reduction and Refinement may lead some households to allocate savings from buying less animal protein towards purchasing meat with higher welfare standards.

However, for other households, consuming less animal protein might reduce the urgency to buy better meat. Similarly, opting for meat with higher welfare standards may lessen the perceived need for Absolute Replacement. Overall, these interactions offer various options for meat-eating consumers interested in improving animal welfare. Consumer reactions depend on how they interpret these options and their priorities alongside animal welfare. For consumers prioritizing animal welfare, choosing meals with less animal protein (Reduction and Absolute Replacement) might be the simplest. If this conflicts with other goals, opting for "less but better" meat (Reduction and Refinement) could be an alternative. Alternatively, choosing "less worrisome" meat (Relative Replacement and Refinement) might be preferred.

Recent experimental studies have explored ways to reduce meat consumption by appealing to animal welfare concerns. These interventions include displaying images of meat dishes alongside pictures of the animals they come from, showing graphic depictions of welfare conditions in factory farms, sharing information about animal-friendly behaviours and providing suggestions for reducing meat consumption [50]. Another significant development is the maturation of the market for plant-based alternatives. There's now a wide variety of convenient products made from novel protein ingredients, including plant-based and cell-based options, with items like burgers gaining popularity [51].

7 Resource Conservation in Meat Production

Meat production has a significantly larger environmental and climate footprint compared to plant-based foods, utilizing a substantial amount of land and water; contributing to greenhouse gas emissions. Animal-based foods are responsible for nitrogen and phosphorus pollution too, in addition to a quantum of freshwater footprint and food-related greenhouse gas emissions. It is said that meat production is a major driver of deforestation and biodiversity loss, leading to threat the global ecosystem functions [52]. At local levels, intensive meat production systems, particularly feedlots, have negative externalities such as poor water and air quality, which can harm local communities' health. The water footprints of meat products are notably larger than those of most plant-based products, primarily due to feed production. Ruminants in extensive grazing systems have large water footprints, although much of this water is rainwater with low competition for other uses. However, issues of water scarcity and pollution mainly relate to groundwater and surface water, making extensive grazing systems preferable over intensive production systems. Overall, the comparison between meat and plant-based foods depends on the production methods used. For instance, intensively produced pork and beef have larger water footprints than cereals and pulses, while beef, sheep and goat meat produced in extensive grazing systems have smaller water footprints [53].

Poultry processing facilities primarily rely on water for carcass cooling post-evisceration and to administer antimicrobials for bacterial control. Water chillers, some as large as 200,000 gallons, are utilized for broiler chickens and turkeys, with significant water usage occurring from evisceration to carcass chilling. Opportunities for water conservation exist in various areas, including reusing water from chill tanks for scalding tanks [54]. Studies in Brazil propose strategies such as reusing effluent from cleaning and cooling towers, resulting in substantial reductions in water consumption and effluent generation [55]. Incorporating water-saving technologies like pressure-activated taps and infrared devices can further enhance conservation efforts. Beef processing facilities also emphasize water conservation, with potential savings of up to 10% reported. In pork and beef processing, air chilling is primary, but water spray chill systems are also used. While water management consulting services are available to the food industry, practicality and economic feasibility of water reuse technologies vary for small slaughter establishments. Overall, implementing water audits and adopting good practices can help identify and address inefficiencies, leading to significant water conservation and cost savings in processing environments [56].

Genetically modified organisms (GMOs), produced by global corporate companies in developed countries, have become increasingly prevalent in food production. These organisms are genetically engineered to enhance desirable traits such as pest resistance and drought tolerance. Genetic engineering can be done with plants, animals or bacteria. With genetic engineering, scientists take the gene for a desired trait in one organism and they insert that gene into the DNA of another organism. Genetically modifying crops to improve water efficiency, could help conserve water resources in food production. For instance, the HARDY gene has been identified to enhance water efficiency in rice by increasing photosynthesis rates and strengthening root structures. GMO animals, such as genetically engineered salmon, have also raised concerns about consumer health, animal welfare and environmental impacts. It is essential to prioritize food safety principles throughout the production process to mitigate potential risks associated with GMO consumption [57].

8 Traceability in the Meat Supply Chain

Traceability in supply chains refers to the ability to retrieve detailed information about products throughout the production and marketing chain. It basically involves tracking system involving tracing items from manufacturing to sale; a new trend contributing to transparency. Here the transparency refers to visibility of supply chain points to different stakeholders. It is an important instrument for gaining consumer trust and establishing brand reputation. Traceability enables firms to manage resources efficiently by providing an information trail and improve consumer perceptions of product safety [58]. In any industry successful traceability depends on effective information sharing without which traceability system is difficult to establish. The system of traceability involves maintaining and tracking data records throughout the key players of supply chain; producers, processors, distributors and consumers. Important features of a traceability system include maintenance of permanent identification marks for every batch/product/lot and storing data securely.

For planning a traceability model there are no fixed rules; it depends on the needs of the organization. At the producer level, attention is focused on recording both livestock details and farming activities to ensure meat quality. This includes geotagging farm locations, managing livestock performance, marking animals for processing and recording various reports and certificates related to health, breeding and feeding. Various tracing methods have evolved, from conventional ear tags and tattoos to advanced technologies like RFID (Radio Frequency Identification), DNA-based tracking, and real-time health monitoring. During processing, meat undergoes slaughtering, processing and packing. The ability to link processes to their source is crucial for tracking products. Cut meat is marked, tagged and linked to its source, undergoing quality checks before packing. Consumers now demand meat safety, quality and sustainability. Technological advancements meet these demands through real-time tracking, QR (Quick Response) code scanning and software implementations. Blockchain traceability is becoming increasingly important in ensuring food safety and product integrity. It provides an extra layer of transparency in the meat supply chain, allowing for end-to-end visibility and customer engagement [59].

Livestock traceability systems offer numerous benefits across various aspects. Animal registration and identification, typically done through visible ear tags or RFID devices, facilitate ownership confirmation, streamline transportation clearances and support precision farming, scientific feeding and health management. These identification systems also aid in recording performance parameters for livestock selection in breeding programs, thus enhancing overall breed quality [60]. Premises registration, including farm and abattoirs, enables centralized monitoring for comprehensive quality assurance. Centralized information on animals, farms and abattoirs facilitates effective disease control programs by enabling the tracking of disease origins during meat inspection. This helps in understanding disease epidemiology, preventing spread, and implementing biosecurity measures. Traceability records also assist in identifying sources of chemical contaminants in meat, contributing to comprehensive quality assurance systems that ensure safe and high-quality meat production free from physical, chemical and biological hazards. These systems document food safety hazards and are expected by importing countries, enhancing market access for traceable meat products and boosting export opportunities [61].

Animal identification is crucial for livestock traceability systems, involving either individual identification or collective identification of groups. The identification method must be tamper-proof and retain its code throughout the animal's life. In the European Union, animals are tagged with two sets of ear tags to ensure identification even if one tag is lost. Ideal identification methods should be resistant to environmental conditions, economical, easily applicable and tamper-proof [62].

Various methods are used for animal identification, including tattooing, branding, bar code tags, visual tags, RFID tags etc. Visual tagging is a simple method where a tag of plastic material with a visible number is fixed to an animal's ear. Bar-coded tags are similar one which contain optical labels that are readable with a scanner. RFID devices are more advanced one that offer non-contact electronic data reading from a distance and are even resistant to dirt and dust [63]. Quick response

(QR) code-based tags use matrix bar code systems, easily readable with smartphone apps [64]. These methods provide options for reliable and efficient animal identification, crucial for maintaining traceability and ensuring the safety and quality of livestock products.

9 Conclusion

The focus in meat product development should prioritize assessing the impact of adding functional ingredients on their physical, chemical, and sensory properties to enhance their nutritional value and health benefits. There is a growing trend towards ready-to-eat meat products, requiring efforts to enhance their appeal and healthiness. Nutritional attributes of meat and poultry products encompass both positive and negative aspects, but strategies such as selecting lean cuts, removing fat, altering fatty acid composition and controlling portion sizes can mitigate some of these negative aspects. Consumers are increasingly interested in the nutritional aspects of animal products, such as the fatty acid profile, which can be altered by changes in animal diets. The approaches of producing meat from grassfed or pasture-fed animals certainly improve the nutritional profile. Practices followed for reduction of antibiotic residues in meat or farm practices involving resource management with positive impact on the environment producing a positive impact on the meat industry. The supply chain management system; empowered with transparency and traceability system, is an appreciable tool to produce a safe and healthy meat. Overall a composite and meticulous planning is required with holistic approach following standard practices to produce healthy meat and meat products.

References

1. Teixeira A, Rodrigues S. Consumer perceptions towards healthier meat products. Curr Opin Food Sci. 2021;38:147–54.
2. Dewhurst RJ, Scollan ND, Lee MR, Ougham HJ, Humphreys MO. Forage breeding and management to increase the beneficial fatty acid content of ruminant products. Proc Nutr Soc. 2003;62(2):329–36.
3. Banni S, Murru E, Angioni E, Carta G, Melis MP. Conjugated linoleic acid isomers (CLA): good for everything. Sci Aliments. 2002;22(4):371–80.
4. Jiménez-Colmenero F, Carballo J, Cofrades S. Healthier meat and meat products: their role as functional foods. Meat Sci. 2001;59(1):5–13.
5. Whitney EN, Rolfes SR, Crowe T, Walsh A. Understanding nutrition. 9th ed. Wadsworth/Belmont: Cengage AU; 2002. p. 1–791.
6. Harris WS. International recommendations for consumption of long-chain omega-3 fatty acids. J Cardiovasc Med. 2007;8:S50–2.
7. Kurćubić V, Stajić S, Miletić N, Stanišić N. Healthier meat products are fashionable—consumers love fashion. Appl Sci. 2022;12(19):10129.

8. Prache S, Adamiec C, Astruc T, Baéza-Campone E, Bouillot PE, Clinquart A, Feidt C, Fourat E, Gautron J, Girard A, Guillier L, Kesse-Guyot E, Lebret B, Lefèvre F, Le Perchec S, Martin B, Mirade PS, Pierre F, Raulet M, Rémond D, Santé-Lhoutellier V. Quality of animal-source foods. Animal. 2022;16:100376.
9. Wyness L, Weichselbaum E, O'connor A, Williams EB, Benelam B, Riley H, Stanner S. Red meat in the diet: an update. Nutr Bull. 2011;36(1):34–77.
10. Park SJ, Beak SH, Kim SY, Jeong IH, Piao MY, Kang HJ, Fassah DM, Na SW, Yoo SP, Baik M. Genetic, management and nutritional factors affecting intramuscular fat deposition in beef cattle—a review. Asian Australas J Anim Sci. 2018;31(7):1043–61.
11. Lenighan YM, Nugent AP, Moloney AP, Monahan FJ, Walton J, Flynn A, Roche HM, McNulty BA. A modelling approach to investigate the impact of consumption of three different beef compositions on human dietary fat intakes. Public Health Nutr. 2020;23(13):2373–83.
12. Ribas-Agustí A, Díaz I, Sárraga C, García-Regueiro JA, Castellari M. Nutritional properties of organic and conventional beef meat at retail. J Sci Food Agric. 2019;99(9):4218–25.
13. Łozicki A, Dymnicka M, Arkuszewska E, Pustkowiak H. Effect of pasture or maize silage feeding on the nutritional value of beef. Ann Anim Sci. 2012;12(1):81–93.
14. Vahmani P, Ponnampalam EN, Kraft J, Mapiye C, Bermingham EN, Watkins PJ, Proctor SD, Dugan ME. Bioactivity and health effects of ruminant meat lipids. Meat Sci. 2020;165:108114.
15. Turner TD, Jensen J, Pilfold JL, Prema D, Donkor KK, Cinel B, Thompson DJ, Dugan MER, Church JS. Comparison of fatty acids in beef tissues from conventional, organic and natural feeding systems in western Canada. Can J Anim Sci. 2022;95(1):49–58.
16. Young OA, Baumeister BMB. The effect of diet on the flavour of cooked beef and the odour compounds in beef fat. N Z J Agric Res. 1999;42:297–304.
17. Priolo A, Micol D, Agabriel J. Effects of grass feeding systems on ruminant meat colour and flavour. A review. Anim Res. 2001;50(3):185–200.
18. Renerre M. La couleur de la viande et samesure. Viande et Produits Carnés. 1981;2:10–6.
19. Sheath GW, Coulon JB, Young OA. Grassland management and animal product quality. In: Proceedings of the XIX International Grassland Congress, Sao Paulo (Brazil), 11–21 February 2001. p. 1019–26.
20. Bailey ME, Suzuki J, Joseph HG, Ross CV, Keisler DH, Purchas RW. Volatile compounds and "grassy" flavor of lamb and beef related to feeding and storage. In: Proceedings of the 34th international congress of meat science and technology; 1988; Brisbane, Australia. p. 187–9.
21. Muir PD, Smith NB, Wallace GJ, Cruickshank GJ, Smith DR. The effect of short-term grain feeding on liveweight gain and beef quality. N Z J Agric Res. 1998;41(4):517–26.
22. Van Soest PJ. Nutritional ecology of the ruminant. 2nd ed. Ithaca: Cornell University Press; 1984.
23. Martin B, Priolo A, Valvo MA, Micol D, Coulon JB. Effects of grass feeding on milk, cheese and meat sensory properties. Options Mediterraneennes Serie A, Seminaires Mediterraneens. 2005;67:213–23.
24. Dutta TK, Yadav SK, Chatterjee A. Antibiotics as feed additives for livestock: human health concerns. Indian J Anim Health. 2019;58(2):121–36.
25. Darwish WS, Eldaly EA, El-Abbasy MT, Ikenaka Y, Nakayama S, Ishizuka M. Antibiotic residues in food: the African scenario. Jpn J Vet Res. 2013;61(Supplement):S13–22.
26. Katz SE, Brady MS. Antibiotic residues in food and their significance. Food Biotechnol. 2000;14(3):147–71.
27. Yates AB, deShazo RD. Allergic and nonallergic drug reactions. South Med J. 2003;96(11):1080–8.
28. Nisha AR. Antibiotic residues-a global health hazard. Vet World. 2008;1(12):375.
29. Tian L, Khalil S, Bayen S. Effect of thermal treatments on the degradation of antibiotic residues in food. Crit Rev Food Sci Nutr. 2017;57(17):3760–70.

30. Mehran Mesgari A, Mahboob N, Hossein B, Masoud A, Ashraf OSadat N. Solid-phase extraction and simultaneous determination of tetracycline residues in edible cattle tissues using an HPLC-FL method. Iran J Pharm Res. 2012;11(3):781–7.
31. Van den Bogaard AE, Stobberingh EE. Epidemiology of resistance to antibiotics: links between animals and humans. Int J Antimicrob Agents. 2000;14(4):327–35.
32. Adesokan HK, Agada CA, Adetunji VO. Oxytetracycline and penicillin-G residues in cattle slaughtered in south-western Nigeria: implications for livestock disease management and public health. J S Afr Vet Assoc. 2013;84(1):1–5.
33. Heshmati A. Impact of cooking procedures on antibacterial drug residues in food: a review. JFQHC. 2015 Jun;10(2):33–7.
34. True primal. Animal Welfare and Regenerative Agriculture [Internet]. 2024. Available from: https://trueprimal.com/animal-welfare-and-regenerative-agriculture
35. Smith G. New methane math could take the heat off cows 2021. Available from: https://www.darigold.com/new-methane-math-could-take-the-heat-off-cows/
36. Ridoutt B. Climate impact of Australian livestock production assessed using the GWP* climate metric. Livest Sci. 2021;246:104459.
37. Chang J, Ciais P, Gasser T, Smith P, Herrero M, Havlík P, Obersteiner M, Guenet B, Goll DS, Li W, Naipal V. Climate warming from managed grasslands cancels the cooling effect of carbon sinks in sparsely grazed and natural grasslands. Nat Commun. 2021;12(1):118.
38. Sexton AE, Garnett T, Lorimer J. Vegan food geographies and the rise of Big Veganism. Prog Hum Geogr. 2022;46(2):605–28.
39. Sheryl Karas MA. Research with Exciting Implications for Regenerative Agriculture and Climate Change [Internet]. 2024. Available from: https://www.csuchico.edu/regenerativeagriculture/blog/multi-species-pasture-study.shtml
40. Penn state extension. High Quality Meat Starts with Animal Care [Internet]. 2022. Available from: https://extension.psu.edu/high-quality-meat-starts-with-animal-care
41. Broom DM. Sentience and animal welfare: new thoughts and controversies. Anim Sentience. 2016;1(5):11.
42. Mellor DJ, Beausoleil NJ. Extending the 'Five Domains' model for animal welfare assessment to incorporate positive welfare states. Anim Welf. 2015 Aug;24(3):241–53.
43. OIE. Terrestrial Animal Health Code 2018 [Internet]. Available from: http://www.oie.int/index.php?id=169&L=0&htmfile=sommaire.htm
44. Phillips C, Phillips C. Animal welfare and animal rights. In: The welfare of animals: the silent majority; 2009. p. 55–77.
45. Madzingira O. Animal welfare considerations in food-producing animals. Anim Welf. 2018;99:171–9.
46. Espinosa R, Tago D, Treich N. Infectious diseases and meat production. Environ Resour Econ. 2020;76(4):1019–44.
47. Van Heek J, Arning K, Ziefle M. Reduce, reuse, recycle: Acceptance of CO2-utilization for plastic products. Energy Policy. 2017;105:53–66.
48. Minson JA, Monin B. Do-gooder derogation: Disparaging morally motivated minorities to defuse anticipated reproach. Soc Psychol Personal Sci. 2012;3(2):200–7.
49. De Boo MJ, Rennie AE, Buchanan-Smith HM, Hendriksen CF. The interplay between replacement, reduction and refinement: considerations where the Three Rs interact. Anim Welf. 2005;14(4):327–32.
50. Mathur MB, Peacock J, Reichling DB, Nadler J, Bain PA, Gardner CD, Robinson TN. Interventions to reduce meat consumption by appealing to animal welfare: meta-analysis and evidence-based recommendations. Appetite. 2021;164:105277.
51. He J, Evans NM, Liu H, Shao S. A review of research on plant-based meat alternatives: driving forces, history, manufacturing and consumer attitudes. Compr Rev Food Sci Food Saf. 2020;19(5):2639–56.

52. Domingo NG, Balasubramanian S, Thakrar SK, Clark MA, Adams PJ, Marshall JD, Muller NZ, Pandis SN, Polasky S, Robinson AL, Tessum CW. Air quality–related health damages of food. Proc Natl Acad Sci. 2021;118(20):e2013637118.
53. Gerbens-Leenes PW, Mekonnen MM, Hoekstra AY. The water footprint of poultry, pork and beef: a comparative study in different countries and production systems. Water Resources Industry. 2013;1:25–36.
54. Amorim AK, De Nardi IR, Del Nery V. Water conservation and effluent minimization: case study of a poultry slaughterhouse. Resour Conserv Recycl. 2007;51(1):93–100.
55. Blevins RE, Feye KM, Dittoe DK, Bench L, Bench BJ, Ricke SC. Aerobic plate count, Salmonella and Campylobacter loads of whole bird carcass rinses from pre-chillers with different water management strategies in a commercial poultry processing plant. J Environ Sci Health B. 2020;55(2):155–65.
56. Timmerman H. Economics and management of hygiene in food plants. In: Hygiene in food processing. Woodhead Publishing; 2014. p. 577–89.
57. Karaba A, Dixit S, Greco R, Aharoni A, Trijatmiko KR, Marsch-Martinez N, Krishnan A, Nataraja KN, Udayakumar M, Pereira A. Improvement of water use efficiency in rice by expression of HARDY, an Arabidopsis drought and salt tolerance gene. Proc Natl Acad Sci. 2007;104(39):15270–5.
58. McGrath P, McCarthy L, Marshall D, Rehme J. Tools and technologies of transparency in sustainable global supply chains. Calif Manag Rev. 2021;64(1):67–89.
59. Trace X Technologies. Importance of meat traceability [Internet]. 2022. Available from: https://tracextech.com/importance-of-meattraceability/#:~:text=Traceability%20refers%20to%20the%20ability,processing%2C%20distributing%20and%20finally%20consumption
60. Schroeder TC, Tonsor GT. International cattle ID and traceability: competitive implications for the US. Food Policy. 2012;37(1):31–40.
61. Thakur M, Hurburgh CR. Framework for implementing traceability system in the bulk grain supply chain. J Food Eng. 2009;95(4):617–26.
62. Girish PS, Nagappa K, Saikia T. Farm-to-fork livestock traceability for quality meat production: an overview. J Meat Sci. 2017;12(1):1–10.
63. Falco A, Salmerón JF, Loghin FC, Lugli P, Rivadeneyra A. Fully printed flexible single-chip RFID tag with light detection capabilities. Sensors. 2017;17(3):534.
64. Tarjan L, Šenk I, Tegeltija S, Stankovski S, Ostojic G. A readability analysis for QR code application in a traceability system. Comput Electron Agric. 2014;109:1–11.

Present Status and Future of Healthier Meat Products

Muhammad Umar Ijaz, Ali Akbar, Muhammad Sajid Arshad, and Hafiz Ansar Rasul Suleria

1 Introduction

Nutrition serves as a cornerstone for the existence of living organisms. It is a process by which humans as well as animals acquire and use essential nutrients necessary for their growth, maintenance, and well-being. Proper nutrition provides essential nutrients that fuel our bodies. Moreover, nutrient-rich foods are reported as an integral element in the current era, which is marked by the widespread occurrence of lifestyle-associated illnesses such as cardiovascular diseases (CVDs), diabetes, and obesity [1].

Meat and its derivatives serve as important reservoirs of proteins, vitamins, and minerals. However, it is noteworthy that they also contain elements such as saturated fatty acids (SFAs), salt, & cholesterol. Despite this, meat & its products are remarkably essential for acquiring energy, protein along with vital micronutrients [2]. Nevertheless, the intake of food derived from animals varies significantly not only between different populations but also within the populations, resulting in a diverse range of impacts on human health [3].

M. U. Ijaz · A. Akbar
Department of Zoology, Wildlife and Fisheries, University of Agriculture, Faisalabad, Pakistan

M. S. Arshad (✉)
Department of Food Science, Faculty of Life Sciences, Government College University, Faisalabad, Pakistan

School of Agriculture and Food, Faculty of Veterinary and Agricultural Sciences, The University of Melbourne, Parkville, VIC, Australia
e-mail: msajidarshad@gcuf.edu.pk

H. A. R. Suleria
School of Agriculture and Food, Faculty of Veterinary and Agricultural Sciences, The University of Melbourne, Parkville, VIC, Australia

A. Kumar Pathera et al. (eds.), *Healthier Meat Products*,
https://doi.org/10.1007/978-3-031-78215-2_3

In 2005, the worldwide average consumption of meat per individual was approximately 110 g daily, demonstrating a tenfold difference among populations characterized by high and low levels of meat consumption [3]. The need for food and meat derived from animals is expected to increase significantly increase in developing countries over the next few years, while in high-income countries, there is a possibility that meat consumption could stabilize or decrease in the future [3, 4]. Furthermore, meat as well as meat products hold significant cultural, nutritional, and economic importance worldwide. The meat industry significantly contributes to the global economy, generating significant revenue and creating millions of jobs throughout the distribution network, from farming & manufacturing to delivering & retail [5].

Reportedly, in the United States (US) alone, the annual medical costs linked with major chronic diseases i.e., CVDs, cancer, osteoporosis, diabetes, & obesity, surpass $400 billion [6]. Numerous studies have established direct link between these health disorders and human dietary habits. This suggests that a proactive enhancement of the food supply to promote healthier options could serve as a preventive healthcare strategy, potentially addressing many challenges in healthcare [7]. Therefore, the current chapter will cover the present status and future aspects of healthy meat products.

2 Historical Perspective on Meat Consumption and Evolution

Meat consumption has constituted a fundamental component of human diets for decades, its history is associated with the evolution of human civilization.

2.1 Early Human Diets and Hunting

Early humans were primarily hunters in the ancient age, and their survival was solely dependent on hunting wild animals. They relied solely on meat to supply the essential proteins and elements in their diet [8]. Archaeological specimens and cave paintings from this era show how important hunting was for survival at that time. The first stages of human involvement with meat consumption were defined by the emergence of instruments such as spears and the creation of hunting tactics [9, 10].

2.2 Agricultural Revolution

The significant shift from hunting & gathering to agriculture during the Neolithic Era was a remarkable moment in human history [11]. This period saw the domestication of animals encompassing pigs, cattle, and sheep for meat manufacture [12]. As agriculture flourished, societies became more settled and dependent on controlled meat sources. The development of animal husbandry practices paved the way for the cultivation of healthier meat options compared to relying solely on hunting [13].

2.3 Industrialization and Mass Production

The industrial revolution brought about progressive changes in the meat industry. Technological innovations in transportation, refrigeration, and food processing allowed for large-scale meat production and distribution [14, 15]. After the invention of refrigerator, meat could be transported longer distances without spoiling, leading to the growth of urban populations [16]. The establishment of massive slaughterhouses and the manifestation of processed meat products, i.e., canned meats and sausages, further transformed meat consumption towards advancement [17, 18].

2.4 Shift Towards Healthier Meat Options Over Time

The development of societies as well as their knowledge of the dietary and health impacts of eating meat encourages a shift towards healthier meat options over time.

2.4.1 Nutritional Awareness

In the late nineteenth century, advancements in nutrition science revealed the nutritional value of various food items, meat based nutritional products [19, 20]. This scientific understanding prompted the recognition of the importance of lean cuts of meat, poultry, and fish in a balanced diet. For example, studies showed that fish was rich in omega-3 fatty acids (ω-3 FA), promoting its inclusion due to its benefits for cardiovascular health [21–23].

2.4.2 Dietary Guidelines and Recommendations

It is reported that governments along with health groups established significant dietary standards in the twentieth century. For example, white meat contains less saturated fat than red meat, the American Heart Association endorsed the use of white meat instead of red meat for the benefits of human wellbeing [24]. These recommendations were indispensable in helping consumers making decisions and developing a preference for healthier meat selections [25]. It has been reported that these recommendations led to a notable decline in the intake of red meat & an escalation in the use of chicken as well as fish meat, which lowered the risk of cardiovascular problems by up to 32% [24, 26]. Moreover, approximately three-quarters of the meat consumed in the US in 1971 came from fish as well as poultry, with red meat making up the remaining 70%. This ratio has changed to a more even distribution by the mid-1990s, at that time fish and poultry rised to 45% and red meat dropped to 55%. By 2016, the percentage of total meat used had dropped to little over 40% from red meat, while the percentage from poultry and fish had reached approximately up to 60% [27, 28].

2.4.3 Organic and Sustainable Meat Production

Animal welfare, environmental sustainability as well as the usage of growth enhancer hormones and antibiotics in the meat manufacturing have garnered a significant attention in the late twentieth century. This resulted in the growth of sustainable as well as organic meat choices [29, 30]. For instance, the organic farming movement introduced practices that focused on the humane treatment of animals and the avoidance of synthetic additives [31]. The demand for organic and sustainable meat options upsurged as consumers became increasingly sensitive to the health and environmental aspects of their food choices [32]. By examining the sales and market expansion of organic and sustainable meat products, it is possible to quantify this growing consumer interest. According to data from the Organic Trade Association, sales of organic meat in the US have consistently grown over the course of past 20 years [33, 34].

3 Nutritional Profile and Its Impact

Meat is recognized for its rich protein content of high quality, encompassing essential amino acids indispensable for the body's growth, repair, & maintenance. Proteins play a vital role in muscle development, immune function, enzyme production, and hormonal balance [35]. It is reported that lean meats such as poultry, fish, and specific cuts of beef are particularly rich in protein, which make them an excellent choice for protein-deficient individuals who want to meet their protein requirements without excess fat intake [36]. The amino acid composition of meat is vital because our body cannot make essential amino acids independently. For instance, methionine and lysine are present in meat in adequate proportions and serve as crucial building blocks for protein synthesis [37].

Meat contains different types of fats and not all fats are the same. The kind and amount of fat present in the meat depend on the animal's diet & the specific cut of meat. Healthier meat products often include lean cuts with lower saturated fat, which are linked with cardiovascular health when consumed excessively. Fatty fish, including mackerel, salmon, and sardines, are well-known for their high ω-3 FA acid [38]. ω-3 FA acids are essential for heart health as well as reducing inflammation and regulating cholesterol levels [39].

Meat is a substantial source of essential vitamins & minerals. One key nutrient is vitamin B12, which is predominantly present in meat products. Vitamin B12 is indispensable for nerve health, DNA synthesis, and RBC formation [40]. Various reports have revealed that the presence of vitamin B12 in meat is necessary for a balanced diet, especially for vegan lifestyles. Furthermore, iron is an essential mineral abundantly present in meat, and it occurs in heme form that is readily absorbed by the body. Heme-iron plays an indispensable part in oxygen transport as well as the prevention of iron deficiency anemia. Meat also contains zinc, which supports immune function and wound healing, while selenium (Se) acts as an antioxidant that safeguards cells from oxidative stress (OS) [41].

Meat is highly valued for its role in promoting muscle health. The amino acids, especially leucine, present in meat stimulate muscle protein synthesis, which is a vital dietary component for athletes, regular exercise, and the old people striving to maintain muscle mass. Additionally, meat-derived products, such as whey protein supplements, have gained substantial attention in the fitness industry [42]. Whey protein, a derivative of cheese manufacture contains a significant amount of branched-chain amino acids (BCAAs) including isoleucine. These BCAAs are well-known for their ability to support muscle recovery and growth, which makes it a valuable supplement for individuals with increased protein requirements [43]. An overview of the nutritional profile and its impact on health is depicted in Table 1.

4 Current Insights into Healthy Meat Products

4.1 Healthy Meat Products as a Functional Food

4.1.1 Unsaturated Fatty Acids

In several countries, there is a prevalent trend of excessive intake of SFA & inadequate ingestion of polyunsaturated fatty acids (PUFAs), particularly ω-3 FA acids [44]. The consumption of FA poses a significant concern due to its capacity to influence cholesterol levels, which are correlated with CVDs. Generally, FAs elevate LDL levels in the bloodstream, consequently raising the hazard of CVDs. On the other hand, PUFAs are considerably involved in lowering cholesterol levels [45].

Table 1 Nutritional profile of meat and its health impact

Nutrient	Impact on health	References
Protein	Essential for muscle development, immune function, enzyme production, and hormonal balance	[35–37]
	Lean meats (poultry, fish, specific beef cuts) are rich in protein, suitable for protein-deficient individuals with minimal fat intake	
	Amino acids like methionine and lysine in meat are crucial for protein synthesis	
Fats	Different types of fats depending on the animal's diet and meat cut	[38, 39]
	Healthier meat choices have lower saturated fats, benefiting cardiovascular health	
Vitamins and minerals	Rich source of vitamin B12 essential for nerve health, DNA synthesis, and red blood cell formation	[40, 41]
	Iron in heme form aids in oxygen transport and prevents iron deficiency anemia	
	Zinc supports immune function and wound healing.	
	Selenium acts as an antioxidant, protecting cells from OS	
Muscle health	Amino acids, predominantly leucine, trigger muscle protein synthesis	[42, 43]
	Vital for athletes, exercisers, and the elderly maintaining muscle mass	
	Meat-derived products such as whey protein supplements rich in BCAAs (leucine, isoleucine, valine) support muscle recovery and growth	

There is a growing attention in integrating ω-3 *FA* acids into functional foods. This is determined by the fact that numerous consumers presently have insufficient intake of *w*-3 FA acids, and an increase in consumption could be beneficial. This increased intake may potentially lead to beneficial effects such as reduced blood triacylglycerols, decreased risk of abrupt cardiac death, alleviation of depression, and mitigation of the symptoms of arthritis [46].

The FA profile of meat derived from ruminants significantly exhibits a higher saturation level. This phenomenon is primarily attributed to the biohydrogenation that occurs in the rumen, where UFAs undergo transformative variations. In the rumen, microbial activity leads to the conversion of UFAs to more saturated forms, as depicted in Fig. 1. In contrast, animals, including poultry as well as fish that lack a rumen, coordinate dietary modifications to modify the FA composition of their muscle tissues [47]. This procedure was previously evident in specialized products, as seen in the case of Iberian hams. These hams have a higher content of oleic acid, a monounsaturated FA, owing to the acorn-rich diet of the animals [48]. Similarly, in the field of aquaculture, salmon are provided with fish oils rich in ω-3 FA acids; this is a dietary strategy that enhances the omega-3 concentration of the resulting salmon meat [49]. Nevertheless, these practices face limitations, as elevated levels of UFAs have effect on the oxidation of the products derived from meat. In contrast, augmented UFAs through dietary manipulation have proven highly effective in the

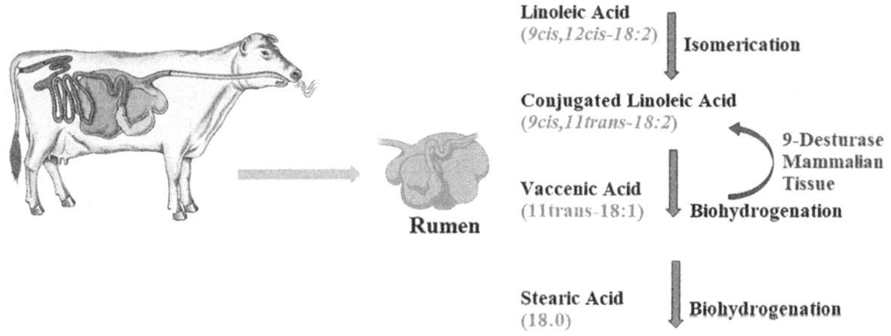

Fig. 1 Rumen bacteria enzymatically bio-hydrogenate linolenic acid, leading to the synthesis of conjugated linolenic acid through endogenous mechanisms

case of eggs. This assertion holds true since eggs inherently possesses antioxidative properties [47]. Increasing UFAs represents a particular challenge in muscle foods due to their elevated levels of prooxidative metals. Additionally, these foods typically have low levels of endogenic antioxidants & undergo operations involved in processing that significantly intensify OS [50].

4.1.2 Minerals

Dietary minerals serve as an important element in various aspects of health, such as nerve function, muscle and bone health, blood pressure maintenance, and the control of blood sugar levels. Consequently, they hold significance in the prevention and management of lethal disorders [45]. Among these minerals, potassium, calcium, and magnesium are frequently identified as minerals consumed in diets [44]. In some populations, iron is also sufficiently disbursed, but incorporating it into meat poses significant challenges. The addition of iron to meat can lead to rapid lipid oxidation (LO) and discoloration, as discussed by Mancini and Hunt [51].

Fortifying processed meats with the above-mentioned minerals has been demonstrated to be feasible without inducing substantial variation in sensory functions [52–54]. An additional benefit of fortifying with these minerals is the potential to overcome Na levels in the meats, thereby enhancing the nutritional level of the product [52]. The inclusion of salts, including potassium lactate, proves beneficial as it hinders LO and curtails microbial proliferation, as reported by Moon et al. [52]. Calcium acts as a remarkable element in activating the protease calpain, which effectively contributes to regulating the tenderness of meat. It is reported that infusion of calcium into beef cattle not only boosts muscle Ca concentrations [55] but also improves beef sensitivity after 14 days [52]. Nevertheless, in certain processed meat products, the presence of calcium, potassium, & magnesium has been found to inhibit proteolytic enzymes that is essential for the flavor profile of dry-cured meats [56]. It is noteworthy that despite this inhibition, these minerals do not adversely affect the salty flavor, as reported by Armenteros et al. [56].

There has been significant attention to supplementing the diets of livestock with Se. Dietary Se supplementation has been shown to increase muscle Se concentrations in beef [57], pork [58], and chicken [59]. This indicates that supplemented animals could potentially provide health benefits for humans. Higher levels of selenium in muscles are often linked to higher levels of glutathione peroxidase, an enzyme that contains selenium and breaks down lipid hydroperoxides. Nevertheless, it is noteworthy that, in the case of beef, the increased activity of GPx did not accelerate the oxidative balance in the muscle [57].

4.1.3 Dietary Fibers

Dietary fiber can be broadly categorized into soluble & insoluble forms, which offer various health-beneficial effects. These include modulating bowel integrity, cholesterol levels which ultimately controls blood sugar levels, & serving as a non-caloric bulking mediator that aids in weight reduction by substituting caloric components i.e., fat. According to the Dietary Guidelines for Americans [44], most adults do not consume an adequate amount of dietary fiber, indicating that fortifying meat products with fiber could confer health benefits. There have been many studies that show adding 2–3 g of fiber per serving to sausages can be done without changing the way they taste [60, 61]. In addition to helping you eat more fiber, adding dietary fibers to meat products has other benefits as well, which make them better at holding water and protecting against oxidation, especially if the fiber source is linked to phenolic antioxidants [61, 62].

4.1.4 Antioxidant Vitamins

Dietary antioxidants including vitamins A, C, and E are reported to show positive effects on immune function, heart health, and cancer prevention. However, many consumers take in inadequate concentrations of these vitamins as compared to the recommended dietary intake levels [44]. Previous studies have reported that tocopherol has been reported to inhibit lipid oxidation in fresh and frozen patties [63]. β-carotene serves as a vital sources of dietary vitamin A. Generally, meats, except chicken, are not significant provider of β-carotene or other carotenoids. However, β-carotene contents in meat could be amplified through nutritional supplementation [47]. Nevertheless, incorporating β-carotene into meats as a food ingredient is challenging owing to its chemical imbalance. Both exogenous accumulation and dietary fortification of β-carotene are restricted, as they can modify the color of the muscle [64].

Meats, particularly poultry and ground beef, play a crucial role as dietary sources of vitamin E. In male, poultry & ground beef rank as the second and sixth most substantial sources of α-tocopherol, while in female, poultry holds the third position [65]. The vitamin E concentration in muscle foods could be amplified through fortification or nutritional supplementation with α-tocopheryl acetate. The addition of the acetate group to tocopherol is beneficial as it protects the molecule from oxidation [50]. In some cases, tocopherols are mixed to meats to constrain LO [66]. However, it's not clear how much these add to the nutritional profile because they may break down during storage and have much lower vitamin E activity than α-tocopherol [67].

The vitamin E level of muscle foods could be enhanced through dietary fortification [68]. Introducing α-tocopherol through the diet allows efficient incorporation into cell membranes. Cell membranes enriched with α-tocopherol exhibit increased resistance to lipid peroxidation. Consequently, α-tocopherol serves a crucial function in safeguarding meat flavor. Additionally, LO in beef muscle leads to myoglobin discoloration, and supplementation with α-tocopherol also contributes to the preservation of meat color. According to Schaefer et al. [68] and Jensen et al. [69], giving α-tocopheryl acetate to animals could have big effects on the qualitative properties of the meat & its nutritious value.

4.1.5 Fundamental Strategies for the Development of Healthy Meat Products as a Functional Food

Numerous studies were aimed at explaining the part of meat in a healthy diet and exploring strategies for producing functional meat-based foods. These strategies mainly involve increasing useful agents & reducing those with undesirable health properties in meat products (Fig. 2). These methods are associated with factors that can impact animal production practices, such as genetics and nutrition, as well as processes in meat transformation including reformulation [70, 71].

Considerations associated to meat processing operations, storage, and ingesting conditions also play a significant role, as they could influence product configuration and consequently, the levels of healthy and, unhealthy agents. In the following sections, we have briefly discussed the possibilities and fundamental approaches for the growth of healthy meat products as functional foods.

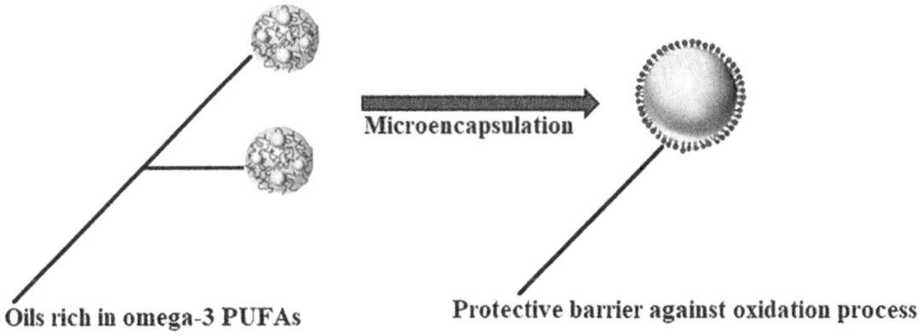

Fig. 2 Microencapsulation protects oils from oxidation process

4.1.5.1 Animal Production Practices

This is the first opportunity to modify the distribution of bioactive components using animal production methods. Animal tissues, including corpses, commercial cuts, and raw meat components, differ depending on the species as well as breed, age, sex, and manner of feeding. There are several methods for causing changes in vivo. These substances must fulfill two crucial requirements: they must be able to be altered in the tissues of animals and exhibit efficacy as bioactive substances. These compounds encompass lipids (FA composition and cholesterol levels) as well as fat-soluble vitamins and minerals [72]. Here, we will describe some of the more relevant procedures.

4.1.5.2 Nutrition and Feeding Management

Lipids have garnered significant attention among bioactive components due to their health implications, especially in the quantitative and qualitative progress of improved meat foods. Animal tissues may exhibit endogenous or exogenous lipid deposition. Numerous animal feeding trials were undertaken to increase the content of PUFA in animal muscles. By raising the amount of MUFA in the animal's diet, the amount of MUFA in meat can be increased. Increasing PUFA levels has been achieved through plant sources like forages, and $n-3$ PUFA-rich plants, as well as marine sources including algae and fish [73–75]. Dietary supplementations are used to enrich chicken, pork, beef, and lamb with conjugated linoleic acid (CLA) [73, 76, 77]. It is possible to selectively lower the amount of cholesterol in living animal's muscle tissue by injecting more dietary copper (up to 20.4% in breast muscle) into chickens [78] or dietary CLA into pigs [79].

4.1.5.3 Genetic Strategies

Genetic improvement is aimed at inducing changes that are economically as well as productively favorable and hold promising prospects in the production of healthy food. Although genetic selection's goals have changed throughout time, they still mostly center on raising fat unsaturation and decreasing fat content.

4.1.5.3.1 Genetic Modifications

Biotechnology in animals, encompassing techniques such as cloning, transgenesis, or a combination of both, represents a more advanced scientific approach. This method has the potential to enhance the quality, yield, and safety of food products through direct genetic manipulation [80]. These options have ramifications for enhancing FA profiles. For example, lack of an n-3 FA desaturase gene makes it difficult for cattle to convert $n-6$ FA into n-3 FA. A $\Delta 12$ FA desaturase gene from spinach has been created and expressed in transgenic pigs [81, 82]. The goal of this genetic alteration is to lower the n − 6/n − 3 PUFA ratio while enhancing the yield of n − 3 PUFAs or facilitating the formation of alpha-linolenic acid (LNA) [82].

4.1.5.3.2 In-Vivo Meat Production

In vitro meat production is proposed as a more humane, safe, and environmental friendly alternative to traditional methods of obtaining animal muscle tissue [83, 84]. The creation of meat products with a variety of flavors, textures as well as nutritional profiles is theoretically possible using methods i.e., co-culturing, medium formulation, or genetic engineering. This novel methodology furthermore permits the modification of bioactive substances particularly FA [84].

4.1.5.4 Reformulation of Healthy Meat Products

Technological strategies employed for the design and development of functional foods, focusing on alterations in meat transformation systems, hold significant promise. Different bioactive components can be removed, reduced, increased, added, or substituted using a variety of methods. Additionally, by altering the meat formulation process, conventional components can be used along with additional substances that have been specially created to have health-promoting qualities [84].

4.1.5.4.1 Optimizing Fatty Acid Profile

Reducing total fat and energy, lowering cholesterol, and changing FA profiles are the three main goals that have been established for controlling fat content in beef reformulation techniques [85]. The use of leaner meat raw materials and the dilution of fat by adding water and other components (gums, protein-, fat-, or carbohydrate-based fat replacers) with low or no calorie content are the two main factors that commonly govern fat reduction. Furthermore, manufacturing and preparation processes need to be taken into account in reformulation plans [86].

The FA composition of meat products plays a crucial role in the diet-health relationship because each FA has distinct effects on plasma lipids. It is reported that the plasma membrane contains various types of lipids. Technological approaches to enhance the FA profile often entail substituting a product's fat with a fat type that better complies with health guidelines. This means that there should be more MUFAs and n − 3 PUFAs, particularly long-chain or CLA, and less SFAs. Better n − 6/n − 3 PUFA ratios and PUFA/SFA ratios are the desired outcomes. Fresh, cooked, or fermented meats can benefit from the addition of natural or processed plant lipids (olive, cottonseed, corn, soybean, peanut) and marine lipids (fish and algae). These can be added in a variety of ways, including as liquid or solid oils, encapsulated, combined with other ingredients to create simple or complex emulsions, or as plant parts [87]. Commercially made CLA isomers were injected into whole muscle [88] and added to meat products such as paté or sausage to get levels high enough for health benefits when eaten in smaller amounts [87]. On the other hand, approaches for cholesterol modulation from raw meat include physicochemical procedures [78], the fermentation of meat products using the cholesterol-lowering bacteria *Eubacterium coprostanoligenes*, or the dilution of fat and lean meat by vegetable oils and plant-based proteins [78].

4.1.5.4.2 Assimilation of Plant Proteins

It has been discovered that plant proteins may be used in meat products for nutritional, cost-saving, technical, and health-promoting reasons. By raising protein levels, balancing amino acid profiles, or lowering calorie and cholesterol contents (when used as fat substitutes), they offer nutritional benefits [85]. In the meat system, proteins from particular plant sources, such as sunflower and walnut, have been used to balance (decrease) lysine/arginine ratios, eventually offering health benefits [89]. Since soy protein has the ability to manage blood cholesterol levels, it has been included as a functional component to a variety of FOSHU (Food for Specified Health Uses) meat products, including low-fat sausage [90].

4.1.5.4.3 Incorporation of Probiotics

Probiotics, owing to their health benefits, provide an encouraging market opportunity for the meat industry. Numerous studies have investigated the addition of probiotic bacteria encompassing *Lactobacillus acidophilus* to fermented sausages [85]. Microorganisms that are resistant to technological processing factors, such as the presence of nitrite and sodium chloride, pH, mechanical and thermal treatments, and the environment of the gastrointestinal tract, are considered suitable as probiotic strains in this fermented matrix [89, 91].

5 Future Aspects of Healthy Meat Products

5.1 Strategies for Integrating Plant-Based Oils into Processed Meat Items

Incorporating plant-based oils into processed meats poses a challenge due to inherent differences in physicochemical characteristics compared to commonly utilized animal fats, such as beef fat [85, 91]. Various methods employed for partial or complete substitution of animal fat with plant-based oils include microencapsulation, direct mixing, and emulsions [92, 93].

The utilization of these plant-based oils necessitates adjustments in meat-processing conditions to achieve the desired quality attributes in the reformulated products. Maintaining the integrity as well as qualities of the finished processed meat products is essential to ensure that they satisfy customer preferences along with anticipated requirements [85, 91]. Thus, it becomes indispensible to investigate novel approaches for the effective incorporation of plant-based oils into processed meat products without sacrificing their overall quality.

5.1.1 Microencapsulation

By encapsulating microscopic particles in a membrane and embedding them in a homogeneous or heterogeneous matrix, a process known as microencapsulation produces tiny capsules. This technique allows the encapsulation of gases, solids as well as liquids in small capsules. The contents of the capsule progressively pervade its wall, allowing for controlled release under certain circumstances [94, 95].

Because microencapsulation safeguards sensitive oils which are high in omega-3 PUFAs from oxidation, it is very effective at delivering these FAs [96] (Fig. 2). Oils that contain ω-3 PUFA are chemically unstable and hence prone to oxidative breakdown. Oxidation can occur when meat products that have been fortified are exposed to moisture, air, light, or heat. This can result in the creation of volatile chemicals and the loss of taste and sensory appeal [94–97].

A number of meat products, such as pate, burgers, and frankfurters, have been the subject of recent research investigating the usage of microparticles [98, 99]. For items containing cooked meat, it is not advised to use spray drying to obtain microparticles. The reason for this is that materials usually used for walls, including starch, gum Arabic, or maltodextrin, can break down when exposed to temperatures higher than 50 °C. As such, microparticles used for goods containing cooked meat need to be made to withstand higher temperatures [100].

5.1.2 Direct Mixing

The incorporation of oil directly into meat products presents a significant drawback that are due to increased vulnerability to oxidation reaction, diminished solubility & increased issues of hardness due to the formation of small fat globules [101]. Recent practices involve the direct addition of vegetable oils to various meat products, such as chicken batters, meatballs, burgers, restructured ham-like products, fermented sausages and chicken nuggets [102]. While these strategies have demonstrated improvements in the FA profile, they often impact certain quality attributes such as flavor, color and sensory acceptance of the reformulated products. To address these challenges, the modification of liquid oils through techniques such as emulsions and organogelation becomes necessary. This modification aims to create a plastic fat with solid-like properties that can effectively substitute for animal fats. Mixing liquid oils with meat batters, particularly after combining them with protein-stabilized emulsions, has been found to mitigate flavor degradations. Moreover, different antioxidants can be added directly to meat products or through emulsion-type delivery methods to lessen oxidation processes and improve the overall stability of the reformed items [85, 103].

5.1.3 Emulsions

An emulsion is a mixture of two entirely immiscible liquids particularly water & oil, where one of the liquids is dispersed on the other in the form of tiny spherical droplets [104]. The interface is the line between the two phases that make up an emulsion: the dispersed phase and the continuous phase [105]. Conventional emulsions come in two primary kinds: water-in-oil (W/O) and oil-in-water (O/W). Despite being thermodynamically unstable and prone to phase separation attributes including aggregation, flocculation, and creaming, O/W emulsions are widely employed in the food company operations. To address these issues, appropriate emulsifiers are added to enhance stability [106].

Emulsifiers, characterized by a hydrophilic part and a lipophilic part, are surface-active substances, when added to an emulsion, reduce the surface tension between aqueous and oil phases, promoting a thermodynamically stable and homogeneous preparation [107, 108]. It is revealed that emulsion preparation technique, the existence of additional functional components, the intended product qualities, and the chemical properties of each phase influence the choice of emulsifier [109]. Food manufacturers identify natural emulsifiers to satisfy consumer demand for natural components, such as proteins, polysaccharides, and phospholipids [110, 111].

Soy proteins, carrageenan and inulin are commonly used natural emulsifiers in food products due to their cost-effectiveness, availability and non-toxic nature. Soy protein, available as isolates or concentrates, is widely used in emulsion formulations to enhance the gel network and cooking yield of final products [111, 112]. Carrageenan, an algal-based polysaccharide, influences emulsion syneresis and hardness with its concentration affecting these properties [113]. Inulin, a dietary fiber which is used in meat products to improve FA profiles without influencing juiciness and texture [112].

While emulsions face limitations in food applications such as alterations in rheological behaviors, physical instability, and textural issues, the development of structured emulsions with innovative functional properties has become popular. These structured emulsions are broadly categorized into hydrogelled and organogelled emulsions. Hydrogelled emulsions are particularly intriguing for replacing animal fat. This interest arises from the ability to incorporate hydrophilic and lipophilic antioxidants, thereby enhancing oxidative stability. This is especially beneficial because the oil content generally does not exceed 50% of the total gel [114].

5.1.4 Processed Meat Products

Processed meat products include items that have undergone modifications such as smoking, salting and curing to enhance quality and prolong shelf life. The range of these products varies based on processing techniques and nutritional composition, including fat and salt content [115] (Fig. 3). Determined by consumer preferences, quality (taste, flavor, texture & nutritional profile) and a growing trend, processed meat products are increasingly becoming popular in the developed world [115,

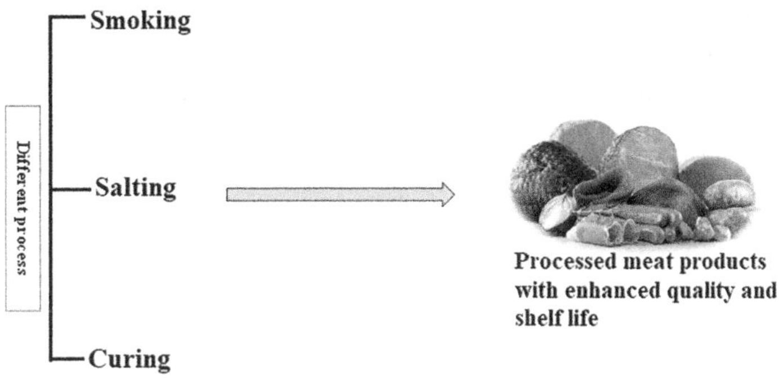

Fig. 3 Different process involved in increasing quality and shelf life of meat

116]. Consumer demand for healthier processed meat products with reduced salt and fat content, especially saturated fat (SFA), has become a dominant concern for the processed meat industry [117].

Processed meat products can be categorized based on the processing method, such as emulsified (e.g., bologna sausages, frankfurters, pâtés, meat batters) and mixed-type (e.g., fermented sausages, meatballs, patties). Emulsion-type meat products usually have oil-in-water (O/W) emulsion systems, where proteins keep the lipid droplets from moving around [104, 118]. Because proteins are amphiphilic, they stick to the surface of each oil droplet and form a layer. This layer lowers the interfacial tension and keeps the emulsion stable [118]. Muscle myofibrillar proteins (MP) and sarcoplasmic proteins are commonly used in meat emulsions due to their high emulsifying capacity [119].

Animal fat plays a crucial role in emulsion-type processed meat products, contributing to stability, product binding, texture improvement, flavor enhancement, and increased juiciness [112]. Attaining extremely high emulsion stability depends on the size of the fat globules and the capacity of correctly solubilized proteins to adsorb across the interface [120]. According to recent research, formulations with higher fat contents provide better emulsion stability and water retaining capability [121]. Conversely, a decrease in fat content has been associated to increased product hardness and cooking loss [122]. It is revealed that approximately 30% of fat can be added to processed meat products to enhance texture and yield [123]. However, consuming large amounts of animal fat and saturated fats may increase the likelihood of developing CVDs, obesity, hypertension, and coronary heart disease [124].

6 Conclusion

Taken together, this chapter discusses the indispensable role of meat in providing essential nutrients for health as well as muscle development. It discovers the evolving status of healthy meat products, highlighting their emergence as significant functional foods to meet growing demand all over the world. The fundamental approaches discussed, from production practices to innovative reformulation, analyze consumer concerns about the health impact of meat consumption. The integration of plant-based oils employing advanced methods provides promising evidence for modulating the nutritional profile of meat products. Overall, this chapter contributes to the continuing discussion about the future of meat production, endorsing a balanced and sustainable approach in line with changing consumer preferences and health-conscious choices.

References

1. Brimblecombe J, Maypilama E, Colles S, Scarlett M, Dhurrkay JG, Ritchie J, O'Dea K. Factors influencing food choice in an Australian Aboriginal community. Qual Health Res. 2014;24:387–400.
2. McAfee AJ, McSorley EM, Cuskelly GJ, Moss BW, Wallace JM, Bonham MP, Fearon AM. Red meat consumption: an overview of the risks and benefits. Meat Sci. 2010;84:1–3.
3. FOOD O. The state of food and agriculture. Clim Change Agri 2016.
4. Vranken L, Avermaete T, Petalios D, Mathijs E. Curbing global meat consumption: emerging evidence of a second nutrition transition. Environ Sci Policy. 2014;39:95–106.
5. Cawthorn DM, Hoffman LC. The role of traditional and non-traditional meat animals in feeding a growing and evolving world. Anim Front. 2014;4:6–12.
6. DHHS. Chronic disease cost. 2010. http://health.nv.gov/CD_ChronicDisease_Costs
7. Trovato GM. Behavior, nutrition and lifestyle in a comprehensive health and disease paradigm: skills and knowledge for a predictive, preventive and personalized medicine. EPMA J. 2012;3:1–5.
8. Speth JD. Paleoanthropology and archaeology of big-game hunting. New York: Springer; 2012.
9. Ambrose SH. Coevolution of composite-tool technology, constructive memory, and language: implications for the evolution of modern human behavior. Curr Anthropol. 2010;51:135–47.
10. Tomasello M, Melis AP, Tennie C, Wyman E, Herrmann E. Two key steps in the evolution of human cooperation: the interdependence hypothesis. Curr Anthropol. 2012;53:673–92.
11. Cummings V, Harris O. Animals, people and places: the continuity of hunting and gathering practices across the Mesolithic–Neolithic transition in Britain. Eur J Archaeol. 2011;14:361–93.
12. Peters J, Buitenhuis H, Grupe G, Schmidt K, Pöllath N. The long and winding road: ungulate exploitation and domestication in Early Neolithic Anatolia (10000–7000 cal bc). In: The origins and spread of domestic animals in southwest Asia and Europe. Routledge; 2016. p. 83–114.
13. Simmons AH. The Neolithic revolution in the Near East: transforming the human landscape. University of Arizona Press; 2011.
14. Sun DW, Zheng L. Vacuum cooling technology for the agri-food industry: past, present and future. J Food Eng. 2006;77:203–14.
15. Rodgers S. Innovation in food service technology and its strategic role. Int J Hosp Manag. 2007;26:899–912.
16. Hammond ST, Brown JH, Burger JR, Flanagan TP, Fristoe TS, Mercado-Silva N, Nekola JC, Okie JG. Food spoilage, storage, and transport: implications for a sustainable future. BioSci. 2015;65:758–68.
17. Horowitz R. Putting meat on the American table: taste, technology, transformation. JHU Press; 2006.
18. Lee PY, editor. Meat, modernity, and the rise of the slaughterhouse. UPNE; 2008.
19. Scrinis G. Nutritionism: the science and politics of dietary advice. CUP; 2013.

20. Winson A. The industrial diet: the degradation of food and the struggle for healthy eating. NYU Press; 2014.
21. Pereira PM, Vicente AF. Meat nutritional composition and nutritive role in the human diet. Meat Sci. 2013;93:586–92.
22. Schonfeldt HC, Hall N. "Fish, chicken, lean meat and eggs can be eaten daily": a food-based dietary guideline for South Africa. S Afr J Clin Nutr. 2013;26:66–76.
23. Rimm EB, Appel LJ, Chiuve SE, Djoussé L, Engler MB, Kris-Etherton PM, Mozaffarian D, Siscovick DS, Lichtenstein AH. Seafood long-chain n-3 polyunsaturated fatty acids and cardiovascular disease: a science advisory from the American Heart Association. Circulation. 2018;138:35–47.
24. Van Horn L, Carson JA, Appel LJ, Burke LE, Economos C, Karmally W, Lancaster K, Lichtenstein AH, Johnson RK, Thomas RJ, Vos M. Recommended dietary pattern to achieve adherence to the American Heart Association/American College of Cardiology (AHA/ACC) guidelines: a scientific statement from the American Heart Association. Circulation. 2016;134:505–29.
25. Lichtenstein AH, Appel LJ, Brands M, Carnethon M, Daniels S, Franch HA, Franklin B, Kris-Etherton P, Harris WS, Howard B, Karanja N. Diet and lifestyle recommendations revision 2006: a scientific statement from the American Heart Association Nutrition Committee. Circulation. 2006;114:82–96.
26. Zhong VW, Van Horn L, Greenland P, Carnethon MR, Ning H, Wilkins JT, Lloyd-Jones DM, Allen NB. Associations of processed meat, unprocessed red meat, poultry, or fish intake with incident cardiovascular disease and all-cause mortality. JAMA Intern Med. 2020;180:503–12.
27. Binnie MA, Barlow K, Johnson V, Harrison C. Red meats: time for a paradigm shift in dietary advice. Meat Sci. 2014;98:445–51.
28. Llorca-Jaña M, Nazer R, Morales-Campos D, Navarrete-Montalvo J. Milk and meat consumption and production in Chile, c. 1930–2017: A history of a successful nutrition transition. 2020.
29. Ronquillo MG, Hernandez JC. Antibiotic and synthetic growth promoters in animal diets: review of impact and analytical methods. Food Control. 2017;72:255–67.
30. Kirchhelle C. Pharming animals: a global history of antibiotics in food production (1935–2017). Palgrave Communications; 2018. p. 4.
31. Niggli U. The evolution of organic practice. In: Organic farming: an international history, vol. 14; 2007. p. 73–92.
32. Winter CK, Davis SF. Organic foods. J food Sci. 2006;71:R117–24.
33. Dimitri C. Recent growth patterns in the US organic foods market. US Department of Agriculture, Economic Research Service; 2002.
34. Greene C. Emerging issues in the US organic industry. DIANE Publishing; 2009.
35. Ahmad RS, Imran A, Hussain MB. Nutritional composition of meat. Meat Sci Nutr. 2018;61:61–75.
36. Wood JD. Meat composition and nutritional value. Lawrie's Meat Sci. 2017:635–59.
37. Wu G, Bazer FW, Dai Z, Li D, Wang J, Wu Z. Amino acid nutrition in animals: protein synthesis and beyond. Annu Rev Anim Biosci. 2014;2:387–417.
38. Webb EC, O'neill HA. The animal fat paradox and meat quality. Meat Sci. 2008;80:28–36.
39. Jain AP, Aggarwal KK, Zhang PY. Omega-3 fatty acids and cardiovascular disease. Eur Rev Med Pharmacol Sci. 2015;19:441–5.
40. Pogorzelska-Nowicka E, Atanasov AG, Horbańczuk J, Wierzbicka A. Bioactive compounds in functional meat products. Molecules. 2018;23:307.
41. Kumar SU, Bhat ZF, Kumar PA, Mandal PK. Functional meat and meat products. Anim Prod Technol. 2013:404–55.
42. Lafarga T, Hayes M. Bioactive peptides from meat muscle and by-products: generation, functionality and application as functional ingredients. Meat Sci. 2014;98:227–39.
43. Khan UM, Selamoglu Z. Nutritional and medical perspectives of whey protein: a historical overview. J Pharm Care. 2019:112–7.

44. Dietary guidelines for Americans. U.S. Department of Health and Human Services, & U.S. Department of Agriculture. 2005. http://www.health.gov/dietaryguidelines/dga2005/document/
45. Whitney EN, Rolfes SR. Understanding nutrition. 9th ed. Belmont: Wadsworth; 2002.
46. Harris WS. International recommendations for consumption of long-chain omega-3 fatty acids. J Cardiovasc Med. 2007;8:50–2.
47. Bou R, Codony R, Tres A, Decker EA, Guardiola F. Dietary strategies to improve nutritional value, oxidative stability, and sensory properties of poultry products. Crit Rev Food Sci Nutr. 2009;49:800–22.
48. Narváez-Rivas M, Vicario IM, Constante EG, León-Camacho M. Changes in the fatty acid and triacylglycerol profiles in the subcutaneous fat of Iberian ham during the dry-curing process. J Agric Food Chem. 2008;56:7131–7.
49. Blanchet C, Lucas M, Julien P, Morin R, Gingras S, Dewailly É. Fatty acid composition of wild and farmed Atlantic salmon (Salmo salar) and rainbow trout (Oncorhynchus mykiss). Lipids. 2005;40:529–31.
50. Fennema OR. In: Damodaran S, Parkin KL, editors. Fennema's food chemistry. Boca Raton: CRC Press; 2008.
51. Mancini RA, Hunt M. Current research in meat color. Meat Sci. 2005;71:100–21.
52. Moon SS, Kim YT, Jin SK, Kim IS. Effects of sodium chloride, potassium chloride, potassium lactate and calcium ascorbate on the physico-chemical properties and sensory characteristics of sodium-reduced pork patties. Food Sci Anim Res. 2008;28:567–73.
53. Schoene F, Mnich K, Jahreis G, Kinast C, Greiling A, Kirmse R, Hartung H, Leiterer M. Analysis of meat products, produced with mineral salt. Fleischwirtschaft. 2009;89:149–52.
54. Selgas MD, Salazar P, García ML. Usefulness of calcium lactate, citrate and gluconate for calcium enrichment of dry fermented sausages. Meat Sci. 2009;82:478–80.
55. Dikeman ME, Hunt MC, Addis PB, Schoenbeck HJ, Pullen M, Katsanidis E, Yancey EJ. Effects of postexsanguination vascular infusion of cattle with a solution of saccharides, sodium chloride, and phosphates or with calcium chloride on quality and sensory traits of steaks and ground beef. J Anim Sci. 2003;81:156–66.
56. Armenteros M, Aristoy MC, Toldrá F. Effect of sodium, potassium, calcium and magnesium chloride salts on porcine muscle proteases. Eur Food Res Technol. 2009;229:93–8.
57. Juniper DT, Phipps RH, Ramos-Morales E, Bertin G. Effect of dietary supplementation with selenium-enriched yeast or sodium selenite on selenium tissue distribution and meat quality in beef cattle. J Anim Sci. 2008;86:3100–9.
58. Morel PC, Janz JA, Zou M, Purchas RW, Hendriks WH, Wilkinson BH. The influence of diets supplemented with conjugated linoleic acid, selenium, and vitamin E, with or without animal protein, on the composition of pork from female pigs. J Anim Sci. 2008;86:1145–55.
59. Pan C, Huang K, Zhao Y, Qin S, Chen F, Hu Q. Effect of selenium source and level in hen's diet on tissue selenium deposition and egg selenium concentrations. J Agric Food Chem. 2007;55:1027–32.
60. Yılmaz I, Gecgel U. Effect of inulin addition on physico-chemical and sensory characteristics of meatballs. J Food Sci Technol. 2009;46:473–6.
61. Choi YS, Choi JH, Han DJ, Kim HY, Lee MA, Lee ES, Jeong JY, Paik HD, Kim CJ. Effects of rice bran fiber on quality of low-fat tteokgalbi. Food Sci Biotechnol. 2008;17:959–64.
62. Sáyago-Ayerdi SG, Brenes A, Goñi I. Effect of grape antioxidant dietary fiber on the lipid oxidation of raw and cooked chicken hamburgers. LWT-Food Sci Technol. 2009;42:971–6.
63. Haak L, Raes K, De Smet S. Effect of plant phenolics, tocopherol and ascorbic acid on oxidative stability of pork patties. J Sci Food Agric. 2009;89:1360–5.
64. Torrissen OJ. Dietary delivery of carotenoids. Antioxidants in muscle foods: nutritional strategies to improve quality; 2000. p. 289–313.
65. Ma J, Hampl JS, Betts NM. Antioxidant intakes and smoking status: data from the continuing survey of food intakes by individuals 1994–1996. Am J Clin Nutr. 2000;71:774–80.
66. Mitsumoto M. Dietary delivery versus exogenous addition of antioxidants.

67. Mustacich DJ, Gohil K, Bruno RS, Yan M, Leonard SW, Ho E, Cross CE, Traber MG. Alpha-tocopherol modulates genes involved in hepatic xenobiotic pathways in mice. J Nutr Biochem. 2009;20:469–76.
68. Schaefer DM, Liu Q, Faustman C, Yin MC. Supranutritional administration of vitamins E and C improves oxidative stability of beef. J Nutr. 1995;125:1792–8.
69. Jensen C, Lauridsen C, Bertelsen G. Dietary vitamin E: quality and storage stability of pork and poultry. Trends Food Sci Technol. 1998;9:62–72.
70. Zhang W, Xiao S, Samaraweera H, Lee EJ, Ahn DU. Improving functional value of meat products. Meat Sci. 2010;86:15–31.
71. Jiménez-Colmenero F, Herrero A, Cofrades S, Ruiz-Capillas C. Meat and functional foods. In: Handbook of meat and meat processing, vol. 2; 2012. p. 225–48.
72. Jiménez-Colmenero F. Meat based functional foods. In: Handbook of food products manufacturing; 2007. p. 989–1015.
73. Raes K, De Smet S, Demeyer D. Effect of dietary fatty acids on incorporation of long chain polyunsaturated fatty acids and conjugated linoleic acid in lamb, beef and pork meat: a review. Anim Feed Sci Technol. 2004;113:199–221.
74. Scollan N, Hocquette JF, Nuernberg K, Dannenberger D, Richardson I, Moloney A. Innovations in beef production systems that enhance the nutritional and health value of beef lipids and their relationship with meat quality. Meat Sci. 2006;74:17–33.
75. Wood JD, Enser M, Richardson RI, Whittington FM. Fatty acids in meat and meat products. In: Fatty acids in foods and their health implications. CRC Press; 2017. p. 101–22.
76. Lynch PB, Kerry JP. Utilizing diet to incorporate bioactive compounds and improve the nutritional quality of muscle foods. Antioxidants Muscle Foods. 2000;14:455–80.
77. Schmid A, Collomb M, Sieber R, Bee GJ. Conjugated linoleic acid in meat and meat products: a review. Meat Sci. 2006;73:29–41.
78. Clarke AD. Reduction of cholesterol levels in meat, poultry and fish products. In: Production and processing of healthy meat, poultry and fish products. Boston: Springer; 1997. p. 101–17.
79. Lauridsen C, Mu H, Henckel P. Influence of dietary conjugated linoleic acid (CLA) and age at slaughtering on performance, slaughter-and meat quality, lipoproteins, and tissue deposition of CLA in barrows. Meat Sci. 2005;69:393–9.
80. Solomon MB, Eastridge JS, Paroczay EW. Transgenic farm animals. In: Meat biotechnology. New York: Springer; 2008. p. 3–20.
81. Saeki K, Matsumoto K, Kinoshita M, Suzuki I, Tasaka Y, Kano K, Taguchi Y, Mikami K, Hirabayashi M, Kashiwazaki N, Hosoi Y. Functional expression of a Δ12 fatty acid desaturase gene from spinach in transgenic pigs. Proc Natl Acad Sci. 2004;101:6361–6.
82. Lai L, Kang JX, Li R, Wang J, Witt WT, Yong HY, Hao Y, Wax DM, Murphy CN, Rieke A, Samuel M. Generation of cloned transgenic pigs rich in omega-3 fatty acids. Nat Biotechnol. 2006;24:435–6.
83. Post MJ. Cultured meat from stem cells: challenges and prospects. Meat Sci. 2012;92:297–301.
84. Datar I, Betti M. Possibilities for an in vitro meat production system. Innovative Food Sci Emerg Technol. 2010;11:13–22.
85. Jimenez-Colmenero F, Salcedo-Sandoval L, Bou R, Cofrades S, Herrero AM, Ruiz-Capillas C. Novel applications of oil-structuring methods as a strategy to improve the fat content of meat products. Trends Food Sci Technol. 2015;44:177–88.
86. Kerry JF, Kerry JP. Producing low-fat meat products. InImproving the fat content of foods. Woodhead Publishing; 2006. p. 336–79.
87. Jiménez-Colmenero F. Healthier lipid formulation approaches in meat-based functional foods. Technological options for replacement of meat fats by non-meat fats. Trends Food Sci Technol. 2007;18:567–78.
88. Baublits RT, Pohlman FW, Brown AH Jr, Johnson ZB, Proctor A, Sawyer J, Dias-Morse P, Galloway DL. Injection of conjugated linoleic acid into beef strip loins. Meat Sci. 2007;75:84–93.

89. Jiménez-Colmenero F, Sánchez-Muniz FJ, Olmedilla-Alonso B. Design and development of meat-based functional foods with walnut: technological, nutritional and health impact. Food Chem. 2010;123:959–67.
90. Arihara K. Functional foods. In: Jensen W, Devine C, Dikemann M, editors. Encyclopaedia of meat sciences, vol. 1. London: Elsevier Science Ltd.; 2004. p. 492–9.
91. Arihara K. Strategies for designing novel functional meat products. Meat Sci. 2006;74:219–29.
92. Paglarini CD, Vidal VA, Martini S, Cunha RL, Pollonio MA. Protein-based hydrogelled emulsions and their application as fat replacers in meat products: a review. Crit Rev Food Sci Nutr. 2022;62:640–55.
93. Ruiz-Capillas C, Herrero AM. Development of meat products with healthier lipid content: vibrational spectroscopy. Food Secur. 2021;10:341.
94. Bakry AM, Abbas S, Ali B, Majeed H, Abouelwafa MY, Mousa A, Liang L. Microencapsulation of oils: a comprehensive review of benefits, techniques, and applications. Compr Rev Food Sci Food Saf. 2016;15:143–82.
95. Delshadi R, Bahrami A, Tafti AG, Barba FJ, Williams LL. Micro and nano-encapsulation of vegetable and essential oils to develop functional food products with improved nutritional profiles. Trends Food Sci Technol. 2020;104:72–83.
96. Calvo P, Castaño ÁL, Hernández MT, González-Gómez D. Effects of microcapsule constitution on the quality of microencapsulated walnut oil. Eur J Lipid Sci Technol. 2011;113:1273–80.
97. Gómez B, Barba FJ, Domínguez R, Putnik P, Kovačević DB, Pateiro M, Toldrá F, Lorenzo JM. Microencapsulation of antioxidant compounds through innovative technologies and its specific application in meat processing. Trends Food Sci Technol. 2018;82:135–47.
98. Domínguez R, Pateiro M, Agregán R, Lorenzo JM. Effect of the partial replacement of pork backfat by microencapsulated fish oil or mixed fish and olive oil on the quality of frankfurter type sausage. J Food Sci Technol. 2017;54:26–37.
99. Vargas-Ramella M, Munekata PE, Pateiro M, Franco D, Campagnol PC, Tomasevic I, Domínguez R, Lorenzo JM. Physicochemical composition and nutritional properties of deer burger enhanced with healthier oils. Food Secur. 2020;9:571.
100. Heck RT, Lorenzo JM, Dos Santos BA, Cichoski AJ, de Menezes CR, Campagnol PC. Microencapsulation of healthier oils: an efficient strategy to improve the lipid profile of meat products. Curr Opin Food Sci. 2021;40:6–12.
101. Zetzl AK, Marangoni AG, Barbut S. Mechanical properties of ethylcellulose oleogels and their potential for saturated fat reduction in frankfurters. Food Funct. 2012;3:327–37.
102. Wirkowska-Wojdyła M, Chmiel M, Ostrowska-Ligęza E, Górska A, Bryś J, Słowiński M, Czerniszewska A. The influence of interesterification on the thermal and technological properties of milkfat-rapeseed oil mixture and its potential use in incorporation of model meat batters. Appl Sci. 2020;11:350.
103. Weiss J, Gibis M, Schuh V, Salminen H. Advances in ingredient and processing systems for meat and meat products. Meat Sci. 2010;86:196–213.
104. Kim TK, Yong HI, Jung S, Kim YB, Choi YS. Effects of replacing pork fat with grape seed oil and gelatine/alginate for meat emulsions. Meat Sci. 2020;163:108079.
105. Loi CC, Eyres GT, Birch EJ. Effect of milk protein composition on physicochemical properties, creaming stability and volatile profile of a protein-stabilised oil-in-water emulsion. Food Res Int. 2019;120:83–91.
106. Liu H, Han G, Zhang H, Liu Q, Kong B. Improving the physical and oxidative stability of emulsions based on the interfacial electrostatic effects between porcine bone protein hydrolysates and porcine bone protein hydrolysate-rutin conjugates. Food Hydrocol. 2019;94:418–27.
107. McClements DJ. Emulsion design to improve the delivery of functional lipophilic components. Annu Rev Food Sci Technol. 2010;1:241–69.
108. Wi G, Bae J, Kim H, Cho Y, Choi MJ. Evaluation of the physicochemical and structural properties and the sensory characteristics of meat analogues prepared with various non-animal based liquid additives. Food Secur. 2020;9:461.

109. McClements DJ, Gumus CE. Natural emulsifiers – biosurfactants, phospholipids, biopolymers, and colloidal particles: molecular and physicochemical basis of functional performance. Adv Colloid Interf Sci. 2016;234:3–26.
110. Given PS Jr. Encapsulation of flavors in emulsions for beverages. Curr Opin Colloid Interface Sci. 2009;14:43–7.
111. de Souza PC, Martini S, Pollonio MA. Using emulsion gels made with sonicated soy protein isolate dispersions to replace fat in frankfurters. LWT. 2019;99:453–9.
112. Utama DT, Jeong HS, Kim J, Barido FH, Lee SK. Fatty acid composition and quality properties of chicken sausage formulated with pre-emulsified perilla-canola oil as an animal fat replacer. Poult Sci. 2019;98:3059–66.
113. Poyato C, Ansorena D, Berasategi I, Navarro-Blasco Í, Astiasarán I. Optimization of a gelled emulsion intended to supply ω-3 fatty acids into meat products by means of response surface methodology. Meat Sci. 2014;98:615–21.
114. Domínguez R, Bohrer B, Munekata PE, Pateiro M, Lorenzo JM. Recent discoveries in the field of lipid bio-based ingredients for meat processing. Molecules. 2021;26:190.
115. Shan LC, De Brún A, Henchion M, Li C, Murrin C, Wall PG, Monahan FJ. Consumer evaluations of processed meat products reformulated to be healthier – a conjoint analysis study. Meat Sci. 2017;131:82–9.
116. Yeung CK, Huang SC. Effects of food proteins on sensory and physico-chemical properties of emulsified pork meatballs. J Food Nutr Res. 2018;6:8–12.
117. Plasek B, Lakner Z, Temesi Á. Factors that influence the perceived healthiness of food. Nutrients. 2020;12:1881.
118. Choi YS, Sung JM, Jeong TJ, Hwang KE, Song DH, Ham YK, Kim HW, Kim YB, Kim CJ. Effect of irradiated pork on physicochemical properties of meat emulsions. Radiat Phys Chem. 2016;119:279–81.
119. Genccelep H, Saricaoglu FT, Anil M, Agar B, Turhan S. The effect of starch modification and concentration on steady-state and dynamic rheology of meat emulsions. Food Hydrocol. 2015;48:135–48.
120. Santhi D, Kalaikannan A, Sureshkumar S. Factors influencing meat emulsion properties and product texture: a review. Crit Rev Food Sci Nutr. 2017;57:2021–7.
121. Guerra IC, Félex SS, Meireles BR, Dalmás PS, Moreira RT, Honório VG, Morgano MA, Milani RF, Benevides SD, Queiroga RC, Madruga MS. Evaluation of goat mortadella prepared with different levels of fat and goat meat from discarded animals. Small Rumin Res. 2011;98:59–63.
122. Álvarez D, Barbut S. Effect of inulin, β-glucan and their mixtures on emulsion stability, color and textural parameters of cooked meat batters. Meat Sci. 2013;94:320–7.
123. Choi YS, Choi JH, Han DJ, Kim HY, Lee MA, Kim HW, Lee JW, Chung HJ, Kim CJ. Optimization of replacing pork back fat with grape seed oil and rice bran fiber for reduced-fat meat emulsion systems. Meat Sci. 2010;84:212–8.
124. Chen Y, Jia X, Sun F, Jiang S, Liu H, Liu Q, Kong B. Using a stable pre-emulsified canola oil system that includes porcine plasma protein hydrolysates and oxidized tannic acid to partially replace pork fat in frankfurters. Meat Sci. 2020;160:107968.

Technological Strategies in Healthier Meat Processing

Moufida Chaari, Teresa D'Amore, Diyar Salahuddin Ali, Mohammad Ali Shariati, and Slim Smaoui

1 Introduction

Meat products are widely consumed in global diets, owing to their rich nutritional components and organoleptic attributes. Meat is a good source of high-quality protein, presenting all essential amino acids, n−3 fatty acids, and docosapentaenoic acid (DPA) [1]. It also includes essential minerals such as iron, zinc, selenium, and vitamins, predominantly B6 and B12 [2]. Along with being essential, meat is important for human evolution since it promotes good health, growth, and intellectual and economic progress [1].

Despite the nutritional advantages, certain studies have highlighted an association between elevated red meat consumption and an increased risk of a wide spectrum of chronic illnesses, including obesity, cardiovascular disease, type-2 diabetes, and various cancers, particularly colorectal cancer [3]. For instance, red meat, such as beef, pork, and lamb, is generally high in saturated fatty acids and cholesterol,

M. Chaari · S. Smaoui (✉)
Laboratory of Microbial and Enzymatic Biotechnologies and Biomolecules (LBMEB), Center of Biotechnology of Sfax, University of Sfax, Sfax, Tunisia
e-mail: slim.smaoui@cbs.rnrt.tn

T. D'Amore
IRCCS CROB, Centro di Riferimento Oncologico della Basilicata, Rionero in Vulture, Italy

D. S. Ali
Department of Chemistry, College of Science, Salahaddin University-Erbil, Erbil, Kurdistan Region, Iraq

Department of Pharmacy, College of Pharmacy, Knowledge University, Erbil, Kurdistan Region, Iraq

M. A. Shariati
Semey Branch of the Institute, Kazakh Research Institute of Processing and Food Industry, Almaty, Republic of Kazakhstan

A. Kumar Pathera et al. (eds.), *Healthier Meat Products*,
https://doi.org/10.1007/978-3-031-78215-2_4

which can raise LDL (low-density lipoprotein) cholesterol levels and contribute to the development of heart disease. Consuming processed meat has also been connected to an increased risk of heart disease and other metabolic conditions including diabetes and weight gain [4, 5]. Because processed meat has a lot of salt and saturated fatty acids, it can contain elements that can cause cancer and mutations. During the processing stages, biogenic amines, N-nitroso compounds, heterocyclic aromatic amines, and polycyclic aromatic hydrocarbons can also form [6, 7].

As a result, there is a growing market for meat products that are upgraded nutritionally and are healthier. In response to this need, researchers have been working on the possibility of making reformulated meat products. Their goal is to introduce more healthful nutrients and unsaturated fatty acids while reducing the number of unhealthy substances like cholesterol, fats, nitrite, or sodium chloride [8]. Concerning fat, the elevated presence of saturated fatty acids prompts numerous consumers to restrict their consumption of meat products. These recommendations stipulate that total fat should not surpass 30% of total energy consumed, with saturated fats constituting less than 10% of the total caloric intake and trans fats less than 1% [9].

In this vein, the need for healthier meat products with functional properties has driven research toward identifying new ingredients capable of reducing or substituting problematic compounds. In light of this situation, the integration of functional components into meat products has garnered considerable attention in recent years, allowing processors to enhance the functional and nutritional aspects of their offerings. Additionally, food technologists tend to develop functional meat products with low fat, natural antimicrobials and antioxidants, lesser sodium content, and enriched with dietary fibers, ω-6 and ω-3 fatty acids.

Regarding the reduction and modification of fat, the focal point of this review, diverse strategies are employed, incorporating the use of fat replacers characterized by superior fatty acid profiles, particularly those enriched in monounsaturated and polyunsaturated fatty acids, to stabilize the products. Numerous studies explore the total or partial substitution of fat with dietary fibers [10], grains [11], or various combinations. Alternatively, some approaches involve the substitution of animal fat with oils boasting optimal nutritional profiles, such as vegetable oils [12, 13], which can be immobilized using gels. Moreover, using various microorganisms, including bacteria, *Lactobacillales*, commonly referred to as LAB (Lactic Acid Bacteria), has garnered significant attention in research. LAB's appealing attributes include the generation of bioactive compounds, some micronutrients, and their recognized safety with a GRAS (Generally Recognized As Safe) status [14, 15].

In fact, the incorporation of these ingredients helps in modifying sensorial characteristics and the overall technology of a meat system, such as fat-holding capacity, water-holding capacity and texture properties. In summary, incorporating functional ingredients in meat products represents a promising strategy for developing healthier and more functional offerings. By eliminating or reducing unhealthy components and introducing beneficial elements, meat processors can contribute to the overall health and well-being of consumers (Fig. 1).

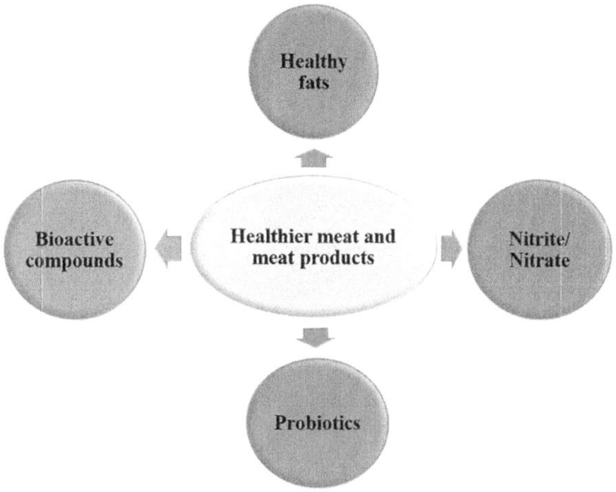

Fig. 1 Functional ingredients for healthier meat and meat products

2 Functional Ingredients

2.1 Healthy Fats

The prevalent use of plant-based lipids is notable in this approach, contributing to the enrichment of meat products with unsaturated fatty acids in the production of low-fat alternatives [16]. Among healthy fats, oils have been employed as ingredients in meat emulsions by directly incorporating them in the form of liquid oils as replacements for fat [12]. Various vegetable oils are utilized for fat replacement (Table 1), each yielding distinct technological and physicochemical attributes in the final meat products. The use of vegetable oils as fat substitutes is a straightforward technique for forming healthier meat products, requiring no pre-processing before incorporation into the meat products. Directly replacing animal fat with specific amounts of vegetable oils has demonstrated a good impact on the quality of meat products [17]. Different types of oils, including canola and linseed oil, safflower oil, and sunflower oil, have been exploited to fully or partially substitute animal fat in meat products [13, 18–20] (Table 1). They integrate plant oils containing polyunsaturated fatty acids (PUFAs) directly, leading to the formation of smaller fat globules within meat emulsions. This process enhances the interaction between fat and protein, potentially allowing the oil to act as filler within the gel matrix of meat products. It may position itself within cavities of the protein gel matrix, restricting matrix movement. Incorporating oils with animal fat has demonstrated an enhancement in the nutritional characteristics of the product. Notably, substituting pork fat with canola oil resulted in a substantial reduction in the saturated fatty acid (SFA) content of chicken meatballs [19]. Incorporating olive oil into meat emulsions resulted in a notable rise in the n−6/n−3 fatty acid ratios, inducing modifications in lipid profiles, as indicated by [21]. Sausages formulated by substituting fat with pumpkin seed oil demonstrated a softer texture and a lighter color when compared

Table 1 Application of functional ingredients in meat products

Functional ingredients	Developed products	Effect on meat product quality	References
Healthy fats			
Sunflower oil	Frankfurter	↗ polyunsaturated fatty acid (PUFA) levels	[39]
		Absence of impact on physicochemical, textural, and sensory attributes	
Linseed oil	Frankfurter	↗ of fatty acid profile	[18]
		↘ of saturated fatty acid (SFA)	
		↘ of cholesterol content	
Canola oil	Liver Pâté	The sensory information yielded highly positive results	[20]
		Textural and physical qualities close to those of the control samples	
Canola oil	Chicken meatball	Elevated carbohydrate content, pH, monounsaturated fatty acids and odor	[19]
		↘ fat content, saturated fatty acids, protein, L* values and texture	
Safflower oil	Wieners	Elevated TBARS, L* value, and b* value	[13]
		↘ cholesterol content and a* value	
Chitosan and Golden flaxseed flour	Hamburgers	↘ of fat content demonstrated positive physical-chemical attributes	[40]
Corn starch	Buffalo calf meat sausages	Low-fat sausages	[41]
		43% lower calorie content	
		Improved quality	
Chicory roots powder	Beef burger	Enhanced protein content, fiber and cooking yield	[42]
		↘ shrinkage, cooking loss, and sensory impact	
		↗ the nutritional properties of the burger	
Ground sesame seed	Chicken sausages	↗ values of all fatty acid profiles, physicochemical and sensory analysis	[11]
		High crude fiber, crude protein, total ash, polyunsaturated fatty acids (PUFA) and PUFA/SFA (saturated fatty acids) ratio	
		Low moisture, crude fat, SFA and cholesterol	
Fibers			
Jabuticaba skin flour	Ham	↗ levels of phenolic compounds	[27]
		Improvement of adhesiveness, cohesiveness, flexibility, and chewiness	
Inulin	Sausages	↘ of fat and energy content	[10]
Inulin, carrot and cellulose fibers	Raw and fried chicken meatballs	Enhancement of color properties, specifically lightness and yellowness values	[25]
		Improvement of overall technological properties	

(continued)

Table 1 (continued)

Functional ingredients	Developed products	Effect on meat product quality	References
Citrus fiber	Bologna sausage	Substitution of fat and sodium tripolyphosphate Acceptable technological parameters	[43]
Chia flour	Chicken nuggets	↗ the overall quantity of dietary fibers Reducing fat and energy content	[44]
Citrus fiber	Turkish meatballs	↗ cooking yield and water holding capacity	[45]
Flax seed and tomato paste	Beef patties	Enhancement of fatty acid profile and nutritional characteristics	[46]
Soy and citrus flavonoid	Meat in broilers	Improvement of crude protein contents ↘ in hepatic malondialdehyde concentration and muscle fat	[47]
Natural preservative			
Encapsulation			
Pomegranate Peel extract-chitosan/polyethylene oxide	Beef meat	Extent of shelf life of meat Enhance the sensory quality	[28]
Garlic essential oil- Maltodextrin and gum Arabic	Beef meat	Encapsulated EO at 20% can be effectively employed in meat products as a natural food additive, serving as a viable alternative to synthetic additives, particularly due to its potent antimicrobial properties	[29]
Laurus nobilis leaf extract with nano-liposome	Minced beef meat	Inhibition of bacterial spoilage (*S. aureus* and *E. coli*) Slowing down the oxidation process	[48]
Tarragon essential oil nano-encapsulated	Pork slices	Improvement of antibacterial and antioxidant properties Enhancement of sensory attributes	[49]
Essential oil of *Thymus capitatus*	Red tilapia fillets	Inhibition effect on Gram (+) bacteria was noted at concentrations of 1000 and 2000 mg/L Extending shelf life for 26 days	[50]
Phlorotannin (Ph) in Sodium alginate (SA) and polyethylene oxide	Chicken meat	Antibacterial activity against *S. enteritidis* in chicken Extending the shelf life of chicken meats	[30]
Clove essential oils encapsulated by chitosan-myristic acid nanogel	Beef cutlets	↘ of *Salmonella* counts in refrigerated beef meat cutlets	[51]
Plant antioxidants			
Rosemary and oregano extract	Hamburger	Maintaining oxidative stability Preserving sensorial characteristics of burgers stored at −18 °C for 30 days	[52]

(continued)

Table 1 (continued)

Functional ingredients	Developed products	Effect on meat product quality	References
Guarana seed extracts	Pork patties	↘ of TBARS and carbonyl formation	[31]
Pomegranate peel extract	Minced beef meat	↘ of MetMb and carbonyl groups accumulation	[53]
		The disappearance of sulphydryl proteins	
Ephedra alata extract	Minced beef meat	Delaying the formation of thiobarbituric acidreactive substances (TBARs) and carbonyls	[54]
Origanum vulgare extract	Sheep sausages	↘ of lipid and protein autoxidation (carbonyls content and TBARS values)	[55]
Salvia officinalis L. extract	Fresh pork sausages	Inhibition of microbial growth	[56]
		↘ of TBARS values	
		Enhancement of sensory properties of sausages	
Prunus salicina peel	Breast chicken patties	↘ in 50% of the formation of TBARs in raw patties during 10-day storage at 4 °C	[57]
Cumin	Ground lean meat	Inhibition of microbial growth and decrease of lipid oxidation	[58]
Cardamom		Enhancement of sensory characteristics	
Clove powder		Prolonging the shelf life of meat during storage at both ambient and refrigeration temperatures for 25 days	
Propolis extract	Fresh patties	↗ of the oxidative stability during storage	[59]
Beetroot peel extract	Raw beef meat	Delaying lipid/protein oxidation processes and microbial growth	[60]
		Enhancement in sensory attributes	
Pitanga leaf extracts	Pork burger	↘ of lipid oxidation	[32]
		Improvement of the redness of burgers	
Date palm seed extract	Raw beef meat	Delaying lipid/protein oxidation processes	[61]
		Delaying microbial growth	
Organic acids/salts			
Citric and malic acids	Sheep meat	↘ of the pH level and moisture content	[36]
		↘ of lipid oxidation in sheep meat	
Lactic acid	Broiler fillets	↘ the *Enterobacteriaceae* and psychrotrophic bacteria counts of broiler fillet	[38]
		↗ of the acceptability of broiler fillet	
		↘ of the biogenic amines formation	
		Preservation of refrigerated broiler fillets up to 12 or 15 days at 4 °C	
Peracetic acid	Raw poultry meat	Reduction in *Salmonella* Typhimurium count	[62]

(continued)

Table 1 (continued)

Functional ingredients	Developed products	Effect on meat product quality	References
Lactic acid and sodium metasilicate	Beef, pork and deli meats	Effective against *E. coli* O157H7 and *Salmonella* spp. on beef bottom rounds and pork ham steaks	[63]
Lactic, malic and fumaric acids	Chicken meat	↘ *Salmonella* counts	[64]
		Improve sensory characteristics and quality of chicken meat	
Peroxy acetic acid	Chicken breast fillets	↘ of *Salmonella* and *Campylobacter* loads	[65]
Peroxy acetic acid	Chicken wings	↘ of *Salmonella* and *Campylobacter* and *E. coli* counts	[66]
Lactic acid	Buffalo meat	Low aerobic plate count	[67]
		Improvement of instrumental color	

to sausages utilizing beef fat, as observed by [17]. When plant oils are added directly to the meat emulsion, they get mixed into the myofibrillar protein matrix. This makes the fat particles' surface area bigger and changes their colour, especially the redness that is important for emulsified meat products [17]. Utilizing safflower oil as a substitute for animal fat in wieners results in the creation of acceptable and healthier products, with no discernible impact on sensory properties [13]. Consequently, integrating healthy fats into meat and meat products is crucial for maintaining product quality and flavour, while simultaneously augmenting the overall health benefits for consumers.

2.2 *Dietary Fibers*

Dietary fibers (DF) are derived from various sources, including fruits, vegetables and grains [22]. Fruits, vegetables, soy, oat bran and psyllium husk are rich in soluble dietary fiber (SDF), while whole grains are a prominent source of insoluble dietary fiber (IDF) [23]. The integration of DF into meat products holds the potential to improve emulsion stability, enhance water and fat retention, and especially minimize cooking losses. DFs demonstrate significant efficacy in improving the texture and juiciness of meat products, as mentioned by Madane et al. [24]. The strategic use of DFs as substitutes for animal fat in meat products proves highly effective in addressing technological and sensory concerns arising from fat reduction. In particular, inulin emerges as a key DF utilized as an animal fat alternative in meat derivatives [10, 25, 26]. Souza Paglarini et al. [26], stated that incorporating a high-fiber emulsion gel with inulin in Bologna sausages exhibited elevated fibre content, a favourable texture profile, and a more cohesive microstructure. These authors observed a 31% reduction in fat content and satisfactory sensory acceptance in Bologna sausages. Notably, jabuticaba skin flour, sourced from jabuticaba fruit cultivation in Brazil, serves as a significant DF reservoir. Alves et al. [27] illustrated that restructured hams containing 0.5% jabuticaba skin flour experienced minimal changes in their physicochemical properties, except for an increase in phenolic compounds and fibre, reduced brightness, and heightened hardness.

Furthermore, a number of studies have been conducted regarding the use of combinations of various types of DF in meat products. The influence of adding inulin fibers, cellulose and carrot to chicken meatballs was examined by Kurt and Kılınççeker [25]. According to the L, a, and b values, the researchers noticed that the inclusion of fibers in meatballs has the potential to improve product quality and color properties. Notably, in the fried samples, cellulose and carrot fibres prevented diameter reduction while increasing yield and moisture absorption values. Combinations of different fibre additions are being investigated in meat compositions through ongoing research, indicating that these products are expected to become more frequent in the market.

2.3 *Natural Preservative*

2.3.1 Encapsulation

Surendhiran et al. [28] investigated chitosan and poly (ethylene oxide) (Ch/PEO) to develop active nanofibers incorporating pomegranate peel extract (PE). Notable thermal, physicomechanical, and antibacterial capabilities were exhibited by the generated nanofibers. Furthermore, laboratory testing verified that the gradually released encapsulated extract extended the shelf life of beef meat compared to untreated samples and possessed a sizable inhibitory zone against *E. coli*. In addition, Najjaa et al. [29] focused on the preservation of minced beef meat by encapsulating the essential oil of *Allium sativum* at 5, 10, 15, and 20%. Over 6 days in cold storage, the treated meat samples revealed an inhibition of *E. coli* growth and the quality remained acceptable and adequate. In another study, Surendhiran et al. [30] generated nanofibers by electrospinning a mixture of sodium alginate and PEO to encapsulate phlorotannin (Ph). The ability of these nanofibers to effectively preserve chicken against *Salmonella enteritidis* was assessed. The *Salmonella enteritidis* count was reduced by 2.92 and 6.27 log CFU/g at temperatures of 4 and 25 °C, respectively. The organoleptic evaluation showed that the application of SA/PEO/Ph nanofibers greatly increased the shelf life of chickens without developing any detectable sensory changes when compared to sodium nitrite.

2.3.2 Plant Antioxidants

In recent years, the use of extracts from vegetables, fruits, and spices as meat product antioxidants has been the subject of several research. An overview of these investigations is given in Table 1, which also lists the individual meat items that the extracts were applied to, the vegetal species from which they were derived, and the major conclusions that were achieved. The quality of meat products deteriorates due in large part to oxidation processes. The use of naturally sourced plant extracts is becoming a noteworthy substitute for managing oxidative alterations in meat products while they are being stored. For example, Pateiro et al. [31] studied the impact of guarana seed extract on pork patties. The patties were subjected to an 18-day analysis period while being refrigerated at 2 °C for the research. The findings showed that the guarana seed extract, at 250 and 500 mg/kg, improved color, stabilized lipids and proteins, and reduced lipid oxidation in comparison to BHT (200 mg/kg). Tyrosol and other low molecular weight phenolics are prominently present in the guarana seed extract, which is renowned for its abundance of phenolic chemicals. In another investigation, Lorenzo et al. [32] investigated the impact of pitanga leaf extract (at 250, 500, and 1000 mg/kg) on protein and lipid oxidation in pork burgers. The main phenolic compounds found in the extract were tyrosol, alkylmethoxyphenols, hydroxycoumarins, hydroxyphenylpropenes, and hydroxycinnamic acid (cinnamic and caffeic acids were the most common). The high antioxidant activity of the extract was due in part to these compounds. They caused lipid and protein oxidation in a way that was similar to BHT (200 mg/kg). Pitanga leaf extract and BHT both kept TBARS readings below 0.5 mg MDA/kg for the duration of the storage period, as was predicted. On the 7th, the antioxidant-free control group, on the other hand, surpassed this value. Additionally, in a research by Šojić et al. [33], thyme by-products—rich in flavonoids, terpenoids, tannins, and other bioactive compounds—were sustainably obtained using supercritical fluid extraction and added to minced pork patties. The researchers used CO_2 as the only solvent and produced two extracts under various supercritical fluid extraction settings, including pressures (100 and 350 bar) and extraction temperatures (40 and 50 °C). In both extracts, thymol, α-terpineol, and carvacrol were shown to be the main terpenoids. The extracts were added at concentrations ranging from 0.075 to 0.150 μL/g to the patties. On the second day of storage, the control group (without antioxidants) had lipid oxidation values higher than 0.5 mg MDA/kg. On the other hand, patties with thyme by-product extract added did not reach this TBARS value until the third day. The extracts prevented discoloration of the samples, confirming their efficacy as antioxidants in ground pork patties.

2.3.3 Organic Acids/Salts

Organic acids and their derivatives function as antioxidants, preventing the spoilage of meat products. Their widespread availability and evident efficacy make them highly promising for inclusion in processed meat items. In fact, the salts derived from organic acids primarily exhibit bacteriostatic effects, while organic acids demonstrate both bactericidal and bacteriostatic properties [34]. The bactericidal impact is primarily attributed to the uncharged and undissociated state of organic acids [35]. According to Albuquerque et al. [36], the introduction of citric acid into sheep "Buchada" led to a decrease in microbial proliferation (thermotolerant coliforms,

Staphylococcus sp., and *Salmonella* sp.), as well as a reduction in pH, moisture, and TBARS value. Importantly, it had no adverse impact on sensory characteristics such as odor, flavor, appearance, and tenderness. Additionally, the study by Abd Ellatif et al. [37] investigated the antibacterial impact of spray treatments containing lactic and acetic acid at concentrations of 1, 1.5, and 2% on the surface of fresh sheep carcasses. The target was the reduction of aerobic plate count, *Enterobacteriaceae*, coliform, and *Staphylococcus* counts. The results revealed notable decreases in the microbial growth of the examined microorganisms following exposure of sheep meat to organic acids, with the 2% lactic acid spray exhibiting a particularly pronounced effect. Moreover, the use of 1% lactic acid exhibited preservation outcomes for refrigerated chicken broiler fillets, according to Hussein et al. [38]. It positively influences the freshness with a significant decrease in total volatile basic nitrogen (TVB-N) and TBARS values. Also, it effectively controlled microbial aspects, leading to a reduction in total psychrotrophic bacteria and *Enterobacteriaceae* loads. Furthermore, it successfully inhibited the formation of biogenic amines, contributing to the enhanced acceptability of refrigerated chicken meat for up to 15 days at 4 °C.

3 Reduction Approach in Meat Processing

3.1 Sodium Content Reduction

NaCl retains a significant role in handling meat and meat products for sensory and technological targets [68]. In addition, salt could enhance the texture and meat flavor supplying detailed processing features through the water-binding properties and solubilization of actin and myosin (myofibrillar proteins). This fact could ensure microbiological security by inhibiting spoilage microorganisms [69]. Recently, researchers have been interested in developing strategies for producing salt analogs.

One of the potentials for reducing the level of the added salt is the stealthy method. This process includes a progressive reduction, over a long period, of the salt in meat products, and therefore the consumers will not observe the changes in the saltiness. Delgado-Pando et al. [70] assessed the lowest suitable salt content in bacon and cooked ham. Outcomes showed that salt levels could be reduced by up to 34% in bacon and 19% in ham; in addition, the microbiological, physicochemical, and sensory properties of products were not affected. Aaslyng et al. [71] reported that a modest salt decrease in the conventional formula is possible for ham and sausages at 2.3–1.8% and 2.2–1.7%, respectively. Remarkably, there was no change in sensory traits, and the shelf life was improved.

An unconventional method for salt reduction in meat products involves replacing NaCl with another mineral. The most frequent kinds of salt substitutes are KCl, $CaCl_2$, $MgCl_2$, $KC_3H_5O_3$, $MgSO_4$, and $C_{12}H_{14}CaO_{12}$ [72, 73]. KCl is the most commonly employed NaCl substitute because of its comparable behaviour concerning the inhibition of protease activity, protein solubilization, and an equivalent preservative effect [74]. Since at great levels KCl can decrease saline taste, generate bitter and metallic tastes, and significant flavor defect, this salt should be employed at a level to 40–50% of NaCl [75, 76]. Some investigations displayed that sensorial flaws of meat products, triggered by NaCl replacement with KCl, can be abridged by the addition of different flavour garnishes, such as amino acids (Lys, His, Arg), $C_{10}H_{11}N_4Na_2O_8P$ and $C_{10}H_{12}N_5Na_2O_8P$ [77], yeast extracts, protein hydrolysate produced from vegetables and $C_5H_8NNaO_4$ [75]. Da Silva et al. [78] confirmed that Arg and His at 1 and 0.2%, respectively, alone or blended, could enhance the sensory features, TPA, and emulsion stability affected by the replacement of 60% NaCl with KCl in emulsified meat products. Edible algae was applied in numerous studies as salt substitutes in reformulated meat products. Choi et al. [79] stated that the addition of edible seaweeds in frankfurters at 60% increased salt, better colour, flavour, tenderness, and juiciness of the newly formatted meat product. Similar trends were observed by Fellendorf et al. [80]. These authors used the Wakame in low-salt black puddings. The research explored the impact of 4 edible seaweeds' (*Himanthalia elongate*, *Undaria pinnatifida*, *Porphyra umbilicalis*, and *Palmaria palmata*) in frankfurters with 50% less NaCl. Compared to untreated samples, treated ones showed that the algae addition provoked: (i) darker colour; (ii) flavour modification; and (iii) a reduction in hardness and chewiness in the final meat product [81].

3.2 Nitrate/Nitrite Content Reduction

For meat and meat derivative preservation, nitrite ($NO^-{}_2$) and nitrate ($NO^-{}_3$) are essential since they provide a (i) vivid colour, (ii) delay the chemical oxidation reaction, and (iii) maintain microbiological properties and sensory traits. Remarkably, in meat processing, NO_2^- salts and NO_3^- salts, like those of sodium and potassium, are employed, which could delay the rancidity onset, produce cured meat flavour or smell, and stabilize the meat's red colour [82]. Biochemically, NO_2^- and NO_3^- can generate pinkish-red colour through interaction with the myoglobin in muscle.

Researchers are interested in substituting the colour formation ingredients, covering natural nitrite or nitrate, with vegetable extracts and bacteria. Zhu et al. [83] applied Lactobacillus *plantarum* to incompletely substitute standard nitrite in Chinese fermented sausages. 50 mg/kg $NaNO_2$ blended with 7 log10 of *Lb. plantarum* has similar effects as 100 mg/kg $NaNO_2$ alone. The rheological assessment displayed that the used strain enhanced the viscoelasticity and gel properties of the treated samples. In addition, protein gels were amended by changes of C-H elongating and bending vibrations, reduction of α-helix, increase of β-sheet, and random coil in secondary structure variation. Furthermore, analysis of biogenic amines showed that the use of *Lactobacillus plantarum* decreased the hazard from biogenic amines by reducing tyramine content. Ozaki et al. [84] incorporated radish powder at 0.5 and 1.0%, and oregano essential oil at 100 mg/kg on nitrite-free fermented cooked sausages [84]. These authors reported that the vegetable powder led to a satisfactory development of color, nitrite development, a_w (0.91–0.92), weight loss (35.8–37.7%), textural properties, and a decrease in total viable count bacteria. Kim and Chin [85] assessed the effects of paprika oleoresin, sunflower seed oil, and $NaNO_2$ on emulsified pork sausage. Findings specified that 7.5×10^{-5} NaNO/0.1% paprika oleoresin solutions could enhance the instrumental redness.

On the other hand, the antimicrobial effects of NO^{-}_2 and NO^{-}_3 are measured to be very significant in meat and meat products. The minimum inhibition concentration (MIC) of NO^{-}_2 that could prevent *Clostridium botulinum* is 4×10^{-5} to 8×10^{-5} g/g [86]. Patarata et al. [87] explored the impact of red wine and garlic on the behaviour of *C. sporogenes* and *Salmonella* in a dry-cured chouriço sausage. These authors concluded that the red wine 7.5% and garlic (1%) could inhibit *C. sporogenes* and *Salmonella* during the processing of chouriço. In order to substitute nitrite in the preparation of cooked beef sausage, [88] applied lupulon–xanthohumol-loaded nanoliposomes (L–X-NL). Formulations produced by 30 ppm of nitrite and 150 ppm of L–X-NL can stabilize the microbiological parameters through 30 days of storage at 4 °C. Additionally, L–X-NL could delay the chemical oxidation and maintain the sensory traits.

As another option, among the functions of nitrite and nitrate in meat products are the improvement of flavour. In this regard, Perea-Sanz et al. [89] inspected the effect of *Debaryomyces hansenii* inoculation on the aroma chemicals of dry fermented sausage. These authors confirmed that D. *hansenii* could produce ethyl ester and 3-methylbutanal. Zeng et al. [90] applied *Lb. fermentum* RC4 in order to reduce the nitrite level in salted meat. Thus, 73 volatile contents in total were described in the fermented group, compared to 67 volatile contents in the control group.

3.3 Unhealthy Fat Reduction

With a view to attaining muscle foods with reduced animal fat, the use of substitutes from plant sources played a dominant role in the stability and formation of these products. A natural high content of biopolymers with technological and functional properties excels these plant materials as fat substitutes in meat products.

Exceptionally, the addition of inulin is connected to advanced stages of moisture and carbohydrates and inferior content of lipids [91–93]. Additionally, the texture was touched by inulin incorporation in meat products, wherein hardness and chewiness can be decreased [91, 94]. In the same way, meat colour could also be impacted by inulin. Similarly, in a dry-fermented sausage, inulin application with water or linseed oil decreased L*, increased a*, and had no impact on b* [94]. Contrariwise, the use of inulin gel/soy protein isolate or soybean oil produced an important level of L*, reduction in a*, and had no effect in b* on a dry-fermented sausage in a reduced salt Bologna sausage [95]. In Brazilian-cooked sausage, the joint usage of inulin and oat fiber did not touch the sensory attributes, excluding texture [92].

Konjac glucommanan was an alternative and applicable option to substitute fat in meat products. In the case of the addition of konjac without other ingredients in a Northeastern Tai fermented sausage, Sorapukdee et al. [96] confirmed a noteworthy decrease in lipid oxidation and pH along with an elevation in texture parameters, a* and scores for sensory attributes. Furthermore, no substantial changes in colour juiciness or sensorial texture among treatments. Similar findings were reported by Kim et al. [97]. Konjac addition impacted the moisture content, as did the Aloe vera, cactus pear, and wheat sprout powders in a frankfurter-type sausage [97].

Oats were likewise considered as potential animal fat substitutes in meat products. The addition of a commercial oat hull product to beef burgers increased the content of moisture, ash, and carbohydrates It also increased the cooking yield, the percentage of polyunsaturated fatty acids (PUFAs), and a* and L* [98]. Equally, the effect of oat beta-glucan with marine collagen peptide as an animal fat replacer was assessed in pork sausage [99]. Compared to untreated samples, treated sausages had higher moisture and chewiness/springiness, carbohydrate content, and b*/juiciness. Furthermore, the L* and a* as well as the flavour and colour, were decreased as a result of the oat β-glucan-marine collagen peptide's incorporation. Another study reported by Guedes-Oliveira et al. [100] confirmed that cashew apple fiber could be a good fat substitute in chicken patties. The moisture, protein, and ash content of the patties were affected by this fiber in this investigation, but the cooking yield and shear force increased.

4 Probiotic/Fermentation Products

Starter cultures, whether single or mixed microbial cultures, serve the purpose of promoting the fermentation process in meat products. These include bacteria, especially lactic acid bacteria (LAB), as well as yeasts and moulds, which contribute significantly to enhancing the safety of fermented meat products. Additionally, the utilization of starters aids in standardizing product characteristics and reducing the ripening duration of fermented meat products. Recognized as Generally Regarded As Safe (GRAS) by the US Food and Drug Administration (FDA), starter cultures possess the capability to delay the growth of pathogenic and spoilage microorganisms [101]. In fact, LAB plays a pivotal role in enhancing the texture and flavor of fermented meat products, contributing to product preservation, and ultimately extending shelf life [102]. The inherent strong buffering capacity and low carbohydrate content of fresh meat facilitate the fermentation process without altering the organoleptic characteristics [103].

Various studies have reported on the bio-preservation of meat and meat products using starter LAB, as outlined in Table 2. For instance, Ayyash et al. [104] studied the impact of using a commercial starter culture (*Staphylococcus carnosus* and *Pediococcus pentosaceus*) individually and in combination with *Lactobacillus plantarum* KX881772, along with variations in meat source (camel meat vs. beef), in the production of bioactive fermented sausages. The study revealed that the choice of meat source influenced the antioxidant activity of sausages, with camel meat exhibiting greater activity than beef, regardless of the starter culture composition.

Table 2 Application of starter cultures in meat products

Starter cultures	Meat products	Biological effects	References
L. sakei JCM1157 and no. 23 and *L. curvatus* NBRC15884 and no. 28 (10^8 CFU/mL)	Fermented beef sausage	↗ inhibition of ACE (angiotensin-converting enzyme of all starters activity	[109]
Commercial starter (*Pediococcus pentosaceus* and *Staphylococcus carnosus*) and *Lactobacillus plantarum* KX881772 (10^7–10^8 CFU/kg)	Fermented camel or beef sausage	↗ degree of hydrolysis and antioxidant ↗ inhibition of α-amylase, α-glucosidase, and ACE activities	[104]
L. plantarum CD101 and *Staphylococcus simulans* NJ201 (10^7 CFU/g)	Fermented pork sausage	↗ degree of hydrolysis, inhibition of ACE activity, free amino acids and antioxidant activity	[105]
Staphylococci xylosus and *L. plantarum*, *Staphylococci xylosus* with *L. pentosus*, or *Staphylococci carnosus* and *L. sakei* (inoculum: 10^7 CFU/g)	Fermented camel sausages	↗ antioxidant and inhibition of ACE activities	[110]
Lactobacillus plantarum AB20–961 and DSM2601 (10^8 CFU/kg)	Fermented beef sausage	Highest content was obtained by DSM2601 (7.5 mg CLA/g fat)	[111]
L. acidophilus CRL1014	Fermented sausages	Enhancement of physicochemical and microbiological characteristics ↗ in oleic acid and ↘ in saturated acid content of sausage	[112]
L. rhamnosus LOCK900 (10^6 CFU/g)	Dry fermented pork loins and sausages	Loins and sausages can be stored in refrigerated, anaerobic conditions for 12 weeks	[106]

Additionally, the antidiabetic potential was influenced by the strains in the starter culture. Sausages containing *Lactobacillus plantarum* KX881772, whether used alone or in combination with the commercial starter culture, demonstrated the most substantial inhibitory effects on α-glucosidase and α-amylase. In another study, Kong et al. [105] combined *Staphylococcus simulans* NJ201 and *Lactobacillus plantarum* CD101, and the resulting sausages exhibited heightened inhibitory activity against angiotensin-converting enzyme and demonstrated elevated antioxidant activity. The study also noted an augmentation in the degree of protein hydrolysis and the concentration of free amino acids. Functional dry-cured loins with probiotics were developed using *L. rhamnosus* LOCK900 as a starter culture, as reported by Neffe-Skocińska et al. [106]. The application of this starter culture led to a decrease in pH and improved oxidative stability throughout the processing and storage periods (up to 4 months at 4 °C). Furthermore, the treated loins exhibited

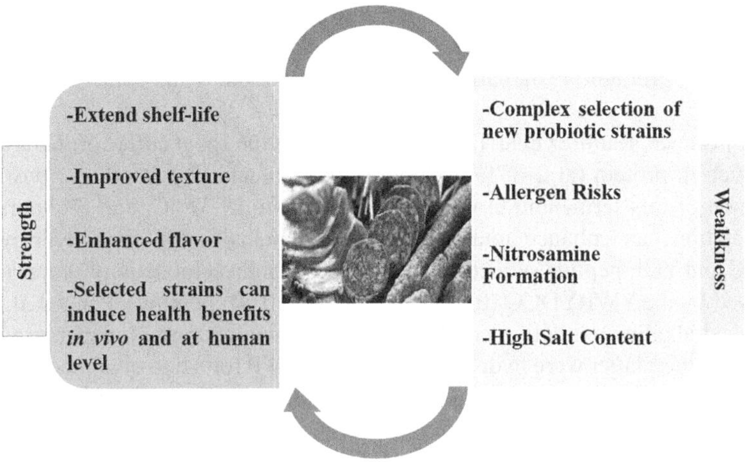

Fig. 2 Strengths and weakness of fermented meat products

elevated counts of LAB, reduced counts of *Enterobacteriaceae,* and the absence of *Salmonella* spp. and *L. monocytogenes*. A parallel investigation by the same research team, involving *L. rhamnosus* LOCK900 and *Bifidobacterium animalis* subsp. *lactis* BB12, demonstrated that this combination of starter cultures yielded fermented and dry-cured loins with sensory qualities deemed acceptable [107]. Additionally, Nikodinoska et al. [108] explored the influence of *L. plantarum* PSC20 on the microbial quality, specifically *L. monocytogenes* and *Salmonella,* in chorizo sausage. Their results revealed a significant decrease in *L. monocytogenes,* while *Salmonella* did not show a significant impact (Fig. 2).

5 Bioactive Compounds/Peptides

Bioactive peptides (BPs), specific protein fragments comprised of 2–20 amino acids, could be produced from animal, plant, fermentation, or enzymatic hydrolysis. Moreover, BAs can also be generated by chemical synthesis. Interestingly, BPs are safe with alternative impact and selective properties. To fulfill the request of consumers for healthy meat and meat derivatives, Scientists, investigators, and academics put effort into employing the BPs because of their antibacterial/antioxidant potential, making new functional foods [113]. Currently, BPs with antioxidant activity possess a good potential for food preservation. In this line, BPs can be used directly, incorporated into food products, or sprayed on the surface [113].

For food preservation, BPs were isolated from numerous sources. Since a protein is the precursor of BPs, contents with elevated protein substances are plentiful sources [114, 115]. It is revealed that BPs derive from animal food, *viz.*, milk and milk products; animals like duck plasma and porcine bone; plants such as cottonseed and bean; and bacteria like *Bacillus amyloliquefaciens* and *Lactobacillus*. Currently, some agricultural by-products are employed to generate BPs, such as waste coffee grounds [116], baijiu vinasse [117] and scales and epidermal mucus of fish [118].

For instance, Ramírez et al. [116] concluded that the spent coffee grounds (SCG) were rich in protein (at 2–17%) and could be fermented by *Bacillus clausii.* The condition of this fermentation was 1.5×10^8 CFU/mL, 37 °C, and 39 hours. The fermentation can enhance total and soluble proteins and protein hydrolysates. GMCC and YGF peptide sequences displayed the highest bioactivity scores (0.97), followed by the YWRYDCQ (0.65) and RMYRY (0.60) peptides. Peng et al. [117] employed alkaline extraction combined with ultrasonication to recuperate vinasse proteins. These latter were hydrolyzed by Corolase PP for 8 h to obtain peptide fractions (VPH-1, −2, and −3) showing high DPPH radical scavenging activity. Characterized by LC-MS/MS, the BP sequences were: KLPDHPKLPK and VDVPVKVPYS. These two BPs have a high anti-inflammatory activity at 0.25 mg/mL. In order to improve the concentration of soluble peptides (P), antioxidant potential, and % of hydrolysis (DH), Sierra-Lopera and Zapata-Montoya [118] optimized the conditions of enzymatic hydrolysis (kind of enzyme, pH, temperature substrate (S), and enzyme concentration (E). The two studied enzymes are Alcalase® and Flavourzyme® and were used for 3 h in a reactor of 0.5 L. The Alcalase® 2.4 L enzyme demonstrated the highest productivity, achieving optimal S and E concentrations of 45 and 4.4 g/L, respectively. Remarkably, the obtained hydrolysates presented antioxidant potential.

5.1 BPs Antimicrobial Activity

Till now, beyond 3000 BPs have been assembled in the Database of Antimicrobial Peptide. When the host is attacked by microbes, BPs could be prompted to destroy or inhibit microorganisms, except with destructive consequences [119]. The majority of BPs could inhibit the microorganisms across the membrane. In this context, BPs are hydrophilic and hydrophobic; on the other hand, a few BPs disturb the normal physiological function of microbial cells by pointing to intracellular components, provoking cell death [120]. Generally, there are four mechanisms implicated: (i) Binding to DNA or RNA: BPs have a direct effect on genetic material (DNA or RNA) [121] (ii) Suppression of protein synthesis: BPs can prevent protein translation or processing modification and perform on the substance or site needed for the synthesis protein [122] (iii) Inhibition of cell wall synthesis: BPs inhibit the formation of cell wall by connecting to key enzymes (iv) Stimulation of reactive oxygen species (ROS) overproduction [123]. A huge number of investigations have shown that the antibacterial activity of BPs is associated with their net positive charge, secondary structure, hydrophobicity and amphipathic structure. It has been suggested that increasing the net positive charge of BPs can improve antibacterial activity. For instance, increasing the number of basic amino acids (Arg, Lys, and His) with + charge in the sequence can successfully increase the BPs antibacterial activity [124]. Hydrophobicity is an additional significant characteristic of the antibacterial activity of BPs. Amino acids like Gly, Val, Ala, Leu, Pro, Ile, Phe, Met and Trp can develop the BPs antibacterial activity.

5.2 BPs Antioxidant Activity

The majority of antioxidant peptides are made of 2–12 amino acids [125]. They can exercise their impacts by (i) scavenging free radicals, (ii) preventing lipid peroxidation, and (iii) networking with metal ions [113]. The antioxidant peptides of BPs covering tyrosine residues scavenge free radicals essentially through the Hydrogen Atom Transfer (HAT) mechanism. Regarding BPs containing Trp, His and Cys residues scavenge free radicals primarily through the electron transfer (ET) mechanism [113].

Practically, the antioxidant BPs affect fatty foods. In this line, BPs are well diffused on the surface of the oil molecules, stopping the oil molecules from communicating with the air, and thus preventing lipid peroxidation. Moreover, antioxidant BPs can decrease lipase activity, blocking the chain reaction. Elsewhere, antioxidant BPs can interrelate with metal, preventing the oxidation-promoting reaction [126]. Some studies have illustrated that antioxidant BPs with a low molecular weight (<5 KDa) usually have higher antioxidant activity. For example, from Pixian broad bean paste, Li et al. [127] isolated four novel BPs: RGLSK, DNLLN, NKGPR and TPCPPQ. RGLSK exhibited the highest ACE inhibitory activity (IC_{50} ~ 87 μmol/L), and TPCPPQ exhibited the greatest antioxidant capacity. Via molecular docking, these authors reported that RGLSK designed H bonds with S1 active sites (Ala354, Tyr523, Glu384) and made coordinate bonds with Zn^{2+} (Glu411, His383) of ACE. In the lipid food matrix, hydrophobic amino acids can develop the BPs solubility and increase the relations with unsaturated fatty acids and free radicals [128]. Moreover, amino acid residues like His, Asp, and Glu displayed antioxidant capacity by virtue of scavenging free radicals and chelating metal ions [129].

5.3 Practical Application of BPs in Meat Product Preservation

Recent investigations have reported the potential of BPs to inhibit microbial growth and limit lipid oxidation in meat products. Jabeen and Khanum [130] investigated the antibacterial activity of BPs from different fractions (pulp, seeds and skin) of bitter melon using *S. aureus*, *E. coli*, *S. typhi* and *P. aeruginosa as* target bacteria. These authors examined the purified peptide from the seed fraction with a molecular mass ~10 kDa for minced meat preservation. In fact, at 40 μg peptide, a decrease of *S. aureus* from 8.0 to 3.77 log CFU/g was reported. Nisin, Nisaplin®, is a polypeptide at 34 amino acids and a 3.5 kDa of molecular mass and has antimicrobial potential *Vs.* an extensive range of Gram (−) and (+) bacteria. The nisin produced from the *L. lactis* 537 strain was evaluated for the *L. monocytogenes* in ready-to-eat (RTE) sliced ham. The addition of the fermentate to RTE-sliced ham resulted in a total decrease of target strains from 3 log CFU/g to below the detection limit stored at 4 °C (20 CFU/g) [131].

Pediocin, a bacteriocin group of class IIa with a molecular mass (<5 kDa) included less than 50 amino acids [132]. Remarkably, the ability of pediocin to suppress Listeria spp. in order to preserve meat has been studied. The antibacterial activity of *L. monocytogenes* and *C. perfringens* was assessed against the pediocin PA-1-producing strain of *P. acidilactici* MCH14 in Spanish dry-fermented sausages and frankfurters [133]. *L. monocytogenes* was reduced by 2 and 0.6 log CFU/g in frankfurters treated with 5000 U/mL of the pediocin PA-1 generated by *P. acidilactici* MCH14 following storage at 4 °C for 60 days and at 15 °C for 30 days, respectively.

Ben Braïek et al. [134] examined the *Ec. lactis* Q1 enterocin P-producing ability to inhibit *L. monocytogenes* in raw beef meat. In fact, a remarkable reduction of *L. monocytogenes* (6.47 log units after 7 days and 7.25 log units below the control after 14 days) was observed. In addition, the organoleptic properties of treated samples were retained or enhanced. In another study, Yan et al. [122], the impact of the bacteriocin BM1829 isolated from *Lactobacillus crustorum* MN047 was examined. The purified bacteriocin displayed antimicrobial activity against Gram (+) and (−) bacteria, with MIC values arranged between 1–16 mg/L. the mechanism of action examined by these authors proved that BM1829 damaged membrane integrity, leading to the leak of cytoplasmic content. In addition, throughout beef storage, *E. coli* and *S. aureus* loads were reduced by using BM1829 [135]. isolated a BP (mytichitin-CB) and investigated its potential for pork preservation. During storage at 4 °C studied BP at 6 and 12 mg/L successfully repressed the microbial growth, decreased the pH, TVB-N, and thiobarbituric acid reactive substances (TBARS), and maintained the sensory features of fresh pork. Besides, the mixture of several preservatives- 4 mg/L mytichitin-CB, 10 g/L chitosan, and 300 mg/L nisin, can further enhance the protective impact on fresh pork.

From the bovine hemoglobin hydrolysate, Przybylski et al. [136] isolated small BP and examined its impact on ground beef meat. The outcomes exhibited that the BP could decrease lipid oxidation by ~60%, and inhibit microbial growth. Hyldgaard et al. [137] investigated the impact of ε-poly-L-lysine (ε-PL) and isoeugenol as natural preservatives on raw turkey meat. Compared to untreated samples, the antimicrobial mixture was qualified to decrease the charge load of *Pseudomonas* spp. by one log 10 of CFU/g of meat. On the other hand, ε-PL-treated samples proceeded from a *Pseudomonas* spp. to *Enterobacteriaceae*-controlled food matrix, while treatment with the antimicrobial mixture yielded an increased relative abundance of *Hafnia* spp., *Photobacterium* spp. and *Enterococcaceae* Marketably called Inovapure®, lysozyme is an antimicrobial enzyme employed to inhibit spoilage microorganisms and food-borne pathogens to extend the shelf-life of raw and processed meat [138]. Cegielska-Radziejewska et al. [139] investigated the role of modified lysozyme on the physicochemical, microbial, and sensory characteristics of pork loin during 28 days of storage. L* and a* were not impacted by the studied BP, and trials with the modified lysozyme possess the maximum score for aroma. Compared to the monomer, the modified lysozyme presented an anti-*Pseudomonas* and anti-*Enterobacteriaceae* activity. Mastromatteo et al. [140] examined the impact of lysozyme at 250 ppm, nisin at 250 ppm, and EDTA at 20 mM against *Listeria monocytogenes* and meat-borne spoilage bacteria in ostrich patties. The blend lysozyme preparations reduced *L. monocytogenes* below 2 log CFU/g of strich meat patties. After 2 days of storage, the treated ones presented a reduction in total viable counts by 1 log CFU/g. In minced lamb meat, Rao et al. [141] confirmed that lysozyme/chitooligosaccharide led to a 3–4 log CFU/g reduction of inoculated *E. coli*, *B. cereus* and *Pseudomonas fluorescens* However, *S. aureus* was not fully removed.

5.4 Factors Touching the Stability of BPs in Meat Preservation

In the process of meat processing and storage, the stability of BPs may be determined by their own structure properties and external environmental factors, *viz.* temperature, salt concentration, pH, meat model matrix, and meat processing. These factors led to bioactivity variations.

5.4.1 Structure of BPs

Some structural changes can enhance BPs stability in food preservation. Particularly the C-terminal amino acid or the penultimate amino acid can achieve a pivotal function in BP stability. Mirzaei et al. [142] substituted the penultimate Pro with Cys Tyr, and His, and elucidated the effects of the amino acid structure on the stability of BP on the antioxidant potential, *vis.* heat, NaCl, and pH. Salt and heat treatment stability was lowest for BPs containing Cys, and highest for residues containing Pro. To endorse the formation of disulfide bonds, Mandel et al. [143] bound L-Cys to the N and C-terminals of the natural linear peptide. Through this bridge creation, a cyclic OMN6, AMP was obtained, which BPs at high stability.

5.4.2 External Environmental Factors

Temperature

Most BPs have high thermal stability since they possess Cys that can be linked by disulfide bonds to form rigid structures [144]. Though some BPs without a firm rigid structure are induced by a high T°, this may lead to protein denaturation or aggregation and cause a change in structure and activity. BPs covering Cys are more sensitive to heat-provoked aggregation shaped by disulfide bonds and dimers or trimers, resulting in lower stability. Moreover, when subjected to an elevated temperature, BPs may denature, touching the secondary structure and decreasing their activity [145].

Salt Concentration
A high level of NaCl can reduce the activity of BPs; however, lower ones don't have a significant impact. Increasing the positive charge can facilitate BPs to substitute the salt ions nearby the bacterial membrane via an ion-exchange mechanism, ultimately overcoming the charge effect [145]. BPs containing Arg, Tyr, and Pro residues presented notable salt tolerance due to the capacity of these amino acids to resist the charge-shielding effect caused by salt ions [146]. It is concluded that the deletion of the N or C-t or both terminals resulted in the loss of BPs sensitivity to salt [146]. On the other hand, the activity of antioxidant BPs rehabilitated little in a low concentration of salt but was diminished in the presence of a high concentration. Similarly, the presence of a high concentration of salt can reduce the solubility of antioxidant peptides, consequently reducing their activity [146].

pH
At a neutral pH, the BPs were high. However, low or high pH will lead to a decreased activity; particularly, high pH has a greater influence on BPs [147]. The main cause could be attributed to the hydrolysis of BPs into inactive residues when exposed to highly alkaline or acidic environments [129, 147]. In addition, the decrease in BP activity could also be the consequence of the racemic reaction. Each BP possesses an optimal pH range for its efficacy. When subjected to various pH conditions, the degradation process varies, leading to alterations in both the structure and functionality of the peptide. The antibacterial activity of cationic BPs against Gram (–) bacteria could reach its maximum at acidic conditions, however, it is reduced under neutral or alkaline conditions [129].

6 Conclusion

Recent years have witnessed the emergence of a number of alternative techniques with many desirable features. In this regard, plant-derived products and microbial metabolites are viewed as preservatives, functional, and nutraceutical substances with a diversity of applications in the meat industry. These substances could be an encouraging solution that can contribute to addressing a number of environmental, economic, and social demands to preserve meat and meat products. This chapter displayed that natural preservatives, isolated from plants, animals and microorganisms, are being investigated for applications in meat and meat products to maintain their quality and extend their shelf-life. In addition, this study focused on bioactive and functional compounds and their fundamental impact on meeting different needs in human health, meat safety, and the global economy.

In the immediate future, new applications and advancements in the incorporation of plant-derived products, microbial metabolites meat and meat products, and active packaging are anticipated. It is necessary to highlight that the usage of these plant-derived products, and microbial metabolites is of particular interest because of the recent global demand for sustainability and a circular bio-economy.

On the other hand, it is also important to consider safety issues by removing toxic substances, and other hazards, and anti-nutritional elements. The sensory quality of the food materials and consumer acceptance are other important restrictions on the use of plant-derived products, and microbial metabolites and by-products as functional food ingredients for developing new food products. Sustainability and safety will have a significant impact on how natural antioxidants are used in food items in the future.

In the upcoming years, it is anticipated that research on plant and animal-derived products, and microbial metabolites will continue to integrate recent innovations and technological advances, such as cutting-edge computational and machine-learning approaches.

References

1. De Smet S, Vossen E. Meat: the balance between nutrition and health. A review. Meat Sci. 2016;120:145–56.
2. Baltic MZ, Boskovic M. When man met meat: meat in human nutrition from ancient times till today. Proc Food Sci. 2015;5:6–9.
3. McBey D, Watts D, Johnstone AM. Nudging, formulating new products, and the lifecourse: a qualitative assessment of the viability of three methods for reducing Scottish meat consumption for health, ethical, and environmental reasons. Appetite. 2019;142:104349.
4. Domingo JL, Nadal M. Carcinogenicity of consumption of red meat and processed meat: a review of scientific news since the IARC decision. Food Chem Toxicol. 2017;105:256–61.
5. Godfray HCJ, Aveyard P, Garnett T, et al. Meat consumption, health, and the environment. Science. 2018;361:eaam5324.
6. Bouvard V, Loomis D, Guyton KZ, et al. Carcinogenicity of consumption of red and processed meat. Lancet Oncol. 2015;16:1599–600.
7. Wolk A. Potential health hazards of eating red meat. J Intern Med. 2017;281:106–22.
8. Gagaoua M, Picard B. Current advances in meat nutritional, sensory and physical quality improvement. Foods. 2020;9:321.
9. Domínguez R, Lorenzo JM, Pateiro M, et al. Main animal fat replacers for the manufacture of healthy processed meat products. Crit Rev Food Sci Nutr. 2022:1–20.
10. Berizi E, Shekarforoush SS, Mohammadinezhad S, et al. The use of inulin as fat replacer and its effect on texture and sensory properties of emulsion type sausages. Iran J Vet Res. 2017;18:253–7.
11. Indumathi J, Shashikumar M, Reddy GVB, et al. Influence of incorporation of ground sesame seed as fat replacer on the functionality of spent broiler breeder hen chicken sausages. J Pharmacogn Phytochem. 2020;9:676–82.
12. Dias MF, Guimarães AS, Benevenuto Júnior AA, et al. Canola and olive oil gelled emulsions as pork fat replacers in beef burgers. Br Food J. 2021;124:50–60.
13. Kaynakci E, Kiliç B. Effect of replacing beef fat with safflower oil on physicochemical, nutritional and oxidative stability characteristics of wieners. Food Sci Technol. 2020;41:52–9.
14. Smaoui S, Echegaray N, Kumar M, et al. Beyond conventional meat preservation: saddling the control of Bacteriocin and lactic acid bacteria for clean label and functional meat products. Appl Biochem Biotechnol. 2023; https://doi.org/10.1007/s12010-023-04680-x.

15. Hernández-González JC, Martínez-Tapia A, Lazcano-Hernández G, et al. Bacteriocins from lactic acid bacteria. A powerful alternative as antimicrobials, probiotics, and Immunomodulators in veterinary medicine. Animals. 2021;11:979.
16. Badar IH, Liu H, Chen Q, et al. Future trends of processed meat products concerning perceived healthiness: a review. Compr Rev Food Sci Food Saf. 2021;20:4739–78.
17. Uzlaşır T, Aktaş N, Gerçekaslan KE. Pumpkin seed oil as a partial animal fat replacer in Bologna-type sausages. Food Sci Anim Resour. 2020;40:551–62.
18. Franco D, Martins AJ, López-Pedrouso M, et al. Strategy towards replacing pork backfat with a linseed oleogel in frankfurter sausages and its evaluation on physicochemical, nutritional, and sensory characteristics. Food Secur. 2019;8:366.
19. Vieira ASP, Souza XR, Rodrigues EC, et al. Replacement of animal fat by canola oil in chicken meatball. Braz J Poult Sci. 2019;21:eRBCA.
20. Barbut S, Marangoni AG, Thode U, et al. Using canola oil organogels as fat replacement in liver Pâté. J Food Sci. 2019;84:2646–51.
21. Nieto G, Lorenzo JM. Use of olive oil as fat replacer in meat emulsions. Curr Opin Food Sci. 2021;40:179–86.
22. Singh A, Kaur V, Kaler RSS. A review on dietary fiber in cereals and its characterization. J Appl Nat Sci. 2018;10:1216–25.
23. Kshirsagar SB, Takarkhede S, Jha AG, et al. A comprehensive review on dietary fiber and their functional properties in human body. World J Biol Pharm Health Sci. 2020;4:059–76.
24. Madane P, Das AK, Pateiro M, et al. Drumstick (Moringa oleifera) flower as an antioxidant dietary fibre in chicken meat nuggets. Food Secur. 2019;8:307.
25. Kurt Ş, Kılınççeker O. Effects of inulin, carrot and cellulose fibres on the properties of raw and fried chicken meatballs. S Afr J Anim Sci. 2018;48:39–47.
26. de Souza Paglarini C, Vidal VAS, Midori Ozaki M, et al. Inulin gelled emulsion as a fat replacer and fiber carrier in healthier Bologna sausage. 2022; https://doi.org/10.1177/1082013220980586.
27. de Alves APC, Marques TR, de Carvalho TCL, et al. Elaboration and acceptability of restructured hams added with jabuticaba skin. Food Sci Technol. 2017;37:232–8.
28. Surendhiran D, Li C, Cui H, et al. Fabrication of high stability active nanofibers encapsulated with pomegranate peel extract using chitosan/PEO for meat preservation. Food Packag Shelf Life. 2020;23:100439.
29. Najjaa H, Chekki R, Elfalleh W, et al. Freeze-dried, oven-dried, and microencapsulation of essential oil from Allium sativum as potential preservative agents of minced meat. Food Sci Nutr. 2020;8:1995–2003.
30. Surendhiran D, Cui H, Lin L. Encapsulation of Phlorotannin in alginate/PEO blended nanofibers to preserve chicken meat from Salmonella contaminations. Food Packag Shelf Life. 2019;21:100346.
31. Pateiro M, Vargas FC, Chincha AAIA, et al. Guarana seed extracts as a useful strategy to extend the shelf life of pork patties: UHPLC-ESI/QTOF phenolic profile and impact on microbial inactivation, lipid and protein oxidation and antioxidant capacity. Food Res Int. 2018;114:55–63.
32. Lorenzo JM, Vargas FC, Strozzi I, et al. Influence of Pitanga leaf extracts on lipid and protein oxidation of pork burger during shelf-life. Food Res Int. 2018;114:47–54.
33. Šojić B, Tomović V, Kocić-Tanackov S, et al. Supercritical extracts of wild thyme (Thymus serpyllum L.) by-product as natural antioxidants in ground pork patties. LWT. 2020;130:109661.
34. Lim C, Lückstädt C, Webster CD, et al. Organic acids and their salts. In: Dietary nutrients, additives, and fish health. Wiley. p. 305–19.
35. Mir SA, Masoodi FA. Use of organic acids for preservation and safety of traditional meat products. J Food Saf. 2018;38:e12514.
36. Albuquerque GN, Costa RG, Barba FJ, et al. Effect of organic acids on the quality of sheep "buchada": from food safety to physicochemical, nutritional, and sensorial evaluation. J Food Process Preserv. 2019;43:e13877.

37. Abd Ellatif ZA, Saad SM, Hassanin FS, et al. Efficiency of some organic acids as decontaminants in sheep carcasses. Benha Vet Med J. 2020;38:116–9.
38. Hussein MA, El-Ghareeb WR, Nasr MAF. The effect of rosemary extract and lactic acid on the quality of refrigerated broiler fillets. J Food Sci Technol. 2018;55:5025–34.
39. Kouzounis D, Lazaridou A, Katsanidis E. Partial replacement of animal fat by oleogels structured with monoglycerides and phytosterols in frankfurter sausages. Meat Sci. 2017;130:38–46.
40. Hautrive TP, Piccolo J, Rodrigues AS, et al. Effect of fat replacement by chitosan and golden flaxseed flour (wholemeal and defatted) on the quality of hamburgers. LWT. 2019;102:403–10.
41. Jairath J, Sharma DP, Dabur RS, et al. Standardization of corn starch as a fat replacer in buffalo calf meat sausages and its effect on the quality attributes. IJAR. 2017; https://doi.org/10.18805/ijar.B-3381.
42. Zeny E, Younis Essa R, Bisar B, et al. Effect of using chicory roots powder as a fat replacer on beef burger quality. Slov Vet Res. 2019;56:509–23.
43. Powell MJ, Sebranek JG, Prusa KJ, et al. Evaluation of citrus fiber as a natural replacer of sodium phosphate in alternatively-cured all-pork Bologna sausage. Meat Sci. 2019;157:107883.
44. Barros JC, Munekata PES, Pires MA, et al. Omega-3- and fibre-enriched chicken nuggets by replacement of chicken skin with chia (Salvia hispanica L.) flour. LWT. 2018;90:283–9.
45. Gedikoğlu A, Clarke AD. Quality attributes of citrus fiber added ground beef and consumer acceptance of citrus fiber added turkish meat-balls. Food Health. 2019:205–14.
46. Valenzuela-Melendres M, Camou JP, Torrentera-Olivera NG, et al. Nutritional quality of beef patties with added flaxseed and tomato paste. CyTA J Food. 2018;16:263–70.
47. Kamboh AA, Memon AM, Mughal MJ, et al. Dietary effects of soy and citrus flavonoid on antioxidation and microbial quality of meat in broilers. J Anim Physiol Anim Nutr. 2018;102:235–40.
48. Tometri SS, Ahmady M, Ariaii P, et al. Extraction and encapsulation of Laurus nobilis leaf extract with nano-liposome and its effect on oxidative, microbial, bacterial and sensory properties of minced beef. Food Measure. 2020;14:3333–44.
49. Zhang H, Liang Y, Li X, et al. Effect of chitosan-gelatin coating containing nano-encapsulated tarragon essential oil on the preservation of pork slices. Meat Sci. 2020;166:108137.
50. Valencia Junca MA, Valencia C, Flórez López E, et al. Chitosan beads incorporated with essential oil of thymus capitatus: stability studies on red tilapia fillets. Biomol Ther. 2019;9:458.
51. Rajaei A, Hadian M, Mohsenifar A, et al. A coating based on clove essential oils encapsulated by chitosan-myristic acid nanogel efficiently enhanced the shelf-life of beef cutlets. Food Packag Shelf Life. 2017;14:137–45.
52. Manhani MR, Nicoletti MA, Barretto ACDS, et al. Antioxidant action of rosemary and oregano extract in pre-cooked meat hamburger. FNS. 2018;09:806–17.
53. Fourati M, Smaoui S, Ben Hlima H, et al. Synchronised interrelationship between lipid/protein oxidation analysis and sensory attributes in refrigerated minced beef meat formulated with *Punica granatum* peel extract. Int J Food Sci Technol. 2020;55:1080–7.
54. Elhadef K, Smaoui S, Ben Hlima H, et al. Effects of Ephedra alata extract on the quality of minced beef meat during refrigerated storage: a chemometric approach. Meat Sci. 2020;170:108246.
55. Fernandes RPP, Trindade MA, Lorenzo JM, et al. Assessment of the stability of sheep sausages with the addition of different concentrations of Origanum vulgare extract during storage. Meat Sci. 2018;137:244–57.
56. Šojić B, Pavlić B, Zeković Z, et al. The effect of essential oil and extract from sage (Salvia officinalis L.) herbal dust (food industry by-product) on the oxidative and microbiological stability of fresh pork sausages. LWT. 2018;89:749–55.
57. Basanta MF, Rizzo SA, Szerman N, et al. Plum (Prunus salicina) peel and pulp microparticles as natural antioxidant additives in breast chicken patties. Food Res Int. 2018;106:1086–94.

58. Ahmad Mir S, Ahmad Masoodi F, Raja J. Influence of natural antioxidants on microbial load, lipid oxidation and sensorial quality of rista – a traditional meat product of India. Food Biosci. 2017;20:79–87.
59. Vargas-Sánchez RD, Torrescano-Urrutia GR, del Torres-Martínez BM, et al. Propolis extract as antioxidant to improve oxidative stability of fresh patties during refrigerated storage. Food Secur. 2019;8:614.
60. Chaari M, Elhadef K, Akermi S, et al. Potentials of beetroot (Beta vulgaris L.) peel extract for quality enhancement of refrigerated beef meat. Qual Assur Saf Crops Foods. 2023;15:99–115.
61. Elhadef K, Chaari M, Akermi S, et al. Production of functional raw chicken meat by incorporation of date palm seed extract: an assessment of microbiological, chemical and sensory properties. J Food Meas Charact. 2023;17:5117–33.
62. Chaplot S, Yadav B, Jeon B, et al. Atmospheric cold plasma and peracetic acid–based hurdle intervention to reduce salmonella on raw poultry meat. J Food Prot. 2019;82:878–88.
63. DeGeer SL, Wang L, Hill GN, et al. Optimizing application parameters for lactic acid and sodium metasilicate against pathogens on fresh beef, pork and deli meats. Meat Sci. 2016;118:28–33.
64. Fernández M, Rodríguez A, Fulco M, et al. Effects of lactic, malic and fumaric acids on Salmonella spp. counts and on chicken meat quality and sensory characteristics. J Food Sci Technol. 2021;58:3817–24.
65. Kumar S, Singh M, Cosby DE, et al. Efficacy of peroxy acetic acid in reducing Salmonella and Campylobacter spp. populations on chicken breast fillets. Poult Sci. 2020;99:2655–61.
66. Vaddu S, Kataria J, Rama EN, et al. Impact of pH on efficacy of peroxy acetic acid against Salmonella, Campylobacter, and Escherichia coli on chicken wings. Poult Sci. 2021;100:256–62.
67. Manzoor A, Jaspal MH, Yaqub T, et al. Effect of lactic acid spray on microbial and quality parameters of buffalo meat. Meat Sci. 2020;159:107923.
68. Petit G, Jury V, de Lamballerie M, et al. Salt intake from processed meat products: benefits, risks and evolving practices. Compr Rev Food Sci Food Saf. 2019;18:1453–73.
69. Chen J, Hu Y, Wen R, et al. Effect of NaCl substitutes on the physical, microbial and sensory characteristics of Harbin dry sausage. Meat Sci. 2019;156:205–13.
70. Delgado-Pando G, Fischer E, Allen P, et al. Salt content and minimum acceptable levels in whole-muscle cured meat products. Meat Sci. 2018;139:179–86.
71. Aaslyng MD, Vestergaard C, Koch AG. The effect of salt reduction on sensory quality and microbial growth in hotdog sausages, bacon, ham and salami. Meat Sci. 2014;96:47–55.
72. Fellendorf S, Kerry JP, Hamill RM, et al. Impact on the physicochemical and sensory properties of salt reduced corned beef formulated with and without the use of salt replacers. LWT. 2018;92:584–92.
73. Gaudette NJ, Pietrasik Z. The sensory impact of salt replacers and flavor enhancer in reduced sodium processed meats is matrix dependent. J Sens Stud. 2017;32:e12247.
74. Nachtigall FM, Vidal VAS, Pyarasani RD, et al. Substitution effects of NaCl by KCl and CaCl2 on lipolysis of salted meat. Food Secur. 2019;8:595.
75. Inguglia ES, Zhang Z, Tiwari BK, et al. Salt reduction strategies in processed meat products – a review. Trends Food Sci Technol. 2017;59:70–8.
76. Armenteros M, Aristoy M-C, Barat JM, et al. Biochemical and sensory changes in dry-cured ham salted with partial replacements of NaCl by other chloride salts. Meat Sci. 2012;90:361–7.
77. dos Santos BA, Campagnol PCB, Morgano MA, et al. Monosodium glutamate, disodium inosinate, disodium guanylate, lysine and taurine improve the sensory quality of fermented cooked sausages with 50% and 75% replacement of NaCl with KCl. Meat Sci. 2014;96:509–13.
78. da Silva SL, Lorenzo JM, Machado JM, et al. Application of arginine and histidine to improve the technological and sensory properties of low-fat and low-sodium bologna-type sausages produced with high levels of KCl. Meat Sci. 2020;159:107939.

79. Choi Y-S, Kum J-S, Jeon K-H, et al. Effects of edible seaweed on physicochemical and sensory characteristics of reduced-salt frankfurters. Korean J Food Sci Anim Resour. 2015;35:748.
80. Fellendorf S, O'Sullivan MG, Kerry JP. Impact of ingredient replacers on the physicochemical properties and sensory quality of reduced salt and fat black puddings. Meat Sci. 2016;113:17–25.
81. Vilar EG, Ouyang H, O'Sullivan MG, et al. Effect of salt reduction and inclusion of 1% edible seaweeds on the chemical, sensory and volatile component profile of reformulated frankfurters. Meat Sci. 2020;161:108001.
82. Ferysiuk K, Wójciak KM. Reduction of nitrite in meat products through the application of various plant-based ingredients. Antioxidants. 2020;9:711.
83. Zhu Y, Guo L, Yang Q. Partial replacement of nitrite with a novel probiotic Lactobacillus plantarum on nitrate, color, biogenic amines and gel properties of Chinese fermented sausages. Food Res Int. 2020;137:109351.
84. Ozaki MM, dos Santos M, Ribeiro WO, et al. Radish powder and oregano essential oil as nitrite substitutes in fermented cooked sausages. Food Res Int. 2021;140:109855.
85. Kim GH, Chin KB. Evaluation of quality characteristics of low-nitrite pork sausages with paprika oleoresin solution during refrigerated storage. Food Sci Anim Resour. 2021;41:428–39.
86. Rivera N, Bunning M, Martin J. Uncured-labeled meat products produced using plant-derived nitrates and nitrites: chemistry, safety, and regulatory considerations. J Agric Food Chem. 2019;67:8074–84.
87. Patarata L, Martins S, Silva JA, et al. Red wine and garlic as a possible alternative to minimize the use of nitrite for controlling clostridium sporogenes and Salmonella in a cured sausage: safety and sensory implications. Food Secur. 2020;9:206.
88. Khatib N, Varidi MJ, Mohebbi M, et al. Replacement of nitrite with lupulon–xanthohumol loaded nanoliposome in cooked beef-sausage: experimental and model based study. J Food Sci Technol. 2020;57:2629–39.
89. Perea-Sanz L, López-Díez JJ, Belloch C, et al. Counteracting the effect of reducing nitrate/nitrite levels on dry fermented sausage aroma by Debaryomyces hansenii inoculation. Meat Sci. 2020;164:108103.
90. Zeng X, Pan Q, Guo Y, et al. Potential mechanism of nitrite degradation by Lactobacillus fermentum RC4 based on proteomic analysis. J Proteome. 2019;194:70–8.
91. Alaei F, Hojjatoleslamy M, Hashemi Dehkordi SM. The effect of inulin as a fat substitute on the physicochemical and sensory properties of chicken sausages. Food Sci Nutr. 2018;6:512–9.
92. Souza CVB, Bellucci ERB, Lorenzo JM, et al. Low-fat Brazilian cooked sausage-*Paio* – with added oat fiber and inulin as a fat substitute: effect on the technological properties and sensory acceptance. Food Sci Technol. 2019;39:295–303.
93. Karwowska M, Stadnik J, Stasiak DM, et al. Strategies to improve the nutritional value of meat products: incorporation of bioactive compounds, reduction or elimination of harmful components and alternative technologies. Int J Food Sci Technol. 2021;56:6142–56.
94. Glisic M, Baltic M, Glisic M, et al. Inulin-based emulsion-filled gel as a fat replacer in prebiotic- and PUFA-enriched dry fermented sausages. Int J Food Sci Technol. 2019;54:787–97.
95. de Souza PC, Vidal VA, Ribeiro W, et al. Using inulin-based emulsion gels as fat substitute in salt reduced Bologna sausage. J Sci Food Agric. 2021;101:505–17.
96. Sorapukdee S, Jansa S, Tangwatcharin P. Partial replacement of pork backfat with konjac gel in Northeastern Thai fermented sausage (Sai Krok E-san) to produce the healthier product. Asian Australas J Anim Sci. 2019;32:1763–75.
97. Kim DH, Shin DM, Seo HG, et al. Effects of konjac gel with vegetable powders as fat replacers in frankfurter-type sausage. Asian Australas J Anim Sci. 2019;32:1195–204.
98. Summo C, De Angelis D, Difonzo G, et al. Effectiveness of oat-hull-based ingredient as fat replacer to produce low fat burger with high beta-glucans content. Food Secur. 2020;9:1057.

99. Fan R, Zhou D, Cao X. Evaluation of oat β-glucan-marine collagen peptide mixed gel and its application as the fat replacer in the sausage products. PLoS One. 2020;15:e0233447.
100. Guedes-Oliveira JM, Salgado RL, Costa-Lima BRC, et al. Washed cashew apple fiber (Anacardium occidentale L.) as fat replacer in chicken patties. LWT Food Sci Technol. 2016;71:268–73.
101. Laranjo M, Potes ME, Elias M. Role of starter cultures on the safety of fermented meat products. Front Microbiol. 2019;10:853.
102. Wang Y, Han J, Wang D, et al. Research update on the impact of lactic acid bacteria on the substance metabolism, flavor, and quality characteristics of fermented meat products. Food Secur. 2022;11:2090.
103. Favaro L, Todorov SD. Bacteriocinogenic LAB strains for fermented meat preservation: perspectives, challenges, and limitations. Probiotics Antimicrob Prot. 2017;9:444–58.
104. Ayyash M, Liu S-Q, Al Mheiri A, et al. In vitro investigation of health-promoting benefits of fermented camel sausage by novel probiotic Lactobacillus plantarum: a comparative study with beef sausages. LWT. 2019;99:346–54.
105. Kong Y, Feng M, Sun J. Effects of Lactobacillus plantarum CD101 and Staphylococcus simulans NJ201 on proteolytic changes and bioactivities (antioxidant and antihypertensive activities) in fermented pork sausage. LWT. 2020;133:109985.
106. Neffe-Skocińska K, Okoń A, Zielińska D, et al. The possibility of using the probiotic starter culture Lacticaseibacillus rhamnosus LOCK900 in dry fermented pork loins and sausages produced under industrial conditions. Appl Sci. 2020;10:4311.
107. Neffe-Skocińska K, Okoń A, Kołożyn-Krajewska D, et al. Amino acid profile and sensory characteristics of dry fermented pork loins produced with a mixture of probiotic starter cultures. J Sci Food Agric. 2017;97:2953–60.
108. Nikodinoska I, Baffoni L, Di Gioia D, et al. Protective cultures against foodborne pathogens in a nitrite reduced fermented meat product. LWT. 2019;101:293–9.
109. Takeda S, Matsufuji H, Nakade K, et al. Investigation of lactic acid bacterial strains for meat fermentation and the product's antioxidant and angiotensin-I-converting-enzyme inhibitory activities. Anim Sci J. 2017;88:507–16.
110. Mejri L, Vásquez-Villanueva R, Hassouna M, et al. Identification of peptides with antioxidant and antihypertensive capacities by RP-HPLC-Q-TOF-MS in dry fermented camel sausages inoculated with different starter cultures and ripening times. Food Res Int. 2017;100:708–16.
111. Özer CO, Kılıç B. Utilization of optimized processing conditions for high yield synthesis of conjugated linoleic acid by L. plantarum AB20–961 and L. plantarum DSM2601 in semi-dry fermented sausage. Meat Sci. 2020;169:108218.
112. Roselino MN, de Almeida JF, Canaan JMM, et al. Safety of a low-fat fermented sausage produced with Enterococcus faecium CRL 183 and Lactobacillus acidophilus CRL1014 probiotic strains. Int Food Res J. 2017;24:2694–704.
113. Tkaczewska J. Peptides and protein hydrolysates as food preservatives and bioactive components of edible films and coatings – a review. Trends Food Sci Technol. 2020;106:298–311.
114. Wang L, Ma M, Yu Z, et al. Preparation and identification of antioxidant peptides from cottonseed proteins. Food Chem. 2021;352:129399.
115. Luz C, Izzo L, Ritieni A, et al. Antifungal and antimycotoxigenic activity of hydrolyzed goat whey on Penicillium spp: an application as biopreservation agent in pita bread. LWT. 2020;118:108717.
116. Ramírez K, Pineda-Hidalgo KV, Rochín-Medina JJ. Fermentation of spent coffee grounds by Bacillus clausii induces release of potentially bioactive peptides. LWT. 2021;138:110685.
117. Peng L, Kong X, Wang Z, et al. Baijiu vinasse as a new source of bioactive peptides with antioxidant and anti-inflammatory activity. Food Chem. 2021;339:128159.
118. Sierra-Lopera LM, Zapata-Montoya JE. Optimization of enzymatic hydrolysis of red tilapia scales (Oreochromis sp.) to obtain bioactive peptides. Biotechnol Rep. 2021;30:e00611.
119. Hussain MA, Sumon TA, Mazumder SK, et al. Essential oils and chitosan as alternatives to chemical preservatives for fish and fisheries products: a review. Food Control. 2021;129:108244.

120. Wang Z, Zhu X, Su Y, et al. Dimethyl phthalate damaged the cell membrane of Escherichia coli K12. Ecotoxicol Environ Saf. 2019;180:208–14.
121. Xu X, Lai R. The chemistry and biological activities of peptides from amphibian skin secretions. Chem Rev. 2015;115:1760–846.
122. Yan H, Lu Y, Li X, et al. Action mode of bacteriocin BM1829 against Escherichia coli and Staphylococcus aureus. Food Biosci. 2021;39:100794.
123. Shwaiki LN, Arendt EK, Lynch KM, et al. Inhibitory effect of four novel synthetic peptides on food spoilage yeasts. Int J Food Microbiol. 2019;300:43–52.
124. Liscano Y, Salamanca CH, Vargas L, et al. Increases in hydrophilicity and charge on the polar face of Alyteserin 1c helix change its selectivity towards Gram-positive bacteria. Antibiotics. 2019;8:238.
125. Sierra L, Fan H, Zapata J, et al. Antioxidant peptides derived from hydrolysates of red tilapia (Oreochromis sp.) scale. LWT. 2021;146:111631.
126. Sun H, Yao X, Wang X, et al. Chemical composition and in vitro antioxidant property of peptides produced from cottonseed meal by solid-state fermentation. CyTA J Food. 2015;13:264–72.
127. Li M, Fan W, Xu Y. Identification of angiotensin converting enzyme (ACE) inhibitory and antioxidant peptides derived from Pixian broad bean paste. LWT. 2021;151:112221.
128. Li Y, Yu J. Research progress in structure-activity relationship of bioactive peptides. J Med Food. 2015;18:147–56.
129. Zhang S, Luo L, Sun X, et al. Bioactive peptides: a promising alternative to chemical preservatives for food preservation. J Agric Food Chem. 2021;69:12369–84.
130. Jabeen U, Khanum A. Isolation and characterization of potential food preservative peptide from Momordica charantia L. Arab J Chem. 2017;10:S3982–9.
131. Dong A, Malo A, Leong M, et al. Control of Listeria monocytogenes on ready-to-eat ham and fresh cut iceberg lettuce using a nisin containing Lactococcus lactis fermentate. Food Control. 2021;119:107420.
132. Papagianni M, Anastasiadou S. Pediocins: the bacteriocins of Pediococci. Sources, production, properties and applications. Microb Cell Factories. 2009;8:3.
133. Nieto-Lozano JC, Reguera-Useros JI, del Peláez-Martínez MC, et al. The effect of the pediocin PA-1 produced by Pediococcus acidilactici against Listeria monocytogenes and Clostridium perfringens in Spanish dry-fermented sausages and frankfurters. Food Control. 2010;21:679–85.
134. Ben Braïek O, Smaoui S, Ennouri K, et al. In situ Listeria monocytogenes biocontrol and sensory attributes enhancement in raw beef meat by Enterococcus lactis. J Food Process Preserv. 2020;44:e14633.
135. Meng D-M, Sun S-N, Shi L-Y, et al. Application of antimicrobial peptide mytichitin-CB in pork preservation during cold storage. Food Control. 2021;125:108041.
136. Przybylski R, Bazinet L, Firdaous L, et al. Harnessing slaughterhouse by-products: from wastes to high-added value natural food preservative. Food Chem. 2020;304:125448.
137. Hyldgaard M, Meyer RL, Peng M, et al. Binary combination of epsilon-poly-l-lysine and isoeugenol affect progression of spoilage microbiota in fresh Turkey meat, and delay onset of spoilage in Pseudomonas putida challenged meat. Int J Food Microbiol. 2015;215:131–42.
138. Lee N-K, Paik H-D. Status, antimicrobial mechanism, and regulation of natural preservatives in livestock food systems. Korean J Food Sci Anim Resour. 2016;36:547–57.
139. Cegielska-Radziejewska R, Szablewski T, Radziejewska-Kubzdela E, et al. The effect of modified lysozyme treatment on the microflora, physicochemical and sensory characteristics of pork packaged in preservative gas atmospheres. Coatings. 2021;11:488.
140. Mastromatteo M, Lucera A, Sinigaglia M, et al. Synergic antimicrobial activity of lysozyme, nisin, and EDTA against Listeria monocytogenes in ostrich meat patties. J Food Sci. 2010;75:M422–9.
141. Rao MS, Chander R, Sharma A. Synergistic effect of chitooligosaccharides and lysozyme for meat preservation. LWT Food Sci Technol. 2008;41:1995–2001.

142. Mirzaei M, Mirdamadi S, Safavi M, et al. The stability of antioxidant and ACE-inhibitory peptides as influenced by peptide sequences. LWT. 2020;130:109710.
143. Mandel S, Michaeli J, Nur N, et al. OMN6 a novel bioengineered peptide for the treatment of multidrug resistant Gram negative bacteria. Sci Rep. 2021;11:6603.
144. Silveira RF, Roque-Borda CA, Vicente EF. Antimicrobial peptides as a feed additive alternative to animal production, food safety and public health implications: an overview. Anim Nutr. 2021;7:896–904.
145. Wang X, Yu H, Xing R, et al. Characterization, preparation, and purification of marine bioactive peptides. Biomed Res Int. 2017;2017:e9746720.
146. Barati M, Javanmardi F, Mousavi Jazayeri SMH, et al. Techniques, perspectives, and challenges of bioactive peptide generation: a comprehensive systematic review. Compr Rev Food Sci Food Saf. 2020;19:1488–520.
147. Fadimu GJ, Le TT, Gill H, et al. Enhancing the biological activities of food protein-derived peptides using non-thermal technologies: a review. Food Secur. 2022;11:1823.

Meat Products Enriched with Antimicrobials

Harsh Kumar, Shivani Guleria, Neetika Kimta, Rajni Dhalaria, Eugenie Nepovimova, and Kamil Kuča

1 Introduction

Meat is recognized as a significant source of protein with a high biological value, meaning it has a well-balanced combination of amino acids. Additionally, it provides essential micronutrients such as iron, zinc, selenium, vitamin B12, and vitamin D [1]. As an illustration, a health claim has been authorized for beef as a facilitator of enhanced iron absorption (Commission Regulation, EU, No. 432/2012). Furthermore, people particularly appreciate meat products due to their sensory qualities [2]. Meat, with its elevated water activity, abundant nutritious components, and pH range of 5.5–6.5, provides an optimal habitat for the proliferation of microbes, particularly bacteria. Microorganisms commonly detected in meat and meat products include *Pseudomonas* sp., *Enterococcus* sp., *Enterobacter* sp., *Acinetobacter* sp., *Moraxella* sp., *Aeromonas* sp., *Psychrobacter* sp., *Serratia* sp., and *Enterobacteriaceae* [3–5]. In the meat industry, chemical preservatives play a

H. Kumar
Centre of Advanced Technologies, Faculty of Science, University of Hradec Kralove, Hradec Kralove, Czech Republic

S. Guleria
Department of Biotechnology, TIFAC-Centre of Relevance and Excellence in Agro and Industrial Biotechnology (CORE), Thapar Institute of Engineering and Technology, Patiala, India

N. Kimta · R. Dhalaria
School of Biological and Environmental Sciences, Shoolini University of Biotechnology and Management Sciences, Solan, India

E. Nepovimova · K. Kuča (✉)
Department of Chemistry, Faculty of Science, University of Hradec Kralove, Hradec Kralove, Czech Republic
e-mail: eugenie.nepovimova@uhk.cz; kamil.kuca@uhk.cz

A. Kumar Pathera et al. (eds.), *Healthier Meat Products*,
https://doi.org/10.1007/978-3-031-78215-2_5

critical role. The two main types of nitrates that are commonly used as meat preservatives are potassium nitrate (E249) and sodium nitrate (E250) [6]. They are either used in addition to salt or in substitute of salt. By reducing the oxidation of meat components and preventing the growth of dangerous pathogens such *Clostridium botulinum*, nitrates (III) increase the shelf life of meat products [6]. Moreover, they improve the meat's sensory qualities and give it a unique pink-red color. The meat business requires the use of nitrates (III) due to their ability to suppress harmful microorganisms, particularly *C. botulinum* [6]. Nevertheless, they have the potential to pose risks to human well-being. The low pH of meat products and high temperature operations (over 130 °C) such as frying or grilling allow nitrates (III) to react with amino molecules, including proteins, amino acids, and amines [6]. Consequently, cancer-causing N-nitrosamines are emitted. Furthermore, nitrates (III) enhance the production of methemoglobin, a molecule that is incapable of oxygen transportation. Therefore, there is a potential danger of developing methemoglobinemia [6].

To reduce the utilization of chemical preservatives, particularly nitrates (III), in meat and meat products, it is necessary to develop alternative techniques of natural preservation [6]. The food industry is currently paying close attention to plant extracts because of their potential as antioxidants and antimicrobials. They are regarded as Generally Recognized as Safe (GRAS) and shouldn't be expected to negatively affect sensory attributes like flavor, odor, or color. They also work well at low concentrations (0.001–0.01%), mix well with a variety of foods, are simple to use, stable during processing and shelf life, and are reasonably priced [7, 8]. Crucially, these components and their byproducts must not be harmful when consumed in larger amounts than what is often eaten in conventional meals [9]. The present chapter centers on the utilization of natural plant extracts for regulating the microbial activities in meat-based food items.

2 Ready-to-Eat (RTE) Meat Products

According to 9 Code of Federal Regulations, Part 430 [10], RTE meat is any kind of meat or poultry product that can be eaten without the requirement for any further precautions to guarantee its safety. To improve its flavor and attractiveness, the product—such as frozen or chilled meat and poultry products—may go through additional processing. These can be broadly (though not exclusively) divided into two groups: hot dogs and deli meats. The delicatessen counter sells deli meats, or you can buy them already wrapped. These meats consist of, but are not limited to, bologna, turkey ham, roast beef, corned beef, baked or boiled ham, and chicken rolls [10]. These products come in a variety of forms, such as presliced, entire as a loaf, combined with salad dressing, or divided into pieces. The popularity of RTE (Ready-to-Eat) meat products has increased because of its tasty flavor, easy preparation, and time-saving advantages. RTE products are typically produced at moderate temperatures, with final product temperatures typically falling between 65 °C and

75 °C [10]. They are not the same as canned goods, which are processed at a high temperature in order to achieve sterility and total eradication of germs [11]. Therefore, ready-to-eat (RTE) foods, particularly those made from muscle, necessitate refrigeration and are commonly known as cold cuts. Despite being ingested in large quantities, RTE meat and poultry are prone to foodborne outbreaks brought on by pathogens such as Listeria monocytogenes and spoiling microorganisms such lactic acid bacteria. Foodborne infections are caused by microbial contamination in ready-to-eat (RTE) products, which mainly occurs after cooking and lasts the duration of the product's shelf life [11]. *L. monocytogenes* is capable of thriving in cold temperatures and has the ability to withstand salts and nitrites [12, 13]. Ready-to-eat (RTE) meats have undergone routine microbiological testing by the Food Safety and Inspection Service (FSIS) of the United States Department of Agriculture (USDA) since 1983 to ensure their safety for human consumption [10]. The global worry arises from the risk of epidemics of *L. monocytogenes* due to the consumption of ready-to-eat meat and poultry products. To improve microbiological safety, several procedures are used during the processing of meat and meat products. This includes the technology of high-pressure processing. Applying pressure of more than 300 megapascals (MPa) to meat samples leads to the creation of secondary lipid oxidation products. These products have an impact on the quality and shelf-life characteristics of ready-to-eat (RTE) meat products [10]. Significant endeavors have been undertaken in the food business to develop methods that lower microbiological contamination and improve the quality of beef products that are ready to eat (RTE). The meat industry is showing interest in plant extracts because of its beneficial inherent antibacterial and antioxidant properties when incorporated into product formulations.

3 Antimicrobial Activity of Plant Extracts in Brief

Plant extracts are a rich source of polyphenolic compounds, which are also known as natural antibacterial agents and are well known for their antioxidant qualities [6]. They show inhibitory effects on microorganisms that are spread through food as well as germs that cause food degradation. Their potential as food preservatives is therefore quite interesting. Microbes' sensitivity to polyphenols depends on the strain and species they belong to, in addition to the molecular makeup of the phenolic chemicals [6]. Important variables are the extract's concentration and composition. Examining the effect of plant extracts on the growth of common bacteria in meat and meat products is the meat industry's main concern [6].

Studies have indicated that the growth of microorganisms is more significantly impacted by mixtures of polyphenols, such as plant extracts, than by individual polyphenols [14, 15]. According to Serra et al. [14], grape extract with a 20 mg/L quercetin content is more effective against Bacillus cereus than the same amount of pure quercetin that has been produced. These bacteria were entirely inhibited from growing by the combination, but they were still able to multiply when quercetin was present. A wide variety of polyphenols, including flavonoids from the anthocyanin group, are abundant in berry fruits [16]. The antibacterial properties of red fruit extracts made from plums, Italian red grapes, and elderberry fruits were investigated by Coman et al. [17]. The pathogens *B. cereus*, *Staphylococcus aureus*, and *Escherichia coli* were all susceptible to the extracts' antibacterial properties. For the most part, *Candida albicans* showed resistance. Surprisingly, the extracts encouraged probiotic bacterial growth. Three isolates of *Cronobacter sakazakii* were tested by Yemis et al. [18] to determine the antibacterial potential of a phenolic extract made from pomegranate peel. The extract had bacteriostatic and bactericidal properties that were concentration-dependent and increased with a drop in the extract's pH and incubation temperature. Polyphenolic components from *Sanguisorba offcinalis*, both in crude and refined extracts, have been shown to exhibit antibacterial effects by Zhu et al. [19]. The investigation showed that in comparison to the crude extract group (15.18, 13.85, and 14.78 mm), the inhibition zones of three Gram-positive bacteria, specifically *S. aureus, B. subtilis*, and *L. monocytogenes*, were twice as large in the purified extract group (29.41, 27.15, and 24.66 mm). The bactericidal qualities of dried spice extracts, particularly those of clove, cinnamon, oregano, and mustard, were investigated by Krishnan et al. [20]. Cloves, cinnamon, oregano, and mustard had the highest antibacterial activity, according to the agar well diffusion experiment. The bacterium with the highest susceptibility was *Lactococcus lactis*. Additionally, the values of the minimal inhibitory concentration (MIC) were computed. With a minimum inhibitory concentration (MIC) of 10–20 mg/mL, *Pseudomonas fluorescens, E. coli*, and *Shawanella putrifaciens* were the bacteria that displayed the greatest susceptibility. However, *Salmonella typhimurium*, with a MIC of 20–30 mg/mL, showed the highest level of resistance.

4 Antimicrobial Mechanism of Plant Extracts

The polyphenols found in plant extracts interact with a variety of nonspecific forces, including as hydrophobic and lipophilic interactions, covalent bond formation, and hydrogen bonding. As illustrated in Fig. 1, these interactions are related to adhesins, enzymes, cell envelope transport proteins, and microbial membranes [21, 22]. Because polyphenols may bind to iron, which is required for all bacteria to survive, it is thought that they have antibacterial qualities [23].

4.1 Interaction with Cell Wall and Cell Membrane

Different traits can be found in the cell walls of bacteria classified as Gram-positive and Gram-negative. For instance, the cell wall of gram-negative bacteria is composed of a thin layer of peptidoglycan and an outer membrane (OM) [23]. The outer

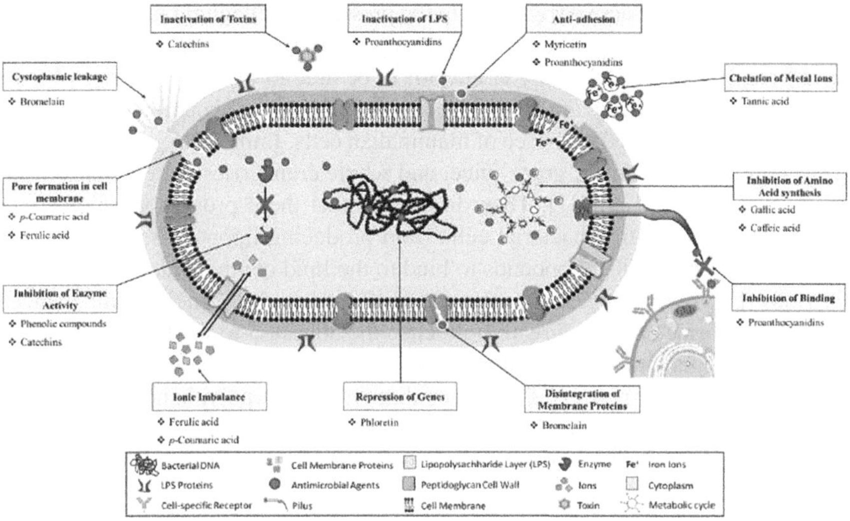

Fig. 1 The antibacterial mechanism of plant polyphenols and their target locations [23] Creative Commons Attribution (CC BY) license

membrane (OM) consists of protein and a phospholipid bilayer, with lipopolysaccharides (LPS) present in the outer leaflet of the membrane. Conversely, the cell wall of Gram-positive bacteria lacks an outer membrane (OM) and is instead composed of a significant layer of peptidoglycan and lipoteichoic acid [24]. Additionally, it's been stated that osmotic protection is mostly dependent on the cell walls of both Gram-positive and Gram-negative bacteria. Therefore, it is claimed that bacterial cell wall impairment reduces the bacteria's ability to withstand high ionic strength and low osmotic pressure. Previous studies have demonstrated that flavanols, flavonoids, and flavones possess potent antibacterial properties against several harmful microorganisms [25–27].

Zhao et al. [27] discovered that epigallocatechin gallate (EGCG) has a direct binding affinity to the peptidoglycan of *S. aureus*. The structural integrity of the cell is altered as a result of this interaction, and its resistance to high ionic strength and low osmotic pressure is reduced. Furthermore, Yoda et al. [28] beheld at the antibacterial capabilities of EGCG against a variety of *Staphylococcus strains* and Gram-negative rods. There have been reports of varying EGCG sensitivities, which have been attributed to differences in EGCG's affinity for different cell wall components. In addition, it has been documented that extracts obtained from cloudberry, blueberry, cranberry, and black currant can, like EDTA, cause *Salmonella enterica* serovar Typhimurium VTT E-981151 and *Salmonella enterica* serovar Infantis VTT E-97738 to emit lipicida [29]. By binding to divalent cations, the phenolic extracts of raspberry and cloudberry were found by Nohynek and colleagues [29] to have the ability to degrade Salmonella cultures' outer membranes. Furthermore, they claimed that the antibacterial activity was due to the presence of ellagitannin in cloudberries and raspberries and ellagic acid in cranberries. Proanthocyanidins from cranberries have been shown by Delehanty et al. [30] to be able to bind to bacterial LPS and neutralize its electrical charge. This process provides an explanation for the inhibition of LPS binding to the surface of mammalian cells. Immobilized proanthocyanidins from cranberry juice, grape juice, and whole cranberries were employed in a study by Johnson et al. [31]. They discovered that these proanthocyanidins were successful in preventing bacterial cells from producing lipopolysaccharide (LPS). The ability of phenolic compounds to bind to the lipid components of lipopolysaccharides (LPS) was suggested as the cause of this activity. Crucial components of bacteria, fimbriae and pili, greatly influence the adhesion to the host tissue. These protein filaments attach to matching carbohydrate receptors on the host cell tissue through a lectin-like process. It has been discovered that polyphenols, in particular proanthocyanidins, prevent pili from adhering to certain cell receptors [32]. Consider proanthocyanidins as receptor mimics to understand this polyphenol's competitive inhibitory function.

4.2 Interaction with Enzymes

Recent research utilizing polyphenols has indicated that these bioactive compounds can impede the action of cyclic di-AMP synthase, an enzyme that facilitates the production of cyclic-di-AMP and is implicated in multiple cellular processes [33–36]. Xiao et al. [37] utilized comparative genomics and molecular modeling to examine 19 flavonoid targets in *E. coli*. Dihydroorotate dehydrogenase, DNA gyrase subunit, fumarate reductase flavoprotein NADH-dependent enoyl-ACP reductase, and dihydrofolate reductase were the enzymes found in the results.

4.3 Interaction with Protein

Phloretin, a flavonoid found in apples, has been shown to inhibit the formation of biofilms in *E. coli* O157:H7. This is achieved by suppressing the expression of curli genes, specifically csgA and csgB, which are responsible for the creation of fimbriae [38]. Fresh pineapples contain significant quantities of bromelain, an enzyme that breaks down proteins. This enzyme is commonly utilized for the purpose of meat tenderization [39]. Various investigations have indicated that bromelain weakens the external membrane of Gram-negative bacteria by breaking down the surface membrane protein. This results in cell leakage, swelling, and damage [40].

5 Antimicrobial Activity of Plant Extracts in Meat-Based Food Products

A significant concern with beef products that are ready to eat is the possibility of bacterial contamination, which can lead to foodborne diseases and product deterioration. Although these substances are frequently combined with antibacterial agents like lactates and diacetates to prevent the growth of bacteria, studies have shown that using them alone is not efficient in inhibiting the growth of Listeria over time [11, 41]. Therefore, in order to prevent the growth of harmful microorganisms (such as *Listeria*), it is customary to employ methods that use post-processing methods including steam pasteurization, adjust pH, and decrease water activity. Pre- or post-heating treatment can involve the application of natural antibacterial agents. Table 1 displays various plant extracts that enhance the quality of different ready-to-eat meat-based products by exerting antimicrobial activities.

The study examined at how two antimicrobial preservatives, potassium lactate (1.5%) and sodium diacetate (0.15%), as well as natural extracts from grape seed (0.22%) and green tea (0.35%) affected the growth and suppression of infections on hot dogs [41]. The highest level of growth inhibition (about 2.0 CFU/g) was achieved by using a combination of chemical preservatives and plant extracts. However, using antimicrobial preservatives alone did not prove to be beneficial. In addition, post-packaging heat treatment of chicken hotdog samples was performed at 65 °C for 104 s. Plant extracts and chemical preservatives were also used, led to a stronger inhibitory effect on the proliferation of *L. monocytogenes* survivors. Hence, the utilization of natural plant extracts in conjunction with heat treatment can serve as a partial substitute for chemical preservatives. Price et al.'s study [42] looked at how sodium ascorbate and grape seed and green tea extracts affected the bacterial degradation of cooked pork meatballs. The meatballs were packaged aerobically and kept in a 4 °C refrigerator. The storage conditions were set to replicate retail display conditions for 0, 4, 8, 12, and 16 days. Throughout the storage period, the total viable count showed inconsistent patterns. Among all the samples, it was observed that green tea extracts displayed noteworthy antibacterial action ($p < .05$) starting from day 12. An additional investigation looked at the effects of rooibos (tea extract), potassium lactate (PL), and sodium diacetate (SDA) on the development of vegetative cells and spores of Clostridium perfringens in RTE Jokbal (pig's trotters) [43], specifically focusing on its antimicrobial activities. Rooibos includes many constituents, including aspalathin, isoorientin, orientin, and rutin, which have the potential to exhibit antibacterial properties [44]. The researchers found that a mixture of 10% rooibos and 4% PL + SDA effectively decreased the number of *C. perfringens* bacteria in the Jokbal, a Korean dish, while stored at a temperature of 10 °C. Throughout the storage period, the combination of 10% rooibos and 4% PL + SDA effectively inhibited the germination and proliferation of *C. perfringens* spores in the Jokbal. Hence, it has been demonstrated that rooibos tea possesses a significant ability to hinder the growth of *C. perfringens*, as well as impede the germination and expansion of *C. perfringens* spores. Recently, Odedina et al. [45]

Table 1 Utilization of botanical extracts on meat-based products and their antibacterial properties [10] (License No. 5777660025753)

Product type	Plant extract type	Foodborne pathogens	Main findings
Beef hot dogs	*Hibiscus* flower	*L. monocytogenes*, *S. aureus*	Increased concentrations of extracts, extended rinse durations, and longer storage periods resulted in enhanced prevention of *L. monocytogenes*. Rinsing hot dogs with extracts at a concentration of 120 mg/ml for 60 min, followed by a 24-h storage, also led to a reduction of around 2 log CFU/g for both bacteria
Hot dogs	Essential oils (thyme, clove, and pimento)	*L. monocytogenes*	No significant increase in antibacterial inhibitory action was seen when employing a larger concentration of extracts. The efficiency of the essential oils was shown to decrease with an increase in the fat level of the hot dog
Cooked pork meatballs; Jokbal (pig's trotters)	Grape seed, green tea; rooibos (tea extract)	Mould and yeast and coliform counts; *Clostridium perfringens*	Green tea extracts shown notable antibacterial activity starting from day 12. The combination of 10% rooibos and 4% potassium lactate + sodium diacetate effectively reduced the levels of *C. perfringens* in the Jokbal at a temperature of 10 °C
Ham steak	Green and black teas	*L. monocytogenes*	Applying a 4% concentration of green tea extract reduced the initial counts from 3.2 to 2.65 log CFU/cm^2 when stored at ambient temperature, and from 3.2 to 1–1.5 log CFU/cm^2 when stored in the refrigerator
Beef muscle slices	Oregano and pimento essential oils	*Pseudomonas* spp., *E. coli* O157:H7	Among all the extracts, oregano had the most significant impact, resulting in a drop in *Pseudomonas* spp. level by 0.95 logs and a decrease of *E. coli* O157:H7 by 1.12 logs
Mortadella sausage	*Zataria multiflora* Boiss essential oil (ZEO), grape seed extract (GSE)	Lactic acid bacteria, aerobic mesophiles, and *L. monocytogenes*	The aerobic mesophiles exhibited the highest sensitivity and resistance among the microorganisms, with a decrease of 0.1–1.1 log cycles. On the other hand, lactic acid bacteria showed a reduction of 0.1–0.7 log cycles. The combination of ZEO and GSE effectively suppressed the growth of *L. monocytogenes*

(continued)

Table 1 (continued)

Product type	Plant extract type	Foodborne pathogens	Main findings
Bologna slices	Oregano essential oil	*E. coli* O157:H7	The utilization of chitosan-based edible film containing 1% and 2% oregano essential oil led to a significant decrease of 4 logarithmic units in the population of *E. coli* O157:H7
Turkey ham	*Alpinia galanga* (Linn.) Swartz Flower Extract GFE)	*L. monocytogenes* and *S. aureus*	The use of GFE led to a reduction in the population of *S. aureus* and *L. monocytogenes* by up to 1 log CFU/g
Cooked shrimp	Galangal (*Alpinia galanga*), rosemary (*Rosmarinus officinalis*), and lemon iron bark (*Eucalyptus staigerana*)	*L. monocytogenes* and *S. aureus*, aerobic bacteria and lactic acid bacteria	The cumulative application of all extracts led to a noteworthy decline in the quantities of aerobic bacteria (1.6 and 1.59) and lactic acid bacteria (log CFU/g) at 4 °C on day 12. However, there was no substantial drop found in the levels of *L. monocytogenes or S. aureus*
Meat paté	Pomegranate extract's (PE)	*L. monocytogenes*	PE suppressed the proliferation of *L. monocytogenes* by 4.1 logarithmic colony forming units per gram (log CFU/g) for a period of 46 days at a temperature of 4 °C, resulting in a final count of 9.2 log CFU/g on the 18th day
Asian sea bass fish (*Lates calcarifer*)	Thyme (*Thymus vulgaris*) and oregano (*Origanum vulgare*)	Spoilage microflora	Thyme and oregano extracts effectively decreased the quantities of spoilage microorganisms. The initial total bacteriological counts decreased to 1.7 × 103 CFU/cm^2
Cooked cured chicken breasts	Oriental mustard	*L. monocytogenes*, Lactic acid bacteria	The use of the extract yielded favorable results in regulating the viability of *L. monocytogenes* and lactic acid bacteria, as well as in inhibiting the growth of possible spoilage bacteria
Chicken breasts	Oriental mustard	*Salmonella* and *C. jejuni*	The viability of microorganisms was lowered by using k-carrageenan/chitosan coated with 1% (v/v) acetic acid and 50 ml/g AITC or 250–300 mg/g oriental mustard extract
Turkey frankfurters	Grape seed	*L. monocytogenes*	After 21 days, the presence of both nisin and grape seed resulted in a substantial decrease in the population of *L. monocytogenes* to levels that could not be detected

(continued)

Table 1 (continued)

Product type	Plant extract type	Foodborne pathogens	Main findings
Bologna slices	Oriental mustard	*L. monocytogenes*	The use of a polymeric film containing oriental mustard extract with 5% (w/w) sinigrin effectively reduced the number of *L. monocytogenes* cells to a level that could not be detected (<1.6 log10 CFU/g) after 52 days at 4 °C
Turkey frankfurters	Grape seed (GSE), green tea (GTE)	*L. monocytogenes*	An considerable reduction in the population of *L. monocytogenes* (7.1 CFU/g) by more than 2 log cycles was found after 28 days at both 4 °C and 10 °C in the samples that contained nisin (10,000 IU) mixed with either GSE (1%) or GTE (1%)
Turkey frankfurters	Grape seed, green tea	*L. monocytogenes*, *E. coli* O157:H7, and *Salmonella typhimurium*	Following a 28-day period at a temperature of 4 °C, a notable decrease in *L. monocytogenes* levels was seen, dropping from 5.5 log/g to 2.3 log/g. this reduction was achieved by combining nisin (6000 IU/g) with GSE (0.5%) and malic acid (1.0%)

examined the impact of an ethanolic leaf extract from *Rhodomyrtus tomentosa* on the control of *L. monocytogenes* infection in cooked chicken meat after cooking. The antilisterial activity of the plant extract was significantly improved when subjected to microwave conditions with an internal temperature of 80 °C. Furthermore, raising the storage temperature from 4 to 37 °C improved the activity much more. The extract effectively reduced the population of *L. monocytogenes* in cooked samples, regardless of the low (104 CFU/g) or high (106 CFU/g) inoculum levels, utilizing both rinse and injection procedures. After being rinsed for 5 minutes in an 8% weight/volume solution of *R. tomentosa* extract, the bacterial count decreased by at least 2 logarithmic units before storage and at least 3 logarithmic units after being stored at 4 °C for 5 days. Before and after being stored at 4 °C for 5 days, the injection of *R. tomentosa* extract at a concentration of 0.4% w/w reduced the number of cells by about two logarithmic units. A different study had shown how allyl isothiocyanate (AITC) can help control the growth of potential spoilage bacteria on cooked, cured chicken breasts as well as the survival of Lactobacillus monocytogenes and lactic acid bacteria [46]. AITC can be derived from the hydrolysis of the glucosinolate sinigrin, which is found in Oriental mustard. This process occurs when the plant's own myrosinase enzyme breaks down sinigrin during physical injury, in the presence of moisture [47]. According to a number of studies, AITC is extremely deadly to common foodborne pathogens like *Salmonella* spp., *L. monocytogenes, E. coli* O157:H7, and *Campylobacter jejuni* [46, 48, 49]. The antibacterial activity of grape seed extract (GSE; 1%), nisin (6400 IU/ml), and a combination of nisin and GSE in tryptic soy broth with 0.6% yeast extract (TSBYE) against *L. monocytogenes* on the surface of full-fat turkey frankfurters was studied by Sivarooban et al. [50]. The combination of nisin and GSE had the greatest inhibitory effect on TSBYE and turkey frankfurters, completely eliminating *L. monocytogenes* after 15 h and 21 days, respectively. Nisin did not show any discernible effect after 72 h in TSBYE. A different study [51] revealed that nisin was not as efficient against *L. monocytogenes* as other bacteriocins, like pediocin. In order to determine how well nisin, grape seed extract (GSE), green tea extract (GTE), and their combinations (nisin combined with either GSE or GTE) worked against *L. monocytogenes* on turkey frankfurters, Theivendran et al. [52] conducted a study. The frankfurters that were treated with a vaccine were immersed in solutions containing soy protein that can create a film. Some of these solutions contained antimicrobial agents such as 1% grapefruit seed extract (GSE), 1% green tea extract (GTE), or 10,000 international units (IU) of nisin. Some solutions contained mixtures of these antimicrobial agents. The findings demonstrated a significant reduction in the population of *L. monocytogenes* (7.1 CFU/g) by more than 2 log cycles after 28 days at both 4 °C and 10 °C in the samples that contained nisin (10,000 IU) mixed with either GSE (1%) or GTE (1%). A different study looked at how well protein isolate (WPI) coatings containing nisin (N), malic acid (MA), grape seed extract (GSE), and ethylenediaminetetraacetic acid (EDTA) eliminated *Salmonella typhimurium*, *E. coli* O157:H7, and *L. monocytogenes* in Turkey frankfurters [59]. The vaccination-treated frankfurters were submerged in WPI-based film-forming solutions. While some of these solutions did not contain antimicrobial agents, others did (GSE, MA, N, and EDTA, or their combinations). When nisin (6000 IU/g) was combined with GSE (0.5%) and MA (1.0%), the number of *L. monocytogenes* bacteria decreased noticeably after 28 days at a temperature of 4 °C, from 5.5 log/g to 2.3 log/g. The application of WPI including a blend of N, MA, GSE, and EDTA resulted in a one logarithmic unit decrease in the *S. typhimurium* population. Throughout the course of 28 days, the population of *E. coli* O157:H7 was successfully reduced by 4.6 log cycles by using a WPI coating including N, MA, and EDTA.

6 Conclusion

Natural antimicrobials offer a clean label substitute that considerably improves the durability and safety of meat products. Nevertheless, it is crucial to customize the extra dosage to maximize antibacterial effectiveness without adversely affecting the sensory attributes. Furthermore, during the production of meat products, especially during high-temperature, high-pressure, or extrusion procedures, bioactive substances may experience chemical degradation that reduces their efficacy. It is important to recognize that using a lot of natural antimicrobials probably gives beef products a better preservation effect. The processes by which these antimicrobials operate against bacteria work together synergistically, guaranteeing the safety and quality of the meat products.

References

1. Cocking C, Walton J, Kehoe L, Cashman KD, Flynn A. The role of meat in the European diet: current state of knowledge on dietary recommendations, intakes and contribution to energy and nutrient intakes and status. Nutr Res Rev. 2020;33:1–9.
2. Asgar MA, Fazilah A, Huda N, Bhat R, Karim AA. Nonmeat protein alternatives as meat extenders and meat analogs. Compr Rev Food Sci. 2010;9:513–29.
3. Dolan AC, Burgess M, Barry TB, Fanning S, Duffy G. A novel quantitative reverse-transcription PCR (qRT-PCR) for the enumeration of total bacteria, using meat micro-flora as a model. J Microbiol Methods. 2009;77(1):1–7.
4. Stoops J, Ruyters S, Busschaert P, Spaepen R, Verreth C, Claes J, Lievens B, Campenhout LV. Bacterial community dynamics during cold storage of minced meat packaged under modified atmosphere and supplemented with different preservatives. Food Microbiol. 2015;48:192–9.
5. Pennacchia C, Ercolini D, Villani F. Spoilage-related microbiota associated with chilled beef stored in air or vacuum pack. Food Microbiol. 2011;28(1):84–93.
6. Efenberger-Szmechtyk M, Nowak A, Czyzowska A. Plant extracts rich in polyphenols: antibacterial agents and natural preservatives for meat and meat products. Crit Rev Food Sci Nutr. 2021;61(1):149–78.
7. Vinceković M, Viskić M, Jurić S, Giacometti J, Kovačević DB, Putnik P, Donsì F, Barbra FJ, Jambrak AR. Innovative technologies for encapsulation of Mediterranean plants extracts. Trends Food Sci Technol. 2017;69:1–12.
8. Kim SJ, Cho AR, Han J. Antioxidant and antimicrobial activities of leafy green vegetable extracts and their applications to meat product preservation. Food Control. 2013;29(1):112–20.
9. Lorenzo JM, Pateiro M, Domínguez R, Barba FJ, Putnik P, Kovačević DB, Shpigelman A, Granato D, Franco D. Berries extracts as natural antioxidants in meat products: a review. Food Res Int. 2018;106:1095–104.
10. Nikmaram N, Budaraju S, Barbra FJ, Lorenzo JM, Cox RB, Mallikarjunan K, Roohinejad S. Application of plant extracts to improve the shelf-life, nutritional and health-related properties of ready-to-eat meat products. Meat Sci. 2018;145:245–55.
11. Horita CN, Baptista RC, Caturla MYR, Lorenzo JM, Barba FJ, Sant'Ana AS. Combining reformulation, active packaging and non-thermal post-packaging decontamination technologies to increase the microbiological quality and safety of cooked ready-to-eat meat products. Trends Food Sci Technol. 2018;72:45–61.
12. Ivy RA, Wiedmann M, Boor KJ. *Listeria monocytogenes* grown at 7°C shows reduced acid survival and an altered transcriptional response to acid shock compared to *L. monocytogenes* grown at 37°C. Appl Environ Microbiol. 2012;78(11):3824–36.
13. McClure PJ, Beaumont AL, Sutherland JP, Roberts TA. Predictive modelling of growth of *Listeria monocytogenes* the effects on growth of NaCl, pH, storage temperature and $NaNO_2$. Int J Food Microbiol. 1997;34(3):221–32.
14. Serra AT, Matias AA, Nunes AV, Leitão MC, Brito D, Bronze R, Silva S, Pires A, Crespo MT, San Romão MV, Duarte CM. In vitro evaluation of olive-and grape-based natural extracts as potential preservatives for food. Innov Food Sci Emerg Technol. 2008;9(3):311–9.
15. Puupponen-Pimiä R, Nohynek LC, Meier M, Kähkönen M, Heinonen AH, Oksman-Caldentey KM. Antimicrobial properties of phenolic compounds from berries. J Appl Microbiol. 2001;90(4):494–507.
16. Ma H, Johnson SL, Liu W, DaSilva NA, Meschwitz S, Dain JA, Seeram NP. Evaluation of polyphenol anthocyanin-enriched extracts of blackberry, black raspberry, blueberry, cranberry, red raspberry, and strawberry for free radical scavenging, reactive carbonyl species trapping, anti-glycation, anti-b-amyloid aggregation, and microglial neuroprotective effects. Int J Mol Sci. 2018;19(2):461.

17. Coman MM, Oancea AM, Verdenelli MC, Cecchini C, Bahrim GE, Orpianesi C, Cresci A, Silvi S. Polyphenol content and in vitro evaluation of antioxidant, antimicrobial and prebiotic properties of red fruit extracts. Eur Food Res Technol. 2017;244(4):1–11.
18. Yemis GP, Bach S, Delaquis P. Antibacterial activity of polyphenol-rich pomegranate peel extract against *Cronobacter sakazakii*. Int J Food Prop. 2019;22(1):985–93.
19. Zhu HL, Chen G, Chen SN, Wang QR, Wan L, Jian SP. Characterization of polyphenolic constituents from *Sanguisorba officinalis* L. and its antibacterial activity. Eur Food Res Technol. 2019;245(7):1487–98.
20. Krishnan KR, Babuskin S, Babu PAS, Sasikala M, Sabina K, Archana G, Sivarajan M, Sukumar M. Antimicrobial and antioxidant effects of spice extracts on the shelf life extension of raw chicken meat. Int J Food Microbiol. 2014;171:32–40.
21. Cowan MM. Plant products as antimicrobial agents. Clin Microbiol Rev. 1999;12:564–82.
22. Kumar S, Pandey AK. Chemistry and biological activities of flavonoids: an overview. World Sci J. 2013:162750.
23. Kumar H, Bhardwaj K, Cruz-Martins N, Nepovimova E, Oleksak P, Dhanjal DS, Bhardwaj S, Singh R, Chopra C, Verma R, Chauhan PP, Kumar D, Kuča K. Applications of fruit polyphenols and their functionalized nanoparticles against foodborne bacteria: a mini review. 2021;26:3447.
24. Brown L, Wolf JM, Prados-Rosales R, Casadevall A. Through the wall: extracellular vesicles in Gram-positive bacteria, mycobacteria and fungi. Nat Rev Microbiol. 2015;13:620–30.
25. Takahashi O, Cai Z, Toda M, Hara Y, Shimamura T. Appearance of antibacterial activity of oxacillin against methicillin resistant *Staphylococcus aureus* (MRSA) in the presence of catechin. Kansenshogaku Zasshi. 1995;69:1126–34.
26. Zhao WH, Hu ZQ, Okubo S, Hara Y, Shimamura T. Mechanism of synergy between epigallocatechin gallate and β-lactams against methicillin-resistant *Staphylococcus aureus*. Antimicrob Agents Chemother. 2001;45:1737–42.
27. Zhao WH, Hu ZQ, Hara Y, Shimamura T. Inhibition of penicillinase by epigallocatechin gallate resulting in restoration of antibacterial activity of penicillin against penicillinase-producing *Staphylococcus aureus*. Antimicrob Agents Chemother. 2002;46:2266–8.
28. Yoda Y, Hu ZQ, Shimamura T, Zhao WH. Different susceptibilities of *Staphylococcus* and Gram-negative rods to epigallocatechin gallate. J Infect Chemother. 2004;10:55–8.
29. Nohynek LJ, Alakomi HL, Kähkönen MP, Heinonen M, Helander IM, Oksman-Caldentey KM, Puupponen-Pimiä RH. Berryphenolics: antimicrobial properties and mechanisms of action against severe human pathogens. Nutr Cancer. 2006;54:18–32.
30. Delehanty JB, Johnson BJ, Hickey TE, Pons T, Ligler FS. Binding and neutralization of lipopolysaccharides by plant proanthocyanidins. J Nat Prod. 2007;70:1718–24.
31. Johnson BJ, Delehanty J, Lin B, Ligler FS. Immobilized proanthocyanidins for the capture of bacterial lipopolysaccharides. Anal Chem. 2008;80:2113–7.
32. Hisano M, Bruschini H, Nicodemo AC, Srougi M. Cranberries and lower urinary tract infection prevention. Clinics. 2012;67:661–8.
33. Opoku-Temeng C, Sintim H. Inhibition of cyclic diadenylate cyclase, DisA, by polyphenols. Sci Rep. 2016;6:25445.
34. Witte CE, Whiteley AT, Burke TP, Sauer JD, Portnoy DA, Woodward JJ. Cyclicdi-AMP is critical for *Listeria monocytogenes* growth, cell wall homeostasis, and establishment of infection. MBio. 2013;4:e00282–13.
35. Sureka K, Choi PH, Precit M, Delince M, Pensinger DA, Huynh TN, Jurado AR, Goo YA, Sadilek M, Iavarone AT, Sauer JD, Tong L, Woodward JJ. The cyclic dinucleotide c-di-AMP is an allosteric regulator of metabolic enzyme function. Cell. 2014;158:1389–401.
36. Huynh TN, Luo S, Pensinger D, Sauer JD, Tong L, Woodward JJ. An HD-domain phosphodiesterase mediates cooperative hydrolysis of c-di-AMP to affect bacterial growth and virulence. Proc Natl Acad Sci USA. 2015;112:E747–E56.
37. Xiao ZT, Zhu Q, Zhang HY. Identifying antibacterial targets of flavonoids by comparative genomics and molecular modeling. Open J Genomics. 2014;3:1–8.

38. Lee JH, Regmi SC, Kim JA, Cho MH, Yun H, Lee CS, Lee J. Apple flavonoid phloretin inhibits *Escherichia coli* O157:H7 biofilm formation and ameliorates colon inflammation in rats. Infect Immun. 2011;79:4819–27.
39. Hossain MF, Akhtar S, Anwar M. Nutritional value and medicinal benefits of pineapple. Int J Nutr Food Sci. 2015;4:84–8.
40. Zharfan RS, Purwono PB, Mustika A. Antimicrobial activity of pineapple (*Ananas cosmosus* L. Merr) extract against multidrug-resistant of *Pseudomonas aeruginosa*: an in vitro study. Indones J Trop Infect Dis. 2017;6:118–23.
41. Perumalla AVS, Hettiarachchy NS, Over K, Ricke SC, Slavik MF, Gbur E, Davis B, Acosta S. Effect of partial replacement of potassium lactate and sodium diacetate by natural green tea and grape seed extracts and postpackaging thermal treatment on the growth of *Listeria monocytogenes* in hotdog model system. Int J Food Sci Technol. 2013;48(5):918–26.
42. Price A, Díaz P, Bañón S, Garrido MD. Natural extracts versus sodium ascorbate to extend the shelf life of meat-based ready-to-eat meals. Food Sci Technol Int. 2013;19(5):427–38.
43. Park HJ, Park KC, Yoon KS. Effect of rooibos (*Aspalathus linearis*) on growth control of *Clostridium perfringens* and lipid oxidation of ready-to-eat jokbal (pig's trotters). J Food Sci. 2014;79(12):M2507–15.
44. Almajano MP, Carbó R, Delgado ME, Gordon MH. Effect of pH on the antimicrobial activity and oxidative stability of oil-in-water emulsions containing caffeic acid. J Food Sci. 2007;72(5):C258–C63.
45. Odedina GF, Vongkamjan K, Voravuthikunchai SP. Use of *Rhodomyrtus tomentosa* ethanolic leaf extract for the bio-control of *Listeria monocytogenes* post-cooking contamination in cooked chicken meat. J Food Sci Technol. 2016;53(12):4234–43.
46. Olaimat AN, Holley RA. Effects of changes in pH and temperature on the inhibition of *Salmonella* and *Listeria monocytogenes* by Allyl isothiocyanate. Food Control. 2013;34(2):414–9.
47. Barba FJ, Nikmaram N, Roohinejad S, Khelfa A, Zhu Z, Koubaa M. Bioavailability of glucosinolates and their breakdown products: impact of processing. Nutr Food Sci Technol. 2016;3:24.
48. Luciano FB, Holley RA. Enzymatic inhibition by allyl isothiocyanate and factors affecting its antimicrobial action against *Escherichia coli* O157:H7. Int J Food Microbiol. 2009;131(2–3):240–5.
49. Olaimat AN, Fang Y, Holley RA. Inhibition of *Campylobacter jejuni* on fresh chicken breasts by κ-carrageenan/chitosan-based coatings containing allyl isothiocyanate or deodorized oriental mustard extract. Int J Food Microbiol. 2014;187:77–82.
50. Sivarooban T, Hettiarachchy NS, Johnson MG. Inhibition of *Listeria monocytogenes* using nisin with grape seed extract on Turkey frankfurters stored at 4 and 10 degrees C. J Food Prot. 2007;70(4):1017–20.
51. Montville TJ, Chen Y. Mechanistic action of pediocin and nisin: recent progress and unresolved questions. Appl Microbiol Biotechnol. 1998;50(5):511–9.
52. Theivendran S, Hettiarachchy NS, Johnson MG. Inhibition of *Listeria monocytogenes* by nisin combined with grape seed extract or green tea extract in soy protein film coated on Turkey frankfurters. J Food Sci. 2006;71(2):M39–44.

Meat Products Enriched with Antioxidants

Sao-Mai Dam, Duyen-Anh Le, and Thien-Hoang Ho

1 Introduction

Recent years have seen a significant development in food science and technology, with an increasing focus on enhancing meat products' quality and nutritional profile. The food industry has expanded into new frontiers in response to the changing desires of consumers, with an emphasis on improving the health and sensory aspects of meat products. A novel approach in this pursuit is the strategic incorporation of antioxidants, essential in mitigating the harmful consequences of oxidative stress on the human organism.

Lipids play a crucial role in shaping the desirable characteristics of various meat types, contributing significantly to their flavour, aroma, tenderness, and juiciness. However, it's essential to recognize that lipid oxidation is the main cause of deterioration of meat, leading to reduced shelf life [1]. This oxidative process adversely impacts several critical aspects of meat quality, including colour, texture, nutritional value, taste, and aroma. The consequences of lipid oxidation include the development of rancidity, which introduces off-flavours and an undesirable taste profile, ultimately leading to consumer rejection [2, 3]. Like lipid oxidation, protein oxidation affects the quality of meat during storage and processing. Protein oxidation may compromise functional attributes such as gelatinousness, emulsification, and sensory qualities, including colour, taste, and softness [4]. The concentration of hydrogen peroxide directly affects protein oxidation, as well as the brightness, ability to retain water, texture, and elasticity of the myofibrillar protein gel. Myofibrillar protein's carbonyl value increased, and protein oxidation became more intense as the peroxide concentration increased. Additionally, increased oxidation concentrations led to more severe damage to the myofibril structure and decreased water

S.-M. Dam · D.-A. Le · T.-H. Ho (✉)
Industrial University of Ho Chi Minh City, Ho Chi Minh City, Vietnam
e-mail: hothienhoang@iuh.edu.vn

A. Kumar Pathera et al. (eds.), *Healthier Meat Products*,
https://doi.org/10.1007/978-3-031-78215-2_6

retention [5]. Protein oxidation is usually the cause of disulphide bond formation, amino acid residue oxidation, and peptide bond breaking, leading to a change in meat's molecular weight and structure.

Antioxidant compounds have been recognized for their ability to eliminate free radicals, which are very reactive chemicals linked to ageing and chronic illnesses and have the potential to harm cells. A variety of chemical antioxidants are commonly used meat products including: (i) Phenol derivatives, such as butylated hydroxy anisole (BHA) and butylated hydroxytoluene (BHT), which remove or neutralize free-radical scavenging [6]. However, synthesis of phenol derivatives is considered to cause mutation and damage to DNA, posing an associated risk to lipid metabolism and cancer [7]. (ii) Phosphates are commonly used as additives and antioxidant agent because they can bind to water and enhance texture of meat. They also bind to divalent metal cation and interact with protein, thereby delaying rancidity and oxidation of protein. (iii) Acids such as sorbic acid, benzoic acid, nitrites, and sulfites can inhibit the protease family (calpain, cathepsins, and aminopeptidase enzymes), preventing the degradation of meat and meat products [8].

Presently, there is a lot of uncertainty over the health impacts of synthetic antioxidants, including potential risks such as allergies, effects on the health of women and children, and the potential for carcinogenicity [9]. There is a growing trend to substitute synthetic antioxidants with natural alternatives to prevent the oxidation of meat products. Oilseeds, cereals, legumes, fruits, vegetables, herbs, and spices have long been utilized for their essential natural antioxidants. These sources are perfect substitutes to fulfil the growing need to create commercial antioxidants.

This chapter presents a comprehensive analysis of the correlation between antioxidants and the quality of meat, covering a range of subjects, including how antioxidants are applied and function in meat products, what influences their effectiveness, whether they are present in raw or cooked meat products, how eating quality is affected, the health benefits of antioxidant-enriched meat, problems and considerations, and the most recent findings and advancements in the field.

2 Antioxidants and Meat Quality

Lipid oxidation is a primary factor in the deterioration of food quality. Lipid oxidation is a complicated chain of chemical processes that occurs when lipids, such as fats and oils, are exposed to oxygen. This process, often called lipid peroxidation, plays a significant role in food deterioration, affecting the flavour, aroma, nutritional value, and overall meat quality. The mechanism of lipid oxidation involves several key steps: (i) initiation, (ii) propagation, (iii) termination, (iv) hydroperoxide decomposition, (v) formation of secondary products (Fig. 1).

(i) Initiation: The process begins with the initiation phase, where oxygen reacts with unsaturated fatty acids, typically present in triglycerides (the main constituents of fats and oils). Free radicals, such as hydroxyl radicals ($^{\cdot}OH$) or lipid

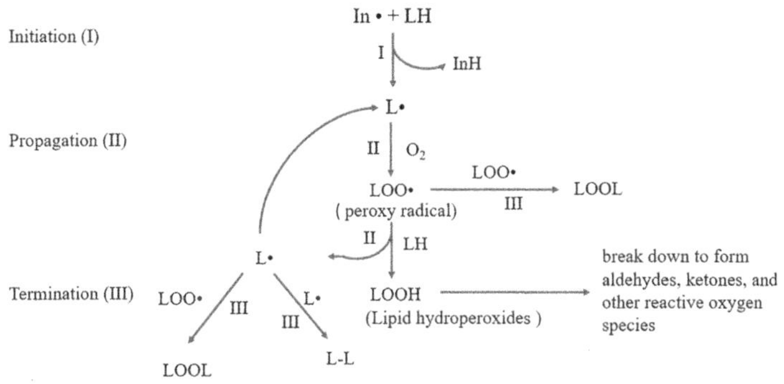

I:Initiation, II:Propagation, III: Termination, LH:attack lipids , L•: lipid radical, LOO•: peroxy radical, LOOL: organic peroxide, LOOH:Lipid hydroperoxides , O_2 : Oxygen, L-L: non-radical product

Fig. 1 Mechanism of lipid oxidation [10]

peroxyl radicals ($LOO^{\cdot}$), are formed during this initiation step. These free radicals are highly reactive and set off a chain reaction.

(ii) Propagation: In the propagation phase, the free radicals produced in the initiation step react with nearby unsaturated fatty acids, creating new lipid radicals. These new radicals can, in turn, react with oxygen to form lipid peroxides. This chain reaction continues, leading to the production of a cascade of lipid peroxides.

(iii) Termination: The termination phase involves deactivating free radicals, often through reactions involving antioxidants. Antioxidants act as scavengers, neutralizing free radicals and breaking the chain reaction. Termination steps are crucial in controlling and slowing down the overall oxidation process.

(iv) Hydroperoxide Decomposition: The lipid peroxides formed during the propagation phase can decompose to form secondary products, including aldehydes, ketones, and other volatile compounds. These decomposition products contribute to the development of off-flavours, odours, and alterations in the sensory properties of the food item.

(v) Formation of Secondary Products: The degradation of lipid hydroperoxides leads to the formation of reactive aldehydes, such as 4-hydroxynonenal (HNE), and malondialdehyde (MDA). These compounds are frequently used as signs of oxidative stress.

Reactive oxygen species (ROS) and reactive nitrogen species (RNS) have a crucial role in initiating protein oxidation. Oxygen-containing compounds, including superoxide ($O_2^{\cdot-}$), hydroxyl radical ($^{\cdot}OH$), alkyl radical ($RO^{\cdot}$), and peroxyl radical ($RO_2^{\cdot}$) are among the key contributors to this process [11]. Additionally, non-radical derivatives of oxygen, such as hydrogen peroxide (H_2O_2), hypochlorous acid (HClO), and ozone (O_3) have been observed to significant influence this particular oxidative pathway [12]. Nitrogen-containing compounds, including nitric oxide ($^{\cdot}NO$), Peroxynitrite (ONOO–), and ($^{\cdot}NO_2$) are also formed during oxidative stress, contribution to oxidation of protein. [13, 14]. The level of protein oxidation is closely related to exposure to reactive oxygen species (ROS) and reactive nitrogen species (RNS). It is regulated by factors such as heat treatment, heme iron content, the presence of oxygen (O_2), and hydrogen peroxide (H_2O_2) [15].

Carbonylation is regarded as an indicator of protein modifications resulting from oxidative stress within cells, resulting in permanent alterations in protein structure. Protein carbonylation is influenced by various mechanisms, including the direct deamination of specific alkaline amino acids such as arginine, lysine and proline [16]. Especially, lysíne constitutes approximately 70% of the total protein carbonyls found in cells [17]. Additionally, The Maillard reaction, formed between reducing sugar molecules and amino acids, also leads to protein carbonylation [18]. Furthermore, non-protein carbonyl groups, such as secondary lipid oxidation products, can covalently bind to proteins, promoting carbonylation [19]. Lastly, the structural and functional aspects of proteins can be altered when the protein backbone undergoes oxidative cleavage, and the oxidation of the glutamyl side chain leads to carbonyl formation, contributing to protein dysfunction and cellular damage. Gaining a comprehensive understanding of these complex processes offers significant insights into the complicated nature of protein oxidation and its possible ramifications on food quality. Nevertheless, a prior study suggested that the process of carbonylation in meat proteins was mainly attributed to the direct oxidation of protein side chains [20]. In contrast, an alternate mechanism was hypothesized by another research, positing that the generation of protein carbonyls might potentially occur through the interaction between water-soluble proteins and myoglobin.

The overall impact of lipid oxidation and protein oxidation on meat product quality includes changes in sensory attributes, nutritional quality, and shelf life of meats. The development of unwanted smells and aromas in meat products can be attributed to lipid oxidation, which forms volatile chemicals. Rancidity, which is distinguished by the presence of undesirable tastes, is a frequently seen outcome. The process of protein oxidation has the potential to impact the sensory attributes of taste and scent, hence exerting an influence on the entire sensory profile. The alteration of meat product colour is influenced by both lipid and protein oxidation. The process of lipid oxidation has the potential to cause a colour change.

In contrast, protein oxidation can give rise to browning processes, affecting the product's visual attractiveness and customer acceptability. The texture of meat products can be affected by protein oxidation, resulting in alterations in both tenderness and juiciness. The structural integrity of the meat matrix can be influenced by the process of cross-linking and fragmentation of proteins. The nutritional content and value of meat products may be determined through the oxidation of both lipids and proteins, which leads to the breakdown of important elements. Lipid oxidation has the potential to result in the degradation of advantageous fatty acids. The process of protein oxidation can lead to the creation of carbonyl molecules, which serve as indicators of permanent alterations in protein structure and have the potential to decrease the bioavailability of crucial amino acids. The generation of detrimental chemicals can occur through lipid oxidation, producing malondialdehyde (MDA). Consumption of MDA has been associated with potential adverse health effects.

Moreover, the association between protein oxidation products, namely advanced glycation end products (AGEs), and various health issues has been shown. The overarching duration of product viability: Rancidity and spoilage in meat predominantly result from lipid oxidation, leading to the development of off-flavors and a shortened shelf life for food products. The stability of fats and oils in meat products is compromised by the development of oxidation products. While protein oxidation may not be the primary catalyst for spoilage, it can indeed contribute to the gradual decline in the quality of a product. The occurrence of microbial growth and deterioration might be facilitated by oxidation processes, hence affecting the stability of the system. The process of lipid and protein degradation can result in the release of essential nutrients that enable the growth of microorganisms, influencing the safety and duration of meat products' storage.

3 Mode of Application and Functionality of Antioxidants in Meat Products

In meat processing, the use of antioxidants plays an important role in diversifying and enhancing quality, as well as ensuring consumer safety. This goal and strategy require further research and attention. The many methods of administration and the functional properties of antioxidants are crucial factors in determining the ultimate characteristics of these products.

Direct incorporation of antioxidants in meat products The direct inclusion of antioxidants into meat products is an essential and diverse approach aimed at enhancing the nutritional value of these goods with substances that reduce oxidative processes. This approach entails directly incorporating substances abundant in antioxidants into the formulation, ensuring their homogeneous dispersion within the meat matrix. This technique not only serves a practical purpose in terms of preserving the food but also contributes to the enhancement of its flavour profile through the addition of several layers of complexity. However, the incorporation of essential oils from plants as antioxidants into meat products as preservatives may face limitations, particularly in terms of their taste profile, especially when used in high quantities [21, 22]. Table 1 presents several uses of natural antioxidants for prolonging the durability of the meat products (Table 1).

Topical application of antioxidants in meat products The topical application of antioxidants in meat products is a strategic approach that protects against oxidative reactions while enhancing flavour and visual attractiveness (Table 2). In contrast to the method of direct cooperation, wherein antioxidants are mixed into the meat composition, topical application applies to forming a protective coating on the external surface of the meat. This approach is effective when applied to recently made incisions, minced meat, and other perishable goods prone to fast oxidation. Spraying, dipping, and immersion are standard techniques used for meat products.

Table 1 The utilization of natural components to prolonging the durability of the meat products

Sources	Bioactive compounds	Application	Meat products	References
Acerola fruit extract	Anthocyanins/ Phenolic acids/ Vitamin C	The lifespan can be prolonged by minimum of 3 days, lipid and colour were stability. Acerola fruit extract did not affect the microbial population of beef patties.	Beef patties	[32]
Red sweet cayenne/Hot cayenne	Carotenoids/ Capsacinoids	*Capsicum* peppers improved red color, inhibited lipid oxidation and bacterial growth.	Fresh pork sausages	[33]
Black pepper/ White pepper	β-Caryophylline/ Limonene/β-Pinene/ Sabinene Capsacinoids/ Caratenoids.	Black pepper/white pepper effectively put off change in color of pork sausages and prevented the creation of off-odors		
Grape seed extract	Flavonoids (Catechins/ Proanthocyanidins)	Green tea and grape seed extracts inhibited lipid oxidation and reduced microbial growth better than sodium ascorbate	Cooked pork meatballs	[34]
Green tea extract	Flavonoids (Catechins)			
Grape seed extract	Proanthocyanidins	Proanthocyanidins mixed into Pork patties decrease bacterial and lower volatile basic nitrogen (VBN).	Pork patties	[35–37]
Oleoresin rosemary	Carnosic acid	Oleoresin rosemary added to meat in oil-soluble form is more effective than in water-soluble form.	Chicken sausage, Chicken meat	
Rosemary extract/ Vitamin E/ Irradiation	Carnosic acid rosmarinus and caffeic acids.	Treated samples with irradiation and a combination of extracts decreased metmyoglobin values and inhibited lipid oxidation	Overwrapped minced meat	[38]

Marination of antioxidants in meat products Marinated meat is the process of using ingredients such as acids, spices, and enzymes that are submerged in uncooked or cooked meat to enhance flavor, colour, texture or soften the meat's surface [23, 24]. Tenderizing meat is a culinary practice involving biological, chemical, and physical techniques [25]. Throughout history, marinades have evolved to include a diverse range of ingredients, resulting in a shift in their primary purpose from preservation or curing to improvement of tenderness and taste [23]. Various biological marinades are used to prepare meat, including kefir, yoghurt, butter, turmeric, torch ginger, lemongrass, curry leaves, Karuk juice, chilli, pepper, garlic, and ginger

Table 2 Topical application of antioxidants in meat products

Natural ingredients	Major bioactive ingredients	Oxidative stability	Experimental findings	Techniques	Meat products	References
Black pepper (Piper nigrum)	Black pepper essential oil	TBARSc color stability	0.5% essential oil treatment has the lowest TBARSc value and MetMb formation on 3, 6, and 9 days	Spraying	Fresh pork	[39]
Sesame	Ascorbic acid, α-tocopherol, sesamol	Volatile compounds, color stability	The ingredient slows down lipid oxidation until 7 d of storage	Spraying	Beef chuck	[40]

[26–28]. According to Manful et al. [29], marinade solutions including wine, vinegar, olive oil, garlic, pepper, onion, mustard, and salt exhibit antioxidant properties throughout the marination process. Consequently, the use of herbs and spices in marinades has been found to significantly enhance the quality of meat and impact health outcomes through the regulation and reduction of lipid oxidation [30].

Encapsulation of antioxidants in meat products Encapsulation is a novel methodology employed in meat processing wherein active components, such as antioxidants or taste compounds, are enveloped inside a safeguarding carrier system. This technique improves the stability of such constituents, guarantees regulated release throughout cooking or digesting, and contributes to the overall excellence and efficacy of meat-based goods [31].

4 Common Antioxidants Used in Meat Products

The use of antioxidants in meat products has become an accepted standard in the food industry, based on the aim to improve product quality, prolong shelf life, and tackle the issues associated with oxidative reactions (Table 3). A wide range of antioxidants derived from natural and synthetic sources are included in meat compositions to mitigate the adverse consequences of oxidation. The process by which natural antioxidants exert their effects varies depending on the specific molecule. Donating a hydrogen atom to chelate metal ions inhibits hydroperoxide degradation [41]. Phenolic compounds are the primary class of natural antioxidants, with notable examples being tocopherols, flavonoids, and phenolic acids. The compounds above exhibit a noteworthy capacity for hydrogen donation and the ability to scavenge free radicals [41, 42]. Oswell et al. [43] have identified several substances that exhibit antioxidant properties in meat products, including carotenoids, hydroxycinnamic acid, flavonoids, terpenes, and antioxidant vitamins.

Table 3 Common natural antioxidants used in meat products

Source	Bioactive compounds	Application	References
Sesame seeds	Sesamin and seasamolin	Sesame oil contributes antioxidant compounds to dishes in various cuisines and for oil extraction.	[45]
Flaxseeds	Lignans, secoisolariciresinol, omega-3 fatty acids.	Ground flaxseeds are incorporated into various foods, such as meatballs and meat products.	[46, 47]
Sunflower seeds	Vitamin E (tocopherols)	Contributing antioxidant properties.	[48]
Whole grains	Phenolic compounds, ferulic acid.	Brown rice and whole wheat contain antioxidants,	[49]
Oat	Avenanthramides and phenolic acids	Using oats and soybeans in beef, chicken, and sheep	[50]
Soybeans	Isoflavones (genistein, daidzein)	Soy sauce decreases lipid oxidation	[51]
Lentils	Polyphenols, flavonoids	The flour of lentils inhibited lipid oxidation in raw burgers	[52]
Berries (blueberries, strawberries, raspberries):	Anthocyanins, quercetin	Berries powder protected pork, ground meat, and patties against oxidation during storage	[53, 54]
Citrus fruits (oranges, lemons)	Vitamin C (ascorbic acid), flavanones and carboxylic acids	Citrus extract reduced lipid oxidation in cooked Turkey meat	[55]
Spinach	Lutein, β-carotene, phenolics and nitrates	Spinach inhibited thiol oxidation in the meat	[56, 57]
Moringa oleifera and *Bidens pilosa*	Polyphenol, β-carotene.	Leaf extracts inhibited lipid oxidation in meat products	[58]
Turmeric	Curcumin	Tumeric powder added to duck meat burgers protects against lipid oxidation in 6 days of storage	[59]
Cinnamon	Cinnamaldehyde	Cinnamon powder 0.5% reduced the TVC (total viable counts) and TBARS (thiobarbituric acid reactive substances) in ground lamb	[60]
Rosemary	Carnosic acid, rosmarinic acid	Essential oils added to minced beef protects against oxidation of lipid.	[61]
Oregano	Thymol, rosmarinic acid	Essential oils added to minced beef protects against oxidation of lipid.	[61]

(continued)

Table 3 (continued)

Source	Bioactive compounds	Application	References
Walnuts	Polyphenols (catechins, resveratrol, ellagitannins), omega-3 fatty acids.	Walnut leaf powder at 0.5% reduced the TBARS.	[62]
Green tea	Catechins (epigallocatechin gallate, EGCG)	Using green tea extract reduced lipid oxidation.	[63]

Table 4 The most common commercial antioxidants used in meat products

Antioxidant	Type	Benefits	Concerns	References
Rosemary extract	Natural	Antibacterial, antifungal, anti-inflammatory	None known	[68]
Tocopherols	Natural	Protects against cell damage	None known	[69]
Ascorbic acid	Natural	Enhances iron absorption, supports immune function	Can cause stomach upset	[69]
BHT	Synthetic	Highly effective in slowing down lipid oxidation	Could potentially harm the liver and thyroid	[6]
BHA	Synthetic	Highly effective in slowing down lipid oxidation	Could potentially harm the lungs and skin	[67]

Incorporating antioxidants in several forms has been a traditional culinary and medicinal practice going as far as long ago, and this tradition continues to the present day. A broad range of antioxidants derived from many sources, ranging from oilseeds to cereals, legumes, fruits, vegetables, herbs, and spices, play a crucial role in both food preservation and the enhancement of flavor, as well as potentially promoting health benefits [44].

Recently, there has been an increasing attraction to exploiting the inherent advantages of natural antioxidants in food production (Table 4). The reason for this interest comes from a combined dedication to improving the quality of products and satisfying the growing customer need for food alternatives that are both natural and health-conscious. The investigation of natural antioxidants remains a vibrant field of study, with numerous antioxidants already being incorporated into commercial applications, showcasing their efficacy in maintaining the quality of diverse food items.

Rosemary extract has emerged as a significant natural antioxidant within the business. The extract under consideration is obtained from the foliage of the Rosemary plant, and it possesses a notable abundance of phenolic components, including rosmarinic acid and carnosic acid. The bioactive components maintain significant antioxidant activities, substantially mitigating lipid oxidation and prolonging the stability of food items. The use of rosemary extract is widespread due to its capacity to augment the oxidative stability of fats and oils, rendering it a preferred option, especially in meat products [43, 64].

Tocopherols, a collection of vitamin E chemicals, are emerging as a notable natural antioxidant within the food sector. Tocopherols, obtained from vegetable oils and seeds, demonstrate efficacy as potent scavengers of free radicals, hence impeding the initiation of oxidation in diverse dietary matrices. The utilization of tocopherols extends beyond their antioxidant capabilities since they also provide additional nutritional advantages. This aligns with the prevailing industrial trend of fortifying foods with bioactive chemicals [43, 65].

Ascorbic acid (vitamin C) is a multifunctional antioxidant generally acknowledged for its capacity to inhibit oxidative processes. The presence of this compound is inherent in several fruits and vegetables, and its application as a commercial antioxidant is noteworthy due to its remarkable efficacy in suppressing both lipid and non-lipid oxidation processes. The dual functioning of ascorbic acid increases the prolongation of product shelf life and protects color and flavor [43, 66].

Furthermore, apart from naturally occurring antioxidants, the food industry has extensively employed synthetic alternatives such as BHT (Butylated Hydroxytoluene) and BHA (Butylated Hydroxy anisole). Despite being synthetic, these compounds have become essential in the development of different food items due to their ability to inhibit oxidative processes and improve the stability of fats and oils. It is necessary to acknowledge that although synthetic antioxidants have been utilized for many years, there is a trend towards emphasizing natural alternatives in the market. This movement is in reaction to the increasing customer demand for clean-label products [6, 67].

The continuous investigation of natural antioxidants highlights the food sector's dedication to pursuing sustainable and health-oriented approaches. With the advancement of scientific knowledge, an increasing number of natural antioxidants are expected to be included in commercial applications. This will not only enhance preservation capabilities but also fit with the changing preferences and expectations of consumers. The ongoing interaction between research, industry innovation, and consumer demand plays a crucial role in influencing the development and impact of antioxidants in food preservation and quality improvement.

The thermal resistance of antioxidants in the process of meat oxidation is a crucial factor to be considered in food preservation. The action of antioxidants is vital in inhibiting lipid oxidation, a process that can negatively affect the quality and longevity of meat products. The capacity of antioxidants to endure the high temperatures encountered during cooking or processing is crucial for preserving their effectiveness.

Various antioxidants demonstrate different levels of thermal stability. Natural antioxidants from rosemary extract, tocopherols, and specific plant extracts frequently exhibit considerable resistance to thermal degradation. This attribute ensures its efficacy even when exposed to high temperatures during cooking or processing.

However, although several synthetic antioxidants exhibit efficacy in inhibiting oxidation, they may need help with their thermal stability. Elevated temperatures have the potential to induce the deterioration of specific synthetic antioxidants, which may, therefore, undermine their efficacy in safeguarding against lipid oxidation.

Therefore, selecting antioxidants in meat products necessitates a careful balance between their effectiveness in preventing oxidation and their capacity to survive the elevated temperatures observed during cooking or processing. Researchers and food producers persist in their efforts to investigate and create antioxidants that have higher heat resistance, aiding in the preservation of meat products' quality and prolonging their shelf life amidst diverse culinary procedures.

The continuous investigation of natural antioxidants highlights the food sector's dedication to pursuing sustainable and health-oriented approaches. With the advancement of scientific knowledge, an increasing number of natural antioxidants are expected to be included in commercial applications. This will not only enhance preservation capabilities but also fit with the changing preferences and expectations of consumers. Consumers play a crucial role in the development of antioxidants to enhance meat products, leading to new research and industry innovations.

5 Factors That Influence the Efficiency of Antioxidants

The efficacy of antioxidants in meat products is subject to several conditions that affect their capacity to prevent or delay oxidation reactions. Comprehending these factors is essential to optimize the composition of meat products to enhance their shelf-life and overall quality. Numerous important factors affect the effectiveness of antioxidants. Influencing factors include the type of antioxidant, concentration of antioxidant, synergistic effects, pH level, temperature, presence of metals, oxygen levels, packaging, storage conditions, processing methods, and type of meat.

According to Lobo et al. [70], antioxidants may be categorized based on their mode of action into primary and secondary molecules. Primary antioxidants react directly with free radicals and generate new stable products. Due to the basic chemical properties of these compounds, they have the capability of acting as acceptors or scavengers of free radicals, hence impeding or retarding the start phase or disrupting the propagation phase of autoxidation. The small molecular antioxidants can efficiently interact with free radicals in the progression of the system's chain reaction, preventing any potential harm to essential components. Certain antioxidants, such as glutathione, ubiquinol, and uric acid, are endogenously synthesized due to regular chemical reactions within the meat. Vitamin E (α-tocopherol), tannin, anthocyanin, and vitamin C are common antioxidants used to prevent lipid oxidation in meat products [71].

Class II antioxidants are active compounds that act as chelators for metal ions that promote oxidation or catalytic reactions, provide hydrogen ions to primary antioxidants, break down hydroperoxide into non-radical compounds, neutralize singlet oxygen, absorb UV radiation, or serve as scavengers for oxygen. These compounds include citric acid, alpha-lipoic acid, polyphenols from herbs and spices, and carotenoids [72, 73].

The efficacy of meat products is substantially influenced by the concentration of antioxidants present. Developing a suitable balance is crucial, as minimal concentrations may prove ineffective in preventing oxidation, while excessive levels may result in unfavorable alterations in sensory perception. El-Nashi et al. utilized pomegranate peel powder at various concentrations ranging from 1% to 3% in beef sausages. The findings indicate that the sausages can be stored at 4 °C for up to 12 days [74]. In a study carried out by Qin et al. [75], it was shown that the application of pomegranate powder extract at a concentration of 0.02 g extract/100 g of raw ground pig meat had a beneficial impact in preventing lipid oxidation. It has been demonstrated that pomegranate peels can be used in meat preservation at high concentrations without affecting the product's taste. Thyme essential oil has been recognized as a potent natural antioxidant for meat products like chicken and minced beef [76, 77]. However, its inclusion may affect customer preference due to its strong scent and flavor [78].

Lipid oxidation is affected by the synergistic effect of antioxidants. Combining some antioxidants has been seen to provide synergistic effects, increasing overall effectiveness. The formulation of blends of antioxidants has the potential to enhance the efficacy of oxidative defense compared to utilizing individual antioxidants in isolation. According to Khodaei et al. [79], According to Khodaei et al. [79], the combination of clove oil and thyme oil at a ratio of 100:25 extended the storage time of ground pork and chicken at 4 °C by 2–4 times. The sensory evaluation also indicated that the treated samples were moderately better accepted than the control group. Mizi et al. [80] investigated the combined effect of sage along with high-pressure processing on the antibacterial and antioxidant characteristics of beef patties during a 60-day storage period. The study's findings indicate that lipid oxidation exhibited an elevated level in all samples during storage. However, it was observed that the inclusion of sage powder in the beef patties, which were subjected to high-pressure processing, demonstrated a notable antioxidant impact. This effect was shown to delay lipid oxidation effectively throughout a 60-day refrigerated storage period.

Furthermore, the microbiological quality of the burgers was assessed to be within acceptable limits. Utilizing both procedures to achieve burgers with enhanced oxidation and microbiological stability for extended storage periods while maintaining unaffected sensory qualities is a viable alternative. Trindade et al. [81] conducted a study to examine the effects of oregano and rosemary extracts, whether given alone or together, as well as (BHA/BHT) on the lipids oxidation and the fatty acids composition in irradiated beef patties. The experimental findings indicate that the application of rosemary extract, either on its own or in conjunction with oregano extract or BHA/BHT, The findings indicated that the combination of extracts was more effective than using individual extracts alone. Additionally, it was observed that while antioxidants decreased during storage, lipid oxidation increased. Rosemary extract had the most significant antioxidant activity among the other natural additions investigated.

The pH level of meat products influences the ionization state of antioxidants. Some antioxidants are more effective in acidic conditions, while others perform better in alkaline environments. Matching the pH requirements of antioxidants with the product's pH is essential for optimal performance. Natural phenolic antioxidants include a variety of organic molecules that contain oxygen, including phenolic acids, flavonoids, chalcones, and stilbenes. The presence of the phenolic-OH moiety is essential in these compounds. Furthermore, there exist several oxygen-containing functional groups, including the alkoxyl group (–OR), ether bond (–O–), carboxyl group (–COOH), ester group (–COOR), carbonyl group (–C=O), and even sugar residue [82–84]. The presence of oxygen atoms in these phenolic antioxidants renders them either acidic or neutral. Ascorbic acid has a significant degree of instability and reactivity. Under certain circumstances, including exposure to light, heat, transition metal ions, and alkaline conditions (high pH), the compound undergoes a reversible oxidation process, forming dehydroascorbic acid (DHA). In contrast, tocopherol is a fat-soluble vitamin with antioxidant properties. It is stable in alkaline solutions and can protect against lipid oxidation [85]. Carnosic acid, carnosol, and rosmarinic acid are highly potent antioxidants in rosemary. These compounds are commonly utilized to prevent oxidation in meat products and exhibit stability even in alkaline conditions. [86].

Storage conditions have a significant role in promoting lipid oxidation [87]. Various conditions, such as temperature, light, and oxygen, might enhance oxidative reactions [88]. Specific processing techniques can also promote oxidation. These activities (including cutting, deboning, grinding, or boiling) expedite the progression of lipid oxidation [1]. In general, these processes result in the disturbance of the muscular membrane, which stimulates oxidative responses. This pertains to the processing of freed membrane phospholipids, which enhances their exposure to prooxidant elements such as oxygen, enzymes, or metals that facilitate oxidation [89]. Furthermore, including additional components (such as substances that inhibit or promote oxidation) in meat product recipes also impacts the oxidation process. Zuzana Réblová observed that phenolic acids with a high susceptibility to oxidation, such as gallic, gentisic, protocatechuic, and caffeic acids, showed a less pronounced decline in antioxidant activity as the temperature increased, as compared to their activity at 90 °C. The less susceptible acids, such as syringic, ferulic, and sinapic acids, and particularly vanillic acid, exhibited a more significant decrease. As a result, only gallic, gentisic, protocatechuic, and caffeic acids showed notable antioxidant activity at a temperature of 150 °C. However, vanillic acid maintained its activity only at a lower temperature of 90 °C [90]. Recently, high-oxygen modified atmosphere packaging (HiOx-MAP) has been frequently used for red meat to ensure the creation and durability of oxymyoglobin (OxyMb). At the same time, it is being shown for sale [91]. Scientific literature suggests that oxygen percentages ranging from 50% to 80% in HiOx-MAP (High Oxygen Modified Atmosphere Packaging) are successful in maintaining OxyMb (Oxymyoglobin) in fresh lamb [92] and beef [93] when stored under cold conditions. However, studies have shown that HiOx-MAP can expedite the process of lipid and protein oxidation in beef [94], lamb [95], and pig meat cuts [96]. The presence of oxidative processes under these packing settings is evident in changes in meat quality, including reduced water-holding capacity and decreased softness [91]. Furthermore, the accumulation of lipid and protein oxidation products in meat is highlighted to underscore the potential safety issues due to the toxicity of these chemical compounds [97]. Gao et al. [98] carried out a study using butylated hydroxytoluene (BHT, 0.02%), carnosine (1%), and plant extracts, including rosemary (RE, 1%), pine bark (PBE, 1%), and pomegranate (PGE, 1%), to prevent lipid oxidation in chicken meat patties. The conditions included treatment with dielectric barrier discharge (DBD) cold plasma (CP), and a modified atmosphere. The results indicate that using natural plant extracts can enhance the inactivation of microbes and reduce lipid peroxidation in meat processed with CP.

6 Antioxidants in Cooked Meat Products

Cooked meat products undergo a range of activities, such as heating and cooking, that can initiate oxidative reactions, resulting in a decline in quality. Antioxidants are vital in reducing these reactions and maintaining cooked meat's sensory characteristics, nutritional quality, and lifespan.

Zang et al. [99] investigated the effects of antioxidants on the physical and chemical characteristics, as well as the aroma compounds, in heat-treated beef taste and their ability to prevent oxidation under various storage conditions. The results show that L-ascorbyl palmitate (L-AP) and butyl hydroquinone (TBHQ) exhibited superior abilities in inhibiting the generation of primary lipid oxidation products compared to tea polyphenol (TP). Based on the findings from oxidation-reduction potential and DPPH radical scavenging studies, TBHQ demonstrated superior antioxidant capacity compared to L-AP and TP during storage. It is worth mentioning that TBHQ impacted the taste characteristics of HPBF, particularly the volatile odour compounds generated through the breakdown of lipids. The use of TBHQ might help reduce the formation of unpleasant odour molecules. The study found that TBHQ improved lipid oxidation stability and preserved the physical and chemical characteristics and flavour of HPBF during storage. It is proposed that TBHQ might be utilized as a substitute ingredient to enhance the quality of thermal process flavourings associated with HPBF.

Kong et al. [100] used six of the extracts (nutmeg, licorice, cassia bark, rosemary, clove, and round cardamom) to inhibit the formation of TBARS while cooking pork patties. The use of spice extracts in cooked pig patties resulted in a significant decrease in TBARS generation and off-flavour scores. Additionally, the patties exhibited improved color stability throughout the storage period.

Ganhão et al. [101] thoroughly investigated the complex impacts of different fruit extracts on protein oxidation, including the color and structure during cold storage. This detailed analysis explored a wide range of fruits, investigating their ability to impact proteins' oxidative processes. Fruit extracts from arbutus berries (*Arbutus unedo L.*, AU), common hawthorns (*Crataegus monogyna L.*, CM), dog roses (*Rosa canina L.*, RC), and elm-leaf blackberries (*Rubus ulmifolius Schott.*, RU) effectively decreased the development of protein carbonyls. They prevented colour and texture degradation during refrigerated storage. The potential ways protein oxidation might significantly impact burger patties' colour and texture changes during refrigerated storage are examined. Out of the several extracts, RC proved to be the most appropriate choice as a functional component in processed meats. This is because it improved burger patties' oxidative stability, colour, and texture attributes without any noticeable disadvantages.

The production of cytotoxic substances in cooked and processed meats may be associated with the presence of free radicals [102]. The formation of heterocyclic aromatic amines (HAA) 2-amino-3,8-dimethylimidazo (4,5-f) (2-amino-3,8-dimethylimidazo (4,5-f) Quinoxaline (MeIOx) and imidazoquinoxaline (IQ) in fried ground beef are result of the interaction between creatinine and alkylpyridine radicals. These compounds have been associated with cancer and other health issues. Therefore, it is possible to decrease the development of harmful chemicals by cooking and processing meat with spices, herbs, and other plant extracts high in antioxidants. Utilizing natural antioxidants is a novel and intriguing strategy that could be a valuable alternative method for enhancing the health of meat products.

Balogh et al. [102] examined how α-tocopherol (vitamin E) and oleoresin rosemary impact the development of HAA in ground beef patties fried at temperatures of 175, 200, and 225 °C. At fat contents of 1% and 10%, the addition of α-tocopherol to the ground beef patties resulted in a reduction of HAAs, including 2-amino-1-methyl-6-phenylimidazo (4,5-b) pyridine (PhIP), the main HAA found in cooked muscle meals, by 69% and 72% in the cooked patties, respectively. Applying oleoresin rosemary at both 1% and 10% concentrations resulted in a 44% decrease in PhIP production. The study group also showed that Montmorency and Balaton cherry tissue reduced the levels of PhIP by 93% and 87%, respectively [103]. In their study, Puangsombat et al. [104] found that adding 0.2% of rosemary, turmeric, fingerroot, and galangal to fried beef patties cooked at 204 °C reduced the levels of HAAs (MeIQx and PhIP) by 43.5%, 39.2%, 33.5%, and 18.4%, respectively.

7 Effect on Eating Quality of Meat Products

The impact of antioxidants on the eating quality of meat products is an essential characteristic that combines the scientific comprehension of food chemistry with consumers' sensory experiences. Antioxidants significantly impact several factors that contribute to the entire eating experience, including taste, softness, juiciness, and aroma. In this analysis, we explore the complex mechanisms by which antioxidants can influence meat products' sensory characteristics and overall taste.

Preserving the flavour of meat products is crucial for ensuring their quality and consumer satisfaction. Antioxidants protect the complex and unique flavours naturally present in meat. One of the most important functions of antioxidants is inhibiting lipid oxidation [105]. Lipid oxidation produces volatile chemicals and undesirable odours, which significantly affect the taste and scent of meat. Antioxidants function as radical scavengers, effectively neutralizing free radicals and inhibiting the cascade of processes that result in lipid oxidation. The suppression of enzymatic activity in meat helps to preserve the natural and desired tastes by maintaining the integrity of the fats [106]. Rancidity, defined by the appearance of undesirable and smelly tastes, is frequently the result of lipid oxidation. Antioxidants inhibit rancidity development, especially those obtained from natural sources such as rosemary extract or tocopherols (vitamin E). Antioxidants have a role in preserving the fresh and unique flavours of meat by preventing the formation of unwanted chemicals that result from the breakdown of lipids. A combination of α-tocopherol and rosemary extract, with concentrations of 0.035% each, was employed to suppress the oxidation of fish lipids induced by Fe^{2+} or hemoprotein [107]. In a study conducted by [108], it was shown that adding cinnamon, cloves, fennel, pepper, and star anise at a concentration of 0.5% led to a decrease in TBA (Thiobarbituric Acid) levels. The spice taste had a significant negative relationship ($P < 0.01$) with both rancid odour and flavour, with correlation values of −0.57 and −0.61, respectively.

The Maillard process, which is responsible for forming delicate tastes and browning in cooked meat, is affected by antioxidants. While this process is favourable for improving flavour, excessive oxidation can result in undesirable flavours. Antioxidants assist in maintaining an appropriate balance, allowing the Maillard reaction to enhance taste positively without straying towards unpleasant flavour profiles. The Maillard reaction can generate various chemicals, such as advanced glycation end products (AGEs), that might cause taste problems and have health effects. Antioxidants can prevent the development of undesired chemicals during the Maillard process by neutralizing free radicals and lowering oxidative stress. This ensures the Maillard reaction enhances taste without causing undesirable sensory effects [109]. In addition, extensive research has been conducted on targeting crucial intermediates, such as GSH, vitamins (pyridoxamine, vitamin C, and E), low molecular weight chitosan, and polyphenols. These substances are used to competitively react with reducing sugars, thereby reducing the formation of Amadori products. They also act to trap α-dicarbonyl compounds and respond with toxic substances for their elimination [110]. Antioxidants, especially those obtained from herbs and spices, can protect and improve aromatic compounds. The Maillard process enhances flavour and adds depth and richness to the sensory experience. The research was done to examine the influence of spices on the amounts and compositions of volatile chemicals produced when cooking beef patties, with each patty having 0.5% of a particular spice (nutmeg, onion, garlic, or ginger powder, by weight), which were cooked using an electric pan until they reached an internal temperature of 75 °C. The researchers employed purge-and-trap gas chromatography-mass spectrometry (GC/MS) to identify 46 volatile chemicals in the cooked beef patties. These compounds comprised 6 alcohols, 6 aldehydes, 5 hydrocarbons, 6 ketones, 9 sulphur compounds, and 14 terpenes [111]. The comparative study demonstrated that the combination of nutmeg, onion, or ginger powder resulted in a considerable reduction in the generation of volatile chemicals due to lipid oxidation in the cooked beef patties compared to the control group. Furthermore, the incorporation of nutmeg and garlic powder into the beef patties led to an increased production of terpenes and sulphur volatile compounds, respectively. The results emphasize the impact of different spices on the primary chemical categories, including alcohols, aldehydes, hydrocarbons, ketones, sulphur compounds, and terpenes, in the volatile component composition of cooked beef patties.

8 Health Benefits of Antioxidant-Enriched Meat

Adding antioxidants to meat products offers several health advantages that go beyond the conventional nutritional components of consuming meat. Despite some encouraging findings from laboratory and field investigations, the efficacy of plant-derived antioxidants in preventing lipid and protein oxidation has yet to be well proven. Furthermore, the effectiveness, user-friendliness, commercial viability, and possible therapeutic advantages of natural antioxidants have yet to be definitively shown [53]. Plant extracts in meat processing have been shown to integrate several antioxidant chemicals that possess therapeutic properties. Extensive documentation exists regarding their multifaceted roles in enhancing human health and preventing physiological diseases and pathological situations. According to [112], the consumption of meat products that are abundant in natural antioxidants has been demonstrated to enhance the inherent antioxidant capacity in the body, hence protecting against oxidative stress, tissue damage, and degenerative illnesses caused by reactive oxygen species (ROS). Using natural phenolic antioxidants as food additives has positively impacted overall health. Notably, these antioxidants can maintain the gastrointestinal system's health, as they do not need absorption [113].

Free radicals generated in the body can damage cellular structures, including DNA, proteins, and lipids. Antioxidants act as defenders against this damage by neutralizing free radicals. Incorporating antioxidant-enriched meat provides an additional layer of cellular protection, promoting cells' overall health and longevity [114]. In our body, cells require a delicate balance achieved by utilizing both antioxidant enzymes and non-enzymatic antioxidant compounds. The body has several antioxidant defense enzymes, which include superoxide dismutase, catalase, peroxidase, glutathione peroxidase, and glutathione reductase. In biological systems, the antioxidant enzymes and reducing agents operate coordinated rather than functioning independently to eliminate reactive oxygen species such as superoxide ($O_2^{\cdot -}$), hydrogen peroxide (H_2O_2), and other oxidizing agents. The group of non-enzymatic antioxidants with low molecular weight consists of tocopherols, carotenoids, ubiquinones, ascorbate, glutathione, phosphate, uric acid, and citric acid, as mentioned by [115]. Certain low-molecular-weight antioxidants, such as vitamin E, beta-carotene, and coenzyme Q, exhibit fat solubility and are found within the cellular membrane.

Conversely, water-soluble antioxidants like ascorbate and citric acid also function within the cytoplasm [116]. Although the body has many antioxidants to balance free radicals, most are susceptible to easy destruction. Incorporating antioxidants into meat could reduce the risk of various illnesses and physiological dysfunctions inside the human body [117].

Oxidative stress is a critical factor in the development of chronic illnesses, including cardiovascular diseases, diabetes, neurological disorders, and cancer [118]. The presence of antioxidants in meat products has the potential to have a beneficial effect on cardiovascular health. Antioxidants are crucial in inhibiting the oxidation process of low-density lipoprotein (LDL) cholesterol, commonly known as "bad" cholesterol. Oxidized low-density lipoprotein (LDL) cholesterol has been identified as a significant risk factor for the development of cardiovascular illnesses. The presence of antioxidants has been seen to have a role in mitigating the oxidation of LDL cholesterol, contributing to overall cardiovascular health. A healthy immune system is crucial for maintaining general well-being and bolstering resilience against diseases. The correct functioning of the immune system is supported by antioxidants, which neutralize free radicals and reduce oxidative stress. The inclusion of antioxidants in meat products has the potential to enhance the immune system's overall resistance. The incorporation of antioxidants into meat products has the potential to improve the preservation of the meat's nutritional integrity. Antioxidants have a crucial role in preventing the oxidation process of vital components, including vitamins and fatty acids, preserving their bioavailability, and augmenting the overall nutritional quality of meat. The correlation between antioxidants and cancer prevention is intricate; nonetheless, some studies propose that antioxidants could offer protection against specific forms of cancer by counteracting the detrimental effects of free radicals, which can facilitate the development and advancement of cancerous cells. Neurodegenerative illnesses, such as Alzheimer's and Parkinson's, have been associated with the occurrence of oxidative stress. The potential involvement of antioxidants in promoting cognitive health is suggested by their ability to mitigate oxidative damage inside the brain. The possible contribution of including antioxidant-enriched meat in one's dietary regimen towards preserving cognitive function is worth considering [118].

9 Challenges and Considerations

Incorporating antioxidants into meat products, although having the potential to improve both nutritional content and life expectancy, presents a few challenges and factors that need to be considered. As the food business explores this novel field, several issues are considered, requiring careful management to guarantee the effective enrichment of meat products with antioxidants.

The stability of antioxidants might differ, and their compatibility with the meat matrix is of tremendous significance. Certain antioxidants have the potential to undergo degradation or experience a decrease in effectiveness throughout the processes of manufacturing or storage, thereby affecting their capacity to provide beneficial effects. Developing a harmonic equilibrium that ensures the stability of antioxidants throughout each step of the manufacturing cycle and in the ultimate meat product presents a substantial obstacle. The study demonstrated that the antioxidant activity of fruits decreases after heat processing and prolonged storage [119]. Various cooking techniques found the food's antioxidant content to be reduced [120, 121]. The vitamin C content and free radical scavenging activity of raspberry fruit were drastically reduced after being stored in frozen conditions for

12 months [122]. Furthermore, the antioxidant capacity of black pepper, allspice, and oregano was diminished after exposure to a temperature of 130 °C for 5 minutes. However, it is worth noting that black pepper exhibited an elevated concentration of phenolic compounds [123].

One crucial factor to consider is the possible influence on sensory characteristics. Antioxidants can modify the sensory attributes of meat-based food items, such as flavour, color, and scent. Achieving a careful equilibrium between optimizing nutritional content and maintaining the intended sensory characteristics is crucial to guaranteeing consumer approval and contentment. The task at hand is harmonizing both features while ensuring neither aspect is compromised.

The determination of the optimal dosage of antioxidants is a complex task. There needs to be more quantity to result in the advantages becoming insignificant. At the same time, too many might give rise to undesired effects, altering sensory attributes such as taste, texture, and overall product quality. The main objective of research and the utilization of antioxidants is to achieve a twofold outcome: enhancing the efficacy of antioxidants while maintaining the product's sensory attributes without compromising any aspects.

Meat products consist of a diverse combination of many components. Understanding how antioxidants interact with preservatives, taste enhancers, and other elements is essential. The product's general stability and functioning may be affected by unexpected interactions. The task at present involves the comprehension and successful management of these relationships.

The perception of consumers assumes significance as a critical aspect. It is crucial to educate customers on the beneficial attributes of antioxidants. Misconceptions or a lack of knowledge might impact the level of acceptance. Effectively communicating the nutritional advantages while conforming to customer expectations and preferences is a strategic challenge.

The use of antioxidants has the potential to influence the entire economic aspect of meat production. Achieving a balance between the possible advantages and the associated costs necessitates the adoption of a strategic methodology. Maintaining competitive pricing in the market presents a complex challenge. The food sector is now exploring the incorporation of antioxidants into meat products, which offers a complex landscape characterized by several obstacles and factors to be considered. Effectively manoeuvring across this terrain necessitates a comprehensive strategy that incorporates scientific knowledge, consumer consciousness, and adherence to industry regulations. Within the complex interaction of nutrition, taste, and market dynamics, the pursuit of achieving optimal antioxidant enrichment in meat products emerges as a varied effort.

10 Research and Innovations

The field of antioxidant-enriched meat has recently made substantial advances due to continuous investigation and the implementation of novel solutions. Researchers are now investigating numerous topics to extend the shelf-life of meat products and enrich them with antioxidants that promote health.

Antioxidants are essential in mitigating the detrimental impacts of free radicals, and scientists are extensively exploring many sources to uncover highly effective bioactive molecules. The scope of investigation encompasses a wide range of agricultural products, including oilseeds, grains, legumes, fruits, vegetables, herbs, and spices. Scientists can optimize nutritional benefits by customizing enrichment procedures based on their understanding of the distinctive antioxidant profiles exhibited by different sources. The study on natural antioxidants has been motivated by a significant change in customer tastes, which now favor clean labels and natural goods. The primary emphasis lies in examining plant extracts, herbs, and spices renowned for their antioxidative characteristics. The trend to natural and minimally processed foods is reflected in the decision to remove synthetic antioxidants.

Nanoencapsulation has recently emerged as a pioneering method for increasing the stability and bioavailability of antioxidants in meat products. The proposed approach entails encapsulating antioxidant chemicals into nanoscale carriers, preserving them from degradation throughout all stages of processing and storage. Scientists are now investigating the potential of nanoencapsulation to enhance the delivery of antioxidants significantly, thereby maximizing their effectiveness.

The investigation of intelligent packaging methods is underway to achieve a controlled release of antioxidants. The preservation of meat extends its longevity and guarantees a consistent presence of antioxidants, offering a dual advantage in both freshness and nutritional enrichment.

Incorporating antioxidants into meat products is transforming its conventional role as a stable food item, as it is increasingly recognized as a functional food or nutraceutical. Researchers are now investigating additional health advantages beyond oxidative stability, including possible anti-inflammatory characteristics and cardiovascular health promotion. This is consistent with the increasing consumer demand for food products that offer functional advantages beyond just nutritional value.

Incorporating antioxidants presents challenges and opportunities for the quality of meat. Academics are doing thorough investigations into the effects of antioxidants on attributes such as colour, texture, flavour, and general acceptability. Achieving an optimal equilibrium between prolonging the duration of product viability and preserving or augmenting sensory attributes is a critical deliberation within this pioneering domain.

Finally, with the increasing emphasis on sustainability, scientists are actively investigating ecologically conscious methods for obtaining antioxidants. This encompasses the enhancement of industrial processes, the reduction of waste, and the evaluation of the ecological consequences associated with various antioxidant-rich constituents. Incorporating sustainability has become a fundamental component within the innovation process of antioxidant-enriched meat manufacturing. Furthermore, rigorous in vitro and in vivo studies are underway to substantiate the potential health benefits of antioxidant-enriched meat. Understanding how antioxidants interact within the human body and contribute to health outcomes is crucial. These studies bridge the gap between laboratory findings and real-world health implications.

References

1. Min B, Ahn D. Mechanism of lipid peroxidation in meat and meat products-a review. Food Sci Biotechnol. 2005;14(1):152–63.
2. Luna A, Labaque M, Zygadlo J, Marin R. Effects of thymol and carvacrol feed supplementation on lipid oxidation in broiler meat. Poult Sci. 2010;89(2):366–70.
3. Śmiecińska K, Daszkiewicz T. Lipid oxidation and color changes in beef stored under different modified atmospheres. J Food Process Preserv. 2021;45(3):e15263.
4. Shui S, Yan H, Tu C, Benjakul S, Aubourg SP, Zhang B. Cold-induced denaturation of muscle proteins in hairtail (Trichiurus lepturus) during storage: physicochemical and label-free based proteomics analyses. Food Chem X. 2022;16:100479.
5. Zhang B, Qi X-e, Mao J-l, Ying X-g. Trehalose and alginate oligosaccharides affect the stability of myosin in whiteleg shrimp (Litopenaeus vannamei): the water-replacement mechanism confirmed by molecular dynamic simulation. LWT. 2020;127:109393.
6. Yehye WA, Rahman NA, Ariffin A, Abd Hamid SB, Alhadi AA, Kadir FA, et al. Understanding the chemistry behind the antioxidant activities of butylated hydroxytoluene (BHT): a review. Eur J Med Chem. 2015;101:295–312.
7. Zhang XJ, Diao MN, Zhang YF. A review of the occurrence, metabolites and health risks of butylated hydroxyanisole (BHA). J Sci Food Agric. 2023;103(13):6150–66.
8. Dave D, Ghaly AE. Meat spoilage mechanisms and preservation techniques: a critical review. Am J Agric Biol Sci. 2011;6(4):486–510.
9. Kumari P, Akhila S, Rao YS, Devi BR. Alternative to artificial preservatives. Syst Rev Pharm. 2019;10:99–102.
10. Halliwell B, Chirico S. Lipid peroxidation: its mechanism, measurement, and significance. Am J Clin Nutr. 1993;57(5):715S–25S.
11. Baron CP, Andersen HJ. Myoglobin-induced lipid oxidation. A review. J Agric Food Chem. 2002;50(14):3887–97.
12. Bao Y, Ertbjerg P. Effects of protein oxidation on the texture and water-holding of meat: a review. Crit Rev Food Sci Nutr. 2019;59(22):3564–78.
13. Soladoye O, Juárez M, Aalhus J, Shand P, Estévez M. Protein oxidation in processed meat: mechanisms and potential implications on human health. Compr Rev Food Sci Food Saf. 2015;14(2):106–22.
14. Nawaz A, Irshad S, Khan IA, Khalifa I, Walayat N, Aadil RM, et al. Protein oxidation in muscle-based products: effects on physicochemical properties, quality concerns, and challenges to food industry. Food Res Int. 2022;157:111322.
15. Macho-González A, Garcimartín A, López-Oliva ME, Bastida S, Benedí J, Ros G, et al. Can meat and meat-products induce oxidative stress? Antioxidants. 2020;9(7):638.
16. Jongberg S, Lund MN, Skibsted LH. Protein oxidation in meat and meat products. Challenges for antioxidative protection. In: Global food security and wellness. New York: Springer; 2017. p. 315–37.

17. Requena JR, Chao C-C, Levine RL, Stadtman ER. Glutamic and aminoadipic semialdehydes are the main carbonyl products of metal-catalyzed oxidation of proteins. Proc Natl Acad Sci. 2001;98(1):69–74.
18. Akagawa M, Sasaki D, Kurota Y, Suyama K. Formation of α-aminoadipic and γ-glutamic semialdehydes in proteins by the maillard reaction. Ann N Y Acad Sci. 2005;1043(1):129–34.
19. Stadtman E, Levine R. Free radical-mediated oxidation of free amino acids and amino acid residues in proteins. Amino Acids. 2003;25:207–18.
20. Estévez M. Protein carbonyls in meat systems: a review. Meat Sci. 2011;89(3):259–79.
21. Dussault D, Benoit C, Lacroix M. Combined effect of γ-irradiation and bacterial-fermented dextrose on microbiological quality of refrigerated pork sausages. Radiat Phys Chem. 2012;81(8):1098–102.
22. Aziz M, Karboune S. Natural antimicrobial/antioxidant agents in meat and poultry products as well as fruits and vegetables: a review. Crit Rev Food Sci Nutr. 2018;58(3):486–511.
23. Lopes SM, da Silva DC, Tondo EC. Bactericidal effect of marinades on meats against different pathogens: a review. Crit Rev Food Sci Nutr. 2022;62(27):7650–8.
24. Ehsanur Rahman SM, Islam S, Pan J, Kong D, Xi Q, Du Q, et al. Marination ingredients on meat quality and safety—a review. Food Qual Saf. 2023;7:fyad027.
25. Gómez I, Janardhanan R, Ibañez FC, Beriain MJ. The effects of processing and preservation technologies on meat quality: sensory and nutritional aspects. Foods. 2020;9(10):1416.
26. Latoch A. Effect of meat marinating in kefir, yoghurt and buttermilk on the texture and color of pork steaks cooked sous-vide. Ann Agric Sci. 2020;65(2):129–36.
27. Latoch A, Libera J, Stasiak DM. Physicochemical properties of pork loin marinated in kefir, yoghurt or buttermilk and cooked sous vide. Acta Sci Pol Technol Aliment. 2019;18(2):163–71.
28. Iqbal SZ, Talib NH, Hasnol N. Heterocyclic aromatic amines in deep fried lamb meat: the influence of spices marination and sensory quality. J Food Sci Technol. 2016;53:1411–7.
29. Manful CF, Pham TH, Nadeem M, Wheeler E, Warren KJ, Vidal NP, et al. Assessing unfiltered beer-based marinades effects on ether and ester linked phosphatidylcholines and phosphatidylethanolamines in grilled beef and moose meat. Meat Sci. 2021;171:108271.
30. Istrati D, Constantin O, Ionescu A, Vizireanu C, Dinică R. Study of the combined effect of spices and marination on beef meat vacuum packaged. Ann Uni Dunarea de Jos Galati Fasc VI-Food Technol. 2011;35(2):75–85.
31. Smaoui S, Hlima HB, Braïek OB, Ennouri K, Mellouli L, Khaneghah AM. Recent advancements in encapsulation of bioactive compounds as a promising technique for meat preservation. Meat Sci. 2021;181:108585.
32. Liu F, Xu Q, Dai R, Ni Y. Effects of natural antioxidants on colour stability, lipid oxidation and metmyoglobin reducing activity in raw beef patties. Acta Sci Pol Technol Aliment. 2015;14(1):37–44.
33. Martínez L, Cilla I, Antonio Beltrán J, Roncalés P. Effect of Capsicum annuum (red sweet and cayenne) and Piper nigrum (black and white) pepper powders on the shelf life of fresh pork sausages packaged in modified atmosphere. J Food Sci. 2006;71(1):S48–53.
34. Price A, Díaz P, Bañón S, Garrido MD. Natural extracts versus sodium ascorbate to extend the shelf life of meat-based ready-to-eat meals. Food Sci Technol Int. 2013;19(5):427–38.
35. Rojas MC, Brewer MS. Effect of natural antioxidants on oxidative stability of frozen, vacuum-packaged beef and pork. J Food Qual. 2008;31(2):173–88.
36. Keokamnerd T, Acton J, Han I, Dawson P. Effect of commercial rosemary oleoresin preparations on ground chicken thigh meat quality packaged in a high-oxygen atmosphere. Poult Sci. 2008;87(1):170–9.
37. Jeong JY, Seol K-H, Seong P-N, Park B-Y, Kim HW. Effects of procyanidin on meat quality and shelf-life for preserving pork patties during chilled storage. Korean J Food Sci Anim Resour. 2015;35(4):564.
38. Formanek Z, Lynch A, Galvin K, Farkas J, Kerry J. Combined effects of irradiation and the use of natural antioxidants on the shelf-life stability of overwrapped minced beef. Meat Sci. 2003;63(4):433–40.

39. Zhang J, Wang Y, Pan D-D, Cao J-X, Shao X-F, Chen Y-J, et al. Effect of black pepper essential oil on the quality of fresh pork during storage. Meat Sci. 2016;117:130–6.
40. Ismail H, Lee EJ, Ko K, Paik H, Ahn D. Effect of antioxidant application methods on the color, lipid oxidation, and volatiles of irradiated ground beef. J Food Sci. 2009;74(1):C25–32.
41. Nikmaram N, Budaraju S, Barba FJ, Lorenzo JM, Cox RB, Mallikarjunan K, et al. Application of plant extracts to improve the shelf-life, nutritional and health-related properties of ready-to-eat meat products. Meat Sci. 2018;145:245–55.
42. Kumar Y, Yadav DN, Ahmad T, Narsaiah K. Recent trends in the use of natural antioxidants for meat and meat products. Compr Rev Food Sci Food Saf. 2015;14(6):796–812.
43. Oswell NJ, Thippareddi H, Pegg RB. Practical use of natural antioxidants in meat products in the US: a review. Meat Sci. 2018;145:469–79.
44. Bellucci ERB, Bis-Souza CV, Domínguez R, Bermúdez R, Barretto ACS. Addition of natural extracts with antioxidant function to preserve the quality of meat products. Biomolecules. 2022;12(10):1506.
45. Dravie EE, Kortei NK, Essuman EK, Tettey CO, Boakye AA, Hunkpe G. Antioxidant, phytochemical and physicochemical properties of sesame seed (Sesamum indicum L). Sci Afr. 2020;8:e00349.
46. Waszkowiak K, Rudzińska M. Effect of flaxseed meals and extracts on lipid stability in a stored meat product. J Am Oil Chem Soc. 2014;91:979–87.
47. Imran M, Ahmad N, Anjum FM, Khan MK, Mushtaq Z, Nadeem M, et al. Potential protective properties of flax lignan secoisolariciresinol diglucoside. Nutr J. 2015;14(1):1–7.
48. Karamać M, Kosińska A, Estrella I, Hernández T, Duenas M. Antioxidant activity of phenolic compounds identified in sunflower seeds. Eur Food Res Technol. 2012;235:221–30.
49. Zhang H, Shao Y, Bao J, Beta T. Phenolic compounds and antioxidant properties of breeding lines between the white and black rice. Food Chem. 2015;172:630–9.
50. Saleh HM, Jameel FR, Hadi ST, Hamid MM. Study of effects oat and soybean on the microbial and sensory analysis of burgers (beef, chicken and sheep). Rev Bionat. 2022;7(1):23.
51. Ham Y-K, Hwang K-E, Song D-H, Choi J-H, Choi Y-S, Kim H-W. Relationship between the antioxidant capacity of soy sauces and its impact on lipid oxidation of beef patties. Meat Sci. 2019;158:107907.
52. Li H. The lentil-meat system: investigating the antioxidant effect of lentil on colour and lipid oxidation of raw beef burgers. University of Saskatchewan; 2017.
53. Manessis G, Kalogianni AI, Lazou T, Moschovas M, Bossis I, Gelasakis AI. Plant-derived natural antioxidants in meat and meat products. Antioxidants. 2020;9(12):1215.
54. Lee CH, Reed JD, Richards MP. Ability of various polyphenolic classes from cranberry to inhibit lipid oxidation in mechanically separated Turkey and cooked ground pork. J Muscle Foods. 2006;17(3):248–66.
55. Contini C, Álvarez R, O'sullivan M, Dowling DP, Gargan SÓ, Monahan FJ. Effect of an active packaging with citrus extract on lipid oxidation and sensory quality of cooked Turkey meat. Meat Sci. 2014;96(3):1171–6.
56. Becerra MO, Contreras LM, Lo MH, Díaz JM, Herrera GC. Lutein as a functional food ingredient: stability and bioavailability. J Funct Foods. 2020;66:103771.
57. Martínez L, Jongberg S, Ros G, Skibsted LH, Nieto G. Plant derived ingredients rich in nitrates or phenolics for protection of pork against protein oxidation. Food Res Int. 2020;129:108789.
58. Kumar P, Chatli MK, Mehta N, Kumar D, Sahoo J. Oxidative stability and quality attributes of emu meat nuggets incorporated with selected levels of broccoli (Brassica oleracea) powder. J Meat Sci Technol. 2013;1(3):83–90.
59. Augustyńska-Prejsnar A, Topczewska J, Ormian M, Saletnik A, Sokołowicz Z, Lechowska J. The effect of the addition turmeric on selected quality characteristics of duck burgers stored under refrigeration. Appl Sci. 2022;12(2):805.
60. Hussain Z, Li X, Ijaz M, Xiao X, Hou C, Zheng X, et al. Effect of Chinese cinnamon powder on the quality and storage properties of ground lamb meat during refrigerated storage. Food Sci Anim Resour. 2020;40(3):311.

61. Ünal K, Babaoglu AS, Karakaya M. Effect of oregano, sage and rosemary essential oils on lipid oxidation and color properties of minced beef during refrigerated storage. J Essent Oil Bear Plants. 2014;17(5):797–805.
62. Boruzi AI, Nour V. Walnut (Juglans regia L.) leaf powder as a natural antioxidant in cooked pork patties. CyTA-J Food. 2019;17(1):431–8.
63. Yoon J, Bae SM, Gwak SH, Jeong JY. Use of green tea extract and rosemary extract in naturally cured pork sausages with white kimchi powder. Food Sci Anim Resour. 2021;41(5):840.
64. Yang Y, Song X, Sui X, Qi B, Wang Z, Li Y, et al. Rosemary extract can be used as a synthetic antioxidant to improve vegetable oil oxidative stability. Ind Crop Prod. 2016;80:141–7.
65. Jessica Elizabeth DLT, Gassara F, Kouassi AP, Brar SK, Belkacemi K. Spice use in food: properties and benefits. Crit Rev Food Sci Nutr. 2017;57(6):1078–88.
66. Nimse SB, Pal D. Free radicals, natural antioxidants, and their reaction mechanisms. RSC Adv. 2015;5(35):27986–8006.
67. EFSA Panel on Additives and Products or Substances used in Animal Feed (FEEDAP), Rychen G, Aquilina G, Azimonti G, Bampidis V, et al. Safety and efficacy of butylated hydroxyanisole (BHA) as a feed additive for all animal species. EFSA J. 2018;16(3):e05215.
68. del Pilar Sánchez-Camargo A, Herrero M. Rosemary (Rosmarinus officinalis) as a functional ingredient: recent scientific evidence. Curr Opin Food Sci. 2017;14:13–9.
69. Gęgotek A, Skrzydlewska E. Antioxidative and anti-inflammatory activity of ascorbic acid. Antioxidants. 2022;11(10):1993.
70. Lobo V, Patil A, Phatak A, Chandra N. Free radicals, antioxidants and functional foods: impact on human health. Pharmacogn Rev. 2010;4(8):118.
71. Shi H, Noguchi N, Niki E. Comparative study on dynamics of antioxidative action of α-tocopheryl hydroquinone, ubiquinol, and α-tocopherol against lipid peroxidation. Free Radic Biol Med. 1999;27(3–4):334–46.
72. Lorenzo JM, Pateiro M, Domínguez R, Barba FJ, Putnik P, Kovačević DB, et al. Berries extracts as natural antioxidants in meat products: a review. Food Res Int. 2018;106:1095–104.
73. Choe E, Min DB. Mechanisms and factors for edible oil oxidation. Compr Rev Food Sci Food Saf. 2006;5(4):169–86.
74. El-Nashi HB, Fattah AFAKA, Rahman NRA, Abd El-Razik M. Quality characteristics of beef sausage containing pomegranate peels during refrigerated storage. Ann Agric Sci. 2015;60(2):403–12.
75. Qin Y-Y, Zhang Z-H, Li L, Xiong W, Shi J-Y, Zhao T-R, et al. Antioxidant effect of pomegranate rind powder extract, pomegranate juice, and pomegranate seed powder extract as antioxidants in raw ground pork meat. Food Sci Biotechnol. 2013;22:1063–9.
76. Fratianni F, De Martino L, Melone A, De Feo V, Coppola R, Nazzaro F. Preservation of chicken breast meat treated with thyme and balm essential oils. J Food Sci. 2010;75(8):M528–M35.
77. Zengin H, Baysal AH. Antioxidant and antimicrobial activities of thyme and clove essential oils and application in minced beef. J Food Process Preserv. 2015;39(6):1261–71.
78. Brewer M. Natural antioxidants: sources, compounds, mechanisms of action, and potential applications. Compr Rev Food Sci Food Saf. 2011;10(4):221–47.
79. Khodaei N, Houde M, Bayen S, Karboune S. Exploring the synergistic effects of essential oil and plant extract combinations to extend the shelf life and the sensory acceptance of meat products: multi-antioxidant systems. J Food Sci Technol. 2023;60(2):679–91.
80. Mizi L, Cofrades S, Bou R, Pintado T, López-Caballero M, Zaidi F, et al. Antimicrobial and antioxidant effects of combined high pressure processing and sage in beef burgers during prolonged chilled storage. Innovative Food Sci Emerg Technol. 2019;51:32–40.
81. Trindade R, Mancini-Filho J, Villavicencio ALCH. Natural antioxidants protecting irradiated beef burgers from lipid oxidation. LWT-Food Sci Technol. 2010;43(1):98–104.
82. Abelskov AK, Smith AT, Rasmussen CB, Dunford HB, Welinder KG. pH dependence and structural interpretation of the reactions of Coprinus cinereus peroxidase with hydrogen peroxide, ferulic acid, and 2, 2′-Azinobis (3-ethylbenzthiazoline-6-sulfonic acid). Biochemistry. 1997;36(31):9453–63.

83. Li X, Jiang Q, Wang T, Liu J, Chen D. Comparison of the antioxidant effects of quercitrin and isoquercitrin: understanding the role of the 6″-OH group. Molecules. 2016;21(9):1246.
84. Li X, Chen B, Xie H, He Y, Zhong D, Chen D. Antioxidant structure–activity relationship analysis of five dihydrochalcones. Molecules. 2018;23(5):1162.
85. Han H, Owens F. Degradation of tocopherol and tocopheryl acetate by ruminal contents in vitro. Anim Sci Res Rep. 1999:119–25.
86. Rashidaie Abandansarie SS, Ariaii P, Charmchian Langerodi M. Effects of encapsulated rosemary extract on oxidative and microbiological stability of beef meat during refrigerated storage. Food Sci Nutr. 2019;7(12):3969–78.
87. Munekata PES, Domínguez R, Campagnol PCB, Franco D, Trindade MA, Lorenzo JM. Effect of natural antioxidants on physicochemical properties and lipid stability of pork liver pâté manufactured with healthy oils during refrigerated storage. J Food Sci Technol. 2017;54:4324–34.
88. Wsowicz E, Gramza A, Hêœ M, Jeleñ HH, Korczak J, Maecka M, et al. Oxidation of lipids in food. Pol J Food Nutr Sci. 2004;13(54):87–100.
89. Chaijan M, Panpipat W. Natural antioxidants. Apple Academic Press; 2017.
90. Réblová Z. Effect of temperature on the antioxidant activity of phenolic acids. Czech J Food Sci. 2012;30(2):171–5.
91. Holman BW, Kerry JP, Hopkins DL. Meat packaging solutions to current industry challenges: a review. Meat Sci. 2018;144:159–68.
92. Morcuende D, Vallejo-Torres C, Ventanas S, Martínez S, Ruiz S, Estévez M. Effectiveness of sprayed bioactive fruit extracts in counteracting protein oxidation in lamb cutlets subjected to a high-oxygen MAP. Foods. 2020;9(11):1715.
93. Shen Y, Guo X, Li X, Wang W, Wang S, Pan J, et al. Effect of cooking temperatures on meat quality, protein carbonylation and protein cross-linking of beef packed in high oxygen atmosphere. LWT. 2022;154:112633.
94. Bao Y, Puolanne E, Ertbjerg P. Effect of oxygen concentration in modified atmosphere packaging on color and texture of beef patties cooked to different temperatures. Meat Sci. 2016;121:189–95.
95. Kim YHB, Bødker S, Rosenvold K. Influence of lamb age and high-oxygen modified atmosphere packaging on protein polymerization of long-term aged lamb loins. Food Chem. 2012;135(1):122–6.
96. Wang C, Wang H, Li X, Zhang C. Effects of oxygen concentration in modified atmosphere packaging on water holding capacity of pork steaks. Meat Sci. 2019;148:189–97.
97. Estévez M, Xiong Y. Intake of oxidized proteins and amino acids and causative oxidative stress and disease: recent scientific evidences and hypotheses. J Food Sci. 2019;84(3):387–96.
98. Gao Y, Yeh H-Y, Bowker B, Zhuang H. Effects of different antioxidants on quality of meat patties treated with in-package cold plasma. Innovative Food Sci Emerg Technol. 2021;70:102690.
99. Zhang Z, Meng F, Wang B, Cao Y. Effects of antioxidants on physicochemical properties and odorants in heat processed beef flavor and their antioxidant activity under different storage conditions. Front Nutr. 2022;9:966697.
100. Kong B, Zhang H, Xiong YL. Antioxidant activity of spice extracts in a liposome system and in cooked pork patties and the possible mode of action. Meat Sci. 2010;85(4):772–8.
101. Ganhão R, Morcuende D, Estévez M. Protein oxidation in emulsified cooked burger patties with added fruit extracts: influence on colour and texture deterioration during chill storage. Meat Sci. 2010;85(3):402–9.
102. Chen C, Pearson A, Gray J. Meat mutagens. Adv Food Nutr Res. 1990;34:387–449.
103. Britt C, Gomaa EA, Gray JI, Booren AM. Influence of cherry tissue on lipid oxidation and heterocyclic aromatic amine formation in ground beef patties. J Agric Food Chem. 1998;46(12):4891–7.
104. Puangsombat K, Jirapakkul W, Smith JS. Inhibitory activity of Asian spices on heterocyclic amines formation in cooked beef patties. J Food Sci. 2011;76(8):T174–T80.

105. Christian KTR, Carole DFM, Alain KY, Dahouenon-Ahoussi E, Avlessi F, Sohounhloue D, et al. Essential oils as natural antioxidants for the control of food preservation. Food Chem Adv. 2023:100312.
106. Decker EA, Elias RJ, McClements DJ. Oxidation in foods and beverages and antioxidant applications: understanding mechanisms of oxidation and antioxidant activity. Elsevier; 2010.
107. Fang X, Wada S. Enhancing the antioxidant effect of α-tocopherol with rosemary in inhibiting catalyzed oxidation caused by Fe2+ and hemoprotein. Food Res Int. 1993;26(6):405–11.
108. Dwivedi S, Vasavada MN, Cornforth D. Evaluation of antioxidant effects and sensory attributes of Chinese 5-spice ingredients in cooked ground beef. J Food Sci. 2006;71(1):C12–C7.
109. Shakoor A, Zhang C, Xie J, Yang X. Maillard reaction chemistry in formation of critical intermediates and flavour compounds and their antioxidant properties. Food Chem. 2022;393:133416.
110. Liu X, Xia B, Hu LT, Ni ZJ, Thakur K, Wei ZJ. Maillard conjugates and their potential in food and nutritional industries: a review. Food Front. 2020;1(4):382–97.
111. Jung S, Jo C, Kim IS, Nam KC, Ahn DU, Lee KH. The influence of spices on the volatile compounds of cooked beef patty. Korean J Food Sci Anim Resour. 2014;34(2):166.
112. Valenzuela A, Sanhueza J, Nieto S. Natural antioxidants in functional foods: from food safety to health benefits. Grasas Aceites. 2003;54(3):295–303.
113. Halliwell B, Rafter J, Jenner A. Health promotion by flavonoids, tocopherols, tocotrienols, and other phenols: direct or indirect effects? Antioxidant or not? Am J Clin Nutr. 2005;81(1):268S–76S.
114. Birben E, Sahiner UM, Sackesen C, Erzurum S, Kalayci O. Oxidative stress and antioxidant defense. World Allergy Organ J. 2012;5:9–19.
115. Gaucher C, Boudier A, Bonetti J, Clarot I, Leroy P, Parent M. Glutathione: antioxidant properties dedicated to nanotechnologies. Antioxidants. 2018;7(5):62.
116. Gutteridge JM, Halliwell B. Mini-review: oxidative stress, redox stress or redox success? Biochem Biophys Res Commun. 2018;502(2):183–6.
117. Poprac P, Jomova K, Simunkova M, Kollar V, Rhodes CJ, Valko M. Targeting free radicals in oxidative stress-related human diseases. Trends Pharmacol Sci. 2017;38(7):592–607.
118. Sharifi-Rad M, Anil Kumar NV, Zucca P, Varoni EM, Dini L, Panzarini E, et al. Lifestyle, oxidative stress, and antioxidants: Back and forth in the pathophysiology of chronic diseases. Front Physiol. 2020;11:694.
119. Keenan DF, Brunton NP, Gormley TR, Butler F, Tiwari BK, Patras A. Effect of thermal and high hydrostatic pressure processing on antioxidant activity and colour of fruit smoothies. Innovative Food Sci Emerg Technol. 2010;11(4):551–6.
120. Wachtel-Galor S, Wong KW, Benzie IF. The effect of cooking on Brassica vegetables. Food Chem. 2008;110(3):706–10.
121. Somsub W, Kongkachuichai R, Sungpuag P, Charoensiri R. Effects of three conventional cooking methods on vitamin C, tannin, myo-inositol phosphates contents in selected Thai vegetables. J Food Compos Anal. 2008;21(2):187–97.
122. de Ancos B, González EM, Cano MP. Ellagic acid, vitamin C, and total phenolic contents and radical scavenging capacity affected by freezing and frozen storage in raspberry fruit. J Agric Food Chem. 2000;48(10):4565–70.
123. Horvathova J, Suhaj M, Šimko P. Effect of thermal treatment and storage on antioxidant activity of some spices. J Food Nutr Res (Slovak Republic). 2007;46(1):20–7.

Meat Products Enriched with Herbs, Spices and Essential Oil

Harsh Kumar, Shivani Guleria, Neetika Kimta, Rajni Dhalaria, Ipsheta Bose, and Kamil Kuča

1 Introduction

Raw fresh meat enrichment with bioactive agents, and the impact of raw meat substances including creatine, taurine, glutathione, L-carnitine, anserine and carnosine on human health, have been widely researched [1]. When the meat-based product is processed, it produces various bioactive substances: peptides bonds that are synthesized by hydrolysis and fermentation have demonstrated physiological advantages for humans [2]. Biologically active peptides can also be synthesized from flesh-based proteins which can subsequently be added to meat products to enhance their functional characteristics [3].

H. Kumar
Centre of Advanced Technologies, Faculty of Science, University of Hradec Kralove, Hradec Kralove, Czech Republic

S. Guleria
Department of Biotechnology, TIFAC-Centre of Relevance and Excellence in Agro and Industrial Biotechnology (CORE), Thapar Institute of Engineering and Technology, Patiala, India

N. Kimta · R. Dhalaria
School of Biological and Environmental Sciences, Shoolini University of Biotechnology and Management Sciences, Solan, India

I. Bose
School of Bioengineering and Food Technology, Shoolini University of Biotechnology and Management Sciences, Solan, India

K. Kuča (✉)
Department of Chemistry, Faculty of Science, University of Hradec Kralove, Hradec Kralove, Czech Republic
e-mail: kamil.kuca@uhk.cz

A. Kumar Pathera et al. (eds.), *Healthier Meat Products*,
https://doi.org/10.1007/978-3-031-78215-2_7

The adoption of functional foods by consumers varies greatly based on their origins in terms of geographic, cultural, economic, political, social, and ethnic factors [2]. Japan was the only country to pioneer the concept of functional foods and has established laws for their use [2]. In 10 years, over 1700 foods with enhanced functional property were founded to the Japanese market, leading to sales of 14 billion dollars in 1999 [4]. The United States is the one of the most vibrant producers for functional foods, in the entire food market it is projected to be between 4% and 6% in 2008 [5]. The European economy for functional foods has been growing gradually, and consumers in Europe are more receptive to them than those in south European countries, where they are more likely to favor natural and fresh food [4].

Lipid degradation is the primary action that degrades the texture, color, flavor and nutritional value of foods [6]. Several synthetic antioxidants, including butylated hydroxyanisole (BHA), butylated hydroxytoluene (BHT) and tertiary-butylhydroquinone, have been employed to hinder the oxidative degradation of foods. However, owing to health concerns, synthetic antioxidants are still not completely accepted by consumers. Thus, many natural substances, such as herbs and spices, have been extensively researched in Asian nations for their potential to act as antioxidants in meat and their products [7]. Herbs and spices include a variety of phytochemicals that may act as natural antioxidants, such as phenolic diterpenes, flavonoids, tannins, and phenolic acids [8]. These substances possess antioxidant, anti-inflammatory, and anticancer properties. In the context of food systems, these substances have the ability to enhance flavor, slow down the deterioration of food caused by lipid oxidation, hinder the growth of microbes, and contribute to reducing the risk of certain diseases [2]. According to reports, clove has the highest free radical scavenging activity among herbs, followed by rose petals, and spices like cinnamon and nutmeg [9]. Moreover, spices possess antimicrobial efficacy mostly attributed to their phenolic components. The potential mechanisms underlying the antibacterial activity of agent compounds encompass altering the permeability of bacterial cells, disrupting membrane activity such as electron transport, nutrient uptake of protein, nutrients and nucleic acid. Additionally, they can interact with membrane proteins, leading to structural and functional deformations, and substituting alkyl groups into the phenol nucleus [2]. The present chapter aims to provide the recent information on the antioxidant activity and antimicrobial properties of regularly used herbs, spices, and their EOs in meat byproducts.

2 Effect of Spices and Herbs in Human Health

Spices and herbs used for culinary purpose have shown wide range of health benefits. There is substantial study data indicating that consuming spices is specifically associated with a decreased risk for mortality from cancer, heart illnesses, respiratory diseases and ischemic [10]. An experimental study conducted in China examined the relationship between the use of savory foods as a regular diet and the overall threats and their reasons of death in a large group of subjects (487,375 individuals), aged 30–79 years, for an average of 7 years, and proved that people consume spicy food with the intention of improving their health [11]. People with spicy food preference had a 10% lower chance of dying, than people who ate spicy food less frequently. Individuals who consumed spicy meals on a daily basis, specifically several times a week, experienced A 14% reduction in the probability of death (the hazard ratios for death in both groups were 0.86). In conclusion, individuals who regularly consumed spicy foods had a 14% decreased mortality risk compared to those who ate spicy foods oftenlly. This observational study indicated a significant correlation between daily-basis consumption of spicy food and a reduced probability of death from respiratory disorders and this was more common among women in comparison to men. People who regularly consumed chili had a decreased danger of death from common diseases [12, 13].

Furthermore, individuals who preferred spicy foods consumed less salt and had lower blood pressure than those who despised spicy foods. Spicy flavors increased sensitivity to salt and decreased preference of salt. Consumption and preference of salt were associated with participants' orbitofrontal cortex (OFC) and insula regional metabolic activity. The administration of capsaicin, a major spicy compound present in chili peppers, elevated the metabolic action of the insula and OFC in reaction to increase-salt stimuli, switching salt intensity-dependent differences, according to a current multicentre, double-blind interventional and observational research study [14]. Thus, it was concluded that eating spicy meals can considerably lower daily salt intake, individual salt preferences and BP by altering the neural stimulation of taste of salt in the brain. The use of savory flavor has the potential to be an effective method for decreasing excessive consumption of salt and lowering blood pressure [10].

3 Enrichment of Spices and Herbs in Meat Products

Aromatic herbs and spices like basil, oregano, dill, parsley, coriander, rosemary, clove, sage, mint, bay, thyme, marjoram and tarragon are commonly used in meat dishes and meat substitute recipes [15]. Different tree parts, such as seeds, branches, and leaf, are utilized in various ways, such as powdered extracts, EOs, pulverized leaf, and water extracts [15]. They are utilized in varying proportions based on the product type to which they are mixed, and have staining, antioxidant and antimicrobial properties, as indicated in Table 1 [15].

A study found that adding clove extract to prepared ground beef reduced lipid degradation, improved storage durability, and improved flavor profile [17]. The hens supplemented with a clove induced diet enhances the microbiological and physical features of the carcass [18]. The usage of dehydrated clove powder in combination with curcumin rhizome has enhanced the moisture holding capacity and retention, tenderness, and sensory attribute of meat [19, 20]. Addition of clove extract to fresh beef patties in a ratio of 1:5 can reduce fat, protein degeneration, and hue angle while securing the quality parameter [21]. Dried cloves treated with 400 millilitres

Table 1 Effect of enrichment of spices and herbs in the meat products [15] (Creative Commons Attribution (CC BY) license)) [16]

Plant common name	Scientific name	Part of plant used	Form used	Quantity added	Meat product	Positive effects
Parsley	*Petroselinum crispum* Hoffm.	Stems	Extract	4.29 g/kg	Mortadella-type sausage	Inhibition of *Listeria monocytogenes* and microbial spoilage during storage time
Dill	*Anethum graveolens* L.	Seeds	A water-soluble polysaccharide named AGP1 was isolated from seeds of Anethum graveolens	0.3%	Turkey sausages	Preservative AGP1 replaced ascorbic acid; reduced lipid peroxidation; preserved pH and color; improved bacterial stability during cold storage at 4 °C for 12 days
Sage	*Salvia officinalis* L.	Leaves	Powder	0.05–0.15%	Chinese style sausage	Improve the oxidative stability of Chinese-style sausage
Rosemary	*Rosmarinus officinalis* L.	Leaves	Extract	0.4%	Beef sausages	Microbial inhibition; rosemary spice can substitute nitrite pickling salt
Rosemary	*Rosmarinus officinalis* L.	Leaves	Extract	0.025–0.05%	Fermented goat meat sausage	Oxidative stability
Clove	*Syzygium aromaticum*	Buds	Powder	0.1%	Ground beef	Decreased lipid oxidation; retarded volatile compound production; enhanced product oxidative stability
Clove	*Syzygium aromaticum*	Buds	Extract	NS	Beef patties	Reduced hue angle, TBARS value, carbonyl content, lipid peroxidation; enhanced redness value, the durability of sensory characteristics; prolonged shelf life up to 10 days, and also quality
Clove	*Syzygium aromaticum*	Buds	Extract	500 ppm	Raw chicken sausage	Reduced microbiological, chemical parameters, physical parameters, sensory attributes; increased antioxidant activity; inhibited lipid oxidation; prolonged shelf life up to 3–6 days

Adopted with License No. 5820200693543

of 95% pure consumable ethanol when applied to Chinese-style sausages declined fat and protein oxidation, improving color, texture, flavor, acceptability, and overall quality [22]. Buffalo meat steaks treated with a 0.5% concentration of black pepper extended the preservation time [23]. The application of curcumin and pepper EOs to chicken meat has been found to effectively minimize fat oxidation and protein degradation, limit microbial growth, and increase the stability and quality of the meat for up to 16 days [24]. When a mixture of 200 ml of acetone and 10 g of turmeric powder is added to chicken and fish meat, it changes the color from yellow to reddish brown, increases water absorption, and decreases water solubility [25]. The use of a powder made from sage and laurel leaves to cook meatballs has been found to decrease the peroxide value, improve sensory characteristics and acceptance, and prolong the shelf life of the product [26]. Minced beef meat infused with onion peel extract retarded amino acid degradation and fat oxidation and suppress the proliferation of microorganism [27]. Addition of fenugreek seed flour to rabbit sausage enhances the antioxidant capacity and reduce lipid oxidation [28].

In irradiation pork meat, the blend of tocopherol and rosemary oleoresin efficiently decreased volatile hexanal levels without affecting sulfur volatile generation [29]. Rosemary extracts improved color retention, as shown by lower metmyoglobin concentrations and higher oxymyoglobin levels after 8 days of storage in irradiation minced beef [30]. Sage extract blended with sodium isoascorbate, decreased pH and water aw, as well as the counts of microbes in raw vacuum-packaged turkey meatballs, but improved the quality of prepared meatballs [31]. Chicken meat which was processed using high pressure had delayed fat oxidation for over two weeks of cold storage [32]. Water based garlic extract prevented the survival of microbiological pollutants, on the surface of refrigerated poultry carcasses [33]. Adding 1 and 3 percent garlic extract to sausage during cold storage reduced peroxide value, residual nitrite, thiobarbituric acid reactive substances (TBARS), and total microbial percent compared to the control [34]. Green tea juice was considered good source for providing antioxidant activity and the concentrations of tyramine and putrescine in a dehydrated and fermented turkey sausage. However, infusing green tea had no direct influence on the color, pH, and overall accecptability of sausages [35]. Green tea powder might partially substitute for nitrite in pig sausages, resulting in reduced TBARS values and volatile basic nitrogen concentrations to those made with nitrite alone [36].

4 Essential Oils (EOs) and Their Enrichment in Meat Products

EOs are aromatic active compounds obtained from different parts of plant, including roots, leaves, flower's petals, branches, and seeds [37]. The worldwide need for EOs is estimated to be more than 200 kilotons in 2020, and the demand is predicted to grow at a 7.5% till 2027, due to consumption practice of natural food product

[38]. EO can be added to meat and their products, and is considered compatible and some EOs have been considered as "GRAS" (Generally Regarded as Safe) by FDA (Food and Drug Administration) [38]. EOs are added to conventional chemical preservatives because of their capacity to delay oxidative reactions, by eliminating free radicals, and terminating microbial growth [39].

Fasseas et al. [40] investigated on uncooked and cooked beef and pork meat treated with oregano and sage EOs and studied the shelf-life activity for 12-day shelf-life study period at 4 °C, the antioxidant estimation of the food sample was evaluated, both raw and cooked. Overall, when treated with EOs, both meat samples had significantly decreased 2,2-diphenyl-1-picrylhydrazyl (DPPH) radical scavenging effects and thiobarbituric acid (TBA) levels during refrigerated storage. TBA values at 532 nm absorbance were estimated to be 0.02 and 0.03 for pork meat treated with essential oils of oregano and sage, respectively, in contrast to 0.12 for the control. When oregano and sage essential oils was added to cooked pork meat the value was 0.02 and 0.05, respectively, compared to 0.36 for the control. Furthermore, raw beef meat treated with essential oils had 0.02 and 0.03 TBA values, while cooked samples had values of 0.03 and 0.05 (Table 2). 0.07 and 0.13 were the estimated TBA values for both raw and cooked phases respectively. In both raw and cooked meat samples, oregano essential oil was more effective than sage essential oil as an antioxidant ($p < 0.001$) [40]. Broiler breast chicken meat samples treated with oregano essential oil showed moderate amounts of malondialdehyde (MDA) when compared to samples with diets fortified with α-tocopheryl, which had the high and least values, respectively. Moreover, chicken meat supplemented at the same rate showed significantly ($p < 0.05$) greater efficacy in postponing lipid degradation during freezing storage temperature at −20 °C. In comparison to a 50 mg/kg feed supplementation level and one basal diet as control, which produced MDA values of 41 and 53 ng/g within 3 months, 48 and 60 ng/g within 6 months, and 70 and 83 ng/g within 9 months, respectively [41]. A research conducted by Estevez et al. [42] discovered that the enrichment of rosemary and sage essential oils at a concentration of 0.1% reduces lipid oxidation in liver pates over 90 days of refrigeration. This is primarily because these essential oils slow down the breakdown of poly-unsaturated fatty acids (PUFAs), which in turn prevents the development of residues like MDA and volatile compounds derived from lipids. Mohamed and Mansour [43] recently investigated the impact of adding rosemary & marjoram EOs (200 mg/kg) to beef patties. The conclusion suggested that the oils effectively decreased the degradation of fats and enhanced the sensory qualities of beef patties while being stored in a frozen state for a duration of 3 months. During three months of frozen preservation study, the TBARS value slightly lower ($p < 0.05$) than those of the standard patties without essential oil.

Furthermore, natural antioxidants were more effective than the synthetic antioxidant (BHT) based on their impact on sensory parameters, as well as their ability to reduce lipid degradation with reduced TBARS percent ($p < 0.05$). Incorporating black pepper oils extracted using supercritical CO_2 method (10 MPa and 60 °C) into minced pork meat led to a lessened elevation in lipid oxidation when stored for 8 days at 4 °C [44]. Dzudie et al. [45] prepared beef meat by combining basilica and

Table 2 Effect of enrichment of EOs in the meat products [15] (Creative Commons Attribution (CC BY) license))

Plant common name	Scientific name	Part of plant used for EOs	Quantity added	Meat product	Positive effects
Basil	*Ocimum basilicum* L.	Leaves	9 μl/ml	Fermented sausage	Reduction in mold growth; antifungal protection
Basil	*Ocimum basilicum* L.	Leaves	0.062–0.25%	Beef burger	Antioxidant; antibacterial activity
Oregan	*Origanum vulgare* L.	Leaves	0.230–0.690 mg/ml	Sausage	Bacteriostatic effect; antimicrobial activity
Oregan	*Origanum vulgare* L.	Leaves	6.25–100 μl/ml	Poultry meat products	Bio-preservative; antimicrobial activity against *Staphylococcus aureus*
Sage	*Salvia officinalis* L.	Leaves	0.2–0.5%	Chicken fat	Effective in respect of hydrolytic rancidity
Coriander	*Coriandrum sativum* L.	Leaves	0.01%	Italian salami	Reduction in lipid oxidation
Coriander	*Coriandrum sativum* L.	Leaves	0.02%	Stored ground beef	Inhibiting the development of unwanted sensory changes and the growth of Enterobacteriaceae
Coriander	*Coriandrum sativum* L.	Leaves	0.075–0.150 μg/ml	Cooked pork sausage	Improved oxidative stability
Marjoram	*Origanum majorana* L.	Leaves	6.25–100 μl/ml	Poultry meat products	Bio-preservative; antimicrobial activity against *Staphylococcus aureus*
Tarragon	*Artemisia dracunculus* L.	Leaves	0.062–0.25%	Beef burger	Natural preservative; flavor enhancer in meat; antioxidant activity; antibacterial effects against *Staphylococcus aureus*
Tarragon	*Artemisia dracunculus* L.	Leaves	0.1%	Frankfurter type sausages	Improving sensory properties
Bay	*Laurus nobilis* L.	Leaves	10%	Fresh lamb meat	Natural preservative; antibacterial activity
Bay	*Laurus nobilis* L.	Leaves	0.05–0.1 g/100 g	Fresh Tuscan sausage	Antibacterial activity; improve the safety and shelf life

(continued)

Table 2 (continued)

Plant common name	Scientific name	Part of plant used for EOs	Quantity added	Meat product	Positive effects
Thyme	*Thymbra capitata* (L.) Cav	Leaves	0.01–3%	Minced beef meat	Antibacterial against *Escherichia coli*, *salmonella typhimurium*, *Staphylococcus aureus*, and *Pseudomonas aeruginosa*
Mint	*Mentha piperita* L.	Leaves	20, 40, and 60 ppm	Cooked sausage	Nitrite partial replacement with *Mentha piperita* essential oil proved oxidative, microbial, and sensory properties

ginger essential oils with maize oil. Over a period of 18 days, the TBA percent of the meat patties containing essential oils were constant. Recent research by Fratianni et al. [46] examined the antioxidant properties of thyme and balm EOs on uncooked chicken breast meat that had been kept for 21 days at 4 °C. During their investigation, it was discovered that both essential oils decreased the generation of radicals in meat as compared to the untreated control. Among the two essential oils, thyme essential oil demonstrated the most effectiveness, with a DPPH radical inhibition percentage ranging from 25% to 30%. The balm oil exhibited a DPPH radical inhibition rate of 15–20%, while the control showed a lower activity of roughly 10%. These findings demonstrate that the use of EOs to fresh chicken breast meat resulted in little lipid oxidation, thus prolonging its shelf life.

5 Working Mechanism of EOs

5.1 Antimicrobial Activity

Meat provides ideal environment for development of microbes due to its nutritional composition, high water activity, and mild acidity. During meat processing, storage factors such as oxygen and temperature levels can alter the kind of microbes present [47]. The antimicrobial properties associated with the EOs of different plant species which differs based on the structure arrangement of the molecule. Terpenes, esters, alcohols, carbonyl compounds, and phenylpropanoids, are the primary metabolites synthesized by plants as essential oils [39]. The mechanism of action in microbial cells is not specific because of various chemical compounds in EO. But one of the most often documented mechanisms entails an elevation in the diffusibility of the cell membrane. The lipophilic characteristic of EOs enables easy diffusion and bind

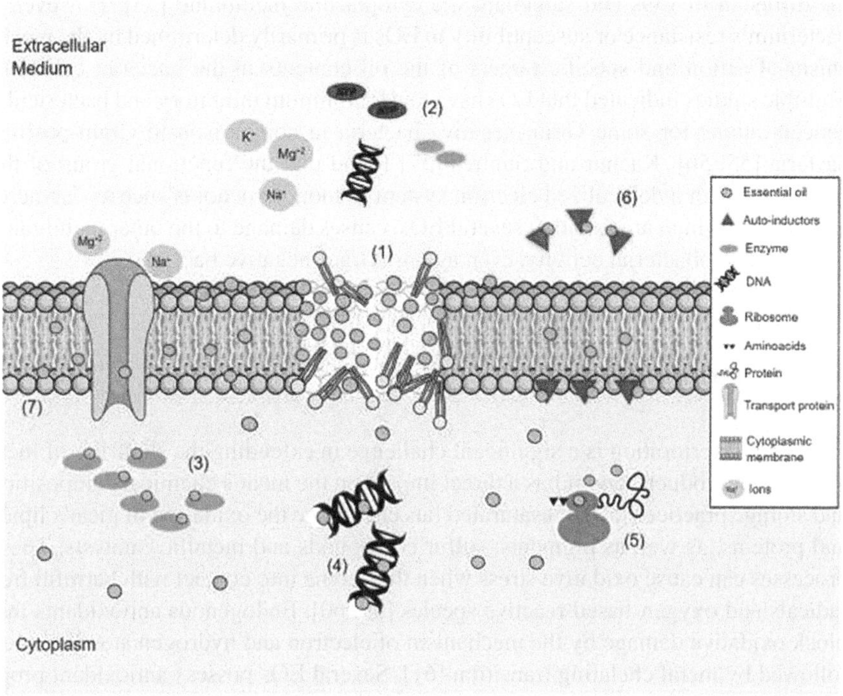

Fig. 1 Working mechanism of EOs on microbial cells. (1) Disruption of cytoplasmic membrane; (2) Release of intracellular components; (3) Impairment of enzymatic activity; (4) DNA degradation; (5) Effect on protein synthesis; (6) Interplay with quorum sensing mechanism; (7) Effect on ATP production [38] (License No. 5815340468100)

with targets in the cytoplasmic layer or within the cell. Enhanced permeability can lead to the outflow of cytoplasmic content and, as a result, cell death (Fig. 1) [48].

Kedia et al. [48] found that Aspergillus flavus showed an elevation in electrolyte loss (such as Na^+, K^+, and Ca^{2+}) after being treated with *Trachyspermum ammi* L. essential oils, and this increase was concentration dependent. Zhang et al. [49] found that cinnamon EO (*Cinnamomum zeylanycum*) increased the amounts of extracellular nucleic acids in viable *Escherichia coli* and *Staphylococcus aureus* cells when used at inhibitory concentrations. The EOs derived from clove (*Syzygium aromaticum*) raised extracellular adenosine triphosphate (ATP) levels and decreased the ATPase enzymatic activity at bactericidal concentrations [50]. An additional mechanism of bacterial DNA replication that has been suggested involves the inhibition of strand DNA transcription into mRNA and mRNA translation for protein synthesis [51]. Apart from the chemical composition of EOs, which is affected by growth of the plant and climatic condition, the microbial activity of the EOs can be influenced by other factors during application, such as oil concentration, microbial concentration, and cell structure [52]. The outer most membrane layer of Gram-negative bacteria consists of lipopolysaccharide, which serves as a barrier to prevent the diffusion of EOs and safeguard the cytoplasmic membrane [53]. However, a bacterium's resistance or susceptibility to EOs is primarily determined by the mechanism of action and specific targets of the oil contents in the bacterial cell [54]. Multiple studies indicated that EOs have lower minimum inhibitory and bactericidal concentrations for some Gram-negative bacteria in comparison to Gram-positive bacteria [55, 56]. Kachur and Suntres [57] found that the functional group of the compound with a delocalized electron system in monoterpenoids such as carvacrol and thymol, which are found in several EOs, causes damage to the outer membrane, increasing antibacterial activity, even against Gram-negative bacteria.

5.2 Antioxidant Activity

Oxidative deterioration is a significant challenge in extending the shelf-life of meat and their byproducts, and it has a direct impact on the meat's chemical composition and storage practices [58]. Unsaturated fats encourage the oxidation of meat's lipids and proteins, as well as pigments, sulfur compounds and metallic catalysts. These processes can cause oxidative stress when they come into contact with harmful free radicals and oxygen-based reactive species [59, 60]. Endogenous antioxidants that block oxidative damage by the mechanism of electron and hydrogen atom transfer, followed by metal chelating transition [61]. Several EOs possess antioxidant properties that are linked to active agents with one or more free secondary and hydroxyl compounds, that can inhibit the oxidative process through mechanisms of inhibition [39]. The EOs obtained from diverse plant types are abundant in phenolic terpenoids and oxygenated monoterpenes, such as menthol, terpenic hydrocarbons, linalool, α-terpineol, citronellal, carvacrol, and thymol and, like myrcene and α-pinene. These compounds possess antioxidant properties and can serve as a natural substitute for meat preservation [38]. Snoussi et al. [62] reported that the EOs from *Mentha spicata*, which are abundant in limonene and carvone, successfully eliminated free radical that was equal to that of BHT, a strong antioxidant that is used in foods. Pirbalouti et al. [63] found that *Ferulago angulata* EOs have comparable antioxidant potential to BHT, with a high concentration of α-pinene and cis-β-ocimenein their structural composition. Semeniuc et al. [64] investigated various EOs from the family of Apiaceae & Lamiaceae and found that thyme EOs had a higher antioxidant potential, which was due to their high phenolic concentration. The EOs of *Ruta chalepensis* obtained from three areas along with diverse chemical composition showed antioxidant property, according to Jaradat et al. [65].

6 Drawbacks of Using EOs in Meat and Meat Byproducts

Investigations based on *In vitro study* with EOs shows a wide range of bioactivity under controlled conditions. However, the effectiveness of EOs decreases when used in food systems, posing a difficulty for their utilization in shelf-life extension of food. The heterogeneous structure of the meat can give rise to particular areas with distinct physico-chemical qualities that can initiate the reaction of EOs with certain targets [66]. A major problem with using EOs in meat and meat byproducts is the dilution factor of the components in the EOs, which decreases their activity in microbial cells [67]. The fat content in raw meat typically averages at 2.5%, however this can vary depending on various components such as the position of the muscle, the species and gender of the animal, its age, genetics, and management [68] EO components can also react with amino acid of meat via hydrophobic or electrostatic interactions, specifically van der Waals forces, which can decrease the bioactivity of the proteins [69]. The interactions between the elements of EOs can also be influenced by the medium's pH. Food commodity with a pH of 5 or lower enhance the hydrophobicity of EOs and can decrease their solubility of meat and their products in the aqueous phase. Typically, the pH level of uncooked meat after slaughtering and stiffening of muscles is approximately 5.5, but in meat byproducts, the pH ranges from 6.0 to 6.5 [66, 70]. Furthermore, the nutritional components present in meat can promote the restoration of microorganisms that have experienced stress because of the antimicrobial properties of EOs [47]. In a study conducted by Shekarforoush et al. [71] Despite the strong antibacterial activity, it was discovered that oregano EOs' inhibited the growth of *Listeria monocytogenes* on chicken meat was only noticeable changes within 1 day of storage at 8 °C. Porter et al. [72] in the research work noted a decline in the growth of anti-Salmonella enterica of white mustard EOs in nutrient broth when added to chilled minced chicken meat. Corrêa et al. [73] conducted a study and revealed that applying cinnamon EOs to sliced ham did not effectively control the development of gram positive and gram-negative bacteria and increasing the concentration of EOs in meat and related byproducts can change sensory thresholds, which will therefore affect consumer acceptance [67]. Smaoui et al. [74] performed a research study in which the intense smell caused by the EOs volatilizing at increased concentrations of Mentha piperita EOs reduced the overall approval rate of grated meat. Silveira et al. [75] observed that bay leaf (*Laurus nobilis*) essential oil infused Tuscan sausage had high sensory acceptance, whereas sausage without EOs was more approved among the consumers. Araújo et al. [76] found that the acceptability of the odor attribute of sausages was reduced by 70% when garlic EOs (*Allium sativum*) were administered at concentrations of 62.6 μl/kg when compared to the sample without EOs. Jaspal et al. [77] also reported similar findings, indicating that the odor and flavor of chicken meat treated with lactic acid remained within approved levels throughout the full storing time. Although, the use of oregano EOs on chicken meat had an adverse impact on the sensory characteristics of meat because of its aromacity.

7 Conclusion

Meat and meat products, being perishable in nature, are very susceptible to lipid degradation, which resulted in safety, security and quality concerns. Spices, herbs, and their essential oils have been identified as the most effective alternative options for preservatives in meat and their products, particularly as free radical scavenging agents, due to specific health concerns, natural substances are substituted for synthetic food additives. Although, research studies on the impact of herbs and spices on meat products are scarce, but their usage is strongly adviced because of the health benefits and flavor enhancements they provide. The EOs derived from oregano, rosemary, thyme, sage, clove, ginger, black pepper, marjoram, basil, and cassia plants have displayed a notable antioxidant property in meat and their products. The primary reason for the antioxidative property of their essential oils is because of the presence of phenolic compounds, which include carvacrol, eugenol, and thymol.

References

1. Arihara K. Functional foods. In: Jensen WK, Devine C, Dikeman M, editors. Encyclopedia of meat sciences. Oxford: Elsevier; 2004. p. 492–9.
2. Zhang W, Xiao S, Samaraweera H, Lee EJ, Ahn DU. Improving functional value of meat products. Meat Sci. 2010;86:15–31.
3. Arihara K. Strategies for designing novel functional meat products. Meat Sci. 2006;74:219–29.
4. Menrad K. Market and marketing of functional food in Europe. J Food Eng. 2003;56:181–8.
5. Benkouider C. Functional foods: a global overview. Int Food Ingred. 2004;5:66–8.
6. Kanner J. Oxidative processes in meat and meat products: quality implications. Meat Sci. 1994;36:169–74.
7. McCarthy TL, Kerry JP, Kerry JF, Lynch PB, Buckley DJ. Assessment of the antioxidant potential of natural food and plant extracts in fresh and previously frozen pork patties. Meat Sci. 2001;57:177–84.
8. Dawidowicz AL, Wianowska D, Baraniak B. The antioxidant properties of alcoholic extracts from *Sambucus nigra* L. (antioxidant properties of extracts). LWT-food. Sci Technol. 2006;39:308–15.
9. Al-Jalay B, Blank G, Mcconnell B, Al-Khayat M. Antioxidant activity of selected spices used in fermented meat sausage. J Food Prot. 1987;50:25–7.
10. Jiang TA. Health benefits of culinary herbs and spices. J AOAC Int. 2019;102:395–411.
11. Lv J, Qi L, Yu C, Yang L, Guo Y, Chen Y, Bian Z, Sun D, Du J, Ge P, Tang Z. Consumption of spicy foods and total and cause specific mortality: population based cohort study. BMJ. 2015;351:h3942.
12. Chopan M, Littenberg B. The association of hot red chili pepper consumption and mortality: a large population-based cohort study. PLoS One. 2017;12:e0169876.
13. Chen YH, Zou XN, Zheng TZ, Zhou Q, Qiu H, Chen YL, He M, Du J, Lei HK, Zhao P. High spicy food intake and risk of cancer: A meta-analysis of case-control studies. Chin Med J. 2017;130:2241–50.
14. Li Q, Cui Y, Jin R, Lang H, Yu H, Sun F, He C, Ma T, Li Y, Zhou X, Liu D. Enjoyment of spicy flavor enhances central salty-taste perception and reduces salt intake and blood pressure. Hypertension. 2017;70:1291–9.

15. Marc RA, Mureşan V, Mureşan AE, Mureşan CC, Tanislav AE, Puşcaş A, Marţiş GS, Ungur RA. Spicy and aromatic plants for meat and meat analogues applications. Plan Theory. 2022;11:960.
16. Mandal D, Sarkar T, Chakraborty R. Critical review on nutritional, bioactive, and medicinal potential of spices and herbs and their application in food fortification and nanotechnology. Appl Biochem Biotechnol. 2023;195:1319–513.
17. Zahid MA, Eom JU, Parvin R, Seo JK, Yang HS. Changes in quality traits and oxidation stability of *Syzygium aromaticum* extract-added cooked ground beef during frozen storage. Antioxidants. 2022;11:534.
18. Suliman GM, Alowaimer AN, Al-Mufarrej SI, Hussein EO, Fazea EH, Naiel MA, Alhotan RA, Swelum AA. The effects of clove seed (*Syzygium aromaticum)* dietary administration on carcass characteristics, meat quality, and sensory attributes of broiler chickens. Poultry Sci. 2021;100(3):100904.
19. Sanwo KA, Adegoke AV, Akinola OS, Njoku CP, Okolo SO, Oladipo NA, Oladejo AS. Meat quality characteristics of improved indigenous chickens (funaab-alpha) fed turmeric (*Curcuma longa*) or clove (*Syzygium aromaticum*) as feed additives. J Agric Sci Environ. 2019;19:102–12.
20. Sanwo KA, Adegoke AV, Egbeyale LT, Abiona JA, Sobayo RA, Obajuluwa OV, Abdulazeez AO. Meat quality and lipid profile of broiler chickens fed diets containing turmeric (*Curcuma longa*) powder and cayenne pepper (*Capsicum frutescens*) powder as antioxidants. J Agric Sci Environ. 2019;19:73–91.
21. Zahid MA, Seo JK, Parvin R, Ko J, Yang HS. Comparison of butylated hydroxytoluene, ascorbic acid, and clove extract as antioxidants in fresh beef patties at refrigerated storage. Food Sci Anim Resour. 2019;39:768.
22. Zhang H, Peng X, Li X, Wu J, Guo X. The application of clove extract protects Chinese-style sausages against oxidation and quality deterioration. Korean J Food Sci Anim Resour. 2017;37:114.
23. Manjunath N, Nayar R, Akkara SS, Sathu T, Vasudevan VN, Rajkumar S. Effect of carrageenan edible film with oleoresins of *Piper nigrum* (black pepper) on quality of buffalo meat steaks. Pharm Innov. 2019;8:397–400.
24. Dalvandi F, Almasi H, Ghanbarzadeh B, Hosseini H, Karimian KN. Effect of vacuum packaging and edible coating containing black pepper seeds and turmeric extracts on shelf life extension of chicken breast fillets. J Food Bioproc Eng. 2020;3:69–78.
25. Erna KH, Felicia WX, Rovina K, Vonnie JM, Huda N. Development of curcumin/rice starch films for sensitive detection of hypoxanthine in chicken and fish meat. Carbohydr Polym Technol Appl. 2022;3:100189.
26. Akcan T, Estévez M, Serdaroğlu M. Antioxidant protection of cooked meatballs during frozen storage by whey protein edible films with phytochemicals from *Laurus nobilis* L. and *Salvia officinalis*. LWT-food. Sci Technol. 2017;77:323–31.
27. Mtibaa AC, Smaoui S, Ben Hlima H, Sellem I, Ennouri K, Mellouli L. Enterocin BacFL31 from a safety *Enterococcus faecium* FL31: natural preservative agent used alone and in combination with aqueous peel onion (*Allium cepa*) extract in ground beef meat storage. Biomed Res Int. 2019;2019:4094890.
28. Zaki E. Incorporation of fenugreek seed powder in the manufacturing of rabbit sausage and its effects on the quality properties during frozen storage. J Adv Food Sci Technol. 2018;5:8–14.
29. Nam KC, Ko KY, Min BR, Ismail H, Lee EJ, Cordray J, Ahn DU. Influence of rosemary–tocopherol/packaging combination on meat quality and the survival of pathogens in restructured irradiated pork loins. Meat Sci. 2006;74:380–7.
30. Formanek Z, Lynch A, Galvin K, Farkas J, Kerry JP. Combined effects of irradiation and the use of natural antioxidants on the shelf-life stability of overwrapped minced beef. Meat Sci. 2003;63:433–40.
31. Karpińska-Tymoszczyk M. Effects of sage extract (*Salvia officinalis* L.) and a mixture of sage extract and sodium isoascorbate on the quality and shelf life of vacuum-packed Turkey meatballs. J Muscle Foods. 2007;18:420–34.

32. Mariutti LR, Orlien V, Bragagnolo N, Skibsted LH. Effect of sage and garlic on lipid oxidation in high-pressure processed chicken meat. Eur Food Res Technol. 2008;227:337–44.
33. Oliveira KADM, Santos-Mendonça RC, Gomide LADM, Vanetti MC. Aqueous garlic extract and microbiological quality of refrigerated poultry meat. J Food Process Pres. 2005;29:98–108.
34. Park WY, Kim YJ. Effect of garlic and onion juice addition on the lipid oxidation, total plate counts and residual nitrite contents of emulsified sausage during cold storage. Food Sci Anim Resour. 2009;29:612–8.
35. Bozkurt H. Utilization of natural antioxidants: green tea extract and *Thymbra spicata* oil in Turkish dry-fermented sausage. Meat Sci. 2006;73:442–50.
36. Choi SH, Kwon HC, An DJ, Park JR, Oh DH. Nitrite contents and storage properties of sausage added with green tea powder. Korean J Food Sci Anim Resour. 2003;23:299–308.
37. Burt S. Essential oils: their antibacterial properties and potential applications in foods-a review. Int J Food Microbiol. 2004;94:223–53.
38. da Silva BD, do Rosário DK, Weitz DA, Conte-Junior CA. Essential oil nanoemulsions: properties, development, and application in meat and meat products. Trends Food Sci Technol. 2022;121:1–13.
39. Bhavaniramya S, Vishnupriya S, Al-Aboody MS, Vijayakumar R, Baskaran D. Role of essential oils in food safety: antimicrobial and antioxidant applications. Grain Oil Sci Technol. 2019;2:49–55.
40. Fasseas MK, Mountzouris KC, Tarantilis PA, Polissiou M, Zervas G. Antioxidant activity in meat treated with oregano and sage essential oils. Food Chem. 2008;106:1188–94.
41. Botsoglou NA, Fletouris DJ, Florou-Paneri P, Christaki E, Spais AB. Inhibition of lipid oxidation in long-term frozen stored chicken meat by dietary oregano essential oil and α-tocopheryl acetate supplementation. Food Res Int. 2003;36:207–13.
42. Estévez M, Ramírez R, Ventanas S, Cava R. Sage and rosemary essential oils versus BHT for the inhibition of lipid oxidative reactions in liver pâté. LWT-Food Sci Technol. 2007;40:58–65.
43. Mohamed HM, Mansour HA. Incorporating essential oils of marjoram and rosemary in the formulation of beef patties manufactured with mechanically deboned poultry meat to improve the lipid stability and sensory attributes. LWT-Food Sci Technol. 2012;45:79–87.
44. Tipsrisukond N, Fernando LN, Clarke AD. Antioxidant effects of essential oil and oleoresin of black pepper from supercritical carbon dioxide extractions in ground pork. J Agric Food Chem. 1998;46:4329–33.
45. Dzudie T, Kouebou CP, Essia-Ngang JJ, Mbofung CM. Lipid sources and essential oils effects on quality and stability of beef patties. J Food Eng. 2004;65:67–72.
46. Fratianni F, De Martino L, Melone A, De Feo V, Coppola R, Nazzaro F. Preservation of chicken breast meat treated with thyme and balm essential oils. J Food Sci. 2010;75:M528–35.
47. Doulgeraki AI, Ercolini D, Villani F, Nychas GJ. Spoilage microbiota associated to the storage of raw meat in different conditions. Int J Food Microbiol. 2012;157:130–41.
48. Kedia A, Prakash B, Mishra PK, Dwivedy AK, Dubey NK. *Trachyspermum ammi* L. essential oil as plant based preservative in food system. Ind crops. Prod. 2015;69:104–9.
49. Zhang Y, Liu X, Wang Y, Jiang P, Quek S. Antibacterial activity and mechanism of cinnamon essential oil against *Escherichia coli* and *Staphylococcus aureus*. Food Control. 2016;59:282–9.
50. Cui H, Zhang C, Li C, Lin L. Antimicrobial mechanism of clove oil on *listeria monocytogenes*. Food Control. 2018;94:140–6.
51. Xu JG, Liu T, Hu QP, Cao XM. Chemical composition, antibacterial properties and mechanism of action of essential oil from clove buds against *Staphylococcus aureus*. Molecules. 2016;21:1194.
52. Siddique S, Parveen Z, Mazhar S. Chemical composition, antibacterial and antioxidant activities of essential oils from leaves of three *melaleuca* species of Pakistani flora. Arab J Chem. 2020;13:67–74.
53. Nazzaro F, Fratianni F, De Martino L, Coppola R, De Feo V. Effect of essential oils on pathogenic bacteria. Pharma. 2013;6:1451–74.

54. Gyawali R, Ibrahim SA. Natural products as antimicrobial agents. Food Control. 2014;46:412–29.
55. Ferreira CD, Nunes IL. Oil nanoencapsulation: development, application, and incorporation into the food market. Nanoscale Res Let. 2019;14:1–3.
56. Gómez-Sequeda N, Cáceres M, Stashenko EE, Hidalgo W, Ortiz C. Antimicrobial and antibiofilm activities of essential oils against *Escherichia coli* O157: H7 and methicillin-resistant *Staphylococcus aureus* (MRSA). Antibiotics. 2020;9:730.
57. Kachur K, Suntres Z. The antibacterial properties of phenolic isomers, carvacrol and thymol. Crit Rev Food Sci Nutr. 2020;60:3042–53.
58. Canto AC, Costa-Lima BR, Suman SP, Monteiro ML, Viana FM, Salim AP, Nair MN, Silva TJ, Conte-Junior CA. Color attributes and oxidative stability of longissimus lumborum and psoas major muscles from Nellore bulls. Meat Sci. 2016;121:19–26.
59. Cunha LC, Monteiro ML, Costa-Lima BR, Guedes-Oliveira JM, Alves VH, Almeida AL, Tonon RV, Rosenthal A, Conte-Junior CA. Effect of microencapsulated extract of pitaya (*Hylocereus costaricensis*) peel on color, texture and oxidative stability of refrigerated ground pork patties submitted to high pressure processing. Innov Food Sci Emerg Technol. 2018;49:1136–45.
60. Falowo AB, Fayemi PO, Muchenje V. Natural antioxidants against lipid–protein oxidative deterioration in meat and meat products: A review. Food Res Int. 2014;64:171–81.
61. Granato D, Shahidi F, Wrolstad R, Kilmartin P, Melton LD, Hidalgo FJ, Miyashita K, van Camp J, Alasalvar C, Ismail AB, Elmore S. Antioxidant activity, total phenolics and flavonoids contents: should we ban in vitro screening methods? Food Chem. 2018;264:471–5.
62. Snoussi M, Noumi E, Trabelsi N, Flamini G, Papetti A, De Feo V. *Mentha spicata* essential oil: chemical composition, antioxidant and antibacterial activities against planktonic and biofilm cultures of *vibrio* spp. strains. Molecules. 2015;20:14402–24.
63. Pirbalouti AG, Izadi A, Malek Poor F, Hamedi B. Chemical composition, antioxidant and antibacterial activities of essential oils from *Ferulago angulata*. Pharm Biol. 2016;54:2515–20.
64. Semeniuc CA, Socaciu MI, Socaci SA, Mureşan V, Fogarasi M, Rotar AM. Chemometric comparison and classification of some essential oils extracted from plants belonging to *Apiaceae* and *Lamiaceae* families based on their chemical composition and biological activities. Molecules. 2018;23:2261.
65. Jaradat N, Adwan L, K'aibni S, Zaid AN, Shtaya MJ, Shraim N, Assali M. Variability of chemical compositions and antimicrobial and antioxidant activities of *Ruta chalepensis* leaf essential oils from three Palestinian regions. Biomed Res Int. 2017;2017:2672689.
66. Gutierrez J, Barry-Ryan C, Bourke P. The antimicrobial efficacy of plant essential oil combinations and interactions with food ingredients. Int J Food Microbiol. 2008;124:91–7.
67. Hyldgaard M, Mygind T, Meyer RL. Essential oils in food preservation: mode of action, synergies, and interactions with food matrix components. Front Microbio. 2012;3:12.
68. López-Bote C. Chemical and biochemical constitution of muscle. In: Lawrie's meat science. Woodhead Publishing; 2017. p. 99–158.
69. Weiss J, Loeffler M, Terjung N. The antimicrobial paradox: why preservatives lose activity in foods. Cur Opin Food Sci. 2015;4:69–75.
70. Juven BJ, Kanner J, Schved F, Weisslowicz H. Factors that interact with the antibacterial action of thyme essential oil and its active constituents. J Appl Bacteriol. 1994;76:626–31.
71. Shekarforoush SS, Basiri S, Ebrahimnejad H, Hosseinzadeh S. Effect of chitosan on spoilage bacteria, *Escherichia coli* and *listeria monocytogenes* in cured chicken meat. Int J Biol Macromol. 2015;76:303–9.
72. Porter JA, Morey A, Monu EA. Antimicrobial efficacy of white mustard essential oil and carvacrol against *Salmonella* in refrigerated ground chicken. Poult Sci. 2020;99:5091–5.
73. Corrêa JA, dos Santos JV, Evangelista AG, Pinto AC, de Macedo RE, Luciano FB. Combined application of phenolic acids and essential oil components against *Salmonella enteritidis* and *Listeria monocytogenes* in vitro and in ready-to-eat cooked ham. LWT-Food Sci Technol. 2021;149:111881.

74. Smaoui S, Hsouna AB, Lahmar A, Ennouri K, Mtibaa-Chakchouk A, Sellem I, Najah S, Bouaziz M, Mellouli L. Bio-preservative effect of the essential oil of the endemic *Mentha piperita* used alone and in combination with BacTN635 in stored minced beef meat. Meat Sci. 2016;117:196–204.
75. Silveira SM, Luciano FB, Fronza N, Cunha A Jr, Scheuermann GN, Vieira CR. Chemical composition and antibacterial activity of *Laurus nobilis* essential oil towards foodborne pathogens and its application in fresh Tuscan sausage stored at 7 C. LWT-Food Sci Technol. 2014;59:86–93.
76. Araújo MK, Gumiela AM, Bordin K, Luciano FB, de Macedo RE. Combination of garlic essential oil, allyl isothiocyanate, and nisin Z as bio-preservatives in fresh sausage. Meat Sci. 2018;143:177–83.
77. Jaspal MH, Ijaz M, ul Haq HA, Yar MK, Asghar B, Manzoor A, Badar IH, Ullah S, Islam MS, Hussain J. Effect of oregano essential oil or lactic acid treatments combined with air and modified atmosphere packaging on the quality and storage properties of chicken breast meat. LWT-food. Sci Technol. 2021;146:111459.

Meat Products with Dietary Fibers

Duyen-Anh Le, Lien Phuong Le Nguyen, and Sao-Mai Dam

1 Introduction

Nowadays, healthy diets and related recommendations are gaining attention. Consumer awareness and interest in the nutritional value and health-promoting effects of food and what they eat daily have increased. The foods must satisfy certain criteria, including food safety and health promotion, to be able to survive in the market. Protein is a crucial macronutrient that plays an important role in facilitating the growth and repair of cells. In addition, protein also helps develop hormones and enzymes, according to the WHO [1]. Meat and meat products are nutritious since they contain all essential amino acids, vitamins, and minerals, while their biological value is high, comparable to egg proteins [2, 3]. The body is able to easily absorb and use those crucial nutrients. In general, the consumption of meat and meat products made of poultry, red meat, and other types of meat is very different by location (regions and countries) around the world. The average consumption of meat, comprising poultry and other types of meat, varies from 4 kilograms/capita/year in some sub-Saharan countries to 100 kilograms/capita/year in the United States [1]. Recently, there has been an emerging trend towards a healthier diet due to the increasing number of nutrition-related illnesses. Many people try to consume diverse foods, including plant-based and animal-sourced foods.

Each product has unique flavor, appearance, and texture characteristics. Consumers have their preferences regarding meat sources and product characteristics, which should be considered carefully while developing technology and

D.-A. Le · S.-M. Dam (✉)
Industrial University of Ho Chi Minh City, Ho Chi Minh City, Vietnam
e-mail: damsaomai@iuh.edu.vn

L. P. Le Nguyen
Institute of Food Science and Technology, Hungarian University of Agriculture and Life Sciences, Budapest, Hungary

A. Kumar Pathera et al. (eds.), *Healthier Meat Products*,
https://doi.org/10.1007/978-3-031-78215-2_8

Table 1 Advantages and health-promoting effects of dietary fibers

Health benefit	Functional property
– Food for probiotic microorganisms. – Reducing the risk of colon cancer. – Accelerating passage in the colon. – Reducing the risk of cardiovascular illness. – Lowering the cholesterol level. – Lowering blood glucose concentration. – Reducing the risk of obesity and related health problems	– Increasing water holding capacity. – Increasing oil holding capacity. – 1Increasing water absorption. – Increasing mineral absorption. – Emulsifier. – Improving texture (hardness, springiness, chewiness, cohesiveness, adhesiveness). – Decreasing cooking loss

recipes. If product change does not meet consumer expectations, the product will fail in the market. The change is often detected by the customers and considered a quality loss, resulting in rejected food. Although red meat and processed red meat have the advantages mentioned above, shifting to plant-based food or enrichment with plant sources is desired to fight against diseases like cancer and cardiovascular illness [1]. Current healthy diet recommendations suggest decreased red and processed meat consumption, with 98–500 g per week for adults. Dietary fibers can offer a transition from traditional red meat consumption, and dietary fibers are already recognized as functional foods or their ingredients [4]. The dietary fibers' health benefits and functional properties are listed below in Table 1.

Comminuted meat products are cut, ground, minced, and shredded. Additional ingredients mixed to form the product may include dietary fibers [3]. These food products can be cooked or raw, such as Bologna-type sausages, fermented sausages, frankfurters, Lyoner-type sausages, meatballs, nuggets, patties, salami, etc. Dietary fibers extracted from plants or byproducts of plant processing (fruit pomace, fruit, vegetable peel, and husk) contain bioactive compounds that help prevent fat oxidation in meat products [4]. Additionally, dietary fibers improve product quality in terms of cooking yield, emulsion stability, juiciness, and shrinkage [5, 6]. In the past few decades, many works have reported the application of dietary fibers in meat products such as banana peel powder [7], bamboo shoot [8], kiwi fruit pomace [9], *Aloe vera* [10], oat fiber [11], rice bran [12], carrot pomace powder [13], pineapple pulp [14], seed pumpkin flour [15], peach fibers [16], citrus byproducts including lemon albedo and orange fiber powder [17].

According to the national food-based guidelines in several countries, balanced diets are highly recommended by the FAO [18]. Studies indicated the vital contribution of the diversity of foods in a healthy diet and the synergy of different kinds of foods [1]. Various components of foods from different sources could complement each other and facilitate the utilization of micronutrients. For example, iron in red meat is readily absorbed from heme iron, which is approximately 1.5–2 times more bioavailable than iron from plant sources. Additionally, the vitamin C consumed with plant material facilitates the utilization of iron. To improve the nutritional value of processed meats, attempts were made to develop new products rich in dietary fibers derived from different sources. In this context, meat products enriched with dietary fibers can satisfy the dietary recommendations and the high demand of the food market nowadays.

2 Dietary Fibers

2.1 Definition of Dietary Fibers

Plant-based sources, particularly dietary fibers, are one of the most important foods for human health. Different approaches define dietary fiber according to its behavior in extraction, chemical composition, and origin. Eben H. Hipsley [19] provided the first definition of dietary fiber by describing the inaccessible carbohydrates in food. The Food and Agriculture Organization of the United Nations (FAO) and the World Health Organization (WHO) similarly define carbohydrates as unavailable after simulated human digestion [18]. The four-hour-long simulated digestion is the recommended measurement method in standard ICC No. 185 [20]. The U.S. Food and Drug Administration (FDA) introduced the term dietary fiber in 2016 for non-digestible carbohydrate food compounds. The nutrition and supplement facts labels can include dietary fibers if they are naturally occurring fibers of plants with confirmed beneficial health effects [21]. Such positive effects can include lowering blood pressure, glucose, and cholesterol concentration, reducing energy intake, facilitating mineral absorption, etc. According to the FDA legislation, dietary fiber listed on label is "non-digestible soluble and insoluble carbohydrates (with three or more monomeric units) and lignin that are intrinsic and intact in plants; isolated or synthetic non-digestible carbohydrates (with three or more monomeric units) determined by the FDA to have physiological effects that are beneficial to human health" [22]. Up to date, seven non-digestible carbohydrates have been approved:

- b-glucan-soluble fiber
- psyllium
- cellulose
- guar gum
- pectin
- locust bean gum
- hydroxypropylmethylcellulose

Additional eight ingredients are pending to receive the same classification presently because of the ongoing scientific research and development of food technology:

- mixed plant cell wall fibers
- arabinoxylan
- alginate
- inulin and inulin-type fructans
- high-amylose starch (resistant starch 2)
- galactooligosaccharide
- polydextrose
- resistant maltodextrin/dextrin

In the above list of candidates, the group of mixed plant cell wall fibers covers numerous ingredients, of which 26 have passed the FDA test [22].

The European Union (EU) registered all food additives and regulated their usage as EC No. 1333/2008 [23]. Dietary fibers are mainly registered in the group of "emulsifiers, stabilizers, thickeners, and gelling agents." The most common dietary fibers as approved food additives in the EU, with name and E-number according to the Food Standard Agency [24]:

- Agar E406
- Alginates E401-E405
- Carrageenan E407
- Plant-based gums E410, E412, E414-E418, E427
- Konjac E425
- Pectins E440
- Cellulose E460
- Methylcellulose E461
- Hydroxypropyl methylcellulose E464
- Carboxymethyl cellulose E466
- Polydextrose E1200

The EU regulation classifies additives based on their functional properties and does not provide distinguished groups for dietary fibers.

Considering the functionality of dietary fiber, two major features can be highlighted to make a classification (Table 2). The most accepted clustering is built using solubility and fermentation behavior. According to the chemical properties of soluble dietary fibers, carbohydrates are fermentable, while insoluble materials are poorly or non-fermentable. Regarding the physiological effect on human digestion, viscosity is the second most important factor in terms of flow behavior affecting bowel movement.

Some animal-origin fibers are also used as dietary fiber, although they are produced by fungi or yeast: chitin, chitosan, chondroitin, and collagen. They are natural compounds of animal-based foods consumed as supplements but not considered food additives [24].

Table 2 Classification of dietary fibers with examples

	Fermentable	Non-fermentable
Viscous	Glucan, glucomannan, gum, pectin, psyllium	n.a.
Non-viscous	Dextrins, fructo oligosaccharides, inulins, oligosaccharides, polydextrose, starch	Cellulose, hemicellulose, lignin

n.a. No data available for that class

2.2 *Application of Dietary Fiber in Meat Products*

Besides the beneficial health effects, dietary fibers also change the food matrix and, therefore, affect the physical, chemical, and sensory properties of meat products. The addition of fibers automatically decreases the concentration of other constituents, for example, fat. Decreased fat content makes healthier products since a lower amount of fat in daily nutrition is generally desired [1]. On the other hand, fat-soluble critical materials, such as cholesterol, will also decrease.

The differences between meat and fibers are apparent in organoleptic characteristics, particularly in texture, which is characterized by hardness, chewiness, gumminess, cohesiveness, and adhesiveness. Material tester instruments with simulated consumption typically assess the latter four parameters. Eating with teeth, cutting, and chewing are simulated with multiple load cycles and relaxation. This method is called Texture Profile Analysis (TPA) and is often used to describe foods [4].

Quality attributes and technological parameters may change with the addition of fibers. Moisture content can decrease. Since water activity (aw) is typically high in meat products, that is the perfect environment for microorganisms. The lower moisture content helps maintain the product's safety. The modified water binding capacity and holding capacity result in altered texture properties and improved cooking loss. While the former is more important in production, the latter is directly experienced by consumers [3].

2.2.1 Advantages of Dietary Fiber Addition in Meat Products

The application of dietary fibers in meat products has potential benefits. The most important one is to increase the nutrient values to prevent the lack of fiber in the human diet. In some cases, dietary fiber is also used as a fat replacement in meat products to promote health [9, 25].

In addition, bioactive compounds in functional dietary fibers, for example, citrus byproducts with antioxidant capacity, could improve the quality and shelf life of meat products [3, 10, 17].

2.2.1.1 Physicochemical Properties of Meat Products Enriched with Dietary Fibers

Adding dietary fibers to meat products can effectively enhance their acceptability to consumers due to their higher quality. The addition of lemon albedo and orange fiber powder at different concentrations to meat products (cooked sausages and dry-cured sausages) obtained promising results in chemical composition and sensory evaluation [17]. The chemical composition tests showed that the residual nitrite level was lower in sausages that had citrus byproducts added (lemon albedo) compared to sausages that did not have citrus byproducts [17]. A decline in the residual nitrite concentration in Bologna-type sausages with lemon albedo obtained health benefits because of the decrease in the probability of the formation of nitrosamines [17].

2.2.1.2 Stability of Meat Products Enriched with Dietary Fibers

Adding antioxidant-rich plant-based sources to meat products could also stop or slow down lipid oxidation and microbial growth, which would make the products last longer and be more stable while they're stored [3, 26]. Many cereals, fruits, and vegetables contain antioxidant dietary fiber due to the presence of polyphenols.

Furthermore, incorporating dietary fiber in meat products could improve the processing characteristics and technology parameters and retain the yield of meat products [5]. During processing, dietary fibers combine with water and fat to form a stable emulsion that retains over the processing and storage time, enhancing the cooking yield, water holding capacity, and juices [3].

2.2.2 Disadvantages of Dietary Fiber Addition in Meat Products

Many plant-based materials from different sources can be used as dietary fibers in various meat products [15]. In some cases, there was no significant difference among evaluated concentrations of dietary fiber in sensory characteristics in terms of appearance, color, taste, odour, texture, and overall acceptability of meat products [15]. However, in other cases, only a certain quantity of dietary fibers can be added to meat products since excess can alter the physicochemical and sensory attributes compared to the control samples. Thus, these changes may affect consumers' perceptions [7, 10]. For example, there was no alteration in sensory profile, including taste, juiciness, color, and acceptance, between the chicken sausage enriched with 2% banana peel powder and the control sample [7].

Nevertheless, those authors also found that adding more than 2% (2–6%) of banana peel powder significantly decreased the aroma and appearance of chicken sausage [7]. Similarly, another report indicated that a beef burger containing 1% *Aloe vera* obtained an acceptable score from the panelists compared to the control group. Still, a sour taste was detected in the beef burger containing 3% *Aloe vera*, and the addition of 5% *Aloe vera* to the meat burger developed a bitter taste [10]. Likewise, adding dietary fiber such as guava powder (at 0.5% and 1.0%) in sheep meat nuggets significantly decreased the stability of the meat emulsion and cooking yield compared to control samples [27]. It could probably be due to the low pH of the emulsion affecting the formation and stability of meat emulsion. Therefore, dietary fiber and its applied concentration should be carefully selected to achieve optimal improvement.

3 Meat Products with Dietary Fiber

Meat products enriched with dietary fiber are becoming more and more popular. Dietary fiber application in meat products is easily found in meatballs, sausages, surimi, meat batter, and many others [5].

3.1 Pork Products with Dietary Fiber

3.1.1 Comminuted Pork Product

Among pork products, sausages, meatballs, and patties are popular and consumed in huge quantities. Since pork meat typically contains 20–30% fat, decreasing the fat in the final product is an active research topic. Han and Bertram [28] compared inulin, carboxymethyl cellulose, cellulose, and chitosan dietary fibers in minced pork products. All dietary fibers were mixed with 2% quantity, substituting ice water (from 8.5% to 6.5%). The water binding capacity of mixtures was measured with cooking loss and expressible water. The texture was assessed with TPA analysis, and parameters of hardness, springiness, cohesiveness, and chewiness were computed from the acquired data. The water-binding capacity was significantly improved by the addition of dietary fiber. The cooking loss decreased from approximately 19% (control) to 4.2% with carboxymethyl cellulose. The amount of expressible water showed a similar result, where carboxymethyl cellulose obtained the best value.

Interestingly, inulin achieved results similar to those of the control group. The TPA test highlighted that dietary fibers affect texture differently. While inulin was identical to the control in hardness, springiness, chewiness, cellulose, and chitosan increased hardness and chewiness, while carboxymethyl cellulose and pectin decreased both. Springiness decreased from 9.5 to 8.0–8.4 mm because of added dietary fiber, but the standard deviation revealed large variations, resulting in no significant difference among the sample groups. Nuclear magnetic resonance relaxometry was performed on heated samples to address the stability of the prepared samples. Han and Bertram [28] concluded that carboxymethyl cellulose showed the best performance in the protection of proteins against heat-induced denaturation. The good results of carboxymethyl cellulose in water binding were attributed to its anionic, water-soluble polymer nature.

3.1.2 Meatballs

Fruit pomace is typically available as a byproduct of juice production. It has the potential to be used as a dietary fiber. Zhao et al. [9] added kiwifruit pomace to low-fat pork meatballs. The experimental procedure for dietary fiber extraction was complicated and lengthy. Fruit pomace was diluted to a 1:10 (w/v) ratio and treated with amylase enzyme for 80 min, followed by protease digestion for 70 min, centrifuge separation, and enzyme inactivation. The prepared material was oven-dried and ground in a planetary ball mill for 12 h. The low-fat meatballs were ready with a 10% reduction in fat content from 30% of the control. Dietary fiber concentration was tested in the range of 0–7%. Besides fat, the recipe consisted of 50% pork meat, 5% eggs, and 15% ice water.

After the batter was measured, additional flavoring substances, such as 2.4% white pepper, 2% salt, and 0.5% sugar, were added. No further additives were used to improve texture or adjust functional properties. This simple recipe is suitable for evaluating the effect of added dietary fiber without potential interactions. The comparison of dietary fiber material before and after grinding revealed that particle size decreased significantly and the surface area increased to approximately four times the value. As a result, small particles with large surface areas are supposed to improve absorption and related attributes. The technological properties, such as water holding capacity, swelling, oil, and fat binding capacity, increased significantly. A slight increase without significant difference was achieved for heavy metal absorption (Lead and Cadmium), and similar behavior was observed for cholesterol and sodium cholate. The authors expected good results for sodium nitrite but figured out that the food pH was higher than the optimal value of nitrite absorption. The most significant difference was achieved in inhibiting pancreatic lipase activity for oil. This supports human health by reducing the blood cholesterol concentration. After the ground kiwifruit fiber was added, several properties of the meatballs changed proportionally with dietary fiber concentration: moisture content, fat content, protein content, and cooking yield.

The emulsion's stability and water-holding capacity were improved with 0.5% added dietary fiber. Although the instrumentally measured color changed with dietary fiber concentration, sensory analysis could not detect an alteration in appearance until 5%. Samples with decreased fat content were evaluated with love scores for flavor, but the addition of kiwifruit fiber could increase this score. Meatballs with 3% kiwifruit dietary fiber and 20% fat achieved the closest taste and texture to control with 30% fat and without fiber. The study reported by Zhao et al. [9] highlighted the potential of fruit pomace in the application of dietary fiber. This material is available in large quantities as a byproduct of the beverage industry. The only challenge is the upscaling of the sophisticated process of dietary fiber production from kiwifruit.

3.1.3 Pork Meat Batters

Bamboo is an ideal plant source of dietary fiber because of its fast-growing and high cellulose and hemicellulose content. Li et al. [8] used bamboo shoot dietary fiber in pork meat batter. The recipe was made with 65% pork leg meat, 32.5% ice water, and 2.5% salt. The added dietary fiber replaced ice water from 1% to 4%. The dietary fiber was added in the very last step of mixing.

Interestingly, the batter was prepared in three steps, and each ingredient was mixed with 1/3 of a portion of ice water. After removing the air bubbles from the batter, samples were cooked at 80 °C for 20 min. Uncooked samples were subjected to 0.1 Hz shearing in oscillatory mode by a rotary rheometer while the temperature increased from 25 to 80 °C using 2 °C/min heating. The rheology measurement revealed that the initial elastic gelation network was formed at 47 °C, the protein network broke at 55 °C, and the stable gel network was finally made at 80 °C. The evaluation of the timely behavior showed that adding dietary fiber increased the gel elasticity and viscosity. Scanning electron micrographs were analyzed and, together with nuclear magnetic resonance data, supported the assumptions regarding the positive effect of bamboo fiber on the gel structure. The texture profile analysis confirmed that hardness and gel strength monotonously increased with dietary fiber concentration. Springiness decreased continuously with increased bamboo fiber concentration. No significant difference was found in cohesiveness. Chewiness did not show a clear tendency but reached the maximum value at 3% bamboo dietary fiber. Li et al. [8] concluded that dietary fiber content affected technological parameters, including water holding capacity and water retention. Bamboo shoot dietary fiber improved the properties of meat butter and proved to be a good candidate for meat product enrichment.

3.1.4 Pork Meat Emulsion

Bamboo fiber was compared with other potential dietary fiber materials to form emulsion gels so that fat content could decrease in the product by dos Santos et al. [29]. Pork skin and canola oil were used to make an essential gel that was enriched with dietary fiber. Different fiber sources of bamboo fiber, inulin, polydextrose, and a-cyclodextrin were combined. According to the recipe, 65% pork meat or chicken meat was used with 20% pork back fat, and 12.75% ice, and the remaining quantity was salt and emulsifier (E452). In this case, it is very important to highlight that chicken meat products also contain pork. People with special diet rules due to their religion cannot consume chicken meat and may not expect to have pork ingredients inside. This product's labeling must be transparent regarding this, or it may be misleading. The pork skin collagen has good gelling and emulsifying characteristics, which were additionally improved with dietary fibers. The emulsions were prepared with 15–25% pork skin gel and 0–3% dietary fiber. During the experiment, emulsions replaced pork fat in the product. Meat products were cooked at 75 °C for 40 min.

The best emulsion stability of the prepared gel mixtures was achieved with bamboo and a-cyclodextrin. On the other hand, they also significantly affected the gel brightness and color (CIE L*, a*, and b*). After these gels were comminuted with meat, the best stability was achieved with bamboo dietary fiber. The scanning electron microscopic images revealed that bamboo had a thickening effect on the gel network, but it became more diffused at the same time. Inulin also had a positive impact, as it made more connections inside the network and filled the gaps. In their study, dos Santos et al. [29] concluded that many parameters of the products made with pork skin gel and dietary fiber were similar to those made with pork back fat. Therefore, the proposed technique can be used to decrease the fat content of the meat. The report lacks sensory analysis and color measurement of the meat product. This might be critical since discoloration was observed on gels but not tested on the final product. From the consumers' point of view, taste, flavor, chewiness, mouth fill, and visual attributes are missing to be able to decide on acceptance.

3.1.5 Pork Salami

Besides the effect of dietary fiber on technological, physical, and chemical parameters, its contribution to fermented meat products is another question. Bis-Souza et al. [30] evaluated inulin, fructooligosaccharides, and a-cyclodextrin dietary fibers in Italian-type salami products. The fat content did not change during the experiment; it was maintained at 10%. Dietary fibers were added up to 2% concentration individually, in pairs, and all together. The fermentation was performed with a 0.025% starter culture of *Pediococcus pentosaceus* and *Staphylococcus xylosus*. Ripening took 13 days at 15–25 °C and 75–95% relative humidity. Those authors focused on the product pH and water activity as technological parameters. Because these are the key factors supporting salami's safety and microbial stability. All salami produced met the food safety requirements and were negative for *Salmonella* and the opportunistic pathogen *Staphylococci*.

No effect was observed on pH or water activity. The final product after ripening showed differences among the dietary fibers or their combinations in terms of moisture and fat content. The weight loss was similar without significant differences. Lipid oxidation was tested, and only the individual application of a single dietary fiber source with the concentration of 2% affected results. It was observed that applying a-cyclodextrin increased the lightness and changed the product's color according to CIE a* and b*.

Additionally, it increased the hardness, which needed to be appreciated in sensory analysis. According to the sensory evaluation performed by untrained panelists, products with 2% inulin and 2% fructooligosaccharides obtained the highest scores in appearance, color, texture, and overall acceptance. The addition of a-cyclodextrin is acceptable in low concentrations if it is used in combination with all others. Although the low-scored texture correlated with hardness, the texture profile analysis found significant differences only for cohesiveness and chewiness. Based on the instrumental assessment of quality parameters and sensory analysis, Bis-Souza et al. [30] concluded that dietary fibers like fermented meat products can be used in salami. Fermentation was not affected by dietary fibers. Unfortunately, the study did not investigate whether metabolic products appeared because of microbial activity and dietary fibers. Control samples were also missing from the experiment. Therefore, dietary fibers were only compared to each other. It would be nice to see which recipe provided the most similar products to the traditional Italian type of salami.

3.2 *Beef Products with Dietary Fiber*

3.2.1 Emulsified and Cooked Meat

Decreasing the fat content is one of the main goals of applying dietary fibers, especially for products including pork meat. A mixed beef and pork meat product was tested by Schmiele et al. [31]. The authors prepared the sample product with 27.82% beef, 27.82% pork meat, 20% pork fat, 19.8% ice, 2% NaCl, 2% maltodextrin, 0.5% potassium tripolyphosphate (E452) as an emulsifier and stabilizer, 0.045% sodium erythorbate (E316) as an antioxidant and 0.015% sodium nitrite (E250) as preservative. The pork fat was replaced with different amounts of amorphous cellulose fiber up to 1.5% quantity. Ingredients were mixed in a cooled cutter at 7 °C, and 50% of the ice with the cellulose fiber was added at the very end. The meat butter was cooked at 180 °C until the cold point reached 72–75 °C. The experimental design was prepared with five levels of pork fat and amorphous cellulose fiber, and results were evaluated with response surface methodology. The beef and pork raw meat, the meat butter, and the cooked product were measured. Both beef and pork meat had a low fat content of below 6%; therefore, they contributed less to the total fat of the final product. Surprisingly, adding various amounts of amorphous cellulose fiber did not affect the cooking loss, but the pork fat content did. However, approximately 15% pork fat with added amorphous cellulose fiber can result in cooking loss similar to the control products. This value was an estimation based on the statistical analysis because the adjusted level of the experiment was 17.1% fat content. The analysis of the cooked product texture revealed that half the amount of pork fat (10%) with 1.3% added fiber achieved similar firmness and hardness to the control product. The same applied to the weight loss of the cooked product within 14 days of cold storage at 5 °C. The color also changed during cold storage regarding CIE L* and b* parameters. According to the color change, products with half a portion of pork fat and 1.07% cellulose fiber can appear similar to the control after 14 days of cold storage. This value is estimated from the response surface model since the adjusted levels of amorphous cellulose fiber were 0.75% and 1.3% around this optimum. The summary of the observations suggests that decreased fat content combined with added cellulose fiber can provide a product with similar characteristics to the known standard regarding firmness, hardness, cooking loss, color, and weight loss in storage. This research offers healthier production of beef and pork meat-based foods with half the amount of fat replaced by approximately 1.3% dietary fiber. On the other hand, consumers will experience the well-known ordinary features.

3.2.2 Beef Hamburger

Essa and Elsebaie [32] used date seed dietary fiber and gelatin to decrease the fat content of hamburger meat. Date seed is commonly treated as a waste from dried fruit production. Its utilization decreases the waste deposited into the environment, and date seeds are available in large amounts. Dietary fiber production from seeds is proposed in a few steps, starting with the drying of clean seeds at 50 °C for 2 days, followed by grinding and defatting with 20% alkaline hydrogen peroxide, filtering, and washing the suspension, which can finally be dried at 55 °C for 12 h. The dried powder shall be processed with amylase, protease, and amyloglucosidase enzymes in consecutive steps to get the dietary fiber. In the presented experiment, the gel was produced with a 30:2 ratio of gelatin and date seed dietary fiber. Hamburger meat patties were produced from beef with 20% back fat as a control, 10% back fat as a low-fat control, and 10% fat content with gelatin and dietary fiber gel. The burgers were grilled at 220 °C for 4 min on each side. The analysis revealed that burgers with dietary fiber had significantly higher moisture content (approximately 10%), while energy (Kcal/100 g) decreased by almost half. Control burgers had a similar color to those with gelatin and date seed fiber in terms of CIE a* and b*. The sensory analysis with 20 trained panelists found significant differences between control samples with 10% and 20% fat already. Burgers with dietary fiber received lower scores, but the difference was insignificant. The dietary fiber-enriched hamburgers were similar to the traditional ones with 20% fat in taste, color, odour, tenderness, and overall acceptability. Only the parameter texture obtained similar values to the low-fat (10%) control. Although panelists were able to detect differences and give more miniature scores to hamburgers made with date seed dietary fiber and gelatin, sensory attributes were in the acceptable range above 8.1 except texture (7.89 average value). The study by Essa and Elsebaie [32] showed that fat replacement in beef hamburgers is possible with gelatin and dietary fiber. Consumers will probably recognize the difference but still like the product according to the sensory scores (close to or above 80%).

3.3 Chicken Products with Dietary Fiber

3.3.1 Chicken Meatballs

Poultry meat is an increasingly popular meat source due to its higher level of unsaturated fatty acids and lower level of cholesterol. Owing to the positive health effects, WHO [1] suggests a change in red meat consumption. Chicken meat is a good choice since it delivers all the amino acids and minerals but is a healthier alternative to pork and beef. Similarly, pork and beef, meatballs, nuggets, and sausages are also produced from chicken meat. Santhi et al. [13] added dietary fiber to chicken meatballs from different fruit and vegetable sources. Pearl millet (*Pennisetum glaucum* L.) flour, wheat (*Triticum aestivum* L.) flour, grape (*Vitis vinifera* L.) pomace powder, pomegranate (*Punica granatum* L.) pomace powder, carrot (*Daucus carota* L.) pomace powder, and beetroot (*Beta vulgaris* L.) pomace powder were used as dietary fiber sources. Pomace application is especially good since these materials are secondary or side products and utilized in low quantities. The control samples were prepared following the recipe of 1 kg of broiler chicken, 100 g of water, 50 g of vegetable oil, 40 g of refined wheat flour (what was removed when wheat flour was added as dietary fiber), 25 g of ginger, 25 g of onion, 25 g garlic, 20 g of salt, and finally 20 g of spices. Parameters of emulsion stability, product yield (cooking loss), energy, moisture, crude protein, crude fiber, total fat, and ash content were measured. When pearl millet and wheat flour were used together, the highest level of pearl millet (7%) obtained the best emulsion stability and cooking loss.

Regarding cooking loss, the pearl millet concentration of 5.25% reached a similar result to 7%. While pearl millet and wheat flour did not change the pH, grape and pomegranate pomace did. The change in pH was primarily affected by the grape pomace content of both the emulsion and the product. No significant difference was observed in the emulsion stability when grape and pomegranate pomace were used together. The cooking loss could not improve, but a significant decrease was detected.

In contrast to these fruits, the addition of carrot and beetroot pomace increased the pH of both the emulsion and the product. However, there needed to be a clear tendency to be able to conclude the effect of these ingredients on the pH value. The parameter emulsion stability did not respond to the added dietary fiber, while product yield increased with decreasing amounts of carrot and beetroot. The best yield was achieved with 0.50% carrot and 0.75% beetroot pomace powder. Regarding the fiber content of the meatballs, covering the daily protein needs with these products can provide 4.75 g of dietary fiber. Depending on the person, this is approximately 15–23% of the recommended daily intake. The results presented by Santhi et al. [13] demonstrated that combining different dietary fiber sources has a synergistic effect, and the optimum solution might be a combination of several plant materials.

3.3.2 Chicken Nuggets

Defatted rice bran was successfully used to decrease the oil content of fried chicken nuggets by Chayawat and Rumpagaporn [12]. Since comminuted and breaded products are typically fried, the reduced fat content of the serving product benefits health. The chicken nuggets were prepared with a mixture of wheat, corn, and rice flour using a ratio of 13:5.4:1, respectively. This mixture was replaced with defatted rice bran, up to 20%. The rice bran had 27.9% total dietary fiber content, of which 23.4% was insoluble. The prepared meat batter's viscosity increased with the rice bran content. Already, 10% added dietary fiber, which approximately doubled the viscosity. The appearance of the batter also changed significantly in terms of the CIE standard color parameters L*, a*, and b*. The batter became darker, and its color changed as well. The most significant amount of dietary fiber resulted in the highest moisture content of 52% and the lowest oil content of the fried nugget with

14%. Chayawat and Rumpagaporn [12] concluded that the dietary fiber-fortified batter made a protective crust against oil penetration during frying and prevented moisture loss from the core. This assumption was supported by a microscopic analysis of the fried products. The crispiness of the product decreased inversely with the increasing dietary fiber content. The sensory analysis could not find differences in oiliness and crispiness among the samples. The overall acceptability and flavor received the same score from tasting panelists (untrained) until 15% rice bran and only the nuggets with 20% dietary fiber were considered different. Due to the significant color changes, the sensory analysis results for color followed the dietary fiber concentration, and the lowest score of 3.28 was assigned to products with 20% rice bran. As a summary of this study, defatted rice bran had promising results, and the positive effects were proportional to the applied concentration. On the other hand, due to the induced changes in appearance and sensory characteristics, there is a threshold value of around 15%. The results published by Chayawat and Rumpagaporn [12] show that product development can find the optimal solution by balancing consumer expectations and the demand for healthier products.

3.3.3 Chicken Sausage

All around the world, researchers try to use the dietary fiber sources they have at hand, mainly focusing on byproducts and plant parts considered waste. A research group in Malaysia tested banana peel powder as dietary fiber in chicken sausage [7]. Zaini et al. [7] prepared the meat batter following the standard recipe, and the homogeneous material was mixed with the banana peel powder in 2–6% quantity. Sausages were steamed at 75 °C for 30 min. It was observed that moisture content decreased monotonously with increasing amounts of additional dietary fiber. The ash content increased to almost double, and the fat content decreased to approximately half with 6% banana peel powder. The protein content changed slowly until 4% added fiber but dramatically dropped to 6% fiber content. The initial protein concentration of 13.73% decreased to 3.39%, which is an extreme change. The tendency was non-linear, and more significant changes were observed with more dietary fiber. Such an accelerated effect was not reported earlier with other nutritional fiber materials. The water holding capacity monotonously increased with the banana peel powder quantity. The cooking yield reached exceptionally high values, but there is confusion between the computation and reported values. According to the instrumentally measured product color, adding banana peel powder resulted in darker sausages with a color change towards blue. The blue color change was detected by decreasing the CIE standard b* parameter. This is unexpected considering the green or yellow peel color.

The analysis of product texture revealed no change in the springiness parameter, but a monotonous tendency was observed for chewiness and hardness with increasing values while decreasing values for cohesiveness—the tendency of the texture parameters correlated with the applied banana peel powder concentration. The sensory analysis showed that the untrained panel was not impressed by chicken sausages. The control products reached 3.73–5.21 points on a seven-point hedonic scale. Hardness and juiciness increased slightly with 2% dietary fiber content and decreased beyond. All other parameters decreased continuously, and the products with 6% dietary fiber achieved 2.26–3.32 points. According to the results published by Zaini et al. [7], banana peel powder can be utilized as dietary fiber and improve chicken sausage's nutritional properties and technological attributes. On the other hand, this product was not popular among university students working in the sensory panel, and the banana peel powder could not help with this issue.

3.4 Turkey Products with Dietary Fiber

3.4.1 Turkey Sausages

Turkey is a popular poultry meat and is considered one of the leanest types due to its low-fat content. Most of the fat is within the skin. Although the United States of America is associated with turkey meat consumption because of the traditional turkey dishes on Thanksgiving, it is not the top consumer country. Surprisingly, Israel had 13 kg per capita in a year, followed by Hungary with 9 kg per capita, and the United States of America was in third place with 7.5 kg per capita, according to the World Population Review [33]. Turkey consumption is observed mainly in Europe and North America, but all continents reported a minimum of 2 kg per capita in their diet, except Asia. Besides the healthy white meat of this bird, processed products, such as sausage, typically contain plant materials as well. Rocchetti et al. [34] evaluated the improvement in a standard turkey sausage recipe, which consisted of 89.53% turkey meat, 4.97% soy, 1.99% pea fiber, 1.99% potato starch, 0.30% ground onion, 0.15% white pepper, and 0.03% nutmeg, and the remaining mass is shared among salts and spices. In their experiment, the meat was substituted with carrot (*Daucus carota* L.) in the range of 9.84–29.84%, decreasing the meat content to 59.69%. The measured total dietary fiber content varied in the range of 2.36–2.80% without significant differences due to the large deviation of samples. The maximum value was obtained with 19.84% carrot.

On the other hand, carbohydrate content decreased significantly with 9.84% carrot in the recipe, and the difference between the control and carrot-enriched samples decreased with increased carrot content. Significant differences in lipid, protein, ash, carbohydrate content, and energy value were found depending on the enrichment. Rocchetti et al. [34] used the Orthogonal Partial Least Squares Discriminant Analysis (OPLS-DA) technique to distinguish samples and identify the discriminating factors. It was found that products suffered changes in color in the first place in terms of the standard CIE a* attribute (red to green), followed by carbohydrates and moisture content, thrombogenic index, energy value, atherogenic index, pH, protein content, etc. in this order. The additional health-promoting effect of adding carrots was decreased sodium in the re-formulated sausages and increased w-3 polyunsaturated fatty acids. As a side effect, the appearance changed, and sausages became more orange-colored. Among the physicochemical parameters, mainly chemical ones and color were measured. Unfortunately, texture analysis was missing, while texture is also an important factor in determining sensory scores and consumer satisfaction.

3.4.2 Vienna-Type Sausages

Pineapple (*Ananas comosus* L.) pomace was also used as dietary fiber in Vienna-type sausages made of pork and turkey meat by Montalvo-González et al. [14]. The dietary fiber concentrate was prepared from pineapple pulp treated at 121 °C and 1.03 bar for 5 min. During the production of fruit fiber concentrate, washed pomace was dried with hot air at 55 °C for 9 h or freeze-dried at 0.08 mbar at −48 °C for 48 h. Sausages were prepared with a mixture of pork, turkey, and fiber, and this mixture was homogenized with a sausage formula in a 1:0.416 mass ratio. Prepared sausages were cooked in water at 80–85 °C. The pressure-steamed pomace with hot air drying was observed to achieve the highest total dietary fiber content and the highest water-holding capacity with the most elevated swelling.

Regarding the chemical analysis, the pressure-steamed pomace with hot air drying obtained the lowest moisture content and the highest total carotenoids expressed in an equivalent mass of b-carotene. The fat content dramatically decreased with the pressure-steamed pomace and reached its lowest value in combination with the freeze-dried concentrate. All measured physical and chemical parameters were sensitive to the meat composition in terms of the pork and turkey meat ratio, and shear force was significantly affected by the amount of dietary fiber concentrate. A lower amount of pineapple fiber resulted in softer sausages. Shrinkage was addressed as a second descriptor of physical properties, but it was less affected by dietary fiber content. Based on the recipe, the authors were able to predict physical and chemical parameters successfully. The concentration of residual nitrites was estimated with the highest R2 of 0.9847, and all chemical attributes reached R2 above 0.82. Shear force was predicted with R2 = 0.9031, and shrinkage resulted in R2 = 0.7731. Compared to fresh fruit pomace, the pressure-steamed pretreatment was the most efficient in contributing dietary fiber, protein, fat, carotenoids, polyphenols content, and antioxidant capacity. Montalvo-González et al. [14] demonstrated that pineapple pomace can be used as dietary fiber in sausages and, at the same time, compared the effectiveness of the dietary fiber preparation methods. Pineapple and probably similar fruits can be utilized following the suggested protocol, which consists of the main steps of high pressure and steam pretreatment, concentration with hot water, and finally, hot air drying.

3.5 *Other Meat Products with Dietary Fiber*

3.5.1 Goat Meat Nuggets

Goat meat is trendy in regions where religion or human health protection do not allow the consumption of pork and beef. Das et al. [35] evaluated the addition of bael (*Aegle marmelos* L.) pulp powder to goat meat nuggets. The powder was prepared from the leftover residue of filtration. It was washed, squeezed to remove the water, and dried at 60 °C. The dried material was ground to produce powder. This protocol is essential since the bioactive compounds of the powder were able to maintain functionality in terms of total phenolics, radical scavenging activity (RSA), and ferric-reducing antioxidant power (FRAP). The recipe for the nuggets was modified to add 0.25% and 0.5% bael pulp powder; meanwhile, the amount of meat decreased from 69.18%. The pulp powder was measured for dietary fiber content, and the insoluble dietary fiber content was 56.48%. This high fiber content is the reason why the total phenolic content of the powder was 15.16 mg GAE/g (gallic acid equivalent per gram) dry weight, far below what has been reported for the whole fruit. The RSA activity of the pulp residue extract varied in the range of 10–46% and increased with increasing concentration. The FRAP followed a similar tendency to the RSA and changed in the range of 0.029–0.089 absorbance at 700 nm. The initial evaluation of the prepared bael pulp powder confirmed the dietary fiber content and revealed additional health-promoting features. Adding 0.5% bael pulp powder improved the observed emulsion stability, while 0.25% concentration significantly improved cooking loss. Together with the significant increase in total dietary fiber of goat nuggets, ash content, and total phenolics, they also increased. The concentration of total phenolics achieved is especially interesting due to the monotonous increase from 855.83 mg/g (control) to 1437.06 mg/g (0.5% bael pulp powder).

Regarding the texture of the nuggets, TPA analysis observed significant changes in chewiness, gumminess, and hardness. Although Das et al. [35] detected significant differences in cohesiveness as well, it did not show a clear tendency like the other parameters. The TPA parameters of chewiness, gumminess, and hardness decreased monotonously with increasing concentrations of bael pulp powder. The sensory analysis performed with trained panelists could not detect differences among samples, except for the appearance affected by brightness and color. This latter conclusion of the sensory analysis was confirmed with instrumental color analysis and Hunter Lab standard parameters *L* (lightness) and *a* (redness). Additionally, sensory parameters such as flavor and juiciness showed a monotonous tendency according to the concentration of bael pulp powder, but groups did not differ significantly. As a result, the new ingredient of bael pulp powder could improve the quality of goat nuggets with bioactive compounds and positively affect their appearance and functional properties.

3.5.2 Fish Sausages

Seafood and fish meat are healthy alternatives to pork and beef. Increased fish consumption is promoted in Europe, North America, and regions without connection to the sea. Meat products made of fish help increase the market share. A recent study by Gore et al. [11] evaluated the addition of oat fiber to minced sausages made of fish. The fiber was incorporated in the 2.5–10% w/w range with cod liver oil (8% w/w) and starch (8% w/w). The crude fiber measurement followed a protocol different from ICC Standard No. 185 [20]. Samples were treated with 1.25% w/v H_2SO_4 solution for 30 min (stirred), and hot water was added to maintain the volume.

The mixture was filtered and washed with hot water to remove acid. The residual material was treated with boiling 1.25% w/v NaOH for 30 min. The mixture was filtered again and dried at 80–100 °C, followed by burning at a high temperature of 550 °C for 3 h. The amount of crude fiber was defined as the difference between the mass of ash and the mass of residue [11]. It was observed that adding oat fiber decreased the moisture content of sausages from 66.21% to 62.94% and the fat content from 7.90% to 6.50% monotonously in the order of fiber content. Crude fiber was measured to increase by up to 7.04%. The water-holding capacity of the product fluctuated with oat fiber concentration, but it was around the value of the control sample (91.21%). Based on scanning electron microscopic images and similar observations from prior studies, the authors concluded that starch did not significantly affect the gel formation of oat fiber. Still, it served as structural material and filled gaps. This statement was supported by the observation that starch does not directly interact with the protein matrix, and starch swelling occurs during cooking. Among the texture properties, only hardness significantly increased from 8.52 N (control) to 13.09 N (10% oat fiber). It must be noted that texture parameters did not have a clear tendency. The uneven homogenization of oat fiber and starch powders probably induced the fluctuation of values. The sensory attributes of the oat-fiber-enriched sausages, assessed by a trained panel, achieved the best scores for 2.5% and 5% fiber content samples. Appearance, odour, taste, and texture were better than others with 2.5% oat fiber. Color, flavor, and overall acceptability reached the maximum with 5% oat fiber. As a result, oat fiber slightly modified the sensory characteristics of the product and increased them in certain aspects. All tested samples were qualified to be suitable for consumption and may attract consumers. The successful experiment proved that fish products can be fortified with dietary fiber.

3.5.3 Surimi

Surimi (*Pagrosomus major*) protein digestibility was evaluated in the presence of dietary fiber of wheat bran (*Triticum aestivum* L.) by Lin et al. [36]. Both insoluble and soluble dietary fibers were used in the range of 0–10% with surimi meat and digested in two steps in an *in vitro* gastrointestinal system. The simulated digestion was achieved in a temperature-controlled Kjeldahl apparatus equipped with a shaker. The wheat bran had 49.39% dietary fiber, of which 40.87% was insoluble fiber. The insoluble fraction was mainly composed of 33.19% hemicellulose, 30.80% lignin, and 12.10% cellulose. The soluble fraction of dietary fiber had 45.09% pectin. Protein digestibility was defined based on the mass and nitrogen content of the input surimi meat and the supernatant of the *in vitro* digestion system. It was observed that protein digestibility decreased with an increased amount of wheat bran dietary fiber. Soluble and insoluble dietary fibers express slightly different behavior. Adding 2% soluble dietary fiber achieved a similar decrease to 6% insoluble fiber. The difference was significant for small amounts below 6%, and protein digestibility values were very close to 10% dietary fiber content. While insoluble fiber absorbed cholesterol, nitrite, and sodium cholate with good efficiency, the soluble dietary fiber showed good antioxidant capacity. According to the chemical analysis, the soluble dietary fiber achieved 2.31 mmol vitamin C equivalent DPPH radical scavenging capacity and 3.34 mmol vitamin C equivalent hydroxyl scavenging capacity. These values exceed the antioxidant capacity of some fruit extracts, therefore suggesting the application of wheat bran dietary fiber as a natural antioxidant as well. This promising feature is probably attributed to the large portion of pectin in the soluble dietary fiber fraction. The study published by Lin et al. [36] demonstrated the benefits of wheat bran dietary fiber and provided details with cutting-edge analytical methods of *in vitro* digestion, Fourier-transform near-infrared spectroscopy, scanning electron microscopy, X-ray diffraction, and measurement of antioxidant and absorption capacity. Besides the impressive analytical methods, no information was provided about the preferences of potential consumers. Sensory analysis and the consumption-simulating texture profile analysis were missing. Finally, we can conclude that adding wheat bran dietary fiber to surimi fortifies the product and has many advantages, but we cannot predict how consumers will evaluate it. This is a critical point in any food product's development, and consumers' decisions rely on appearance (primarily color) and texture in terms of hardness, crispiness, chewiness, and tenderness. The product's palatability must meet market expectations, or it will fail.

4 Conclusions

As the standard of living is increasing daily, consumers pay much more attention to their daily foods because diet is related to health issues. Daily diet intakes that are rich in dietary fibers can decrease the risk of severe diseases and enhance body function regulation. Meat products enriched with dietary fiber are one of the solutions to overcoming fiber deficiency in the daily diet and preventing some serious diseases. Increasing the consumption of legumes and eating fruits and vegetables regularly every day is recommended. Dietary fibers are getting attention due to their health-promoting functions. Meat is also an essential part of daily meals. Therefore, enrichment of meat products with dietary fibers is one of the good options to meet the food market nowadays. In addition, many studies have increasingly emphasized the synergistic effect of nutrients from diverse food sources. A healthy diet could enhance the absorption of bioactive compounds and thus be health-promoting. Besides, increased plant-based food consumption could reduce greenhouse gas emissions and decrease the human environmental footprint.

5 Future Directions

Recently, trends in food science have shifted to healthier diets because unhealthy lifestyles could cause serious illnesses, including diabetes, obesity, and cardiovascular and digestion problems. This transition period is toward healthy and sustainable foods. Moreover, experts and academic groups also suggested increasing the consumption of plant-based foods.

Protein of meat is an essential macronutrient that facilitates the development, maintenance, and repair of the body. In addition, meat protein has highly contributed to several cultures, especially in health, fitness, and dieting. Furthermore, broadening the dietary patterns from different sources, including animal and plant-based foods, is necessary for the synergies within and among foods. Therefore, combining a wide range of meal types, such as meat products, with fruit and vegetables is a current approach to developing healthier foods and reducing environmental impacts (including greenhouse gas emissions, water usage, and pollution) from livestock production. The benefits of incorporating dietary fibers or plant-based foods in traditional meat products suggest more promising applications of plant-derived sources for healthier foods.

Moreover, further research on enrichment with dietary fibers in animal-source foods is needed to develop new, healthier meat products for the food supply chain. The newer products are made, the more options customers will have as well. The diversity of meat products with health-promoting functions may offer multiple choices and easily satisfy the high-demand markets. Simultaneously, additional work is also necessary to widen the application of plant-based sources to improve dietary fibers' economics and practical value, especially plant-byproducts from food processing.

Because of the health benefits and sustainability of the environment, the potential application of some types of plant byproducts and further research, such as the enrichment of meat products with plant-based foods, is recommended. Moreover, the demand for natural ingredients in food processing is increasing significantly due to food safety and health promotion. Some byproducts, such as citrus byproducts from citrus juice production, can cause environmental pollution if disposed of improperly. However, these materials are still rich in dietary fiber and bioactive compounds. Thus, grain, fruit, and vegetable byproducts, including dietary powder and fruit pomace, are valuable sources due to their nutritional properties, and the application of such byproducts in meat products is promising.

References

1. World Health Organization. Red and processed meat in the context of health and the environment: many shades of red and green. Information brief. Geneva: WHO; 2023.
2. Lorenzo JM, Pateiro M. Influence of type of muscles on nutritional value of foal meat. Meat Sci. 2013;93(3):630–8. https://doi.org/10.1016/j.meatsci.2012.11.007.
3. Das AK, Nanda PK, Madane P, Biswas S, Das A, Zhang W, Lorenzo JM. A comprehensive review on antioxidant dietary fibre enriched meat-based functional foods. Trends Food Sci Technol. 2020;99:323–36. https://doi.org/10.1016/j.tifs.2020.03.010.
4. Younis K, Yousuf O, Qadri OS, Jahan K, Osama K, Islam RU. Incorporation of soluble dietary fiber in comminuted meat products: special emphasis on changes in textural properties. Bioact Carbohydr Diet Fibre. 2022;27:100288. https://doi.org/10.1016/j.bcdf.2021.100288.
5. Yang YY, Ma S, Wang XX, Zheng XL. Modification and application of dietary fiber in foods. J Chem. 2017;2017:9340427–8. https://doi.org/10.1155/2017/9340427.
6. Zhang W, Xiao S, Samaraweera H, Lee EJ, Ahn DU. Improving functional value of meat products. Meat Sci. 2010;86(1):15–31. https://doi.org/10.1016/j.meatsci.2010.04.018.
7. Zaini HB, Sintang MD, Pindi W. The roles of banana peel powders to alter technological functionality, sensory and nutritional quality of chicken sausage. Food Sci Nutr. 2020;8(10):5497–507. https://doi.org/10.1002/fsn3.1847.
8. Li K, Liu JY, Fu L, Zhao YY, Zhu H, Zhang YY, Zhang H, Bai YH. Effect of bamboo shoot dietary fiber on gel properties, microstructure and water distribution of pork meat batters. Asian Australas J Anim Sci. 2020;33(7):1180–90. https://doi.org/10.5713/ajas.19.0215.
9. Zhao D, Guo C, Liu X, Xiao C. Effects of insoluble dietary fiber from kiwi fruit pomace on the physicochemical properties and sensory characteristics of low-fat pork meatballs. J Food Sci Technol. 2021;58:1524–37. https://doi.org/10.1007/s13197-020-04665-2.
10. Soltanizadeh N, Ghiasi-Esfahani H. Qualitative improvement of low meat beef burger using Aloe vera. Meat Sci. 2015;99:75–80. https://doi.org/10.1016/j.meatsci.2014.09.002.
11. Gore SB, Xavier KM, Nayak BB, Tandale AT, Balange AK. Technological effect of dietary oat fiber on the quality of minced sausages prepared from Indian major carp (Labeo rohita). Bioact Carbohydr and Diet Fibre. 2022;27:100305. https://doi.org/10.1016/j.bcdf.2021.100305.
12. Chayawat J, Rumpagaporn P. Reducing chicken nugget oil content with fortified defatted rice bran in batter. Food Sci Biotechnol. 2020;29(10):1355–63. https://doi.org/10.1007/s10068-020-00782-y.
13. Santhi D, Kalaikannan A, Natarajan A. Characteristics and composition of emulsion-based functional low-fat chicken meat balls fortified with dietary fiber sources. J Food Process Eng. 2020;43(3):e13333. https://doi.org/10.1111/jfpe.13333.
14. Montalvo-González E, Aguilar-Hernández G, Hernández-Cázares AS, Ruiz-López II, Pérez-Silva A, Hernández-Torres J, Vivar-Vera MD. Production, chemical, physical and technological properties of antioxidant dietary fiber from pineapple pomace and effect as ingredient in sausages. CyTA J Food. 2018;16(1):831–9. https://doi.org/10.1080/19476337.2018.1465125.
15. Mumyapan M, Aktaş N, Gerçekaslan KE. Seed pumpkin flour as a dietary fiber source in Bologna-type sausages. J Food Process Preserv. 2022;46(7):e16586. https://doi.org/10.1111/jfpp.16586.
16. Grigelmo-Miguel N, Martín-Belloso O. Comparison of dietary fibre from by-products of processing fruits and greens and from cereals. LWT-food. Sci Technol. 1999;32(8):503–8. https://doi.org/10.1006/fstl.1999.0587.
17. Fernández-López J, Fernández-Ginés JM, Aleson-Carbonell L, Sendra E, Sayas-Barberá E, Pérez-Alvarez JA. Application of functional citrus by-products to meat products. Trends Food Sci Technol. 2004;15(3–4):176–85. https://doi.org/10.1016/j.tifs.2003.08.007.
18. Food and Agriculture Organization of the United Nation. Proposal for introduction of ICC Standard No. 185 (AOAC method 2017.16) to replace AOAC method 2009.01. NFSDU/41 CRD 6. 2019. Last Accessed 26 Nov 2023.

19. Hipsley EH. Dietary "fibre" and pregnancy toxaemia. Br Med J. 1953;2(4833):420–2. https://doi.org/10.1136/bmj.2.4833.420.
20. International Association for Cereal Science and Technology. ICC Standard No. 185. Measurement of total dietary fibre in cereals, ingredients and food products using the Rapid Integrated TDF Procedure (RINTDF). 2019. https://icc.or.at/publications/icc-standards/standards-overview/185-standard-method. Last Accessed 26 Nov 2023.
21. US. Food and Drug Administration. Questions and answers on dietary fiber. 2023. https://www.fda.gov/food/food-labeling-nutrition/questions-and-answers-dietary-fiber. Last Accessed 26 Nov 2023.
22. US. Food and Drug Administration. The declaration of certain isolated or synthetic non-digestible carbohydrates as dietary fiber on nutrition and supplement facts labels: guidance for industry. 2018. https://www.fda.gov/media/113663/download. Last Accessed 26 Nov 2023.
23. European Union. Regulation (EC) No 1333/2008 of the European Parliament and of the Council of 16 December 2008 on food additives. 2008. Last Accessed 26 Nov 2023.
24. Food Standard Agency [Internet] Approved additives and E numbers. 2023. https://www.food.gov.uk/business-guidance/approved-additives-and-e-numbers. Last Accessed 26 Nov 2023.
25. Verma AK, Banerjee R, Sharma BD. Quality of low fat chicken nuggets: effect of sodium chloride replacement and added chickpea (Cicer arietinum L.) hull flour. Asian-Australas. J Anim Sci. 2012;25(2):291. https://doi.org/10.5713/ajas.2011.11263.
26. Xiong YL. Nonmeat ingredients and additives. In: Handbook of meat and meat processing. CRC Press; 2012. p. 573–88.
27. Verma AK, Rajkumar V, Banerjee R, Biswas S, Das AK. Guava (Psidium guajava L.) powder as an antioxidant dietary fibre in sheep meat nuggets. Asian-Australas J Anim Sci. 2013;26(6):886. https://doi.org/10.5713/ajas.2012.12671.
28. Han M, Bertram HC. Designing healthier comminuted meat products: effect of dietary fibers on water distribution and texture of a fat-reduced meat model system. Meat Sci. 2017;133:159–65. https://doi.org/10.1016/j.meatsci.2017.07.001.
29. dos Santos M, Ozaki MM, Ribeiro WO, de Souza PC, Vidal VA, Campagnol PC, Pollonio MA. Emulsion gels based on pork skin and dietary fibers as animal fat replacers in meat emulsions: an adding value strategy to byproducts. LWT-Food Sci Technol. 2020;120:108895. https://doi.org/10.1016/j.lwt.2019.108895.
30. Bis-Souza CV, Ozaki MM, Vidal VA, Pollonio MA, Penna AL, Barretto AC. Can dietary fiber improve the technological characteristics and sensory acceptance of low-fat Italian type salami? J Food Sci Technol. 2020;57:1003–12. https://doi.org/10.1007/s13197-019-04133-6.
31. Schmiele M, Mascarenhas MC, da Silva Barretto AC, Pollonio MA. Dietary fiber as fat substitute in emulsified and cooked meat model system. LWT-Food Sci Technol. 2015;61(1):105–11. https://doi.org/10.1016/j.lwt.2014.11.037.
32. Essa RY, Elsebaie EM. New fat replacement agent comprised of gelatin and soluble dietary fibers derived from date seed powder in beef burger preparation. LWT-Food Sci Technol. 2022;156:113051. https://doi.org/10.1016/j.lwt.2021.113051.
33. World Population Review. Which country consumes the most Turkeys per capita each year? 2023. https://worldpopulationreview.com/country-rankings/which-country-consumes-the-most-turkeys-each-year-per-capita. Last Accessed 26 Nov 2023.
34. Rocchetti G, Pateiro M, Campagnol PC, Barba FJ, Tomasevic I, Montesano D, Lucini L, Lorenzo JM. Effect of partial replacement of meat by carrot on physicochemical properties and fatty acid profile of fresh Turkey sausages: a chemometric approach. J Sci Food Agric. 2020;100(13):4968–77. https://doi.org/10.1002/jsfa.10560.
35. Das AK, Rajkumar V, Verma AK. Bael pulp residue as a new source of antioxidant dietary fiber in goat meat nuggets. J Food Process Preserv. 2015;39(6):1626–35. https://doi.org/10.1111/jfpp.12392.
36. Lin Y, Chen K, Tu D, Yu X, Dai Z, Shen Q. Characterization of dietary fiber from wheat bran (*Triticum aestivum* L.) and its effect on the digestion of surimi protein. LWT-Food Sci Technol. 2019;102:106–12. https://doi.org/10.1016/j.lwt.2018.12.024.

Fermented Meat Products

Aashruti Singh, Alok Saxena, and Tanushree Maity

1 Introduction

Among the most crucial dietary components for humans are meat and animal products. The traditional production of raw, preserved, and matured meat products, including various types of sausages and smoked meat-based goods, has historically relied on the fermentation process [1]. This procedure encompasses a regulated degradation using a combination of inherent meat enzymes and microbial-derived enzymes. In the conventional preparation of uncooked meat items, natural or supplementary carbohydrates are subjected to fermentation by lactic acid bacteria (LAB). These LAB can naturally exist within the meat or its vicinity, serving as the main source of energy. Ketones, acetic acid, bacteriocins, lactic acid, aldehydes, and alcohols are just a few of the compounds created during fermentation in beef products, that can significantly affect shelf life, quality, and safety [2].

Probiotic foods refer to nutritional items that promote excellent health and have a significant market appeal. Their market shares are on the rise, as they are currently in vogue for promoting consumer wellness and overall health through the creation of innovative functional foods tailored to specific health goals. Over the past four decades, a widely used method for producing fresh-ripened beef products has involved employing starting cultures with highly diverse compositions. It is advisable to utilize bacterial strains possessing probiotic qualities in the production of meat-based probiotic products [3]. These strains, typically found in the human intestines, should have the ability to outcompete bacteria in meat, which is not their natural habitat. Moreover, they must be capable of reaching a size considered

A. Singh · A. Saxena (✉)
Amity Institute of Food Technology, Amity University Uttar Pradesh,
Noida, Uttar Pradesh, India

T. Maity
Defence Research and Development Organization, Timarpur, New Delhi, India

A. Kumar Pathera et al. (eds.), *Healthier Meat Products*,
https://doi.org/10.1007/978-3-031-78215-2_9

beneficial for health and withstand the processes of fermentation, drying, freezing, and storage. Unlike the pasteurization process used for milk, the initial quantity of germs, present in sausages stuffing or on the surface of salami or loin cannot be decreased during the production of meat products [4]. Hence, selecting the right microbial strains is crucial.

There exists a viable technological option to employ microorganisms closely connected to the meat industry environment, possessing desirable attributes and qualities that promote health. Probiotic bacteria can occur naturally in items that have been fermented include hams and dried sausages, or they can be sourced from established cultured commercial meat starters [5]. Alternatively, using starting cultures with well-known probiotic qualities is another approach. As meat items with probiotics represent a comparatively new and undervalued sector within the meat industry, the primary challenge remains striking a balance between technological obstacles, quality, safety, and health advantages. Probiotics are commonly found in various forms, such as capsules, medications, and fermented foods. Identification and detailed characterization of the probiotic strain at several levels, such as species, genus, and strain, constitute the initial stage in the development of a probiotic food product. This comprehensive understanding is essential for the growth process. Subsequent stages encompass researching physiological traits, technological prerequisites, operational strategies, and providing evidence of human benefits.

With the growing demand for meat products containing high-quality ingredients and valuable nutritional content, some traditional meat items are being reformulated. However, it is essential that innovations do not compromise the sensory qualities of the original product from the consumer's perspective. Therefore, exploring novel techniques is crucial to creating wholesome meat products that are not only nutritionally beneficial but also highly appealing to consumers. It is crucial to monitor the current surge in fermented meat products with active fermentation. The goals listed in the 2030 Agenda for Sustainable Development published by the United Nations, particularly those related to promoting excellent health and wellbeing, can be interpreted to include fermented meat products enriched with probiotics [6]. This perspective becomes even more significant in the context of enhancing nations' ability to mitigate health risks and improve health hazard management, ultimately leading to reduced healthcare costs across all age groups. By combining dietary fibres and probiotics, innovative low-fat fermented meat-based diets can be formulated [3]. This amalgamation offers a myriad of potential health benefits to consumers. The primary aim of this chapter is to facilitate the food industry and academic researchers with diverse knowledge of creating distinctive, versatile meat-based products with enhanced nutritional profiles.

1.1 Meat Products in Human Nutrition

Meat food products supply critical amino acids necessary for protein synthesis, making it easier to create and maintain muscles since they have a composition that is comparable to that of human skeletal muscle [7]. It is essential to maintain skeletal muscle mass for metabolic health and continued physical functioning. Age-related decreases in muscle strength and fragility can be prevented by including meat in the diet, which is crucial for older individuals. Meat serves as an excellent source of essential minerals, vitamins, and proteins. For example, an adult consuming an average of 189 g of meat daily receives approximately 19%, 52%, 28%, and 38% of the recommended intakes for iron, zinc, selenium, and phosphorus, respectively, based on diverse group reference values [8]. Meat stands out as a primary source of high-quality, nutritionally significant proteins. The muscle tissue of various animal species contains protein levels ranging from 15% to 20% [9]. Furthermore, meat proteins offer a well-balanced mix of amino acids, encompassing all the essential amino acids vital for the human body. Since only connective tissue proteins contain the non-essential amino acids tryptophan and cysteine, they have minimal biological relevance. Meat and meat products vary in their energy value based on their water and fat levels. Depending on the animal species, area of the carcass, or product type, the fat content might vary greatly. A primary source of several mineral substances, including zinc, copper, phosphorus, and sulphur as well as iron (sometimes known as hem iron, which is particularly easily absorbed), is meat. B-complex vitamins are abundant in meat, and the amounts of certain B group vitamins vary greatly between different animal species [10]. The only food sources of vitamin B_{12} are animal products. Animal products fulfil approximately 70% of the typical diet's vitamin B_{12} requirement. Meat and its products also contain fat-soluble vitamins, primarily vitamins A and D, and to a lesser extent, vitamin E [11]. If meat intake is reduced, it becomes vital to substitute it with a variety of meals that can still provide the nutritional benefits generally found in meat.

1.2 Probiotic Microorganism in Fermented Meat and Their Health Benefits

Probiotic bacteria are vital for the human body as they regulate flatulence, control diarrhoea, prevent and treat it, protect intestinal infections like irritable bowel syndrome, boost the immune system, possess antibacterial properties, inhibit the growth of *Clostridium botulinum* endospores in the digestive system, and strengthen epithelia [12]. Moreover, they aid in lowering cholesterol levels by dissolving bile salts, making it harder for cholesterol to be absorbed, and facilitate the absorption of calcium, iron, and zinc, contributing to overall health.

A microbial strain's probiotic potential depends on the existence of species and genera that are often discovered in the digestive system of humans. Most gram-positive Lactobacillus species, recognized as bifidobacteria or lactic acid bacteria, qualify as probiotics [13]. The genus Bifidobacterium is historically and practically categorised as a kind of LAB, despite having certain physical characteristics in common with conventional LAB. On the contrary, enterococci, due to their potential toxicity and transmission of antibiotic-resistant genes, have not been recommended for human use and have even been suggested as probiotics. Probiotic meals often have beneficial effects on the host's health because they include some lactic acid bacteria strains that, going through the gastrointestinal tract and intestinal tract, may survive in the large intestine.

Probiotics have been chosen over time based on a variety of factors, such as their human origin, capacity to endure the surface GI area (includes resistance to proteolytic enzymes, gastric acid, and bile salts as it travels from the gastrointestinal tract to the colon.), engaging with the local microbiota, which involves adhering to the intestinal epithelium and colon (colonisation without significantly affecting the majority of microbiota), and positive effects [14]. Probiotics can be found in several different forms, most frequently as capsules, tablets, and fermented foods. The probiotic bacterial strain must be identified and characterised at the strain, species, and genus levels in the early phases of manufacturing any probiotic food product. Subsequent stages involve examining the probiotic strain's physiological traits, technological requirements, therapeutic applications, and evidence of its positive impact on human health. Probiotic foods constitute up to 70% of the functional foods market, with a steady increase observed in non-dairy options like fruit and vegetable juices and cultured meats [15]. This growth is particularly evident in dairy-based probiotic meals. Considering the multitude of potential fermentable substrates and diverse environments where LAB strains can demonstrate "functional performance," it is logical to expect ongoing research and development for innovative food-based probiotics in the future.

1.3 Fermented Meats in Human Nutrition

Human nutrition and fermented meat products are closely linked disciplines that investigate the impact of fermented meats on overall health and well-being. Fermented meats, such as sausages, salami, and cured hams, undergo a natural preservation process involving beneficial microorganisms such as lactic acid bacteria and molds [16]. This fermentation not only enhances the taste and texture of meats but also boosts their nutritional value.

Improved digestibility is one of the main benefits of fermented meats. Enzymes and beneficial microbes break down complex carbohydrates and proteins during the fermentation process, increasing the body's ability to absorb nutrients. Moreover, probiotics present in some fermented meats have the potential to improve gut health by promoting healthy gut microbiota, which is necessary for effective digestion and nutrient absorption [17]. Additional sources of these nutrients include the essential amino acids, quality proteins, minerals, and vitamins found in fermented meats.

Because they are frequently simpler to digest, proteins from fermented meats are beneficial for those with sensitive digestive systems. Vital minerals including zinc, iron, and vitamin B complex may also be present in these goods. These minerals are necessary for many biological functions, including energy metabolism and immune system support.

However, it's important to eat fermented meats in moderation because they can be high in saturated fats and sodium, which can lead to health problems like cardiovascular disease and hypertension if ingested in excess [18]. Therefore, to fully use the nutritional benefits of these products, it is imperative to incorporate a variety of foods, including fermented meats, into a diet that is balanced and aware of portion sizes and overall dietary consumption.

1.4 Microorganisms as Starter Cultures for Processed Meat Products

Meat has been preserved for ages by fermenting with indigenous microorganisms. Several compounds, such as alcohols, aldehydes, ketones, lactic acid, pyruvic acid, and carboxylic acids, are produced during these processes and have an impact on the functionality and storage viability of these products. Nowadays, ensuring a consistent level of quality during the fermentation process is commonly achieved by employing starter cultures [19]. These are specific bacterial combinations, that promote the required metabolic processes in the fermentation substrate, whether actively or latent. Although microbes typically thrive on this substrate, that's not always the case. Starter cultures are a type of bacteria capable of fermenting glucose into lactic acid [19]. Pentoses like arabinose and xylose can be utilized to produce lactic and acetic acids, underscoring the importance of selecting appropriate strains for each product. Common LAB strains in early cultures include acid from *Pediococcus acidilactici*, *L. pentosaceus*, *L. plantarum*, *L. sakei*, *L. curvatus*, *L. lactis*, and *L. casei*. Among these, *L. curvatus*, *L. sakei,* and *L. plantarum,* frequently found in fermented sausages, adapt to sausage conditions during fermentation, influencing the sausage's maturation rate and controlling the growth of harmful bacteria [20].

In contrast to European fermented sausages, where initial microorganisms are infrequently isolated, American counterparts intentionally incorporate starter microbes to hasten meat batter acidity. It is important to understand that various LAB strains naturally create varied quantities of lactic acid when selecting a LAB strain for a particular meat product. The rate and quantity of acid production can also be affected by several other factors, such as bacterial growth being faster in warmer environments with higher water flow. Additionally, the sausage's diameter influences lactic acid generation, with lower or higher oxygen levels playing a role [21]. Table 1 discusses some of the fermented meat products and how probiotic cultures affect their quality.

Table 1 Various fermented meat products and effect various probiotic cultures on their quality

Fermented meat product	Probiotic culture used	Effect on quality
Fermented sausage	*E. faecium* CRL183	Boost microbial diversity in the diet [82]
Italian salami sausage	*Bifidobacterium lactis* and *L. acidophilus*	Favourable sensory, physical-chemical, and microbiological attributes [83]
Norwegian fermented sausages	*L. farciminis MF1288* and *L. plantarum* MF1297	Rapid lactic acid production [84]
Fresh pork sausage	*L. sakei* BAS0117	The properties that the injected strains supplied the product helped it during its storage duration [85]
Nostrano salami	*E. faecium UBEF-41* and *L. lactis ssp. lactis* strain 16	Transform the aroma through the conversion of fatty and amino acids, allows the production of fermented products that mature at lower temperatures [3]
Fermented sausage	*L. sakei*	Effective starting cultures for sausage maturing procedures [86].
Spanish fermented sausages	*Companilactobacillus alimentarius*	Demonstrated a promising capacity for development under conditions simulating the salt and temperature levels characteristic of fermented sausages [87]
Fermented beef	*Meyerozyma guilliermondii* and *Debaryomyces hansenii*	Reduced lightness, moisture content, water activity, pH, protein content, nitrite content, and N-nitrosamine content, thereby imparting the desired meat color, mouthfeel, odor, taste, and tissue quality [88]
Fermented sausage	*Limosilactobacillus fermentum 332*	Influence the generation of flavor compounds [89]

1.5 Types of Fermented Meat Products

A preserved fresh sausage is produced from assorted meats that have been cured, encased, and fermented using microorganisms [22]. It is seasoned with spices, predominantly garlic and black pepper, as well as nitrate or nitrite, salt, and sugar. To support the fermentation process and attain the appropriate level of lactic acid, the temperature of the sausage must be kept at a level where lactic acid bacteria may grow (ranging from 27 to 37.8 °C) for a duration of 10–15 h [23]. Throughout this period, bacteria proliferate and convert sugars into lactic acid.

The tradition of making raw-cured sausages has deep roots in the Mediterranean region, dating back to the Roman era [24]. The invention of maturation chambers ensured consistent production of raw-cured sausages regardless of the weather conditions. In modern times, reducing ripening time while preserving essential sensory qualities is crucial in raw cured sausage production [25]. Fast-acidifying starter cultures are employed to expedite the process, resulting in products with a pH between 5.2 and 5.3, or even slightly lower, within 24 h. These sausages can be ready for sale in about 14 days, sometimes even sooner, thanks to the action of bacteria such as *L. sakei* or *L. curvatus*.

Raw cured sausages of the salami type exhibit a wide range of shapes, sizes, maturation rates, appearances, and flavors. The traditional technique for crafting these salami-like sausages, primarily derived from donkey meat, originated in Italy and has been adopted in countries such as France, Germany, Spain, Italy, and Hungary. Poland also produces in small amounts, fresh preserved fermented sausages. Numerous elements affect these fermented sausages' flavour profiles, including the starting culture, processing time, temperature, smoking, and types of ingredients like meat, salt, and spices [26]. The combination of aroma and taste creates the fundamental flavor, primarily shaped by the formation of peptides, free amino acids, and lactic acid.

During the Middle Ages, several cured products from raw gammon were developed, particularly in Central and Southern Europe. One of China's most renowned traditional meats, Jinhua gammon, is also crafted, mostly sliced from a hind leg. Creating raw matured hams is more challenging compared to other pork products [27]. Salt components disperse slower in these hams, which might lead to undesirable microbial growth, product deterioration, or health risks. Hence, careful selection of raw materials is crucial. Additionally, differences in muscle consistency also impact the production process. Quick-ripening raw preserved sausages differ from traditional sausages, especially those with longer ripening periods, as they possess a milder aroma and a lighter color. This distinction arises from alterations caused by Staphylococci, leading to a distinct aroma and color due to the rapid decrease in pH value [28].

2 Technological Characteristics of Probiotic Cultures Used in Meat Products

Probiotic bacteria incorporated into food made of fermented meat, like sausages, must withstand the harsh conditions established during the fermentation process. Probiotic cultures can be introduced either as a starter culture or mixed with the original culture during sausage fermentation [29].

2.1 Lactic Acid Production

Lactobacillus is a species of lactic acid bacteria that is often employed in meat fermentation. Lactobacillus converts carbohydrates into lactic acid during the fermentation process, resulting in a pH drop. The lactic acid released during fermentation improves the meat's sensory properties [30]. This encompasses everything, like flavor, texture, and color. The reduction in pH during fermentation causes the myofibrillar proteins (muscle proteins) in the meat to coagulate. Coagulation affects the texture and structure of the sausage, resulting in greater stiffness and cohesiveness. In addition to protein coagulation, the drop in pH leads to the creation of a red color in the sausage [31]. This is frequently a desirable feature in cured and fermented meat products. The rapid pH drop, and the generation of lactic acid contribute to the safety of the fermented beef product. Lactic acid's acidic environment hinders the growth of dangerous germs, acting as a natural preservative. Rapid pH decreases during sausage fermentation may also help to minimize the accumulation of dangerous biogenic amines [21]. Biogenic amines are chemicals that can arise during protein fermentation and can be toxic in large quantities.

2.2 Resistance to Salt (NaCl) and Nitrite (NO_2)

Traditional methods of preserving fermented sausages include fermenting with the aid of beneficial microbes such as lactic acid bacteria (LAB). These bacteria contribute to the acidification of the sausage, resulting in an environment that limits the growth of dangerous microorganisms. It's unusual to try to maintain fermented sausages fully microbe-free, because the fermentation process relies on certain helpful microorganisms. Sodium chloride (NaCl) is often used in cured meats, notably sausages, to improve flavor, texture, and meat preservation [27]. However, it is rarely employed to manufacture a product free of germs. Sodium nitrite ($NaNO_2$) is a curing salt that is used in small doses to prevent the growth of dangerous bacteria and to color the meat pink [32]. It is an essential component in the curing of some varieties of sausages. While sodium chloride and sodium nitrite quantities can vary based on the type of sausage and desired qualities, the 3% sodium chloride and 200 ppm sodium nitrite amounts indicated above appear reasonable for certain cured sausages.

2.3 Bacteriocin Production in Meat Products

Bacteriocins are proteins or peptides that are generated by bacteria, particularly gram-positive microbes. These chemicals are antibacterial and can prevent the growth of other bacteria. Bacteriocins are specifically designed to prevent the growth of *Listeria monocytogenes*, a pathogenic bacterium that can pose a food safety risk, particularly in meat products [33]. Bacteriocin incorporation into meat-based goods has the potential to improve both product quality and customer health. The safety and shelf life of meat food products can be increased by limiting the growth of dangerous bacteria such as *Listeria*. Bacteriocins are thought to be a natural food preservation strategy. Their usage as antibacterial agents allows for the enhancement of food safety without the use of synthetic chemicals. Bacteriocins are typically considered harmless and do not offer any toxicological risks. As a result, they are a good choice for food preservation because they pose no risk to customers. Bacteriocins are produced by a variety of lactic acid microbes, including *L. curvatus, L. casei, L. sakei, L. brevis,* and *L. plantarum* [23]. These lactic acid bacteria are often utilized in the fermentation of meat products, and their capacity to create bacteriocins provides an extra layer of defense against pathogenic germs.

3 Most Utilized Probiotic Cultures in Meat Products

Probiotic strains are commonly classified within the *Bifidobacterium* and *Lactobacillus* groups. Other species, such as *Saccharomyces, Enterococcus, Propionibacterium, and Lactococcus*, are also considered probiotics due to their potential to improve host health [34]. The application of *Lactobacillus gasseri* in fermented meat products has been studied for its potential benefits, particularly from a microbiological perspective [35]. The success of incorporating *L. gasseri* or any other specific strain into fermented meat products depends on various factors, including the formulation, fermentation conditions, and the specific characteristics desired in the final product. It is also important to consider factors like flavor, texture, and consumer acceptability in general. The interesting field of investigation of meat product fermentation employing probiotic strains such as *Lactobacillus plantarum*, *Lactobacillus rhamnosus*, and *Lactobacillus paracasei* combines traditional food preservation methods with the possible health benefits of probiotics [3]. These Lactobacillus strains are different from one another in ways that can affect the fermentation process and improve the overall quality of the finished product. Probiotic strains including *Lactobacillus rhamnosus GG*, *L. rhamnosus* VTT-97800, *L. rhamnosus* VTT, and *L. rhamnosus* LC-705 are some of the strains that can be used to dry sausage manufacture as an intriguing way to combine the benefits of probiotics with traditional meat fermentation [36].

Studies and experiments in this area could help to develop functional meat products that have both traditional flavor qualities and possible health advantages. Combining a regular starting culture with a particular strain of probiotics can result in a distinctive item that not only ferments but also has potential health advantages. The use of a common starter culture, Bactoferm T SPX from Chr Hansen, in conjunction with potential probiotic bacteria like Lactobacillus casei LC-01 and Bifidobacterium lactis Bb-12, may indicate a desire to create a fermented meat product [37]. The uniqueness of probiotic strains emphasizes the significance of customized techniques for producing fermented meat products. Different strains may contribute distinct properties to the finished product, influencing safety, flavor, texture, and potential health effects. This approach is consistent with the concept of precision fermentation, which selects microbial strains based on their functional qualities for specific purposes.

4 Probiotic Fermented Sausages

While adding probiotics to various food products is not a novel notion, manufacturers rarely use fermented sausages as probiotic carriers. This suggests a potential area of innovation and exploration. Probiotic bacteria can be added to sausages during the process of fermentation [38]. This is especially true for dry fermented pork items that require little or no heat, as elevated temperatures may reduce the vitality of probiotic bacteria. The microbial atmosphere of dry fermented sausages is optimal for beneficial bacteria to grow. The absence of heating ensures the survival of these beneficial microbes. Different microorganisms, each with a distinct function, are mixed to form starting cultures for the process of fermentation. This enables controllable and consistent fermentation. Lactic bacteria are frequently used in the fermentation process of sausages to produce controlled and strong acids. This acidification technique not only improves the overall flavor of the sausages, but also increases product safety and stability [21]. The controlled, strong acidity produced by lactic bacteria serves to limit the proliferation of harmful germs. This is an important part of the safety of food in fermented goods.

5 Probiotic Meat Products in Human Nutrition: Benefits and Risks

5.1 Safety Considerations of Probiotic Meat Products

Incorporating probiotic bacteria into fermented meat products can help to reduce disease development and survival. Probiotic bacteria produce chemicals such as bacteriocins along with other antimicrobials which can kill pathogenic bacteria. Lactobacillus and Lactococcus bacteria have a long history of safe eating. They are generally considered safe (GRAS) [39]. There are potential risks associated with probiotics, including the risk of systemic infections. Adverse metabolic activity is listed as a potential risk associated with probiotics. Probiotics can potentially modulate the immune system, and this is noted as a risk factor. Gene transfer is mentioned as a potential risk associated with probiotics [40]. When it comes to probiotic meat products, ensuring consumer safety involves addressing various factors, including epidemiological data, potential health risks, and considerations related to meat processing. The safety of probiotic meat products involves a comprehensive approach that considers various factors, including consumer characteristics, processing methods, HACCP principles, and advancements in the field [41]. Ongoing research, monitoring, and adherence to safety protocols are essential to mitigate risks and ensure the production of safe and high-quality probiotic meat products. *Enterococcus* species are commonly found in the environment and are known to inhabit the gastrointestinal tracts of humans and animals. While some Enterococcus strains are used as probiotics, certain species within this genus can pose health risks, especially in individuals with compromised immune systems [42].

6 Factors Influencing Probiotic Viability in Fermented Sausages

Several factors, such as oxygen levels, temperature, moisture content, acidity, sugar, salt, levels of organic acid concentrations and additions, the possible existence of additional microorganisms, and hydrogen peroxide, can impact the survivability of probiotic cultures in fermented products [43]. Among these factors, the following are crucial in determining the survival of probiotics in fermented sausages:

6.1 pH Levels

The inherent low pH and intense acidity of fermented sausages create a difficult habitat for probiotic microorganisms. This acidity causes problems for probiotics because they normally require less acidic environments to develop or survive for a lengthy period. Understanding the optimal pH ranges of specific probiotic strains sheds more light on these issues. Bifidobacteria, for example, grow best at pH levels ranging from 6.0 to 7.0 [13]. Another typical probiotic strain, *Lactobacillus acidophilus*, thrives well in a slightly lower pH range of 5.5 to 6.0 [44]. The growth of bifidobacteria is significantly hampered below a pH of 5.5, highlighting the acidic sensitivity of probiotic strains. The pH values of fermented sausages alter dynamically as they age. The action of LAB, which produces lactic acid and other organic acids throughout the fermentation and maturation phases, is principally responsible for this change. The lactic acid and other organic acids induce a significant drop in pH during the first three days of maturation, reaching levels between 4.1 and 4.5. The breakdown of acids stabilizes the pH after the first dip, resulting in a constant level. This stabilization is critical for the fermented sausages' overall quality and safety. Indigenous LAB bacteria with significant acidifying potential proliferate during the fermentation and maturation phases. The prevalence of LAB in the microbial population is distinctive, highlighting their critical role in defining the sensory and safety characteristics of fermented sausage.

6.2 Water Activity

Water accessibility is a significant component impacting the development, activity, and survival of microorganisms in fermented sausages, including probiotic bacteria. Many bacteria require water activity (a_w) of 0.90 or higher for optimal growth and survival, except for halophilic bacteria that flourish in high-salt conditions. Fortunately, fermenting sausages frequently fit this criterion. Water activity naturally diminishes throughout the maturing and drying processes of salami [45]. The steady loss of moisture during the fermentation and maturation phases of sausage preparation causes this drop in water activity. For example, the water activity value may decline from 0.96 at the end of the fermentation phase (7 days) to 0.90 at the end of the maturation phase (28 days). The loss of moisture in the sausage is closely related to the decrease in water activity. Weight reductions of 20% and 40% at specified time points, such as 30 and 60 days, are indicative of the drying process, respectively. This drying procedure affects the water activity levels as well as the desirable texture and flavor of the fermented sausage. Probiotic microorganisms face a problem as water activity drops. Probiotics, like many bacteria, frequently require higher levels of water activity for viability and optimal functionality [46]. If the water activity falls below 0.90, it can have a detrimental effect on the survival and function of probiotic bacteria. As a result, fluctuations in water activity levels during the manufacturing and maturation phases have a substantial impact on probiotics'ability to survive in fermented sausages.

6.3 Type of Probiotic Strain and Bacterial Synergies

When it comes to selecting optimal starters for probiotic meat products, it is critical to carefully evaluate the specific qualities of each probiotic strain. The complex interplay of diverse strains within a mixed beginning culture needs a grasp of how they work both alone and together. The finest fermented sausage strains can efficiently digest carbohydrates in meat while maintaining reasonable acidity levels, which improves both the sensory characteristics and the safety of the product [6]. The ability of the probiotic strains to resist the adverse conditions observed in fermented sausage is an important factor in strain selection. Traditional sausage recipes sometimes have a low pH level, high sodium chloride content, and excessive nitrate/nitrite concentrations [47]. The chosen strains must be resistant to these environments to survive and perform well during the fermentation and maturation procedures. A carefully designed strain selection approach considers potential antagonistic effects between co-cultured strains. To maintain the sensory qualities of the finished product, the probiotic strains must not only complement one another but also resist antagonistic interactions that could compromise flavor, texture, or overall product quality. Another consideration in the context of probiotics meats-starting cultures is their potential to compete with the naturally existing microflora in meat. This competition makes sure these strains establish and rule effectively, contributing to the desirable functional properties of the finished product [3]. The end goal is to offer a probiotic-rich product that benefits the host. This necessitates the presence of many active probiotic cells at the moment of consumption. The strains chosen must be well-adapted to the host's conditions, allowing them to adapt and assert their dominance in the end outcome. In contrast to many dairy products, which may necessitate the addition of probiotic cultures, fermented meat products have a natural symbiotic interaction between certain probiotics and meat-based substrates [15].

6.4 Presence of Nitrite or Nitrate Compounds

In many countries, meat preservation is an important part of food safety, and sodium nitrite and sodium chloride are routinely used for this purpose. It has been demonstrated in traditional meat fermentation that specific starting cultures may resist the addition of these preservation agents. This hardness is essential for maintaining the integrity and nutritional value of preserved meat and meat products. To protect the safety of preserved meat products, Japanese regulations mandate the concentration of sodium nitrite at 200 ppm and salt chloride at 3.3% [48]. Meeting these stringent criteria necessitates the effectiveness of establishing cultures under those conditions. In this context, 202 intestinal lactobacilli strains have been evaluated with a focus on their ability to survive and resistance capabilities under specified conditions. L. rhamnosus PERM P-15120 and L. paracasei FERM P-15121 proved to be particularly well-suited to the Japanese method requirements for non-heat-treated beef products [49]. The strains used were chosen for their extraordinary resistance to hard conditions, such as their capacity to persist at 20 °C—a temperature crucial to the preservation technique - as well as their resistance to high salt concentrations (4% sodium chloride and 200 ppm sodium nitrite). The ability of these selected strains to resist the specified conditions demonstrates their suitability to be utilized as starting cultures in the fermentation of beef products which comply with Japanese safety standards. Their ability to deal with the existence of sodium chloride and sodium nitrite is crucial for assuring the safety, quality, and conformity of the finished fermented meat products. In essence, the selection of these specific lactobacilli strains demonstrates the careful balance required when choosing starter cultures for preserved meat products. The strains not only meet the stringent safety criteria established by regulatory norms, but they also demonstrate resistance and adaptation to the difficult conditions imposed using sodium nitrite and sodium chloride in the preservation process [50].

6.5 Fermentation Temperature

The fermentation temperature is a major parameter in moulding the quality of probiotic-fermented foods, having a significant impact on factors such as probiotic bacteria viability and the total duration of the fermentation process. Temperature management is critical during fermentation to produce optimal conditions for bacterial growth and metabolic activity, providing the required properties of the product [21]. In general, the best temperature range for bacterial development in fermented foods is 40–42 °C [51]. This temperature range is thought to be advantageous for the proliferation and metabolic activities of a wide variety of probiotic strains. Bifidobacteria, a common probiotic genus, grows best at somewhat cooler temperatures, often between 37 and 41 °C [52]. There are, however, some strains that can tolerate higher temperatures. Notably, strains such as *L. sakei, L. curvatus,* and

L. plantarum can survive at temperatures as high as 45 °C [29]. This property can be useful in certain fermentation processes that need higher temperatures. The ability of *L. sakei* to grow at lower temperatures, on the other hand, highlights its adaptability. All *L. sakei* strains tested showed growth at temperatures as low as 15 °C, with many flourishing even at 4 °C. This cold tolerance is especially important in circumstances requiring fermentation or storage at lower temperatures. It is critical to stress that the temperature of incubation has a significant impact on the dynamics of mixed starting cultures. Changing the fermentation temperature can boost the growth of specific strains over others, causing population dynamics to vary. Temperature-mediated selection can lead to the dominance of specific strains within the microbial community, impacting the flavor, texture, and overall features of the fermented product [3].

6.6 Fat and Salt Content

Salt and fat are necessary ingredients in fermented sausages that influence the overall quality, safety, flavour, and texture of the finished product [53]. Fermented sausages have a rich, savory flavor and a melt-in-mouth texture that are greatly enhanced by fat. The result is influenced by the kind and distribution of fat. Belly fat contributes a more robust pig flavor, while back fat usually has a softer flavor [54]. A meat's ability to retain moisture and suppleness is enhanced by fat marbling [55]. The smoothness and general mouthfeel of the sausage are influenced by the fat content. Lower fat varieties may feel rougher and dryer, while higher fat variations may have a creamier, more opulent texture. Fat functions as a naturally occurring preservative by delaying the onset of rancidity and oxidation. This helps fermented sausages keep longer on the shelf. Fat raises the sausage's calorie and saturated fat contents while simultaneously adding flavor and texture [56]. When picking fermented sausages as part of a balanced diet, it's crucial to pay attention to the fat level.

During fermentation and storage, salt inhibits dangerous germs from growing, guaranteeing the product's safety. It also prolongs shelf life and prevents food from spoiling. Salt brings out and harmonizes the flavors of the meat, spices, and other components. It is essential to the development of the distinct savory flavor of fermented sausages [57]. Salt creates a harder and more cohesive result by encouraging the cross-linking of proteins. Additionally, it helps to retain moisture, preventing the sausage from crumbling and drying out [58]. Salt helps avoid unintended browning and preserve the meat's natural color by slowing down oxidation. It's crucial to keep in mind that the salt content of fermented sausages usually makes them rich in sodium. Achieving the ideal ratio of fat to salt is essential for creating premium fermented sausages. A sausage with too much fat will be too greasy, whereas one with too little fat will be tasteless and dry [59]. As too much salt can ruin a sausage's flavor and raise health issues, too little salt might also jeopardize its safety and shorten its shelf life.

There has been an attempt to lower the fat and salt levels of fermented sausages in response to consumer demand for healthier options. This frequently entails using thinner meat portions, adding fat substitutes, and employing different seasoning techniques [60]. Reducing these essential elements, however, may affect the flavor, texture, and shelf life; therefore, careful changes may be necessary to preserve the intended quality. All things considered, fat and salt are crucial in determining the sensory qualities and safety features of fermented sausages. Comprehending their distinct contributions enables one to make knowledgeable decisions while choosing or consuming these delectable and adaptable products.

6.7 Addition of Prebiotic Compounds

Prebiotics are essential to the creation of probiotic-rich fermented sausages since they are inert food ingredients that enhance host health by changing the flora in the digestive tract. These food ingredients, which are often very fermentable, help certain probiotic strains develop quickly, particularly when employed as starter cultures. The combination of probiotics and prebiotics in a single product is known as "synbiotics." To create a beneficial synergy in the digestive system and to optimise the health benefits of both components, they are combined [61]. The effects of prebiotics on intestinally relevant microorganisms have captivated researchers. In synbiotics, scientists investigate the complicated and dynamic interactions that occur within the digestive tract by combining diverse probiotic bacterial strains with prebiotic carbohydrates. The goal is to improve probiotics'survival, proliferation, and overall efficacy in fostering a healthy gut environment. Despite the potential benefits of synbiotics, locating appropriate prebiotic partners for specific probiotic cultures can be difficult [62]. The compatibility and synergy between the probiotic strains and the prebiotic components determine the efficiency of the symbiotic connection. Choosing the best combination requires considering the specific properties of each strain, the fermentability of the prebiotic, and the desired health outcomes [3]. The addition of highly fermentable prebiotic substances to fermented sausages can boost the growth and activity of probiotic bacteria. This, in turn, contributes to the product's overall probiotic content [63]. The synergistic action of probiotics and prebiotics in synbiotics has the potential to improve gut microbial balance, nutritional absorption, and immune function. As the subject of synbiotics evolves, continuing research strives to uncover the specific processes driving probiotic and prebiotic interactions. Understanding this is essential for tailoring synbiotic blends to address certain health issues and optimise the beneficial properties of fermented sausages. Finally, the development of fermented foods high in synbiotics signifies a deeper comprehension of the intricate connection between gut health and overall wellbeing.

6.8 Microencapsulation of Probiotic Cells

Microencapsulation is a method that includes covering bacterial cells with hydrocolloidal polymers to protect them from harsh environmental conditions. This encapsulation procedure is especially advantageous for probiotic bacteria since it allows them to withstand the harsh conditions experienced during sausage manufacture that they would not be able to tolerate otherwise. The application of microencapsulation technology to probiotic bacteria improves their fermentation capacities and competitiveness throughout the meat fermentation process [6]. Bean gum, alginate, xanthan-gellan, whey proteins, alginate-starch, gelatin, and starch have all been successfully used as wall components for probiotic microencapsulation [64]. Microencapsulation based on alginate has been shown to be efficient in guaranteeing the safety and stability of beginning cultures during the fermentation of meat products. The use of alginate microencapsulation has been widely researched to protect beginning cultures during the meat fermentation process [65]. Alginate microencapsulation and mixes of alginate with other polymers have shown various degrees of efficacy in improving bacterial stability. The impact of microencapsulation on the sensory aspects of the food product is an important issue in its application. Fortunately, when alginate particles, which resembled discrete fat particles in size and color, were mixed into the sausage matrix, they remained unnoticed, having no visible effect on the meat's texture at the concentrations utilized. Researchers have also investigated the possibility of using microencapsulation to protect probiotics in beef products. When lyophilized cultures were applied directly to the meat, their survival in the final product was jeopardized because they were promptly released into the meat matrix, resulting in delayed fermentation [23]. This problem was solved by microencapsulating beef starter cultures in alginate before freeze-drying. When immobilized in lyophilized alginate, probiotic microorganisms such as *L. plantarum* and *P. pentosaceus* performed better in the setting of fermented beef products than their unencapsulated counterparts [66].

6.9 Packaging Materials

To ensure the good quality of probiotic sausages, the viability of probiotics, particularly bifidobacteria, must be maintained during storage period [6]. The packaging process, which has a substantial impact on oxygen permeability, complicates matters even more. The viability of probiotics is critical for delivering the intended health benefits to customers. To address the issue of oxygen penetration, it is critical to employ low-oxygen-permeability packaging materials, sometimes known as oxygen-barrier packaging materials [67]. These materials are dense and less breathable, reducing oxygen ingress into the sausage and potentially jeopardizing the life of probiotic microorganisms. An alternate technique is to use active packaging, which involves integrating oxygen scavengers directly into the packaging polymers.

An oxygen-permeable food-contact polymer is stacked behind an oxygen-scavenging component in this system [67]. This design actively removes oxygen from the packing environment, resulting in a regulated atmosphere within the box. Importantly, the materials used in active packaging must be food-grade to assure the product's safety and quality.

Several factors influence oxygen permeability, including storage temperature, relative humidity, and packaging material properties. Temperature is important for both bacterial cell survival and the oxygen permeability of packing materials. Maintaining an optimal storage temperature is therefore critical for keeping the probiotic content of the sausages. While the importance of packing in retaining probiotic viability is obvious, information about how probiotic-fermented sausages are packaged is scarce in the existing literature. As a result, more research is needed to investigate and optimize packaging solutions for probiotic sausages. Understanding the interaction between temperature, packing materials, and probiotic viability is critical for creating effective packaging solutions that can maintain the quality and health advantages of probiotic-fermented sausages across their shelf life [68].

7 Sensory Characteristics of Probiotic Fermented Sausages

When choosing food products such as meat, consumers consider a variety of elements, such as sensory attributes, nutritional value, and health impact. A thorough understanding of sensory characteristics is critical for product creation, market expansion, and quality evaluation. Fermented sausages are subjected to sensory study based on a wide range of parameters to fully capture their distinct qualities [69]. The distinct smells and textures of fermented meats are obtained from byproducts of breakdown such as peptides, amino acids, carbonyls, and volatile chemicals produced by lipolysis and proteolysis [70]. The enzymatic patterns of starter cultures are important in product development because they influence the flavor and aroma of fermented meat products. Ripening during the drying process has a substantial impact on the final product's quality, influencing the nutritional content, safety, and sensory characteristics of dry fermented sausages [71]. The complex interplay of bacterial metabolism, enzymatic processes in muscle and adipose tissue, and dry sausage maturation has technical consequences for enzymatic activities that have been examined. Bacterial enzyme production speeds up the replication of specific meat flavors and broadens flavor diversity [72]. Because of the diverse and active local microbiota, sausages with extended ripening durations typically have more appealing sensory qualities than their mass-produced counterparts employing starters [73]. Choosing strains with high lipolytic and/or proteolytic activity can result in a plethora of aroma components, boosting the entire sensory experience and ensuring high sensory quality in fermented sausages [74].

8 Sensory Properties of Probiotic Fermented Meat Products

While research on the impact of bacteria on meat products is limited, there is an increasing interest in understanding how bacteria influence the sensory quality of probiotic meals. Products containing gut bacteria have different smells, which are frequently characterized by an inexplicable acidic scent [75]. Manufacturers typically improve flavors by including herbs, perfumes, sweeteners, and extra dry mass to minimize any unpleasant sensory characteristics. It is advised that prebiotic saccharides such as galacto-oligosaccharides, lactulose, or fructo-oligosaccharides be included in current fermented foods [76]. These substances encourage the growth of beneficial bifidobacteria in the gastrointestinal tract. Experimenting with probiotic bacteria added to uncooked sausages has been demonstrated to prevent the growth of pathogenic microbiota while retaining the flavor and scent of commercial sausages [21]. Lactic acid bacteria are added to uncooked sausages to create starter cultures and probiotics. The flavor characteristics of samples produced by commercial starter cultures and those fermented by different Lactobacillus strains were found to be very similar [29]. The acidic flavor had a variety of nuances, although being modest. The sensory quality of fermented pork loin was apparently altered by the probiotic strains used in fermentation [69]. Storage procedures had a considerable impact on overall sensory quality, with varying degrees of severity depending on the test samples. Fermented pork loin with *L. acidophilus* Bauer and 0.2% glucose had the best sensory quality when stored [69]. The research investigated how ripening temperature influences sensory quality. Products that were ripened between 16 °C and 20 °C had the best overall quality. Dry, uncooked loins injected with probiotic bacteria matured the best in a laboratory setting at 20 °C over a storage period of 21 days, indicating maximum microbiological and sensory quality [77]. However, because of safety concerns, a temperature of 16 °C is preferred for industrial-scale production.

9 Beneficial Effects Associated with the Consumption of Probiotic Meat Products

Most of the probiotic research on meat products focuses on their persistence inside the meat matrix as well as their technical and sensory impact on the final product. There have been fewer investigations into their implications for consumer health. The lower number of studies in this domain is due to the higher expenses, longer durations, and ethical approval procedures connected with in vivo research. A small number of studies have investigated the effects of consuming 50 g of probiotic salami daily for a few weeks on healthy volunteers' immune systems, blood triglyceride levels, and cholesterol levels [6]. The results were generally positive, with the host's immunity being weakened but plasmatic lipid levels remaining stable. Furthermore, Caco-2 cells derived from colon cancer were employed to assess probiotic cultures' ability to colonize the human gastrointestinal system and their antagonistic interactions with pathogenic organisms in vitro [78]. In the context of meat products, research is primarily focused on how probiotics can lower the likelihood of pathogenic infections, hence improving product safety. The ability of probiotic bacteria to coaggregate, providing a barrier against pathogenic germs, and their autoaggregation, critical for sticking to intestinal epithelial cells, have been major topics of research [79]. Furthermore, certain probiotic strains have been shown to organically synthesize nutraceuticals within meat products. Genetically modified *Lactobacillus plantarum* strains, for example, have produced significant amounts of folate (vitamin B_{11}) [80]. Certain probiotic bacteria also create conjugated linoleic acid, which has anti-obesity, anti-diabetic, anti-atherogenic, anti-cancer, and immune-boosting properties [81]. This opens the door to in-situ fortification of meat products, as certain bacteria can produce micronutrients and nutraceutical components. Other active substances and probiotics, such as dietary fiber, added to meat-based meals have also been studied to see how they interact [3]. These results demonstrate that, in addition to ensuring the safety of meat products, probiotic meat products have the potential to improve their nutritional value and functional qualities.

10 Conclusions and Prospects

The use of probiotic bacteria in the fermentation of meat products has lately gained popularity. However, several technological barriers must be overcome in order to properly manufacture probiotic meat products. These limitations include meat's innate microbiome, the need for low salt, nitrite, and water activity, and the lack of natural carbs. Probiotic strains used to make dry fermented meat products must be able to compete with other bacteria in the product while retaining its sensory features. They must also be able to survive the conditions seen in fermented meals. Ripening and storage have an immediate impact on the finished product, whether it's ham, loin, or sausage. This technique promotes the growth of probiotic microorganisms, which affects the flavor, nutritional value, safety, and other characteristics of the beef product. To affect the consumer's gut flora, probiotics must be present in the product at least 106 times per gram. Because probiotics in meat can work in a variety of ways, further research is needed to discover the specific strains and combinations that deliver the highest health advantages. While studies on the impact of probiotic meat-based products on human well-being have received some scientific backing, the therapeutic benefits of eating foods containing probiotic microorganisms are driving improvements in this developing field of functional foods.

References

1. Bamidele OP, Adeyanju AA, Wokadala OC, Mlambo V. African fermented fish and meat-based products. In: Indigenous fermented foods for the tropics. Elsevier; 2023. p. 117–31.
2. Pellissery AJ, Vinayamohan PG, Amalaradjou MAR, Venkitanarayanan K. Spoilage bacteria and meat quality. In: Meat quality analysis. Elsevier; 2020. p. 307–34.
3. Manassi CF, de Souza SS, Hassemer G d S, Sartor S, Lima CMG, Miotto M, et al. Functional meat products: trends in pro-, pre-, syn-, Para- and post-biotic use. Food Res Int. 2022;154:111035.
4. Cassani L, Gomez-Zavaglia A, Simal-Gandara J. Technological strategies ensuring the safe arrival of beneficial microorganisms to the gut: from food processing and storage to their passage through the gastrointestinal tract. Food Res Int. 2020;129:108852.
5. Bis-Souza CV, Barba FJ, Lorenzo JM, Penna ALB, Barretto ACS. New strategies for the development of innovative fermented meat products: a review regarding the incorporation of probiotics and dietary fibers. Food Rev Intl. 2019;35(5):467–84.
6. Munekata PES, Pateiro M, Tomasevic I, Domínguez R, da Silva Barretto AC, Santos EM, et al. Functional fermented meat products with probiotics—a review. J Appl Microbiol. 2022;133(1):91–103.
7. Li P, He W, Wu G. Composition of amino acids in foodstuffs for humans and animals. 2021. p. 189–210.
8. Geiker NRW, Bertram HC, Mejborn H, Dragsted LO, Kristensen L, Carrascal JR, et al. Meat and human health—current knowledge and research gaps. Food Secur. 2021;10(7):1556.
9. Lee HJ, Yong HI, Kim M, Choi YS, Jo C. Status of meat alternatives and their potential role in the future meat market — a review. Asian Australas J Anim Sci. 2020;33(10):1533–43.
10. Hrubša M, Siatka T, Nejmanová I, Vopršalová M, Kujovská Krčmová L, Matoušová K, et al. Biological properties of vitamins of the B-complex, part 1: vitamins B1, B2, B3, and B5. Nutrients. 2022;14(3):484.
11. Hashemi SMB, Pourmohammadi K, Gholamhosseinpour A, Es I, Ferreira DS, Mousavi KA. Fat-soluble vitamins. In: Innovative thermal and non-thermal processing, bioaccessibility and bioavailability of nutrients and bioactive compounds. Elsevier; 2019. p. 267–89.
12. Zaib S, Hayat A, Khan I. Probiotics and their beneficial health effects. Mini-Rev Med Chem. 2024;24(1):110–25.
13. Liu W, Chen M, Duo L, Wang J, Guo S, Sun H, et al. Characterization of potentially probiotic lactic acid bacteria and bifidobacteria isolated from human colostrum. J Dairy Sci. 2020;103(5):4013–25.
14. Sanders ME, Merenstein DJ, Reid G, Gibson GR, Rastall RA. Probiotics and prebiotics in intestinal health and disease: from biology to the clinic. Nat Rev Gastroenterol Hepatol. 2019;16(10):605–16.
15. Kumar D, Lal MK, Dutt S, Raigond P, Changan SS, Tiwari RK, et al. Functional fermented probiotics, prebiotics, and synbiotics from non-dairy products: a perspective from nutraceutical. Mol Nutr Food Res. 2022;66(14)
16. Cui S, Fan Z. Lactic acid bacteria and fermented meat products. In: Lactic acid bacteria. Singapore: Springer Singapore; 2019. p. 211–25.
17. Casertano M, Fogliano V, Ercolini D. Psychobiotics, gut microbiota and fermented foods can help preserving mental health. Food Res Int. 2022;152:110892.
18. Nestel PJ, Beilin LJ, Clifton PM, Watts GF, Mori TA. Practical guidance for food consumption to prevent cardiovascular disease. Heart Lung Circ. 2021;30(2):163–79.
19. Cruxen CE d S, Funck GD, Haubert L, Dannenberg G d S, Marques J d L, Chaves FC, et al. Selection of native bacterial starter culture in the production of fermented meat sausages: application potential, safety aspects, and emerging technologies. Food Res Int. 2019;122:371–82.
20. Ferrocino I, Franciosa I, Rantsiou K, Cocolin LS. The microbiome of fermented sausages. In: Good microbes in medicine, food production, biotechnology, bioremediation, and agriculture. Wiley; 2022. p. 160–8.

21. Ashaolu TJ, Khalifa I, Mesak MA, Lorenzo JM, Farag MA. A comprehensive review of the role of microorganisms on texture change, flavor and biogenic amines formation in fermented meat with their action mechanisms and safety. Crit Rev Food Sci Nutr. 2023;63(19):3538–55.
22. Leroy F, Charmpi C, De Vuyst L. Meat fermentation at a crossroads: where the age-old interplay of human, animal, and microbial diversity and contemporary markets meet. FEMS Microbiol Rev. 2023;47(2)
23. Barcenilla C, Puente A, Cobo-Díaz JF, Alexa EA, Garcia-Gutierrez E, O'Connor PM, et al. Selection of lactic acid bacteria as biopreservation agents and optimization of their mode of application for the control of Listeria monocytogenes in ready-to-eat cooked meat products. Int J Food Microbiol. 2023;403:110341.
24. Sharifi-Rad J, Rodrigues CF, Stojanović-Radić Z, Dimitrijević M, Aleksić A, Neffe-Skocińska K, et al. Probiotics: versatile bioactive components in promoting human health. Medicina (B Aires). 2020;56(9):433.
25. Fernández M, Ordóñez JA, Bruna JM, Herranz B, de la Hoz L. Accelerated ripening of dry fermented sausages. Trends Food Sci Technol. 2000;11(6):201–9.
26. Flores M, Piornos JA. Fermented meat sausages and the challenge of their plant-based alternatives: a comparative review on aroma-related aspects. Meat Sci. 2021;182:108636.
27. Jia S, Shen H, Wang D, Liu S, Ding Y, Zhou X. Novel NaCl reduction technologies for dry-cured meat products and their mechanisms: a comprehensive review. Food Chem. 2024;431:137142.
28. Hu Y, Wang J, Liu Q, Wang Y, Ren J, Chen Q, et al. Unraveling the difference in flavor characteristics of dry sausages inoculated with different autochthonous lactic acid bacteria. Food Biosci. 2022;47:101778.
29. Najjari A, Boumaiza M, Jaballah S, Boudabous A, Ouzari H. Application of isolated *Lactobacillus sakei* and *Staphylococcus xylosus* strains as a probiotic starter culture during the industrial manufacture of Tunisian dry-fermented sausages. Food Sci Nutr. 2020;8(8):4172–84.
30. Hu Y, Li Y, Li XAO, Zhang H, Chen Q, Kong B. Application of lactic acid bacteria for improving the quality of reduced-salt dry fermented sausage: texture, color, and flavor profiles. LWT. 2022;154:112723.
31. Zhu N, Liang ZS, Zhao B, Rong WQ, Min ZH, Li S, et al. Effect of processing on protein degradation and quality of emulsion sausages. Food Biosci. 2020;37:100685.
32. Guéraud F, Buisson C, Promeyrat A, Naud N, Fouché E, Bézirard V, et al. Effects of sodium nitrite reduction, removal or replacement on cured and cooked meat for microbiological growth, food safety, colon ecosystem, and colorectal carcinogenesis in Fischer 344 rats. NPJ Sci Food. 2023;7(1):53.
33. Nisa M, Dar RA, Fomda BA, Nazir R. Combating food spoilage and pathogenic microbes via bacteriocins: a natural and eco-friendly substitute to antibiotics. Food Control. 2023;149:109710.
34. Yadav M, Mandeep SP. Probiotics of diverse origin and their therapeutic applications: a review. J Am Coll Nutr. 2020;39(5):469–79.
35. Arihara K, Ohata M. Use of probiotics and prebiotics in meat products. In: Processed meats. Elsevier; 2011. p. 403–17.
36. Macedo REF, Luciano FB, Cordeiro RP, Udenigwe CC. Sausages and other fermented meat products. In: Starter cultures in food production. Wiley; 2017. p. 324–54.
37. Työppönen S, Petäjä E, Mattila-Sandholm T. Bioprotectives and probiotics for dry sausages. Int J Food Microbiol. 2003;83(3):233–44.
38. Pavli FG, Argyri AA, Chorianopoulos NG, Nychas GJE, Tassou CC. Effect of Lactobacillus plantarum L125 strain with probiotic potential on physicochemical, microbiological and sensorial characteristics of dry-fermented sausages. LWT. 2020;118:108810.
39. Anadón A, Ares I, Martínez-Larrañaga MR, Martínez MA. Probiotics: safety and toxicity considerations. Nutraceuticals Elsevier. 2021:1081–105.
40. Sotoudegan F, Daniali M, Hassani S, Nikfar S, Abdollahi M. Reappraisal of probiotics' safety in human. Food Chem Toxicol. 2019;129:22–9.

41. Sheng L, Wang L. The microbial safety of fish and fish products: recent advances in understanding its significance, contamination sources, and control strategies. Compr Rev Food Sci Food Saf. 2021;20(1):738–86.
42. Wang X, Yang Y, Huycke MM. Risks associated with enterococci as probiotics. Food Res Int. 2020;129:108788.
43. Marino M, Innocente N, Melchior S, Calligaris S, Maifreni M. Main technological challenges associated with the incorporation of probiotic cultures into foods. In: Advances in probiotics. Elsevier; 2021. p. 479–95.
44. Soares MB, Almada CN, Pereira EPR, Ferreira BM, Balthazar CF, Khorshidian N, et al. Review—Sporeforming probiotic bacteria: characteristics, health benefits, and technological aspects for their applications in foods and beverages. Trends Food Sci Technol. 2023;138:453–69.
45. Mohamed HMH, Burroughs S, Emch AW, Waite-Cusic J. Enhancing the reduction of Salmonella and Listeria monocytogenes during traditional salami processing by adding a finishing phase. Food Control. 2022;131:108432.
46. Gurram S, Jha DK, Shah DS, Kshirsagar MM, Amin PD. Insights on the critical parameters affecting the probiotic viability during stabilization process and formulation development. AAPS PharmSciTech. 2021;22(5):156.
47. Fraqueza MJ, Laranjo M, Elias M, Patarata L. Microbiological hazards associated with salt and nitrite reduction in cured meat products: control strategies based on antimicrobial effect of natural ingredients and protective microbiota. Curr Opin Food Sci. 2021;38:32–9.
48. Zhang Y, Zhang Y, Jia J, Peng H, Qian Q, Pan Z, et al. Nitrite and nitrate in meat processing: functions and alternatives. Curr Res Food Sci. 2023;6:100470.
49. Verdenelli MC, Silvi S, Cecchini C, Orpianesi C, Cresci A. Influence of a combination of two potential probiotic strains, lactobacillus rhamnosus IMC 501® and Lactobacillus paracasei IMC 502® on bowel habits of healthy adults. Lett Appl Microbiol. 2011;52(6):596–602.
50. Mladenović KG, Grujović MŽ, Kiš M, Furmeg S, Tkalec VJ, Stefanović OD, et al. Enterobacteriaceae in food safety with an emphasis on raw milk and meat. Appl Microbiol Biotechnol. 2021;105(23):8615–27.
51. Tjener K, Stahnke LH, Andersen L, Martinussen J. A fermented meat model system for studies of microbial aroma formation. Meat Sci. 2004;66(1):211–8.
52. He BL, Xiong Y, Hu TG, Zong MH, Wu H. *Bifidobacterium* spp. as functional foods: a review of current status, challenges, and strategies. Crit Rev Food Sci Nutr. 2023;63(26):8048–65.
53. Osman OA, Sulieman AME. Factors influence the quality and safety of fermented sausages. In: African fermented food products- new trends. Cham: Springer International Publishing; 2022. p. 219–33.
54. Lebret B, Čandek-Potokar M. Review: pork quality attributes from farm to fork. Part II Processed pork products Animal. 2022;16:100383.
55. Siddiqui SA, Bhowmik S, Afreen M, Ucak İ, Ikram A, Gerini F, et al. Bodybuilders and high-level meat consumers' behavior towards rabbit, beef, chicken, Turkey, and lamb meat: a comparative review. Nutrition. 2024;119:112305.
56. Domínguez R, Lorenzo JM, Pateiro M, Munekata PES, Alves dos Santos B, Basso Pinton M, et al. Main animal fat replacers for the manufacture of healthy processed meat products. Crit Rev Food Sci Nutr. 2022;64:1–20.
57. Sangaré M, Karoui R. Evaluation and monitoring of the quality of sausages by different analytical techniques over the last five years. Crit Rev Food Sci Nutr. 2023;63(26):8136–60.
58. Lu M, Zhang C, Chen B, Ai C, Chen L, Teng H. Improvement of gelation properties of Penaeus vannamei surimi by magnetic field-assisted freezing in combination with curdlan. Int J Biol Macromol. 2024;257:128323.
59. Shao L, Bi J, Li X, Dai R. Effects of vegetable oil and ethylcellulose on the oleogel properties and its application in Harbin red sausage. Int J Biol Macromol. 2023;239:124299.
60. Essa RY, Elsebaie EM. New fat replacement agent comprised of gelatin and soluble dietary fibers derived from date seed powder in beef burger preparation. LWT. 2022;156:113051.

61. Vinicius De Melo Pereira G, De Carvalho Neto DP, Junqueira ACDO, Karp SG, Letti LAJ, Magalhães Júnior AI, et al. A review of selection criteria for starter culture development in the food fermentation industry. Food Rev Intl. 2020;36(2):135–67.
62. Roy NK, Panda S, Dey G. Engineering a sustainable protein revolution: recent advances in cultured meat production. Food Bioengineering. 2023;2(4):301–16.
63. Câmara AKFI, Paglarini C d S, Vidal VAS, dos Santos M, Pollonio MAR. Meat products as prebiotic food carrier. Advances in Food and Nutrition Research. 2020;94:223–65.
64. Radosavljević M, Lević S, Pejin J, Mojović L, Nedović V. Encapsulation technology of lactic acid bacteria in food fermentation. In: Lactic acid bacteria in food biotechnology. Elsevier; 2022. p. 319–47.
65. Jurić S, Tanuwidjaja I, Fuka MM, Vlahoviček-Kahlina K, Marijan M, Boras A, et al. Encapsulation of two fermentation agents, Lactobacillus sakei and calcium ions in microspheres. Colloids Surf B Biointerfaces. 2021;197:111387.
66. Yoha KS, Nida S, Dutta S, Moses JA, Anandharamakrishnan C. Targeted delivery of probiotics: perspectives on research and commercialization. Probiotic Antimicrob Proteins. 2022;14(1):15–48.
67. Ahmed I, Lin H, Zou L, Brody AL, Li Z, Qazi IM, et al. A comprehensive review on the application of active packaging technologies to muscle foods. Food Control. 2017;82:163–78.
68. Sirini N, Loyeau P, Ruiz M, Stegmayer M, Soto L, Werning M, et al. Development of probiotic fermented sausages and viability monitoring of supplemented Lactiplantibacillus plantarum BFL strain. Fermentation. 2022;8(10):526.
69. Sha ZS, Yu WC, Ying HY, Yang L, Cai XB. Enhancement of fermented sausage quality driven by mixed starter cultures: elucidating the perspective of flavor profile and microbial communities. Food Res Int. 2024;178:113951.
70. Afzal A, Saeed F, Afzaal M, Maan AA, Ikram A, Hussain M, et al. The chemistry of flavor formation in meat and meat products in response to different thermal and non-thermal processing techniques: an overview. J Food Process Preserv. 2022;46(10)
71. Balamurugan S, Gemmell C, Lau ATY, Arvaj L, Strange P, Gao A, et al. High pressure processing during drying of fermented sausages can enhance safety and reduce time required to produce a dry fermented product. Food Control. 2020;113:107224.
72. Li Y, Cao Z, Yu Z, Zhu Y, Zhao K. Effect of inoculating mixed starter cultures of Lactobacillus and Staphylococcus on bacterial communities and volatile flavor in fermented sausages. Food Sci Human Wellness. 2023;12(1):200–11.
73. Halagarda M, Wójciak KM. Health and safety aspects of traditional European meat products. A review. Meat Sci. 2022;184:108623.
74. Meijer GW, Lähteenmäki L, Stadler RH, Weiss J. Issues surrounding consumer trust and acceptance of existing and emerging food processing technologies. Crit Rev Food Sci Nutr. 2021;61(1):97–115.
75. Ghandehari-Alavijeh S, Can Karaca A, Akbari-Alavijeh S, Assadpour E, Farzaneh P, Saidi V, et al. Application of encapsulated flavors in food products; opportunities and challenges. Food Chem. 2024;436:137743.
76. Ángeles Stegmayer M, Sirini N, Frizzo LS, Fernández-López J, Pérez Álvarez JÁ, Rosmini MR, et al. Enrichment of foods with prebiotics. In: Strategies to improve the quality of foods. Elsevier; 2024. p. 171–201.
77. Bosse R, Wirth M, Weiss J, Gibis M. Effect of storage temperature on volatile marker compounds in cured loins fermented with Staphylococcus carnosus by brine injection. Eur Food Res Technol. 2021;247(1):233–44.
78. Damodharan K, Palaniyandi SA, Suh JW, Yang SH. Probiotic characterization of Lactobacillus paracasei subsp. paracasei KNI9 inhibiting adherence of Yersinia enterocolitica on Caco-2 cells in vitro. Probiotic Antimicrob Proteins. 2020;12(2):600–7.
79. van Zyl WF, Deane SM, Dicks LMT. Molecular insights into probiotic mechanisms of action employed against intestinal pathogenic bacteria. Gut Microbes. 2020;12(1):1831339.

80. Fatima M, Ghani SM, Khanum S, Mohammad Said Al-Tawaha AR, Karnwal A, Dey A, et al. Lactic acid bacteria as a source of functional ingredients. In: Food microbial sustainability. Singapore: Springer Nature Singapore; 2023. p. 153–72.
81. Bahmid NA, Rahmayanti A, Hastuti HP, Muhpidah KI. Potential role of bioactive compounds from foods and current trends. In: Functional foods. Boca Raton: CRC Press; 2023. p. 279–316.
82. Roselino MN, Sakamoto IK, Tallarico Adorno MA, Márcia Canaan JM, de Valdez GF, Rossi EA, et al. Effect of fermented sausages with probiotic Enterococcus faecium CRL 183 on gut microbiota using dynamic colonic model. LWT. 2020 Oct;132:109876.
83. Ruiz JN, Villanueva NDM, Favaro-Trindade CS, Contreras-Castillo CJ. Physicochemical, microbiological and sensory assessments of Italian salami sausages with probiotic potential. Sci Agric. 2014;71(3):204–11.
84. Klingberg TD, Axelsson L, Naterstad K, Elsser D, Budde BB. Identification of potential probiotic starter cultures for Scandinavian-type fermented sausages. Int J Food Microbiol. 2005;105(3):419–31.
85. Ameer A, Seleshe S, Kim BJ, Kang SN. Inoculation of *Lactobacillus sakei* on quality traits of dry fermented sausages. Prev Nutr Food Sci. 2021;26(4):476–84.
86. Casaburi A, Di Martino V, Ferranti P, Picariello L, Villani F. Technological properties and bacteriocins production by Lactobacillus curvatus 54M16 and its use as starter culture for fermented sausage manufacture. Food Control. 2016;59:31–45.
87. Tabanelli G, Barbieri F, Baños A, Madero JMG, Daza MVB, Cortimiglia C, et al. Companilactobacillus alimentarius: an extensive characterization of strains isolated from spontaneous fermented sausages. Int J Food Microbiol. 2024;410:110489.
88. Zhang Q, Shen J, Meng G, Wang H, Liu C, Zhu C, et al. Selection of yeast strains in naturally fermented cured meat as promising starter cultures for fermented cured beef, a traditional fermented meat product of northern China. J Sci Food Agric. 2024;104(2):883–91.
89. Gu Y, Qiao R, Jin B, He Y, Tian J. Effect of Limosilactobacillus fermentum 332 on physicochemical characteristics, volatile flavor components, and quorum sensing in fermented sausage. Sci Rep. 2023;13(1):3942.

Meat Products with Bioactive Peptides

Shweta Kelkar, Sarang Dilip Pophaly, Awadhesh Kumar Tripathi, Manorama Chauhan, and Jadhav Shanthilal

1 Introduction

Bioactive peptides are compounds derived from amino acids, these are connected by covalent bonds while these connections are commonly referred to as amide linkages or peptide bonds. Majority of bioactive peptides are liberated from the precursor protein after their hydrolysis is completed except a few which are found free in nature. They have a notable impact on preserving human health through performing important metabolic functions and affects the digestive, cardiovascular, endocrine, nervous and immune systems [1]. BAPs are polymers of short length composed of 2–20 amino acids with a particular sequence and a lower molecular weight than the parent protein. FitzGerald and Murray (2006) describe bioactive "peptides exhibiting hormones-or-drug-like activity, which ultimately regulate physiological functions by binding to specific receptors on target cells, eliciting physiological responses [2]. The BAPs can be derived from a variety of foods like grains, dairy, eggs, beans, seaweed, certain animal protein sources such as beef, pork, chicken and seafood [3]. These BAPs are well known for their antioxidant, antimicrobial, opioid, antihypertensive, antidiabetic, antithrombotic and other important physiological properties. Along with exhibiting all the above-mentioned properties, they are scientifically proven to combat several cardiovascular and lifestyle diseases [4]. With this field hypocholesterolaemia, antioxidant, and antithrombotic properties have been identified as regulators of the cardiovascular system. Similarly, peptides with immunomodulatory effects and those capable of binding minerals play roles in the resilience and gastrointestinal functions. Certain peptide groups exhibit

S. Kelkar (✉) · J. Shanthilal
National Dairy Research Institute, Karnal, Haryana, India

S. D. Pophaly · A. K. Tripathi · M. Chauhan
College of Dairy Science and Food Technology, DSVCKV, Raipur, Chhattisgarh, India

A. Kumar Pathera et al. (eds.), *Healthier Meat Products*,
https://doi.org/10.1007/978-3-031-78215-2_10

multifunctionality, such as opioid antagonists affecting the nervous, gastrointestinal, and immune systems. Additionally, antimicrobial peptides demonstrate regulatory abilities in both gastrointestinal and immune system [5].

2 Bioactive Peptides Obtained from Meat Sources

The European Commission (EC) has provided a specific definition for the term "meat" as the product obtained after processing the consumable portion of the carcass of various domestic animals like cattle, sheep, pigs, goats, poultry and more involves extractions [6]. Amino acids present in meat and its products play a crucial role in preventing severe muscle-wasting diseases, maintaining a healthy gut environment, and regulating blood pressure through ACE inhibitory activity in connective tissues. Additionally, meat and its derivatives provides abundant essential minerals and vitamins like iron, folic acid, and vitamin B12 making them valuable nutrient sources. Moreover, they serve as immunity boosters due to antioxidants, phytanic acids, and linoleic acids, which support normal immune function [7]. The BAPs possess specific bioactivity and perform physiological functions which could deliver many health benefits. The most beneficial attributes of BAPs obtained for meat and meat products are associated with various health benefits including their potential antihypertensive, antimicrobial, ACE inhibition, antithrombotic, antidiabetic and anticancer properties. However, some Consumers connect these products with increased risks of obesity, cancer, and other chronic lifestyle diseases due to their content of saturated fatty acids and fats content, thus contributing to a negative perception of these products and cholesterol along with the intake of excess sodium is added to these products and considered major cause of hypertension. This perspective of consumers ignores the fact that these products could have a considerable impact on preserving health. At present, more research studies are being conducted to find the link between nutrients present in the diet and their physiological performance Most food components are recognized as beneficial for human health, either in treating various diseases or enhancing biological effects. Numerous Biologically Active Peptides (BAPs) derived from meat, numerous compounds like carnosine, L-carnitine, anserine, conjugated linoleic acid, taurine, creatine, and glutathione, have undergone thorough scientific investigation due to their beneficial physiological effects, rendering them increasingly appealing for inclusion in the human diet [2].

The formation of these peptides widely employs enzymatic hydrolysis, fermentation procedures or advanced emerging technologies [1]. However, the peptides produced from meat and meat products could also be released during gastrointestinal digestion. Meat proteins are characterized by their high quality, comprising essential amino acids vital for human development and normal bodily functions, thus offering versatile applications in both food and medicinal products [3].

2.1 *Synthesis of Bioactive Peptides*

Research on meat and its derivatives stand out as one of the most extensively studied area within the realm of food products relevant for the extraction of novel BAPs. Harvesting of Biologically Active Peptides (BAPs) from various dietary sources involves in the retrieval of theses compound using diverse methods protein sources is a multifaceted process, and research on this topic is extensively documented in the literature. There are two widely employed methods for the production of BAPs viz., fermentation using desirable microbial cultures and enzymatic hydrolysis. On the other hand, these peptides can be naturally generated during digestion in the GI tract. Certain researchers have documented the production of ACE inhibitory and antimicrobial peptides from beef haemoglobin through pepsin (porcine) action, while other studies have explored the utilization of enzymes present in muscle cells for extracting these peptides. Some researchers have utilized a combination of certain enzymes such as trypsin, pepsin, proteinase K, protease, thermolysin + proteinase A, papain, and tyrosinase for the hydrolysis of beef protein to generate antimicrobial and antihypertensive peptides [1]. The extraction of BAPs from the parent protein is followed by the purification of the resultant peptide mixture by employing one of the appropriate technologies like ultrafiltration, Reversed-Phase High-Performance Liquid Chromatography (RP-HPLC), Gel filtration chromatography (GFC), Size exclusion chromatography (SEC) and Ion exchange chromatography (IEC) [8].

2.1.1 Enzymatic Breakdown

Enzymatic breakdown stands out as the most reliable and efficient method to generating Biologically Active Peptides (BAPs) from meat proteins, showcasing notable antioxidant properties [9]. In various industrial settings, enzymatic hydrolysis is extensively employed for large-scale production of BAPs, with its efficiency contingent upon factors such as pH, temperature, the ratio of enzyme to substrate, duration of reaction, reaction duration, and enzyme deactivation are influential variables. Adjusting these parameters becomes imperative to achieve optimal hydrolysis of meat proteins. Commonly utilized enzymes for animal muscle hydrolysis include papain and pepsin, while food-grade enzymes like Protamex, Alcalase, and Flavourzyme find application in industries. A study highlighted utilizing Alcalase in acquiring peptides with the sequence HNGN to pork blood plasma, demonstrating significant antioxidant activity [8]. The creation of Biologically Active Peptides (BAPs) through enzymatic breakdown relies on use of commercial enzyme sourced from microbial, plant, or animal origins. Notably, industrially significant enzymes like Falvourzyme originate from Aspergillus oryzae, Alcalase from Bacillus licheniformis, and Neutrase from Bacillus subtilis. Additionally, plant-derived enzymes such as bromelain (from pineapple) and papain (from papaya) are traditionally employed for meat tenderization. These enzymes play a crucial role in meat protein

Table 1 BAPs produced through the process of enzymatic hydrolysis

Type of meat product	Process involved	BAP sequence	Biological effect	References
Muscle	Proteolysis, oxidation	APPPPAEVP, IPITAAKASRNIA, LPLGG, FAGGRGG	ACE-I activity	[10]
Cured meat products	Proteolysis	DSGVT, VPSIDDQEELM, IEAEGE DAQEKLE, EELDNALN, SLTA, SAGNPN, ALTA, DLEE, GLAGA	Antioxidant activity	[10]
Fermented meat products	Proteolysis	FGG, DM	Antioxidant activity	[10]
Collagen	Enzymatic hydrolysis	AKGANGAPGIAGAPGFPGARG PSGPQGPSGPP, GAXGLXGP, VGPV, LQGM, LC, PAGNPG ADGQPGAKGANGAP, GPRGF, QGAR, LQGMH	Antioxidant activity and ACE-I	[10]

hydrolysis and are noted for their broader specificity in protein hydrolysis site selection compared to animal-derived enzymes like trypsin and pepsin [10]. Few studies have evaluated food grade microbial enzymes to hydrolyse proteins of connective tissue and myofibril to obtain peptides with angiotensin- converting enzyme inhibiting and antioxidant efficacy. The generation of crude protein hydrolysate followed by the step of purification this included gel permeation chromatography sub fractionation. Following purification, cell-based assays were conducted to assess the cytotoxicity of the obtained hydrolysates. The findings from these procedures reveal that the produced Biologically Active Peptides (BAPs) exhibit non-cytotoxic properties and are expected to remain stable under gastrointestinal tract conditions [9] (Table 1).

2.1.2 Microbial Fermentation

In microbial fermentation to extract information differently of Biologically Active Peptide (BAPs) from meat and its products, a variety of microorganisms are employed instead of purified enzymes [11]. This fermentation process is widely recognized in the meat industry for BAP generation and is valued for its ability to improve storage conditions through the protective effects conferred by the microorganisms. In the method of fermentation to obtain BAPs, the meat products are used as substrates, followed by incubation with specific microorganisms added during fermentation processes. Microorganisms release enzymes during microbial fermentation that are unique to catalysing a variety of chemical processes, resulting in the production of hydrolysates. Meanwhile, several metabolites are also produced. Some researchers manufactured fermented meat products like those containing fermented peptidase, undergo transformations sausages, and other food products, with the addition of bacteria, yeast, and mold as beneficial starter cultures, and these fermentation processes enabled the protein hydrolysis and subsequent formation of BAPs. After fermentation, the resulting Biologically Active Peptides (BAPs) were subjected to evaluation to ascertain their physiological attributes, revealing antioxidant and antihypertensive properties in the finalized fermented products. Generally regarded as a safe, cost-effective, and environmentally friendly approach, microbial fermentation is employed for BAP production. Some researchers advocate its feasibility for deriving BAPs from various fish and byproducts compared to alternative methods; for instance, fermentation produced protein hydrolysates with ACE inhibitory effects from tuna viscera meat. BAPs from fermented fish products have been documented to exhibit antioxidant and antimicrobial activities. Notably, two antioxidant peptides with sequence LLLLLLL and PQLLLLLL, derived from fermented fish byproducts demonstrated potent antioxidant properties. However, factors affecting BAP production, such as the need for stringent control over microbial fermentation in meat and meat products, necessitate additional steps to evaluate the bioactivity of the generated BAPs [12].

2.1.3 Gastrointestinal Digestion

The meat products that consist of endogenous proteins undergo proteolysis to generate BAPs during gastrointestinal digestion. In general, gastrointestinal digestion of meat proteins consists of four stages; in the initial stage, food material undergoes oral processing and allows the subsequent formation of a swallowable lump of partially digested protein fraction. Later, food matrix disruption happens through the combined action of saliva, amylases, and mechanical forces. The gastrointestinal compartment, second phase of digestion occurs as the mass ingested is propelled through the esophagus. Here, pepsin in the intestine acts upon proteins, initiating their hydrolysis under acidic conditions typically ranging between pH 2 and 5.5, resulting in the formation of polypeptides. This process generates an acidic fluid comprising a blend of partially digested food and gastric juice, known as chyme. Chyme progresses from stomach to small intestine, where digestion continues is an environment typically maintained at a pH of 7–8. The hydrolysis of polypeptides occurs through the action of trypsin, chymotrypsin when exposed to bile salts and hormones, various peptidases and enzyme undergo alterations, thereby modifying their activity and the whole process results in the formation of oligopeptides and amino acids. Further, peptides and minerals are initiated to get absorbed by the enterocytes and reach the bloodstream. After their appropriate absorption and reaching the blood stream in intact condition, they act specifically in terms of exhibiting their physiological properties [12].

3 BAPs Derived from Meat and Meat Products

3.1 BAPs Extracted from Eggs

Hen eggs are considered one of the best foods offered by nature and are widely consumed by all age groups in the world. However, eggs are animal-origin products that can be stored in raw or uncooked form, as they tend to be a shelf-stable product at room temperature. Eggs are highly regarded as a significant protein source in the human diet. Their popularity stems from their rich content of essential amino acids, easy digestibility, and nutritional value, making them a favoured food choice [13]. Notably, egg proteins are recognized as promising precursors of Biologically Active Peptides (BAPs) owing to their diverse biological and multifunctional properties. The BAPs produced by egg proteins are shown to demonstrate antimicrobial, anti-hypertensive, antioxidant, anti-diabetic, and anti-inflammatory properties, along with binding actions with iron and calcium ions [14]. The proteins are mostly located within the albumen and yolk, along with a small amount in conjugation with lipids and carbohydrates to form lipoproteins and glycoproteins. The proteins derived from eggs and utilized for the production of BAPs are ovotransferrin, ovomucin, lysozyme, cystatin, ovalbumin, ovoinhibitor, ovomucoid, avidin, lipoprotein and glycoprotein. Egg proteins are esteemed as superior sources for generating Biologically Active Peptides (BAPs) because of their remarkable physiological characteristics which include reducing blood pressure and fighting against microorganisms, antioxidant, antidiabetic, mineral-binding, antiadhesive, and immunomodulatory activities. Additionally, hydrolysates derived from egg proteins hold potential for uncovering high-value BAPs [13] (Table 2).

Table 2 BAPs isolated from egg proteins

Precursor protein	Sequence	Property exhibited	Synthesis	References
Pepsin hydrolysate (egg white)	RADHPFL and YAEERYPIL	Activity against hypertension and inhibition peptides	In vitro GI digestion	[14]
Ovalbumin and egg white	Tyr-Ala-Glu-Glu-Arg-Tyr-Pro-Ile-Leu	Antioxidant activities	Hydrolysed by pepsin	[13]
Egg white lysozyme	KIVSDGNGMNAW VSDGNGM RNTDGSTDYGIL KIVSDGNGM	Antioxidant activities	Hydrolysed by alcalase	[13]
Egg yolk	Peptides consisting phosphorus	Antioxidant activities	Tryptic digestion	[13]
Egg white and yolk	Hydrolysates	Antioxidant activities	Enzymatic hydrolysis	[13]
Egg proteins	RADHPF, IRW, YAEERYPIL, YPI, IQW, RADHP, QIGLF, RADHPFL	ACE-inhibitory activity	Purification techniques	[13]

3.2 BAPs Generated from Poultry Products

The quantity of BAPs derived from poultry products such as chicken meat varies with the breed, type and the cooking technique is utilized. Carnosine, betaine, carnitine, and anserine are among the common Biologically Active Peptides (BAPs) found in chicken meat. Breast meat is noted to contain significantly higher levels of carnosine and anserine compared to leg meat, likely due to the higher demand for these endogenous compounds in breast meat, which possess buffering capacity. However, upon cooking, the amounts of these two BAPs decrease due to their solubility in water and subsequent loss in cooking juices [1]. Chicken is popularly consumed for its high protein content, low fat content, and abundance of unsaturated fatty acids. Its meat is extensively studied for its ease of digestion, with particular attention paid to evaluating the BAPs it produces. Carnosine and anserine are particularly notable for their roles in regulating intracellular pH, neurological functions, and oxidation inhibition. The BAPs physiological displayed by muscles extracts from chickens have been extensively documented and acknowledged for their significant impact like anti-hypertension, antibacterial and antioxidant effects. In commercial chicken meat, the carnosine content typically ranges from 0.3 to 1.83 mg/g, with breast meat containing approximately seven times more carnosine than leg muscle. Notably, in female chickens, carnosine levels were measured at 0.35 mg/g in leg muscle and 1.04 mg/g in breast muscle, while in male chickens, levels were reported at 0.30 mg/g in leg muscle and 1.71 mg/g in breast muscle. A study investigating BAPs derived from chicken collagen identified an octapeptide exhibiting ACE inhibitory activity, which subsequently lowered blood pressure levels in spontaneously hypertensive rats were assessed following the administration of treatment for a duration of 6 hours. Additionally, research has focused on cationic peptides as host defence proteins against pathogens, with studies showing antimicrobial effects against E. coli. Notably, peptides such as CATH3 and β-defensin 4 demonstrated potent antibacterial activity. Apart from these biological properties, the antioxidant activities were also identified in some protein hydrolysates obtained from chicken muscle through hydrolysis with the action of flavorzyme and α-chymotrypsin [15, 16] (Table 3).

3.3 BAPs Derived from Fish and Marine Products

The health advantages associated with fish and marine products stem predominantly from their nutritional compositions, which include polyunsaturated fatty acids, indispensable amino acids, essential nutrients and vital minerals. The BAPs produced from fish have numerous physiological properties, such as being possessing properties that counter inflammation prevent thrombosis, inhibit fat cell formation, combat oxidative stress, deter cancer growth, regulate heart rhythm, safeguard neurons, and aid in arthritis management. Also, these BAPs can find several beneficial

Table 3 BAPs obtained from poultry products (chicken)

BAP with sequence	Property exhibited	References
PVPVEGV	Decreasing power and chelating potential concerning metal ions, DPPH radical scavenging evaluation	[15]
HVTEE	Decreasing the potency and chelating properties concerning metal ions, DPPH radical scavenging assay	[15]
GPRG, GPR, GPRGP	Antithrombotic	[16]
PPGP	Antibacterial	[16]
LVL	Antihypertensive	[16]
GQ	Neuropeptide	[16]
EAE	Immunomodulating	[16]
HY, VAP, AAP, VRP, YW, LVR, PLP	Antihypertensive	[16]
RGD	Embryotoxic and antithrombotic	[16]

utilization in the food and pharmaceutical sectors. Furthermore, these BAPs can be used as functional ingredients due to their useful attributes such as emulsifying, gelling, and foaming properties. The fish industry is growing at a faster rate, and approximately 196 million tons of fish are projected to be processed by 2025. Additionally, this sector produces numerous by-products like head, trimmings, skin, fins, frames, viscera, and occasionally muscles, which are often discarded, not fully utilized, or diverted to create lower-value food item like fish silage and fish meal. The production of fish by-products can account for up to 50% of the total fish yield [17] (Table 4).

3.4 BAPs Obtained from Porcine Meat Products

Numerous Biologically Active Peptides (BAPs) have been obtained from porcine sources, sourced either starting from pork meat or its by-products. Research has successfully isolated bioactive peptides from various porcine protein fractions, including myofibrillar proteins and sarcoplasmic proteins, as well as specific muscle types like longissimus lumborum. Additionally, BAPs have been obtained from porcine by-products such as hematoid fluid, fibrous protein, skeleton structure and prepared swine items such as air-dried pork. On the other hand, extensive studies investigation the DPPH radical scavenging activity of red blood cell fraction hydrolysates derived from pigs using FP400, FPH, or bromelain exhibited elevated levels compared to those from sheep, dear and cattle red blood cell fractions treated with identical methods. Certain researchers have documented the generation of Biologically Active Peptides (BAPs) from porcine myofibrillar proteins, yielding sequences such as DSGVT, IEAEGE, DAQEKLE, EELDNALN, and VPSIDDQEELM. These peptides exhibit remarkable DPPH radical scavenging capacity and demonstrate chelating capabilities toward metal ions [1] (Table 5).

Table 4 BAPs extracted from different species of fish and marine products

Species	BAP Sequence	Property exhibited	References
Pollock (*Theragra chalcogramma*)	NGMTY, WT, NGLAP	Immunoregulatory	[18]
Atlantic salmon (*Salmo salar*)	WM, WA, VW, MW, IW, LW, LF, AW, WV, WL, FL, DW, GIG	ACE inhibitory	[18]
Atlantic salmon (*Salmo salar*)	GPGA, GPGE	DPP IV inhibitory	[18]
Atlantic salmon (*Salmo salar*)	AEVAPAPAAAAPAKAPKKKAAA KPKKAGPS	Antimicrobial	[18]
Atlantic salmon (*Salmo salar*)	VR, AP	ACE inhibitory	[18]
Hoki	VLSGGTTMYASLYAE, GSTVPERTHPACPDFN	Ca-binding and antioxidant	[18]
Pollock (*Theragra chalcogramma*)	SCH	Iron chelating activity	[18]
Pollock (*Theragra chalcogramma*)	GPAGPHGPPG, SGSTG, WR, MLGER, GPAGPR, KIGER, QGLIGPR, AGPAGPR, GPVGHGPPGKDG	Mineral chelating	[18]
Tuna (*Thunnus thynnus*)	GDLGKTTTSNWSPP, VKAGFAWTANQQLS	Antihypertensive, antioxidant	[18]
Sea bream (*Sparus autarus*)	VIY, GY, GF, VY	ACE inhibitory	[18]
Sole (*Limanda aspera*)	MIFPGAGGPEL, N-terminal RPDFDLEPPY	Antihypertensive, antioxidant	[18]
–	FKYDSTHGRFHGEVKAEGGKL (F21)	Ace inhibitory and antioxidant	[19]

3.5 BAPs Derived from Beef Meat Products

Beef stands as a significant reservoir of high-quality dietary protein essential for human health, often harnessed for the extraction of beneficial Biologically Active Peptides (BAPs). The presence of bioactive compounds in beef meat varies considerably with muscle and fiber composition, as well as their functionalities. Muscles engaged in aerobic metabolism contain coenzymes like taurine and Q10, while carnosine levels are typically low. BAPs derived from beef meat proteins exhibit a range of health-promoting biological functions. In a study, researchers identified approximately 452 peptides, of which 62 peptides, each comprising 7–21 amino acids, demonstrated a bioactivity index exceeding 0.5. Consequently, these 62 peptides were scrutinized for their potential in exhibiting the highest biological activity. Notable beef proteins commonly utilized for BAP extraction include myosin, carnosine, taurine, creatine, beta-casein, troponin, and myoglobin [1] (Table 6).

Table 5 Production of BAPs obtained from porcine meat sources

BAP sequences	Property exhibited	References
DSGVT, IEAEGE, DAQEKLE, EELDNALN, VPSIDDQEELM	DPPH radical scavenging capacity	[15]
QYP	Hydroxyl scavenging assay and DPPH	[15]
HNGN	Reducing enhancing potency in capturing metal ions, chelating activity and scavenging DPPH radicals	[15]
FISNHAY	ACE inhibition activity	[15]
NLLPP, DAVLP, DEP	ACE inhibition activity	[15]
ALTA, SLTA, VT		[15]
GKFNV, LPGGGHGDL, LPGGGT and HA	Enhancing metal ion chelation, scavenging hydroxyl radicals, assay, DPPH	[15]
GLAGA, SAGNPN	Reducing power, DPPH radical scavenging activity	[15]
AAATP	Lowering the blood pressure and ACE inhibition activity	[15]
RPR, PTPVP, KAPVA	Lowering the blood pressure and ACE inhibition activity	[15]
DLEE	Chelating activity towards metal ions, radical scavenging assay, DPPH	[15]
RMLGQTPTK	Enzymatic hydrolysis: pepsin	[20]
QGAR, LQGM, LQGMH, HC	–	[15]

4 Properties of BAPs Derived from Meat and Meat Products

4.1 *Antioxidant Activity*

The antioxidant property holds vital significance for human health, safeguarding against reactive oxygen species (ROS) that can inflict damage on membrane lipids, DNA, and proteins [23]. Numerous research endeavours have explored plant and animal protein sources, including milk, sunflower, fish, marine products, and various meats, in search of peptides exhibiting antioxidant capabilities. Hydrolysates derived from egg proteins have demonstrated antioxidant activity, effectively preventing enzymatic and non-enzymatic peroxidation of essential fatty acids. Investigators have isolated and characterized various Biologically Active Peptides (BAPs) with antioxidant properties, such as 11 KIVSDGNGMNAW, Tyr-Ala-Glu-Glu-Arg-Tyr-Pro-Ile-Leu, VSDGNGM, RNTDGSTDYGIL, and KIVSDGNGM released from egg proteins. Common techniques utilized for assessing antioxidant capacity include Capacity for Absorbing Oxygen Radicals, trolox equivalent antioxidant capacity (TEAC), Total Radical-Trapping Antioxidant parameter (TRAP), Total Oxyradical-Scavenging Potency (TOSP), and 2,2-diphenyl-1-picrylhydrazyl (DPPH) free radical scavenging assay radical-scavenging [13]. In general, this activity can be determined by the ability to reduce stable radical 2,2-diphenyl-1-picrylhydrazyl (DPPH assay) due to the action of antioxidant peptides. Several

Table 6 BAPs reported in beef and beef products

BAPs	Sequence	Property exhibited	Synthesis	References
Myosin-1 Limousine cattle (uncured)	FPMNPPKF	Antithrombotic and hypotensive	BAPs released through acid whey	[1]
Myosin-2 Limousine cattle (uncured)	FPMNPPKFD	ACE-I inhibitor and antithrombotic	BAPs released through acid whey	[1]
Ace-inhibitory peptides	INDPFIDLHYM	Ace-I inhibitory	–	[1]
Ace-inhibitory peptides	RGDLGIEPAEKVF	Ace-I inhibitory	–	[1]
Ace inhibitor hexapeptide	VLAGYK	Ace inhibitory	–	[1]
Bioactive peptide (34 amino acid residues)	TALGGILKKKGHHEAEVKHLAESHANKHKIPVKY	ACE inhibitory activity and antimicrobial effects	Enzymatic digestion: pepsin	[19]
Ace-inhibitory peptides	INDPFIDLHYM	ACE inhibitory activity	–	[8]
Ace-inhibitory peptides	RGDLGIEPAEKVF	ACE inhibitory activity	–	[8]
Ace-inhibitory peptides	VLAQYK	ACE inhibitory activity	–	[8]
ACE-I inhibitory peptides	AKGANGAPGIAGAPGFPGARGPSGPQGPSGPP	ACE inhibitory activity	Enzymatically hydrolysed: bacterial collagenase	[20]
ACE-I inhibitory peptides	PAGNPGADGQPGAKGANGAP	ACE inhibitory activity	Enzymatically hydrolysed: bacterial collagenase	[20]
Antimicrobial peptides	TKAVEHLDDLPGALSELSDLHAHKLRVDPVNFKLLSHSLL	Antibacterial activity and ACE inhibition	Hydrolysed by pepsin	[20]

(continued)

Table 6 (continued)

BAPs	Sequence	Property exhibited	Synthesis	References
Antimicrobial peptides	LDDLPG+ALSELSDLHAHKLRVDPVNFKLLSHSL	Antibacterial activity and ACE inhibition	Hydrolysed by pepsin	[20]
Antimicrobial peptides	KLLSHSL	Antibacterial activity and ACE inhibition	Hydrolysed by pepsin	[20]
Antimicrobial peptides	LLSHSL	Antibacterial activity and ACE inhibition	Hydrolysed by pepsin	[20]
Antimicrobial peptides	VTLASHLPSDFTPAVHASLDKFLANVSTVL	Antimicrobial activity	Hydrolysed by pepsin	[20]
Antimicrobial peptide	VNFKLLSHSLLVTLASHL	Antimicrobial activity	–	[20]
Opioid active tetrapeptide	YPWT	Opioid properties	Hydrolysed by intestinal enzymes	[20]
Opioid active	LVVYPWT, LVVYPWTQR and LVVYPWTQRF	Maintain and interact with opioid receptors in the brain and cardiovascular system	–	[20]
Troponin T	PVVPPLIPPKIPEG	DPP IV inhibitor	Fermented by acid whey	[21]
Immunomodulatory peptides	FDKFKHLKTEAEMK	Immunomodulating	Fermented by acid whey	[21]
Beta-enolase	SRYLGKGVLK	Neuropeptide	Fermented by acid whey	[21]
Myoglobulin	AKHPSDFGADAQ	DPPH scavenging activity, FRAP assay, and Fe^{2+} chelating ability	Papain	[22]
Mineral binding peptide	AKHPSDFGADAQA	DPPH scavenging activity, FRAP, and Fe^{2+} chelating capacity	Papain	[22]
Fibrinogen and -globulins	Whole hydrolysate	FRAP and ORAC assays	Papain	[22]
Sarcoplasmic proteins	GFHI, DFHING, FHG, and GLSDGEWQ	High antihypertensive activity, antimicrobial and cancer cell cytotoxic effects	–	[2]

BAPs isolated from meat and meat products exhibit this property in an efficient manner because of their ability to inhibit the oxidation of lipids, the chelation of metal ions, and the inactivation of free radicals and ROS. Most of the antioxidant peptides present in meat and meat products are the dipeptides anserine and carnosine. Some researchers hydrolysed the myofibrillar proteins of porcine with the help of papain and actinase E and obtained a set of five BAPs with specific sequences such as DSGVT, EELDNALN, IEAEGE, DAQEKLE and VPSIDDQEELM, which resembles antioxidant activity. Similarly, four peptides i.e. QGAR, LQGMH, LQGM and LC were isolated from pork collage by the action of plant microbial enzyme sources, namely pepsin, papain, protease derived from bovine pancreas, and bacterial proteases from *Bacillus polymyxa* and *Streptomyces* spp. Two peptides, DM and FGG, were isolated from pork sausages. However, the BAPs derived from finished meat products such as dry-cured ham were also reported to have the highest antioxidant properties revealed in the peptides GLAGA, DLEE and SAGNPN [23].

4.2 Antimicrobial

The BAPs with antimicrobial activity have been recognized through a plethora of the origin's wide variety of sources available in nature, including several plants, animals, and microorganisms. However, the peptides obtained from animal origin were found to be more effective as compared to those of plant and microbial origin and showed inhibitory effects against a large spectrum of bacteria [23]. These BAPs show inhibitory effects against bacteria causing food spoilage as well as a diverse array of pathogens in their natural habitat, comprising bacteria, fungi, viruses, and eukaryotic parasites. Antimicrobial peptides isolated from meat and meat products these substance have the potential to act as antibacterial, antiviral, and antifungal agents. Their effectiveness can be assessed by identifying the lowest concentration necessary for their activity antimicrobial peptides that is required to inhibit the proliferation of spoilage and disease-causing microorganisms. In general, these peptides have some common characteristics, as they are mostly made up of less than 50 amino acids along with 50% hydrophobic amino acids and often present as folded into amphipathic 3D structures. In a study, four BAPs with the codes GFHI, GLSDGEWQ, FHG, and DFHING obtained from bovine meat sources were evaluated to determine their antimicrobial activity against six pathogens: three gram-positive bacteria (*Staphylococcus aureus, Bacillus cereus, and Listeria monocytogenes*) and three gram-negative bacteria (*Salmonella typhimurium, Pseudomonas aeruginosa, and Escherichia coli*). The peptide GLSDGEWQ demonstrated inhibition of B. cereus, S. typhimurium, L. monocytogenes, and E. coli growth, representing the sole peptide to exhibit growth inhibition against both gram-positive and gram-negative pathogens [24]. Conversely, FHG and GFHI peptides inhibited the growth of P. aeruginosa. Eggs harbour specific proteins like ovotransferrin and lysozyme, renowned for their antimicrobial activity against a broad spectrum of bacteria. Additionally, eggs contain natural peptides distributed throughout various egg components, including the shell, vitelline membranes, and albumen. Most of these peptides exhibit amphipathic or cationic characteristics, with a few being hydrophobic, particularly α-helical peptides known for their antimicrobial efficacy [13].

4.3 Anti-diabetic

Diabetes is a metabolic disease often considered a multifactorial lifestyle disorder linked with a disturbance in the metabolism of blood glucose (blood sugar) in the human body. At present, it has become a considerable threat to global health, and in combination with obesity, it has emerged as a combined epidemic and is a significant public health problem all over the world. According to IDF data, the prevalence of diabetes globally is projected to increase by 10.1% by the year 2030. A higher probability of diabetes was identified in association with obesity. In general, diabetes is characterized as an elevated concentration of blood glucose, which is either because of inadequate ability of the pancreas to produce sufficient insulin (type1 diabetes or T1D) or the failure of the human body to utilize the insulin produced by the pancreas for the glucose metabolism (type2 diabetes or T2D). The elevated glucose levels caused by any of the two types of diabetes, namely T1D or T2D, can cause major health complications to the human body, such as diabetic retinopathy, hypertension, foot ulceration, neuropathy, and congestive heart failure [25]. However, there are several anti-diabetic drugs and medications developed commercially for the purpose of treating and curing diabetes, but they are frequently linked with harmful consequences like chromosomal abnormalities and alterations in genetic and metabolic pathways. In general, dipeptidyl peptidase IV (DPP-IV) is a serine proteinase that causes the rapid breakdown of these specific hormones such as GIP (glucose-dependent insulinotropic polypeptide) and glucagon-like peptide-1 (GLP-1), α-amylase, and α-glycosidase, and these consequences lead to the occurrence of diabetes [26]. Therefore, the blocking DPP-IV shows the most potential as a therapeutic approach in averting and management of type-2 diabetes and this has been considered an interesting target for the formulation of antidiabetic drugs. Several BAPs with DPP-IV inhibitory activity were obtained from meat muscle and its byproducts through enzymatic hydrolysis. One study has reported a dipeptide with the sequence ER obtained from bovine serum albumin (BSA) after hydrolysis by the action of papain that demonstrated an antidiabetic effect. Similarly, tripeptides PPG and PPL were isolated from bovine collagen and BSA, respectively, and were reported to inhibit DPP-IV with IC_{50} values of 2.2 and 0.3 mM. Several animal research studies suggested that the hydrolysates, or BAPs, exhibiting DPP-IV inhibitory effects are generally linked to antidiabetic activities after ingestion. A classic example of the above-mentioned suggestion is a study that included the daily administration of BAPs obtained from porcine skin by the action of Flavorzyme, and results demonstrated the improvement of glucose tolerance in diabetic rats (streptozotocin-induced) when evaluated at days 21 and 42. Hence, in this way these BAPs could be used as potential ingredients for the development of functional foods [22].

4.4 Antihypertensive and ACE Inhibitory Peptides

Antihypertensive is a physiological property which is widely studied, and ACE inhibitory BAPs can manage to prevent hypertension, which is considered the major reason for the occurrence of cardiovascular diseases. Since then, hypertension has emerged a very health concern in many developed nations, presenting major risk factors for causing severe cardiovascular ailments. Therefore, there has been burgeoning curiosity in the production of antihypertensive peptides due to their potential in lowering blood pressure and resulting in the possible preventing high blood pressure and blocking ACE activity are crucial maintaining cardiovascular health. These peptides could be used as an alternative to commercial hypertensive drugs. A spike in blood pressure is observed as age increases. People of elderly age have been reported to be severely affected by high blood pressure with 150 mmHg (systolic blood pressure) and 90 mmHg (diastolic blood pressure). In general, enzymes such as rennin and angiotensin-converting enzyme (ACE) play a significant role in the mechanisms of normal blood pressure. The use of commercial or synthetic antihypertensive drugs may cause cough, headache, dizziness and dysgeusia. Hence, we require better alternative antihypertensive drugs for the maintenance of normal blood pressure. Therefore, BAPs derived from meat and meat products can be a possible alternative to being used as an antihypertensive drug [27]. Angiotensin-converting enzyme (ACE), a dipeptidyl carboxypeptidase that plays as a catalyst to convert angiotensin I to angiotensin II (potent vasoconstrictor), This enzyme plays an imperative role in blood pressure regulation. Ovokinin, an egg white-derived BAP with the order FRADHPFL was reported to have great antihypertensive activity when it was emulsified with egg yolk [13]. Few scientists have Peptides originating from porcine skeletal muscle protein were identified in studies, wherein two ACE inhibitory peptides with sequences ITTNP and MNPPK, termed myopentapeptides A and B respectively, were isolated from thermolysin digestion. These myopentapeptides exhibited antihypertensive properties in certain in vivo studies when administered orally. Additionally, six peptides—MNP, PPK, NPP, TTN, ITT, and TNP—were identified as sequence components of myopentapeptides. Myosin, serving as a protein source, was utilized for Biologically Active Peptide (BAP) production, yielding the ACE-inhibitory octapeptide VKKVLGNP, which demonstrated significant antihypertensive activity, establishing its role as a temporally effective hypotensor. In a similar manner, another myosine-derived peptide, KRVIQY, was identified in the pepsin hydrolysate of porcine myosin, and the peptide was shown to retain its activity even after heating at 98 °C for 10 minutes before hydrolysis by pepsin. This step demonstrated that these peptides will have antihypertensive properties even after cooking at elevated temperatures [2]. Some researchers isolated a peptide sequence RMLGQTPTK from porcine troponin C hydrolysed with the help of pepsin and showed great resistance to the digestive proteins, suggesting that this peptide may work well in vivo as a good antihypertensive agent. A similar group of researchers conducted other experiments and identified two unique ACE-I inhibitory BAPs with the same sequence: KRQKYDI and EKERERQ. Among both of these BAPs, KRQKYDI exhibited the strongest ACE inhibitory activity. In a study, chicken muscle was hydrolysed to find ACE inhibitory BAPs and a total of seven such peptides, having the same sequence as LKA, LAP, LKP, FQKPKR, IVGRPRHQG, IKW, and FKGRYYP, through thermolysin treatment followed by *in vitro* evaluation to check their activity. However, FKGRYYP failed to exhibit antihypertensive activity, suggesting that the inhibitory activity against ACE does not always correspond to its in vivo antihypertensive effects. Few research studies have demonstrated the ability to extract BAPs from beef proteins and assayed the antihypertensive activities of many enzymatic hydrolysates of sarcoplasmic proteins from beef sources. A hexapeptide with the sequence VLAQYK was isolated and purified from the above-mentioned source, and the substance was extracted from the previously described source and exhibited the highest level of ACE inhibitory action [2].

4.5 *Antithrombotic Properties*

Certain peptides play a role in inhibiting the aggregation of blood platelets, offering potential benefits in thrombosis prevention [23]. Biologically Active Peptides (BAPs) obtained from meat products and their by-products hold promise as antithrombotic agents. Scientists have isolated a BAP from pork muscle, hydrolysed by papain, which exhibited antithrombotic activity. Subsequent research endeavours have investigated the antithrombotic potential of this peptide using enhanced platelet activity tests in vitro with rat blood (non-anticoagulated) and in vivo through oral administration, assessed via an examination of carotid artery thrombosis in mice induced by a helium-neon laser. The BAP shows an antithrombotic effect equivalent to that of aspirin at a dosage of 50 mg/kg of body weight. In addition, few studies have been conducted to evaluate the effect of papain hydrolysate obtained from the impact of pork consumption on plasma and liver cholesterol level concentrations, and the study results demonstrated that the levels of plasma and liver cholesterol were found to be significantly lower in rats when they were provided a diet consisting of hydrolysed meat, in comparison to untreated meat, undergoes the process where it is broken down using water as compared to the untreated meat [22].

4.6 Anticancer and Cytomodulatory Activity

In cancer prevention and treatment, proteins, peptides, and specific amino acids have demonstrated beneficial effects. Some researchers examined a series of four antimicrobial peptides sourced from bovine origins, revealing cytotoxic effects when tested on cell lines including instances of lung carcinoma (specifically A549) and breast adenocarcinoma are common occurrences in oncology. Notably, the peptide GFHI exhibited potent cytotoxicity against MCF cells and reduced the viability of AGS cells, while GLSDGEWQ notably suppressed the proliferation of AGS cells. A team of scientists evaluated the hydrolysate of dark muscle (tuna) for the estimation of antiproliferative activity by exposing it to the MCF-7 (human breast cancer cell line). They identified two BAPs with the sequences PTAEGVYMVT and LPHVLTPEAGAT from protease and papain hydrolysate, and these peptides demonstrated the highest antiproliferative properties. Several other researchers identified a novel anti-cancer peptide with the sequence ACBP, administered into goat spleen and liver, followed by immunization with gastric protein extract from humans. These study results reported that the ACBP peptide sequence in vitro in a dose-dependent manner inhibited the growth of BGC-823 (human gastric cancer cell line), and along with that, it also demonstrated antitumor activity without any measurable side effects Studies conducted in living organisms have demonstrated that ACB efficiently inhibits the proliferation of human gastric tumours in xenograft models, devoid of any discernible cytotoxic impact on the host. These results collectively imply that both Biologically Active Peptides (BAPs) and ACBP possess substantial promise as potent anticancer agents, operating through mechanisms including the induction of cell apoptosis and arrest of the cell cycle. Similarly, one study analysed a novel antitumor bioactive peptide-3 sequenced as ACBP-3 extracted from the liver of a goat to evaluate the anti- cancer effect on gastric cancer stem cells both *in vitro* and *in vivo,* and this peptide can cause a drastic decrease in CD44+ cells, suppressing the expansion of the SC (spheroid colonies) cells is impeded and their ability to form clones is suppressed [24].

5 The Production of BAPs Associated with the Meat by-Products and Wastes

The livestock sector produces a significant volume of residual meat materials and wastes, such as skin, blood, collagen, bones, and other animal body parts, which can be used as a source of valuable bioactive peptides. These products constitute a good amount of lipids, carbohydrates, vitamins, and fiber, especially the high protein content that offers numerous BAPs that can find possible uses in both the food and pharmaceutical industries. For example, blood contains haemoglobin, which is an excellent source of iron, and iron can be hydrolysed to produce valuable BAPs such as neokyotorphin (α137–141), an antimicrobial BAP. Similarly, plasma as a byproduct from pork, sheep, and other animals can be used to obtain proteases with antioxidant properties. Byproducts from beef, such as beef muscle, contain high-quality proteins for the human diet due to the presence of essential amino acids and can be an excellent source for the generation of BAPs. Collagen derived from meat and meat products can also be utilized to produce BAPs with specific biological functions. The utilization of meat over the past several years, there has been considerable research into the utilization of by- products as a significant reservoir of BAPs. In this context, two major meat by-products, such as blood and collagen, which are very essential by-products obtained by the meat industry and slaughter houses have undergone the most scrutiny. Blood is a great source of proteins, which consist of an iron-containing protein known as haemoglobin. It is significantly produced in large amounts in several countries all over the world but utilized as a potential food ingredient only in a few countries such as Europe, Asia, and Africa. As a potential source of BAPs, extensive research has been conducted on both the cellular component and the plasma component, with findings indicating that their breakdown products possess antimicrobial, antioxidant, ACE inhibitory and opioid properties. Among all the physiological properties mentioned above, antimicrobial activity exhibiting peptides derived from haemoglobin hydrolysates has been the most studied. Collagen is reported as the most abundant protein in various by-products generally obtained from the meat industry. In general, it is the main constituent of many animal body parts, such as skin, bones, hide and cartilage. Though the nutritional significance of collagen is low, it is extensively analysed as a source of BAPs [5].

Therefore, due to the high nutritional attributes of meat by-products, these products should not be discharged directly into the environment; otherwise, the improper discharge can result in several health hazards and other detrimental effects on both the environment as well as the particular industry involved [8] (Table 7).

5.1 Potential Applications of the BAPs Obtained from Meat by-Products

There are several potential ways in which these BAPs can be utilized; the majority of traditional applications falls under the manufacturing of human and animal food products. On the other hand, some applications include rendered fat for the development of useful chemicals and cosmetics, as well as hides for leather. Much of the recent advancements are pertinent to the utilization of proteins, taking advantage of their technological characteristics along with some nutritional properties, and the hydrolysis of the proteins to generate beneficial peptides exhibiting beneficial biological activities. Therefore, it becomes imperative to evaluate these by-products resulting from meat for their nutritional properties in order to find their potential applications in food and nutrition [20]. The cellular fractions obtained from red blood cells, white blood cells, and platelets possess the potential to serve as pigments enhancers in the manufacture of sausages. A heme iron polypeptide can be

Table 7 Several cases where meat remnants are utilized for creating bioactive peptides

By-product	BAP	Property	References
Pork skin	DPP-IV inhibitory peptides	ACE inhibitory, can be used to treat diabetes	[8]
Myofibrillar proteins (porcine)	DSGVT, IEAEGE, DAQEKLE, EELDNALN, and VPSIDDQEELM	Antioxidant	[8]
Liver hydrolysate (porcine)	APAAIGPYSQAVLVDR (unknown protein), GLNQALVDLHALGSAR, ALFQDVQKPSQDEWGK and LSGPQAGLGEYLFER (ferritin) and LGEHNIDVLEGNEQFINAAK (trypsinogen)	Antioxidant	[8]
Collagen (porcine)	QGAR	DPPH free radical scavenging potential was assessed within a linoleic acid oxidation environment and metal chelating assay	[22]
Plasma protein (porcine)	HNGN	Metal chelating and DPPH radical scavenging potential, reducing power and thiobarbituric acid-reactive substances	[22]
Blood (bovine)	STVLTSKYR	Antimicrobial activity	[28]
Myoglobin (bovine)	AKHPSDFGADAQA	Antioxidant activity	[28]
Blood (bovine)	TKAVEHLDDLPGALSELSDLHAHKLR VDPVNFKLLSHSLL	Antihypertensive activity	[28]
Tendon (bovine)	AKGANGAPGIAGAPGFPG ARGPSGPQGPSGPP	–	[28]

generated through the enzymatic hydrolysis of haemoglobin, and that helps with the better absorption of iron. Interesting innovations and technological advancements in the field of food processing can be explored extensively with the aid of blood proteins. Therefore, these proteins have been reported to impart beneficial properties such as gelation and emulsification contributed by immunoglobulins, fibrinogen, and serum albumin, whereas other proteins such as plasma proteins contribute to protein cross-linking, foaming and protein enrichment. Few investigations have demonstrated antimicrobial activity by red blood cell fractions generated from sheep, pigs, cattle and red deer. Similarly, some white blood cells obtained from sheep have been shown to exhibit antimicrobial activity against *E. coli, P.*

aeruginosa, and *S. aureus*. The BAPs resulting from the meat by-products can also be used as flavour enhancers; their sensory attributes largely depend on the balance and composition of small peptides chains and unbound amino acids [29].

6 Potential Applications of these BAPs

6.1 Applications of Meat Derived BAPs in Food Industries

Foods containing bioactive peptides can be divided into a number of groups according to how they are made. Even though traditional foods like meat, milk, fish, and legumes can be beneficial but aren't necessarily the healthiest choices, they can include components that are enhanced with a variety of bioactive peptides to meet specific demands. Under stringent guidelines, bioactive peptides can be incorporated into diets with particular objectives in mind, such as assisting individuals who are susceptible to certain health challenges specific goals, such as aiding vulnerable groups, such as the elderly and children. Antioxidants are often employed compounds that inhibit the oxidation of proteins and lipids. In particular, animal proteins may experience chemical alterations such as nitrosation and oxidation, which would lower their nutritional content and hinder the digestion of proteins. A recent trial using minced beef has further proved the potential of peptides as food antioxidants. These therapeutic peptides are mostly used to enhance or replace type 2 diabetes therapies, which can be burdensome for patients (requiring multiple insulin injections) and may have unintended side effects due to the presence of chemically manufactured molecules. An additional option for utilizing antidiabetic peptides would be to incorporate them as dietary supplements to have a Diabetes-Preventive Effect [12].

At present, due to major changes in lifestyle as well as shifts in regular eating habits, a large part of the global population trusts that convenient foods are more promising as they offer a wide range of BAPs. And upon the delivery of BAPs, they offer beneficial physiological functions beyond better nutrition as well as other health-promoting properties. Therefore, these BAPs have been scientifically proven to exhibit important health benefits, play a major role in human health and maintenance, and could be potentially used as an ingredient for the development of functional and novel foods [26]. In 2019, the world-wide nutraceutical industry reached a valuation of approximately USD 382.51 billion and is projected to witness a compound annual growth rate (CAGR) of 8.3% from 2020 through 2027. With time, consumers' positive attitudes towards functional as well as nutraceutical food products have tremendously increased, as these products provide additional health and wellness benefits beyond providing nutrition to the human body [30].

In several food industries, egg proteins are usually utilized as plasticizers for the reduction of hardness in food products due to their lower glass transition (T_g) than at room temperature. Egg white, as a clarifier, has been added during the manufacturing of alcoholic beverages for the purpose of removing deposits. Similarly, egg lysozyme is used as a natural preservative due to its bacteriolytic activity. At present, the BAPs generated from egg proteins are found in many nutraceuticals and functional foods [13]. BAPs obtained from fish can be potentially used as supplements in the fish meal provided to feed animals for the purpose of cost reduction. In animal feeds, these fish BAPs enhance digestion and absorption as well as strengthen immunity. Furthermore, the presence of essential amino acids in high amounts makes these BAPs useful as dietary supplements. For instance, a protein hydrolysate derived from fish (Atlantic salmon) is deemed both safe and effective due to the fact that it is reported to contain mostly short peptides and free amino acids. Also, because of their great emulsifying, foaming, and dispersing activities, these BAPs can be used as food additives and ingredients. Moreover, these BAPs have also been shown to demonstrate antioxidant properties, which can be used as a preservative to extend the shelf life of various food products. All these properties make fish protein-derived BAPs potential functional food ingredients [17].

With a comprehensive overview of the Biologically Active Peptides (BAPs), including their nutritional significance and health-enhancing attributes, such as anti-hypertensive, antimicrobial, antioxidant, anti-diabetic, ACE-inhibitory, anticancer, anti-inflammatory, and neuroprotective properties, they can be effectively employed in the development of functional foods, pharmaceuticals, and cosmetic products.

6.2 *Applications of Meat Derived BAPs in Pharmaceutical Industries*

BAPs produced for pharmaceutical purposes may be utilized to create novel, less-invasive medication regimens. Therapeutic peptides, in fact, have become an important class of medications that play a variety of clinical roles in nearly every branch of medicine. The strictest rules apply to pharmaceutical uses. In fact, BPs will improve an illness or how the body operates if they are utilized as a medication. Clinical trials must be conducted in order to verify the effects of BPs, and these are costly and time-consuming. Furthermore, relatively few of them are carried out since they present ethical, legal and financial difficulties. Moreover, marketing authorization must be obtained if the intended medication based on BAPs is able to go to the clinical trial stage. As previously said, a peptide needs to reach its target in sufficient quantity without changing in order to effectively exert its physiological function. The patient may receive an intravenous injection of BAPs to do this. Peptides can also be administered subcutaneously to shorten the time required for administration, enhance convenience, and utilize fewer medical resources than are required for intravenous injection. However, the oral route is recommended for peptides due to its ease of administration, increased patient comfort and compliance, reduced discomfort, and cheaper manufacturing costs related to the requirement for sterile production. These findings are especially important since BAPs may be used as targets in the creation of novel, long-term therapies for chronic illnesses. Despite the fact that peptides' poor bioavailability and instability in biological settings limit their therapeutic applications, more research is being done on the creation of novel orally administered BAP delivery systems. Compared to the parenteral route, oral administration facilitates easier absorption into physiological processes, although it may also have peripheral effects. Egg proteins are considered the better source to produce BAPs that can be potentially used in both the pharmaceutical as well as food industries. The BAP generated from egg proteins having the order Arg-Val-Pro-Ser-Leu is reported to exhibit antihypertensive properties. In a similar study, around 60 BAPs were extracted from boiled eggs and they showed antioxidant and anti-inflammatory properties observed in experiments conducted in vitro. A research study conducted by a team of scientists suggested that both fish and meat-derived BAPs were capable of exhibiting several health-promoting properties, such as in living organisms, this substance exhibit microbial, anti-growth, and cardiovascular-protective, *in vivo* and *in vitro* experiment models. Similarly, BAP (Arg-Pro-Arg) derived the highest blood pressure- lowering effects in living organism were observed when pork was combined with Pro-Thr-Pro-Val-Pro and Lys-Ala-Pro-Val-Ala. Therefore, these BAPs can be potentially employed as various pharmaceutical and nutraceutical supplements. Although very few nutritional supplements incorporating BP sourced from meat and readily accessible in the market, much extensive research is required in this field to explore more BAPs [26].

7 Challenges and Limitations

The primary hindrance to the promotion of health possessed by BAPs is their remarkably poor bioavailability. Bioavailability may be decreased by nucleophilic moieties' reactions and interactions with the food matrix [31]. These BAPs may interact with emulsifiers, metal ions, lipids, biopolymers, and other substances to be sequestered and prevented from freely diffusing through the extraction digesting liquid. Consequently, following digestion within the digestive system, carnosine is effectively liberated from the food substrate, enabling optimal bioavailability and absorption. Consequently, peptide transporters attached to protons PepT1 mediate the majority of carnosine absorption in the small intestine [32]. To make it easier for biopeptides to enter the market, the current legal system needs to be improved. Many people currently ingest BAPs made from conventional protein sources such milk, whey, fish, and soybeans. To determine the health advantages of these bioactive peptides, more research including in vitro and animal studies is necessary. Furthermore, it is crucial to confirm the safety of peptides by testing for toxicity, cytotoxicity, and allergen city.

To sum up, although BAPs derived from common protein sources have received interest, more research has to be done to fully realize their potential. In order to facilitate the commercialization of these promising items, the legislative framework must be modified. Additionally, additional study is required to produce strong proof of the health benefits of these medicines. The extensive research studies and interest in the production and application of BAPs is rising year by year, relevant to their utilisation in clinical trials and several efforts as well as initiatives taken by the food and pharmaceutical industries. Many BAPs reported are well known to regulate specific metabolic and cellular functions.

Many research studies conducted in human society have reported the effectiveness of BAPs generated from animal meat and meat products; whereas, they also possess some challenges and limitations that are required to be considered. Several consequences were determined during clinical trials such as smaller sample sizes, short-interval treatment procedures and the yield of results consisting of high variability. Presently, the best evidence is found in the supplementation of BAPs in the prevention of cardiovascular diseases; in general, these BAPs exhibit lipid-lowering or antihypertensive activities. BAPs in the development of nutraceuticals appear valid when in combination with other molecules, establishing their safety profiles, efficacy, and tolerance limits. It is noteworthy that the long-duration clinical trials have not been carried out appropriately to suggest their effectiveness, and the mode of action of these BAPs still needs to be clarified. Many research studies have demonstrated great physiological properties such as antioxidant, antihypertensive, antidiabetic, antimicrobial, anti-inflammatory, and many more. However, it is interesting to determine that the preliminary data of a few studies suggested the potential application of BAPs as anti-cancer agents, but more phase I studies are required. Also, in the case of immunomodulatory effects reported by several BAPs on discussing both innate and adaptive immunity, various research works are currently under investigation in pre-clinical studies [33]. In addition to clinical trials, pharmacokinetic data is also necessary to obtain, which will enable us to ascertain the appropriate dosage and dossing schedule while comprehending the varied differences among individuals in both inter -and intra- personal contexts. The biological effects of BAPs on inter-variables such as age, sex, diseases, concomitant therapy, and other factors need to be considered prior to their prescription and administration. For instance, two tripeptides, (IPP and VPP), demonstrated differences in the effectiveness of blood pressure reduction appear to vary between individuals of European and Asian descent, hinting at a genetic. Therefore, pharmacokinetic studies are necessary to be performed to study the biopharmaceutical aspects of these BAPs as well as to formulate unique strategies such as lipid microparticle systems, emulsions micelle formation and microencapsulation. However, this kind of study is often limited due to the short life of BAPs i.e., not more than 2 hours along with the low concentration of plasma (pmol/mL), thus limiting the measurement of oral bioavailability.

From the perspective of oral bioavailability BAPs reported lower bioavailability, and hence their chemical structure should be considered. In this context, along with the above-mentioned studies, some extensive *in vitro* research is essentially required to determine the bio accessibility along with the stability of several BAPs. In a few studies, it has been observed that, during the digestive process, the BAPs got inactivated, resulting in a major reduction of their physiological effects, and these consequences resulted in such a way that they were unable to reach the bloodstream in sufficient quantities to demonstrate their potential activities. However, recent developments in non-traditional dosage forms have enhanced the bioavailability of BAPs, lowering both chemical and microbial degradation, which was the main demerit of the oral supplementation of BAPs. Another crucial aspect requiring thorough investigation is the determination of the several kinds of transporters taking part in the absorption of BAPs in the intestine. The primary transporter in the duodenum responsible for absorbing most of the BP can become saturated with varying dosages. Consequently, when multiple peptides vie for the same transporter, their availability for absorption can diminish due to competition [33].

8 Conclusion

In recent decades, global meat consumption patterns have undergone significant changes. Meat and its by-products are now recognized as beneficial for human health, playing crucial roles in diet, nutrition, and cognitive development. Notably, they serve as rich sources of high-quality protein, providing essential amino acids necessary for human growth, development, and tissue repair. Additionally, meat contributes vital nutrients including essential vitamins, minerals, and bioactive compounds essential for maintaining overall human health. Meat and its by-products are considered a major as well as beneficial source of BAPs that can be utilized in numerous ways. There are several kinds of animal meats, such as bovine, porcine, poultry, and other meats, that can be potentially used to generate or produce BAPs with beneficial physiological properties, namely, antihypertensive, ACE inhibitory, antimicrobial, anti-diabetic, anticancer, antithrombotic, antioxidant, mineral binding, opioid, and immunomodulatory properties. These peptides with such great biological activities must be extracted from their specific parent proteins. There are some important methodologies that are generally employed to produce BAPs from different meat proteins, such as enzymatic hydrolysis, microbial fermentation, gastrointestinal digestion, acid-base hydrolysis, and non-conventional techniques, namely isoelectric solubilization/precipitation and sub-critical water hydrolysis. In the present scenario, several studies and experiments are focusing on the development of novel and functional food products from meat and its by-products, primarily utilizing the BAPs generated from these meat proteins.

However, there is an urgent requirement for pharmacokinetic and bioactivity studies involving human subject to gain deeper insights into these compounds. Furthermore, it is imperative to conduct long- term, randomized clinical trials to evaluate their effectiveness and assess any potential immunogenic responses. Additionally, standardization and scalability of the isolation, extraction, and production methods of BAP are crucial for their industrial viability. Consequently, an economic evaluation of potential therapeutic utilization of BAP should be undertaken in this context.

References

1. Rezaharsamto B, Subroto E. A review on bioactive peptides derived from various sources of meat and meat by-products. Int J Sci Technol Res. 2019;8(12):3151–6.
2. Department of Meat Technology and Food Quality, University of Life Sciences in Lublin, Poland, Stadnik J, Kęska P. Meat and fermented meat products as a source of bioactive peptides. Acta Sci Pol Technol Aliment. 2015;14(3):181–90.
3. Madhu M, Kumar D, Sirohi R, Tarafdar A, Dhewa T, Aluko RE, et al. Bioactive peptides from meat: current status on production, biological activity, safety, and regulatory framework. Chemosphere. 2022;307:135650.
4. Mustafa K, Kanwal J, Musaddiq S, Khakwani S. Bioactive peptides and their natural sources. In: Egbuna C, Dable Tupas G, editors. Functional foods and nutraceuticals: bioactive components, formulations and innovations [Internet]. Cham: Springer International Publishing; 2020 [cited 2023 Dec 20]. p. 75–97. Available from: https://doi.org/10.1007/978-3-030-42319-3_5.
5. Mine Y, Kovacs-Nolan J. New insights in biologically active proteins and peptides derived from hen egg. Worlds Poult Sci J. 2006;62(1):87–96.
6. Lafarga T, Hayes M. Bioactive peptides from meat muscle and by-products: generation, functionality and application as functional ingredients. Meat Sci. 2014;98(2):227–39.
7. Stephen J, Manoharan D, Radhakrishnan M. Immune boosting functional components of natural foods and its health benefits. Food Prod Process Nutr. 2023;5(1):61.
8. López-Pedrouso M, Zaky AA, Lorenzo JM, Camiña M, Franco D. A review on bioactive peptides derived from meat and by-products: extraction methods, biological activities, applications and limitations. Meat Sci. 2023;204:109278.
9. Sánchez A, Vázquez A. Bioactive peptides: a review. Food Qual Saf. 2017;1(1):29–46.
10. Chernukha IM, Mashentseva NG, Afanasev DA, Vostrikova NL. Biologically active peptides of meat and meat product proteins: a review. Part 1. General information about biologically active peptides of meat and meat products. Теория и практика переработки мяса. 2019;4(4):12–6.
11. Chauhan V, Kanwar SS. Chapter 4 – Bioactive peptides: synthesis, functions and biotechnological applications. In: Verma ML, Chandel AK, editors. Biotechnological production of bioactive compounds [Internet]. Elsevier; 2020 [cited 2023 Dec 12]. p. 107–37. Available from: https://www.sciencedirect.com/science/article/pii/B9780444643230000047.
12. Bechaux J, Gatellier P, Le Page JF, Drillet Y, Sante-Lhoutellier V. A comprehensive review of bioactive peptides obtained from animal byproducts and their applications. Food Funct. 2019;10(10):6244–66.
13. Yu Z, Yin Y, Zhao W, Chen F, Liu J. Application and bioactive properties of proteins and peptides derived from hen eggs: opportunities and challenges. J Sci Food Agric. 2014;94(14):2839–45.
14. Liu YF, Oey I, Bremer P, Carne A, Silcock P. Bioactive peptides derived from egg proteins: a review. Crit Rev Food Sci Nutr [Internet]. 2018 [cited 2023 Dec 4]. Available from: https://www.tandfonline.com/doi/abs/10.1080/10408398.2017.1329704.
15. Xing L, Li G, Toldrá F, Zhang W. Chapter 4 – The physiological activity of bioactive peptides obtained from meat and meat by-products. In: Toldrá F, editor. Advances in food and nutrition research [Internet]. Academic Press; 2021 [cited 2023 Nov 18]. p. 147–85. Available from: https://www.sciencedirect.com/science/article/pii/S1043452621000218.
16. Dziuba J, Minkiewicz P, Plitnik K. Chicken meat proteins as potential precursors of bioactive peptides. Pol J Food Nutr Sci. 1996;5(4):85–96.
17. Gao R, Yu Q, Shen Y, Chu Q, Chen G, Fen S, et al. Production, bioactive properties, and potential applications of fish protein hydrolysates: developments and challenges. Trends Food Sci Technol. 2021;110:687–99.
18. Neves AC, Harnedy PA, FitzGerald RJ. Chapter 4 – Marine processing proteinaceous by-products: a source of biofunctional food ingredients. In: Singh Dhillon G, editor. Protein byproducts [Internet]. Academic Press; 2016 [cited 2023 Dec 24]. p. 63–86. Available from: https://www.sciencedirect.com/science/article/pii/B9780128023914000045.

19. Maky MA, Zendo T. Generation and characterization of novel bioactive peptides from fish and beef hydrolysates. Appl Sci. 2021;11(21):10452.
20. Mora L, Reig M, Toldrá F. Bioactive peptides generated from meat industry by-products. Food Res Int. 2014;65:344–9.
21. Kęska P, Wójciak KM, Stadnik J. Bioactive peptides from beef products fermented by acid whey – *in vitro* and *in silico* study. Sci Agric (Piracicaba, Braz). 2019;76:311–20.
22. Lafarga T, Álvarez C, Hayes M. Bioactive peptides derived from bovine and porcine co-products: A review. J Food Biochem. 2017;41(6):e12418.
23. Biologically active peptides of meat and meat product proteins: a review. Part 2. Functionality of meat bioactive peptides [Internet]. [cited 2023 Dec 8]. Available from: https://cyberleninka.ru/article/n/biologically-active-peptides-of-meat-and-meat-product-proteins-a-review-part-2-functionality-of-meat-bioactive-peptides/viewer.
24. Bhat ZF, Kumar S, Bhat HF. Bioactive peptides of animal origin: a review. J Food Sci Technol. 2015;52(9):5377–92.
25. Mallappa RH, Balasubramaniam C, Amarlapudi MR, Kelkar S, Adewumi GA, Kadyan S, et al. Chapter 22 – Role of probiotics in the prevention and management of diabetes and obesity. In: Dwivedi MK, Amaresan N, Sankaranarayanan A, Kemp EH, editors. Probiotics in the prevention and management of human diseases [Internet]. Academic Press; 2022 [cited 2023 Dec 24]. p. 321–36. Available from: https://www.sciencedirect.com/science/article/pii/B9780128237335000064.
26. Manzoor M, Singh J, Gani A. Exploration of bioactive peptides from various origin as promising nutraceutical treasures: in vitro, in silico and in vivo studies. Food Chem. 2022;373:131395.
27. Onuh JO, Selvamuthukumaran M, Pathak YV. Bioactive peptides: production, bioavailability, health potential, and regulatory issues. CRC Press; 2021. p. 509.
28. Zaky AA, Simal-Gandara J, Eun JB, Shim JH, Abd El-Aty AM. Bioactivities, applications, safety, and health benefits of bioactive peptides from food and by-products: a review. Front Nutr [Internet]. 2022 [cited 2023 Sep 25];8. Available from: https://www.frontiersin.org/articles/10.3389/fnut.2021.815640.
29. Toldrá F, Mora L, Reig M. New insights into meat by-product utilization. Meat Sci. 2016;120:54–9.
30. Wu YHS, Chen YC. Trends and applications of food protein-origin hydrolysates and bioactive peptides. J Food Drug Anal. 2022;30(2):172–84.
31. Udenigwe CC, Fogliano V. Food matrix interaction and bioavailability of bioactive peptides: two faces of the same coin? J Funct Foods. 2017;35:9–12.
32. Marcolini E, Babini E, Bordoni A, Di Nunzio M, Laghi L, Maczó A, et al. Bioaccessibility of the bioactive peptide carnosine during in vitro digestion of cured beef meat. J Agric Food Chem. 2015;63(20):4973–8.
33. Colletti A, Favari E, Grandi E, Cicero AFG. Pharmacodynamics and clinical implications of the main bioactive peptides: a review. Nutraceuticals. 2022;2(4):404–19.

Meat Products with Modified Fatty Acid Profile

Pradeep Kumar Singh, Namrata Agrawal, and Surbhi Yadav

1 Introduction

Meat and its products are significant nutrient sources, providing essential amino acids, minerals, vitamins, and various nutrients [11]. While meat and its products are valuable sources of essential nutrients [54], various scientific evidence has linked their consumption to elevated risks of severe health conditions like coronary heart disease [53, 90]. There's a growing global demand for healthier meat options, characterized by reduced fat, cholesterol, sodium chloride, nitrite levels, improved fatty acid profiles, and the incorporation of health-enhancing ingredients [46]. Consumer acceptance of foods with modified features varies widely due to social, economic, geographical, political, cultural, and ethnic factors [40]. In Europe, the market for foods with enhanced functionality has been steadily increasing, with Central and Northern European consumers showing more preference for functional foods compared to Mediterranean regions, where a preference for fresh and natural foods is more prominent [57].

Reducing fat content while adjusting the fatty acid profile (lipid modification) is a potential strategy. Typically, lipid modification involves decreasing saturated fatty acids (SFAs) while increasing monounsaturated fatty acids (MUFAs) and polyunsaturated fatty acids (PUFAs), particularly $n-3$ PUFAs, aiming to lower the $n-6/n-3$ ratio [79]. In fact, foods with PUFA/SFA ratios of 0.4 to 1.0, $\omega-6{:}\omega-3$ ratios below 7, lower atherogenic and thrombogenic index, are considered to be more healthy, especially for the cardiovascular system [41]. The American Health Association (AHA) recommends for consumption a balanced fatty acid ratio of approximately 1:1:1 for PUFA: MUFA: SFA. Such proportions of fatty acids can promote health and reduce the level of blood LDL-cholesterol [83]. Meat is often

P. K. Singh (✉) · N. Agrawal · S. Yadav
Department of Livestock Products Technology, College of Veterinary Science and Animal Husbandry, Jabalpur, India

A. Kumar Pathera et al. (eds.), *Healthier Meat Products*,
https://doi.org/10.1007/978-3-031-78215-2_11

high in saturated fatty acids (SFAs), leading to the adoption of nutritional strategies aimed at enhancing meat quality from livestock. This strategy involves altering the animal feed to increase the levels of monounsaturated (MUFAs) and polyunsaturated fatty acids (PUFAs), which helps decrease the SFAs content in the meat. Additionally, antioxidants are added to the feed to boost the meat's oxidative stability, as MUFAs and PUFAs are more prone to oxidation, potentially compromising the quality of the meat [62].

An alternative method to improve the lipid profile in meat involves directly enriching the products with monounsaturated (MUFAs) and polyunsaturated fatty acids (PUFAs). However, increasing levels of unsaturated fatty acids (UFAs) can make the products more prone to oxidation and defects such as oiling out. Therefore, employing robust meat processing techniques, such as adding natural antioxidants, becomes a compelling option. Moreover, some replacers have been found to be effective in reducing the fat (SFA) content with the least effect on product quality; inulin has been proposed to enhance water entrapment and improve the lubricant and flow characteristics in various meat products [3], while barley β-glucan has been recognized for its thickening and gelling properties, presenting potential as a fat reducer in diverse foods, including meat products [69].

Fatty acids play a crucial role in various aspects of meat quality. Differences in their melting points significantly influence the firmness or softness of fat in meat, impacting both subcutaneous and intramuscular fat [25]. The composition of fatty acids also affects fat coloration; solidified fats with higher melting points appear whiter than those with lower melting points. Additionally, unsaturated fatty acids, especially those with multiple double bonds, contribute to rapid oxidation, influencing meat shelf life by causing rancidity and color deterioration [92]. However, this oxidation propensity is pivotal in flavor development during cooking. Fatty acids influence meat flavor by generating volatile, odorous lipid oxidation products when cooked [75]. These compounds interact with Maillard reaction products, forming additional volatiles that contribute significantly to the overall aroma and flavor profile of the meat [59].

2 Fatty Acids in Meat Products

2.1 Fatty Acid Composition in Meat

Fatty acids found in foods are typically categorized as saturated fatty acids (SFA, no double bond), monounsaturated fatty acids (MUFA, 1 double bond), or polyunsaturated fatty acids (PUFA, 2 or more double bonds). In red meat, major fatty acids include myristic (14:0), palmitic (16:0), and stearic (18:0) in the SFA category; palmitoleic (16:1) and oleic (18:1) in the MUFA category; and linoleic (18:2), linolenic (18:3), and arachidonic (20:4) in the PUFA category.

Palmitic and stearic acids are the most prevalent among saturated fatty acids (SFAs) and are commonly found in fish oil, milk, body fats of land animals, and various vegetable oils, ranging from 5% to 50% of total fatty acids [96, 97]. Stearic acid tends to be more abundant in animal fat due to biological processes like the elongation of palmityl-coenzyme A and rumen biohydrogenation [97]. Oleic acid (18:1 n − 9cis) is notably present in animal fats and other food sources, serving as a precursor for other n − 9 monoene fatty acids and the n − 9 family of polyene acids [98]. Common C18 polyunsaturated fatty acids (PUFAs) like linoleic and linolenic acids are prevalent in both meats and vegetable oils [100], whereas arachidonic acid (20:4 n − 6) is primarily found in animal fats and is relatively scarce in other sources [99] (Table 1).

2.2 *Health Implications of Fatty Acids*

Certainly, the effects of fatty acids present in meat products on health can be diverse. Here are some key points supported by research:

2.2.1 Saturated Fats

Red and processed meats are major dietary sources of saturated fats. Consuming high amounts of saturated fats can raise LDL cholesterol levels, which may increase the risk of cardiovascular diseases such as heart disease and stroke. To lower the risk of these diseases, the American Heart Association advises reducing saturated fat consumption [58].

2.2.2 Trans Fats

Some processed meat products contain artificial trans fats, which are particularly detrimental to cardiac health. The consumption of trans fats has been linked to increased LDL cholesterol and a higher risk of coronary heart disease. Many health organizations advocate reducing or eliminating trans fats from the diet [61].

2.2.3 Omega-3 and Omega-6 Fatty Acids

While meat contains small amounts of essential fatty acids, they are predominantly found in oily cold-water fish such as salmon, herring, mackerel, anchovies, and sardines. Mackerel, for example, has the highest concentration, with 2.2 grams of $\omega - 3$ PUFAs per 100 grams of fish [91]. It has been proposed to fortify food products with $\omega - 3$ PUFAs like eicosapentaenoic acid (EPA; 20:5) and docosahexaenoic acid (DHA; C22:6), replacing saturated, monounsaturated, and $\omega - 6$

Table 1 Major fatty acids (percentage) in different type of meat

Product		SFA	MUFA	PUFA	Myristic 14:0	Palmitic 16:0	Stearic 18:0	Palmitoleic 16:1	Oleic 18:1	Linoleic 18:2	Linolenic 18:3	Arachidonic 20:4	References
Beef	Concentrate finishing	40.9–47.9	40.1–49.9	2.9–10.7	2.0–3.7	23.8–28.0	12.8–14.8	3.6–5.7	30.5–44.5	2.5–7.7	0.1–0.9	0.2–3.5	[7, 45, 86]
	Grass finishing	47.7–51.5	25.4–47.2	3.6–7.9	2.5–3.0	28.3–30.7	14.2–18.0	2.5–3.5	31.3–42.8	2.12–3.1	0.7–1.7	0.3–0.8	
Lamb		40.8–53.3	37.9–47.3	3.6–8.6	1.9–4.0	20.9–28.8	16.1–23.9	1.4–2.2	35.7–54.3	3.2–5.2	0.6–2.20	0.5–1.1	[7, 86]
Goat		37.2–53.8	33.7–53.9	8.9–16.5	1.6–5.1	17.3–31.4	15.0–17.3	2.1–5.7	28.0–49.2	5.2–11.8	1.1–1.2	3.1–4.7	[7, 86]
Pork		36.9–41.7	26.9–46.8	14.0–32.5	1.1–1.5	23.8–26.7	11.9–14.0	2.5–3.6	31.1–42.8	11.3–23.3	0.2–0.4	1.4–6.5	[7, 86, 95]
Turkey		50.67–52.64	23.79–28.07	21.26–23.57	0.54–1.17	20.12–21.15	7.2–7.86	3.86–5.9	19.93–22.17	19.91–22.21	1.17–1.24	0.12–0.18	[35]
Horse meat		35.8–36.8	27.4–34.6	29.7–35.8	2.3–4.5	24.2–25.9	5.5–7.5	2.9–4.5	22.2–28.1	11.7–22.9	7.9–16.6	0.7–3.0	[29, 84]

MUFA monounsaturated fatty acids, *PUFA* polyunsaturated fatty acids, *SFA* saturated fatty acids

polyunsaturated fatty acids to improve health outcomes [39]. These fatty acids, particularly EPA and DHA, are known for their cardiovascular benefits, such as lowering the risk of heart disease, although the levels in meat are generally lower than those in fish, which may limit their health impact [60].

2.2.4 Conjugated Linoleic Acid (CLA)

This naturally occurring trans fat found in meat and dairy products, particularly in grass-fed animals, has been suggested to have potential health benefits. Research indicates CLA might have anti-cancer properties and could aid in reducing body fat. However, more studies are needed to fully understand its effects on human health [10].

Remember, the health effects of fatty acids in meat can also be influenced by other dietary and lifestyle factors. Moderation and balance in consumption, along with an overall healthy diet, are essential for maintaining good health. The Committee on Medical Aspects of Nutrition Policy (COMA) established guidelines for daily fat intake: total fat should not exceed 35% of daily food energy, with specific limits for saturated fats (SFAs), monounsaturated fats (MUFAs), polyunsaturated fats (cis-PUFAs), and trans-fatty acids. They recommend replacing SFAs with a mix of MUFAs and $n - 6$ PUFAs to balance fatty acid consumption [48]. Unsaturated fatty acids like MUFAs and $n - 6$ PUFAs, found in oils, can reduce LDL cholesterol, lowering cardiovascular disease (CVD) risk. Long chain $n - 3$ PUFAs in fish oils potentially protect against fatal heart disease, but evidence on (alpha linolenic acid) ALNA's role in CVD risk is not very clear. It has been proposed that to obtain the physiological benefits similar to those of DHA and EPA, one must consume about five times the quantity of ALA (Alpha-Linolenic Acid) due to the body's low efficiency in converting ALA to the active forms of DHA and EPA [91]. For diabetes, a low-saturated-fat diet is recommended due to the heightened CVD risk associated with high SFA intake. High-MUFA diets may improve glycaemic control, but there's insufficient evidence for specific recommendations. Certain unsaturated fatty acids, like $n - 3$ PUFAs and MUFAs, might lower the risk of certain cancers, yet authoritative bodies don't provide specific dietary advice. Instead, a balanced diet rich in fruits, vegetables, and high-fiber foods is encouraged. Inflammatory conditions like asthma or arthritis might benefit from $n - 3$ and $n - 6$ PUFA modifications, though the clinical impact remains uncertain without more comprehensive trials. During pregnancy, adequate intake of long chain $n - 3$ PUFAs is crucial for fetal growth and neurological development, but caution is advised regarding excess fish consumption due to potential contaminants. Breastmilk and PUFA-supplemented infant formulas help meet infant PUFA needs. It has been found that long chain PUFAs, particularly DHA and arachidonic acid, are crucial for brain development and cognitive function.

3 Modification of Fatty Acid Profile

3.1 Dietary Modification in Livestock Feeding

The fatty acid composition profoundly influences the quality of ruminant and non-ruminant meat, affecting factors like fat tissue hardness, shelf life, and flavor. Oxidation of unsaturated fatty acids (UFAs) within meat can lead to structural changes in cells and the production of potentially harmful by-products. Various factors, including animal production systems and diet composition, can influence these fatty acids' susceptibility to oxidation. Modifications in beef fatty acid composition, achieved by offering different fat supplements to cattle, led to significant alterations in the meat's DHA and ALA content. However, increased n − 3 PUFA levels correlated with higher lipid oxidation, impacting meat color and sensory attributes.

3.1.1 Vitamin E Supplementation

Vitamin E supplementation played a role in preventing oxidation, but its efficacy varied with the degree of unsaturation in the meat. Grass-fed beef showed lower lipid oxidation and better color retention, attributed to higher tissue levels of vitamin E due to antioxidants present in grass. Vitamin E combined with plant extracts rich in polyphenols effectively protected n − 3 PUFA enriched meat from oxidation during processing. In pork, enriching with n − 3 long chain PUFA can affect shelf life and meat color due to PUFA oxidation, influencing product quality. Studies altering pork's fatty acid content through marine algae supplementation highlighted the potential impact on muscle composition, particularly the selective enrichment of DHA in muscles over subcutaneous fat. Vitamin E supplementation in pork, while beneficial for oxidative stability, showed inconclusive effects on aspects like muscle color and water holding capacity. Studies examining the impact of dietary modifications on cooked meat revealed alterations in flavor attributes, with higher n − 3 PUFA content correlating with specific sensory scores. Lamb meat with varying fatty acid compositions elicited differences in flavor perceptions, with fish oil and marine algae diets displaying oxidative instability and reduced shelf life. Familiarity with specific meat types influenced consumer preferences, with correlations observed between PUFA content and flavor differences. Ultimately, variations in fatty acid composition due to dietary changes significantly affect meat quality and sensory attributes (Woods and Fearon, [94].

3.1.2 Fodder

Numerous studies confirm that ruminants fed on fresh pasture exhibit higher unsaturated fatty acid (UFA) levels in their meat compared to those on cereal-based diets. Grass, particularly rich in n − 3 polyunsaturated fatty acids (PUFA) like linolenic acid (C18:3 n − 3), varies in content based on maturity and variety. Animals raised on pasture have notably higher proportions of linolenic acid in their fat than those stall-fed. Research by French et al. [30] observed that increasing grass intake while reducing concentrate proportion in steers' diets led to a rise in PUFA:SFA ratio and a decrease in saturated fatty acid (SFA) concentration. Animals exclusively fed grass exhibited the highest PUFA concentration in intramuscular fat. Grass and grass silage showed significantly higher-linolenic acid compared to concentrates. Studies by Nuernberg et al. [63]; Demirel et al. [22], and Scerra et al. [77] consistently found higher levels of specific fatty acids like lauric and palmitic acid in animals fed concentrates compared to those on pasture diets, believed to pose a health risk. Other studies, such as Sañudo et al. [78], highlighted that grass-fed lambs and cattle showed higher percentages of linolenic acid and various unsaturated fatty acids in their meat compared to those fed concentrates. This could potentially enhance the nutritional value of meat and shift the negative perception of the saturated nature of beef when finished on pasture.

3.1.3 Oil Seeds and Vegetable Oils

Several studies have explored modifying meat fatty acid composition using whole oilseeds. For instance, replacing soybeans with extruded or crushed linseed in the diets of animals like Belgian Blue young bulls increased linolenic acid content [73]. Similarly, feeding Holstein bulls whole canola or linseed at different lipid levels raised the n − 3 PUFA concentration in muscle [26]. Feeding lambs diets rich in fat and oils like fish oils, kelp, and flax seeds was reported to increase polyunsaturated fatty acids [93]. Other studies have explored rabbit diets containing extruded linseed, showing an increase in specific fatty acids in muscles and cooked meat, resulting in an improved nutritional value of rabbit meat [42].

Research conducted by Crespo and Esteve-García [17] in broiler chickens demonstrated distinct impacts of different plant oils on the fatty acid deposition sites. Supplementation with olive oil resulted in higher C18:1 levels; sunflower oil led to increased linoleic acid proportions (51.1 g/100 g total body FA); and linseed oil contributed the most n − 3 PUFA and a favorable n − 6:n − 3 ratio in the carcass fat. Additionally, Lu et al. [47] study on pork indicated that dietary supplements of linseed and soybean oil significantly raised C18:3 and C18:2 contents in neutral lipids and phospholipids within specific muscles like the longissimus and biceps brachii.

3.1.4 Fish Oil and Marine Algae

Studies involving pigs and their dietary incorporation of n − 3 long chain polyunsaturated fatty acids (LCPUFA) from fish oil revealed time-dependent influences on the transfer of these fatty acids. Introducing fish oil during specific fattening phases significantly affected the incorporation of DHA, with a substantial increase observed when administered during the second phase. Direct dietary intake of DHA was emphasized as essential for its augmentation in animal muscle, indicating that pork's DHA levels were not significantly affected by dietary precursors.

Marine algae, an alternative to fish oil, richly supply n − 3 long chain PUFA (LCPUFA) like EPA and DHA. Studies incorporating marine algae into sheep diets, reported by Cooper [15] and others, highlighted higher percentages of specific PUFAs like DHA in muscle and adipose tissue, compared to diets with fish oil. Similar studies across rabbits, lambs, and pigs also indicated effects on fatty acid composition in muscle tissue.

3.2 *Genetic Modification*

Regarding the nutritional value of animal-based foods, the primary focus lies on altering lipid profiles to elevate n − 3 long-chain PUFA content. The bulk of research exploring genetic engineering methods, particularly transgenesis, in meat composition centres on enhancing this specific aspect. Scientists have been tinkering with animal genes, especially in pigs, to change how their meat stores fat. In an experiment, by Saeki et al. [74], they put a gene (Δ12 fatty-acid desaturase) from spinach into pigs. This gene helps make some important fat LA and ALA (Linoleic Acid and Alpha-Linolenic Acid) that mammals usually can't produce but plants can. These pigs ended up having more of this fat in their bodies. It showed that we can change what fats animals have by using genes from plants. Another study by Lai et al. [44] worked on making pigs that carry a special gene (C. elegans fat-1 gene) found in a round worm. This gene helps make healthy fats found in fish. These pigs had way more of these good fats in their muscles, i.e. substantial increases in n − 3 PUFA levels and reductions in the n − 6/n − 3 PUFA ratio, making their meat healthier to eat. Tang et al. [82] also checked if these changes affected the animals' health. They found that the special genes (the fat-1 gene) didn't mess up the gut bugs in these pigs. They also made sure these new genes didn't spread to other pigs that didn't have them. Other experiments by Chen et al. [14] are looking into genes from fungi and plants to improve the fat in livestock. For example, there's a gene (sdd17 gene) from a fungus (*Saprolegnia diclina*) that can boost healthy fat and make bad fat less in animals, i.e., enhance n − 3 long chain PUFA content and decrease the n − 6/n − 3 PUFA ratio in cells. Similar things happened when Indo et al. [36] used a gene (fad-3 n − 3 fatty-acid desaturase) from a plant called scarlet flax in cows, which led to increased n − 3 PUFA levels and reduced n − 6/n − 3 PUFA ratios in bovine adipocytes. Transgenesis, aimed at improving animal growth performance, has also significantly influenced the lipid composition of meat [72]. Transgenic pigs supplemented with the bovine growth hormone transgene demonstrated reduced fat content in their carcasses and a more optimal SFA: MUFA: PUFA ratio [81].

3.3 *Post-Slaughter Variations Observed in Fatty Acids*

3.3.1 Seasonal Variations

The time of slaughter significantly impacted the levels of linoleic acid C18:2, arachidonic acid C20:4, and CLA in bull carcasses, with higher concentrations observed during the summer [16]. A similar trend was noticed for vaccenic acid and docosapentaenoic acid C22:5 (DPA), but without statistical significance. These fatty acids are linked to health benefits, indicating that beef from summer-slaughtered cattle has a more favorable fatty acid profile compared to winter-slaughtered ones [2]. The type of fat deposits influenced the concentrations of functional fatty acids, with intramuscular fat having the lowest CLA levels but higher concentrations of certain PUFAs compared to other fat types [23]. Bulls slaughtered in the summer showcased higher arachidonic acid and DPA content in intramuscular fat. Overall, beef from summer-slaughtered cattle appears to possess higher nutritional value and health-promoting properties, attributed to seasonal variations in forage quality available to the cattle [18]. Oleic acid was the predominant functional MUFA, significantly higher in intramuscular and subcutaneous fat compared to other fat types [1]. The interaction between slaughter season and fat type influenced the content of C20:4 and C20:5 fatty acids, notably higher in the intramuscular fat of animals slaughtered in summer, possibly due to adaptations in muscle metabolism caused by different grazing periods [80].

3.3.2 Lipid Oxidation

After slaughter, the cessation of blood flow and metabolic activities in animal tissues leads to the oxidation of fatty acids. This oxidation process involves three phases: initiation, propagation, and termination, during which oxygen interacts with the double bonds in fatty acids, forming various byproducts such as hydroperoxides, aldehydes, acids, and ketones [4]. The susceptibility of lipids in meat to oxidation is influenced by the fatty acid composition, the presence of antioxidants like vitamin E, and prooxidants such as free iron in muscle tissues [55]. Polyunsaturated fatty acids, in particular, are more vulnerable to oxidation, which can degrade the quality and nutritional value of meat [34]. In a study, goose meat stored in a high-oxygen modified atmosphere showed an increase in saturated fatty acids and a decrease in polyunsaturated fatty acids, regardless of storage temperature [65]. The breakdown of lipids in meat can happen through enzymatic or non-enzymatic processes. Enzymatic breakdown involves specific enzymes like lipases and phospholipases that split glycerides, forming free fatty acids that can lead to off-flavors known as rancidity. Non-enzymatic breakdown occurs due to heme proteins like hemoglobin and myoglobin, which can produce hydroperoxides contributing to oxidation. This process involves the oxidation of iron and hydrogen peroxide formation. Additionally, after slaughter, natural enzymes within the animal tissues break down complex compounds like carbohydrates, fats, and proteins, leading to changes in meat texture and color. This process, known as autolysis, involves proteolysis and fat hydrolysis, which are essential for further microbial decomposition. The breakdown of proteins by enzymes like calpains, cathepsins, and aminopeptidases contributes to flavor and texture changes in meat, including tenderization. The breakdown of proteins and fats in meat after slaughter, caused by enzymes and oxidation, can affect its quality, flavor, and nutritional value [19].

4 Fat Replacer

Fat replacers are substances designed to mimic the functional and sensory properties of fat, primarily to reduce both the fat and calorie content in foods while minimally affecting their taste and texture [37]. Recent research has focused intensively on their use in meat products. The adoption of low-fat diets is often hindered by a noticeable decline in the texture and flavor of reduced-fat foods. Fat replacers are classified into two main types based on their function and composition [6]. The first type includes fat substitutes such as biomolecules or their derivatives that perform similarly to fats, categorized by their origins into carbohydrate-based (like starches that absorb water and provide a soft texture), protein-based (including albumen, casein, and whey protein), and fat-based substitutes (like Caprenin or Olestra, which are too large to digest and thus contribute minimal calories). Notably, Olestra, made from sucrose and fatty acid esters, has been restricted in use due to potential gastrointestinal effects [38].

The second category encompasses fat mimetics, which partially replicate the sensory qualities of animal fat and include various hydrocolloids (gums, pectins), protein aggregates, and emulsion gels [71]. Selecting the right fat replacers that maintain the sensory integrity of foods is crucial for enhancing the quality of low-fat products [13]. Recent innovations, such as olive and chia oil oleogels and emulsion gels [70], and inulin-based emulsion-filled gels [31], have shown promise in substituting pork backfat in dry fermented sausages, indicating the potential for creating healthier meat products using vegetable oils.

5 Effects on Meat Quality and Sensory Attributes

5.1 Texture, Flavor and Aroma

The burgers had a smoother texture thanks to a mix of inulin, β-glucan, and breadcrumbs. In these burgers, 5.3% of breadcrumbs were replaced by 2.2% β-glucan and 3.1% inulin, which lowered some textural values. The primary part of breadcrumbs is wheat starch, which swells during cooking, making a firm structure with protein gel. Yet, when hydrocolloids like inulin and β-glucan were added, the starch swelling was reduced. This led to less leaching of amylose, disrupting network formation during cooking [9]. β-glucan in the meat system created a tight, porous network that trapped water, preserving moisture and reducing protein cross-linkages during cooking. This made the burgers less hard, cohesive, and springy. However, while β-glucan and inulin (up to 8% w/w) separately affected texture differently, combining them produced a softer texture. This might be due to better moisture retention with the β-glucan/inulin mix in low-fat and traditional burgers.

The beef patties with added olive and linseed oils along with grape seed extract showed better color preservation compared to regular beef patties throughout the display period. This positive impact on color aligns with the instrumental color results obtained. Discoloration scores gradually increased for both types of patties during display, more prominently for conventional patties. This increase in discoloration is usually linked to pigment oxidation in beef [27]. Interestingly, odor change wasn't influenced by the type of formulation during the display period, aligning with findings in low-fat beef patties with higher PUFA levels. For the modified patties, off-odors became noticeable later than the color change, suggesting that odor was the limiting factor. The mixture of olive and linseed oils in these patties didn't negatively impact their sensory quality on the first day of cooking. The proportions of these oils in the mixture were relatively low and didn't significantly affect aroma, flavor, oiliness, or other sensory aspects of the patties.

Assessors did not find significant differences in firmness, moistness, flavour, and overall scores between low fat products with starch or pea fibre addition and high fat control [68]. Carrageenan with pectin gel [12] or with whey protein and tapioca starch [50] can be successfully used without altering the sensory attributes of low fat pork sausages.

5.2 Shelf-Life and Stability

The stability and shelf life of modified fatty acids in meat products depend on various factors, including processing methods, storage conditions, and the type of modified fatty acids incorporated. Research by Delgado-Pando et al. [20] focused on the stability of conjugated linoleic acid (CLA) in cooked meat products during storage. They observed that CLA content was relatively stable when meat products were stored under refrigeration, with minimal degradation over time. Additionally, studies by Offer et al. [64] investigated the impact of different storage conditions on the stability of omega-3 fatty acids in meat products enriched with these lipids. They found that packaging methods, especially vacuum packaging or modified atmosphere packaging, helped maintain the integrity of omega-3 fatty acids, prolonging their shelf life by protecting against oxidation. However, it's important to note that the stability of modified fatty acids in meat products can vary based on the specific lipid modifications, the presence of antioxidants, packaging methods used, and storage temperatures. Monitoring changes in fatty acid composition over time and under different storage conditions is crucial to understanding the shelf life of modified fatty acids in meat products.

6 Consumer Perception and Acceptance

6.1 Consumer Attitudes and Preferences

Consumer attitudes and preferences regarding the modified fatty acid profile of meat products are influenced by various factors, including health considerations, sensory attributes, and the information provided. Several studies have explored consumer perceptions of meat products with altered fatty acid profiles. Research by Grunert et al. [33] indicated that consumers are generally interested in meat products with improved nutritional profiles, such as those with reduced saturated fat and increased levels of omega-3 fatty acids. Health-conscious consumers tend to prefer meat products that align with dietary recommendations for heart health. A study by Verbeke et al. [89] found that consumers perceive omega-3-enriched meat products as healthier and are willing to pay a premium for such products. The positive association of omega-3 fatty acids with heart health contributes to the favorable consumer attitude.

Providing clear information about the modified fatty acid profile on product labels is crucial. Consumers are more likely to choose meat products with modified fatty acids when they are well-informed about their health benefits. Research by Van Wezemael et al. [87] emphasized the importance of transparent labeling. Sensory attributes, especially taste and flavor, play a significant role. Studies, such as that by Grunert et al. [32], suggest that while health considerations are important, consumers are not willing to compromise on the taste and overall sensory quality of meat products. Cultural factors and dietary habits also influence consumer preferences. For example, a study by Pieniak et al. [66] highlighted differences in consumer attitudes towards modified fatty acid profiles in meat products across European countries, indicating the influence of cultural variations. Consumer acceptance is also influenced by the cost of modified meat products. Studies, including those by Lusk and Sullivan [49], have shown that consumers may be willing to pay a premium for perceived health benefits, but there are limits to this willingness.

In summary, consumer attitudes and preferences towards meat products with modified fatty acid profiles are multifaceted, involving considerations related to health, taste, information transparency, cultural factors, and affordability. Successful market acceptance often depends on a balance between improved nutritional benefits and sensory quality.

6.2 Challenges in Acceptance

The acceptance of meat products with modified fatty acid profiles faces several challenges, rooted in consumer perceptions, market dynamics, and industry limitations. These challenges impact the adoption and market success of such products. Consumer understanding of the health benefits associated with modified fatty acids in meat products can be limited. Lack of awareness or misconceptions about the benefits of omega-3 enriched meats might hinder acceptance. Educating consumers about the advantages of these modifications is crucial for acceptance [66, 67]. Altering the fatty acid composition can sometimes lead to changes in taste, texture, or overall sensory attributes of meat products. Consumers are sensitive to these alterations and may resist products that differ significantly from traditional tastes or mouthfeel. Balancing health benefits with maintaining taste and texture is a key challenge [56].

Inadequate or unclear labelling about modified fatty acids may create mistrust or confusion among consumers. Ambiguity in labelling might fail to communicate the nutritional improvements, leading to consumer scepticism or reluctance to try these products [88]. The accessibility and availability of modified fatty acid meat products can be limited in certain regions or markets. Their distribution and availability in supermarkets or retail outlets might be challenging, reducing consumer exposure and thus acceptance [51].The cost of producing meat with modified fatty acid profiles could be higher due to specialized feeding regimes or production methods. This can result in a higher price for these products, limiting their affordability for some consumers. Regulatory frameworks regarding labelling requirements, health claims, and safety standards for modified meat products vary across regions. Compliance with these regulations can pose further challenges for producers and might influence consumer trust. Consumer scepticism about genetically modified organisms (GMOs) or perceived risks associated with altered food might impact acceptance. Negative perceptions about genetic modification or alterations in food could affect consumer trust in such products [8].

Addressing these challenges requires a multifaceted approach involving transparent communication, targeted education, improving sensory quality, regulatory alignment, cost optimization, and ensuring widespread availability. Overcoming these hurdles is crucial for the successful acceptance and integration of meat products with modified fatty acid profiles into the market.

7 Regulatory Considerations

7.1 Labelling and Marketing

Labelling and marketing of modified fatty acid profiles in meat products are crucial for consumer understanding and acceptance. Providing accurate and transparent information on packaging is pivotal. It involves clearly stating the modified fatty acids present in the meat product. The labelling should be straightforward, easy to comprehend, and backed by scientific evidence. This transparency builds trust and helps consumers make informed choices about the products they purchase [88]. Educating consumers through campaigns can significantly impact their understanding and acceptance. These campaigns might involve information about the health benefits of modified fatty acids, explaining how they contribute to a healthier diet, and addressing any misconceptions consumers might have [51]. While promoting the benefits of modified fatty acids, adherence to regulatory guidelines is crucial. Any health claims made on the packaging should be substantiated by scientific evidence and comply with the regulations set by food governing bodies. This ensures that consumers receive accurate information and prevents misleading claims [66, 67].

Visual aids on packaging can simplify complex information. Using symbols, graphics, or labels that indicate improved fatty acid profiles can quickly convey the nutritional benefits to consumers, even those who might not delve deeply into the nutritional information [88]. Engaging consumers through various channels, such as social media, websites, or informational campaigns, can foster a better understanding of modified fatty acids. Interactive platforms allow for direct communication with consumers, addressing their concerns, sharing recipes, and offering information about the sources and benefits of modified fatty acids [52]. These approaches collectively aim to bridge the gap between technical nutritional information and consumer comprehension, fostering a greater acceptance and appreciation of meat products with modified fatty acid profiles.

7.2 Health Claims and Nutritional Information

Health claims associated with modified fatty acid profiles in meat products typically revolve around the potential benefits they offer for human health. These claims often focus on the improved nutritional content compared to traditional meat products. Some common health claims and nutritional information related to modified fatty acids are related to omega-3 fatty acids. Incorporating health claims on product packaging requires adherence to regulatory standards. The claims should be scientifically supported and compliant with the guidelines provided by regulatory bodies such as the FDA (Food and Drug Administration), EFSA (European Food Safety Authority), or similar local authorities. For instance, claims related to reduced saturated fat content or increased omega-3 fatty acids should be substantiated by credible scientific studies [24, 28].

Providing detailed nutritional information is essential for informed consumer choices. This includes specifics about the modified fatty acids present, their quantities, and their health benefits. The nutritional content, especially in terms of fatty acid composition, can be presented in a standardized format, often displayed in a table or chart on the packaging [88]. Supplementing packaging with additional educational material or leaflets can offer a deeper understanding of the health benefits associated with modified fatty acids. These materials can explain the role of different fatty acids in a balanced diet, how they contribute to overall health, and their specific effects on the body. Scientific literature, health organization publications, and government dietary guidelines can be referenced in these educational materials [5, 85]. These strategies collectively aim to ensure that health claims and nutritional information associated with modified fatty acids in meat products are accurate, transparent, and supported by credible scientific evidence.

8 Future Trends

The future trends in meat products with modified fatty acid profiles are poised for several advancements. With growing health consciousness, consumers seek healthier food options. Future trends suggest a rise in demand for meat products boasting healthier fatty acid compositions. This shift in consumer preferences may drive the food industry to focus more on producing and promoting these healthier meat options [43]. Anticipated advancements in biotechnology and genetic engineering could revolutionize the precision and efficiency of modifying fatty acid profiles in meat. This may lead to more tailored compositions that cater to specific dietary needs, potentially offering a broader range of healthier meat choices [78].

There's a likelihood of evolving regulations governing the labelling and marketing of modified fatty acid meat products. This is aimed at ensuring transparency, accuracy, and standardized information for consumers. Regulatory changes might aim to provide clearer guidelines for the nutritional content and health claims on product labels [21]. Continuous research on the health impacts of different fatty acids could significantly influence the formulation of meat products. This could lead to optimized meat compositions that offer enhanced health benefits while retaining desirable taste and texture, potentially expanding the range of healthier meat options available to consumers. These trends suggest a future where meat products with modified fatty acid profiles could become more diverse, transparently labeled, and tailored to meet specific nutritional needs while maintaining consumer appeal and satisfaction.

9 Conclusion

In conclusion, the demand for healthier meat options with modified fatty acid profiles and the incorporation of health-enhancing ingredients is steadily increasing globally. Fatty acids play a crucial role in determining meat quality, from texture and coloration to flavor development and shelf life. The emphasis on creating "healthy" foods has gained significant attention, considering the potential health risks associated with consuming high-fat diets. Reducing fat content while adjusting the fatty acid profile presents a potential strategy to address these concerns. Incorporating prebiotic dietary fibers and utilizing ingredients such as inulin and barley β-glucan can further enhance the nutritional and functional properties of meat products. As the market for functional foods continues to grow, the development of meat products with modified fatty acid profiles and health-enhancing ingredients holds promise in meeting the evolving preferences of consumers for healthier and more nutritious food options.

References

1. Aldai N, Nájera AI, Dugan MER, Celaya R, Osoro K. Characterisation of intramuscular, intermuscular and subcutaneous adipose tissues in yearling bulls of different genetic groups. Meat Sci. 2007;76(4):682–91.
2. Alfaia CM, Castro ML, Martins SI, Portugal AP, Alves SP, Fontes CM, et al. Effect of slaughter season on fatty acid composition, conjugated linoleic acid isomers and nutritional value of intramuscular fat in Barrosã-PDO veal. Meat Sci. 2007;75(1):44–52.
3. Alvarez D, Barbut S. Effect of inulin, β-Glucan and their mixtures on emulsion stability, color and textural parameters of cooked meat batters. Meat Sci. 2013;94(3):320–7.
4. Amaral AB, Silva MVD, Lannes SCDS. Lipid oxidation in meat: mechanisms and protective factors–a review. Food Sci Technol. 2018;38:1–15.
5. American Heart Association. Dietary fats. https://www.heart.org/en/healthy-living/healthy-eating/eat-smart/fats/dietary-fats
6. Asyrul-Izhar AB, Bakar J, Sazili AQ, Meng GY, Ismail-Fitry MR. Incorporation of different physical forms of fat replacers in the production of low-fat/reduced-fat meat products: which is more practical? Food Rev Intl. 2023;39(9):6387–419.
7. Banskalieva V, Sahlu TA, Goetsch AL. Fatty acid composition of goat muscles and fat depots: a review. Small Rumin Res. 2000;37:255–68.
8. Belaunzaran X, Lavín P. Consumer perception of omega-3 enriched meat. Food Res Int. 2018;103:131–6.
9. Bemiller JN. Pasting, paste, and gel properties of starch–hydrocolloid combinations. Carbohyd Polym. 2011;86(2):386–423.
10. Bhattacharya A, Banu J, Rahman M, Causey J, Fernandes G. Biological effects of conjugated linoleic acids in health and disease. J Nutr Biochem. 2006;17(12):789–810.
11. Biesalski HK. Meat as a component of a healthy diet — are there any risks or benefits if meat is avoided in the diet? Meat Sci. 2005;70:509–24.
12. Candogan K, Kolsarici N. Storage stability of low-fat beef frankfurters formulated with carrageenan or carrageenan with pectin. Meat Sci. 2003;64:207–14.
13. Chavan RS, Khedkar CD, Bhatt S. Fat replacer. In: The Encyclopedia of food and health, vol. 2; 2016. p. 589–95.
14. Chen Y, Zhang M, Gou K. SDD17 desaturase can convert arachidonic acid to eicosapentaenoic acid in mammalian cells. Biochem Biophys Res Commun. 2010;394(1):158–62.
15. Cooper SL. Dietary manipulation of the fatty acid Composition of sheep Meat (Doctoral dissertation, The Open University), 2002.
16. Costa P, Roseiro LC, Partidário A, Alves V, Bessa RJB, Calkins CR, Santos C. Influence of slaughter season and sex on fatty acid composition, cholesterol and α-tocopherol contents on different muscles of Barrosã-PDO veal. Meat Sci. 2006;72(1):130–9.
17. Crespo N, Esteve-Garcia E. Nutrient and fatty acid deposition in broilers fed different dietary fatty acid profiles. Poult Sci. 2002;81(10):1533–42.
18. Dannenberger D, Nuernberg G, Scollan N, Schabbel W, Steinhart H, Ender K, Nuernberg K. Effect of diet on the deposition of n-3 fatty acids, conjugated linoleic and C18: 1 trans fatty acid isomers in muscle lipids of German Holstein bulls. J Agric Food Chem. 2004;52(21):6607–15.
19. Dave D, Ghaly AE. Meat spoilage mechanisms and preservation techniques: a critical review. Am J Agric Biol Sci. 2011;6(4):486–510.
20. Delgado-Pando G, Cofrades S, Ruiz-Capillas C, Solas MT, Triki M, Jiménez-Colmenero F. Low-fat frankfurters formulated with a healthier lipid combination as functional ingredient: microstructure, lipid oxidation, nitrite content, microbiological changes and biogenic amine formation. Meat Sci. 2011;89(1):65–71.
21. Delgado-Pando G, Cofrades S, Ruiz-Capillas C, Teresa Solas M, Jiménez-Colmenero F. Healthier lipid combination oil-in-water emulsions prepared with various protein sys-

tems: an approach for development of functional meat products. Eur J Lipid Sci Technol. 2010;112(7):791–801.
22. Demirel G, Ozpinar H, Nazli B, Keser O. Fatty acids of lamb meat from two breeds fed different forage: concentrate ratio. Meat Sci. 2006;72(2):229–35.
23. Duckett SK, Wagner DG, Yates LD, Dolezal HG, May SG. Effects of time on feed on beef nutrient composition. J Anim Sci. 1993;71(8):2079–88.
24. EFSA (European Food Safety Authority). Guidance on the scientific requirements for health claims related to antioxidants, oxidative damage and cardiovascular health. EFSA J. 2011;9(12):2474.
25. Farmer LJ, Mottram DS. Lipid-Maillard interactions in the formation of volatile aroma compounds. In: Developments in food science, vol. 35; 1994.
26. Faustman C, Cassens RG, Schaefer DM, Buege DR, Williams SN, Scheller KK. Improvement of pigment and lipid stability in Holstein steer beef by dietary supplementation with vitamin E. J Food Sci. 1989;54(4):858–62.
27. Faustman C, Sun Q, Mancini R, Suman SP. Myoglobin and lipid oxidation interactions: mechanistic bases and control. Meat Sci. 2010;86(1):86–94.
28. FDA (Food and Drug Administration). Guidance for industry: a food labeling guide (14. Claims). U.S. Department of Health and Human Services, FDA, 2021.
29. Ferjak EN, Cavinder CA, Sukumaran AT, Burnett DD, Lemley CO, Dinh TTN. Fatty acid composition of mesenteric, cardiac, abdominal, intermuscular, and subcutaneous adipose tissues from horses of three body condition scores. Livest Sci. 2019;223:116–23.
30. French P, Stanton C, Lawless F, O'riordan EG, Monahan FJ, Caffrey PJ, Moloney AP. Fatty acid composition, including conjugated linoleic acid, of intramuscular fat from steers offered grazed grass, grass silage, or concentrate-based diets. J Anim Sci. 2000;78(11):2849–55.
31. Glisic M, Baltic M, Glisic M, Trbovic D, Jokanovic M, Parunovic N, Dimitrijevic M, Suvajdzic B, Boskovic M, Vasilev D. Inulin-based emulsion-filled gel as a fat replacer in prebiotic- and PUFA-enriched dry fermented sausages. Int J Food Sci. 2019;54:787–97.
32. Grunert KG. Food quality and safety: consumer perception and demand. Eur Rev Agric Econ. 2005;32(3):369–91.
33. Grunert KG, Verbeke W, Kügler JO, Saeed F, Scholderer J. Use of consumer insight in the new product development process in the meat sector. Meat Sci. 2011;89(3):251–8.
34. Gumułka M, Połtowicz K. Comparison of carcass traits and meat quality of intensively reared geese from a Polish genetic resource flock to those of commercial hybrids. Poult Sci. 2020;99(2):839–47.
35. Igenbayev A, Nurgazezova A, Okuskhanova E, Rebezov Y, Kassymov S, Nurymkhan G, Tazeddinova D, Mironova I, Rebezov M. Fatty acid composition of female Turkey muscles in Kazakhstan. J World Poult Res. 2019;9(2):78–81.
36. Indo Y, Tatemizo A, Abe Y, Suzuki I, Matsumoto K, Hosoi Y, et al. Functional expression of a humanized gene for an ω-3 fatty acid desaturase from scarlet flax in transfected bovine adipocytes and bovine embryos cloned from the cells. Biochim et Biophys Acta Mol Cell Biol Lipids. 2009;1791(3):183–90.
37. Jacqueline MB. Chapter 6—Lipids basics: fats and oils in foods and health: healthy lipid choices, roles and applications in nutrition. In: Culinary nutrition. Cambridge, MA: Academic; 2013. p. 231–77.
38. Jandacek RJ, Letton JC, Procter and Gamble Co. Compositions containing novel solid, non-digestible, fat-like compounds. U.S. Patent No. 4,797,300, 10 January 1989.
39. Jimenez-Colmenero F. Healthier lipid formulation approaches in meat-based functional foods. Technological options for replacement of meat fats by non-meat fats. Trends Food Sci Technol. 2007;18(11):567–78.
40. Jimenez-Colmenero F, Carballo J, Cofrades S. Healthier meat and meat products: their role as functional foods. Meat Sci. 2001;59:5–13.

41. Jones T. A review: value added meat products with modified fatty acid profiles. Adv Food Technol Nutr Sci Open J. 2016;SE(2):S18–S26. https://doi.org/10.17140/AFTNSOJ-SE-2-103.
42. Kouba M, Benatmane F, Blochet JE, Mourot J. Effect of a linseed diet on lipid oxidation, fatty acid composition of muscle, perirenal fat, and raw and cooked rabbit meat. Meat Sci. 2008;80(3):829–34.
43. Kris-Etherton PM, Fleming JA. Emerging nutrition science on fatty acids and cardiovascular disease: nutritionists' perspectives. Adv Nutr. 2015;6(3):326S–37S.
44. Lai L, Kang JX, Li R, Wang J, Witt WT, Yong HY, et al. Generation of cloned transgenic pigs rich in omega-3 fatty acids. Nat Biotechnol. 2006;24(4):435–6.
45. Leheska JM, Thompson LD, Howe JC, Hentges E, Boyce J, Brooks JC, Shriver B, Hoover L, Miller MF. Effects of conventional and grass-feeding systems on the nutrient composition of beef. J Anim Sci. 2008;86:3575–85.
46. Li Y, Mei H, Liu Y, Li Z, Qamar H, Yu M, Ma X. Dietary supplementation with Rutin alters meat quality, fatty acid profile, antioxidant capacity, and expression levels of genes associated with lipid metabolism in breast muscle of Qingyuan partridge chickens. Food Secur. 2023;12(12):2302. https://doi.org/10.3390/foods12122302.
47. Lu P, Zhang LY, Yin JD, Everts AK, Li DF. Effects of soybean oil and linseed oil on fatty acid compositions of muscle lipids and cooked pork flavour. Meat Sci. 2008;80(3):910–8.
48. Lunn J, Theobald HE. The health effects of dietary unsaturated fatty acids. Nutr Bull. 2006;31(3):178–224.
49. Lusk JL, Sullivan P. Consumer acceptance of genetically modified foods. Food Technol. 2002;56(10):32–7.
50. Lyons PH, Kerry JF, Morrissey PA, Buckley DJ. The infl uence of added whey protein/carrageenan gels and tapioca starch on the textural properties of low fat pork sausages. Meat Sci. 1999;51:43–52.
51. Magnusson MK, Arvola A, Koivisto Hursti UK, Aberg L, Sjödén PO. Attitudes towards organic foods among Swedish consumers. Br Food J. 2003;105(3):261–72.
52. Mandell M, Shuster M, Fetterolf D, Wolinsky I. Identifying and understanding consumer preferences for specific beef flavor characteristics. J Sens Stud. 2005;20(2):107–24.
53. Mcafee AJ, Mcsorley EM, Cuskelly GJ, Moss BW, Wallace JMW, Bonham MP, Fearon AM. Red meat consumption: an overview of the risks and benefits. Meat Sci. 2010;84(1):1–13.
54. Mehta N, Ahlawat SS, Sharma DP, Dabur RS. Novel trends in development of dietary fiber rich meat products—a critical review. J Food Sci Technol. 2015;52(2):633–47.
55. Meng H, Matthan NR, Wu D, Li L, Rodríguez-Morató J, Cohen R, et al. Comparison of diets enriched in stearic, oleic, and palmitic acids on inflammation, immune response, cardiometabolic risk factors, and fecal bile acid concentrations in mildly hypercholesterolemic postmenopausal women—randomized crossover trial. Am J Clin Nutr. 2019;110(2):305–15.
56. Menozzi D, Sogari G, Veneziani M, Simoni E, Mora C, Marescotti A. "What is natural is good": the role of the naturalness bias in consumer evaluation of agri-food technologies. Food Qual Prefer. 2017;62:340–8.
57. Menrad K. Market and marketing of functional food in Europe. J Food Eng. 2003;56:181–8.
58. Mensink RP, Zock PL, Kester AD, Katan MB. Effects of dietary fatty acids and carbohydrates on the ratio of serum total to HDL cholesterol and on serum lipids and apolipoproteins: a meta-analysis of 60 controlled trials. Am J Clin Nutr. 2003;77(5):1146–55.
59. Mottram DS. Flavour formation in meat and meat products: a review. Food Chem. 1998;62:415–24.
60. Mozaffarian D, Wu JH. Omega-3 fatty acids and cardiovascular disease: effects on risk factors, molecular pathways, and clinical events. J Am Coll Cardiol. 2011;58(20):2047–67.
61. Mozaffarian D, Katan MB, Ascherio A, Stampfer MJ, Willett WC. Trans fatty acids and cardiovascular disease. N Engl J Med. 2006;354(15):1601–13.
62. Nieto G, Ros G. Modification of fatty acid composition in meat through diet: effect on lipid peroxidation and relationship to nutritional quality—a review. Lipid Peroxid. 2012;12:239–58.

63. Nuernberg K, Fischer A, Nuernberg G, Ender K, Dannenberger D. Meat quality and fatty acid composition of lipids in muscle and fatty tissue of Skudde lambs fed grass versus concentrate. Small Rumin Res. 2008;74(1–3):279–83.
64. Offer GJ, Mermelstein J, Brightman E, Brandon NP. Thermodynamics and kinetics of the interaction of carbon and sulfur with solid oxide fuel cell anodes. J Am Ceram Soc. 2009;92(4):763–80.
65. Orkusz A, Wolańska W, Krajinska U. The assessment of changes in the fatty acid profile and dietary indicators depending on the storage conditions of goose meat. Molecules. 2021;26(17):5122. https://doi.org/10.3390/molecules26175122. PMID: 34500555; PMCID: PMC8433622.
66. Pieniak Z, Verbeke W, Scholderer J. Health-related beliefs and consumer knowledge as determinants of fish consumption. J Hum Nutr Diet. 2010a;23(5):480–8.
67. Pieniak Z, Verbeke W, Scholderer J, Brunsø K. Impact of consumers' health beliefs, attitudes and nutritional knowledge on the use of health claims. Food Qual Prefer. 2010b;21(1):87–96.
68. Pietrasik Z, Janz JAM. Utilization of pea fl our, starch-rich and fiber-rich fractions in low fat bologna. Food Res Int. 2010;43:602–8.
69. Pinero M, Parra K, Huerta-Leiddenz N, Arenas DE, Moreno L, Ferrer M, Areujo S, Barboza Y. Effect of oat's soluble fiber (β-glucan) as a fat replacer on physical, chemical, microbiological and sensory properties of low-fat beef patties. Meat Sci. 2008;80(3):675–80.
70. Pintado T, Cofrades S. Quality characteristics of healthy dry fermented sausages formulated with a mixture of olive and chia oil structured in oleogel or emulsion gel as animal fat replacer. Food Secur. 2020;9:830.
71. Protte K, Weiss J, Hinrichs J, Knaapila A. Thermally stabilised whey protein-pectin complexes modulate the thermodynamic incompatibility in hydrocolloid matrixes: a feasibility-study on sensory and rheological characteristics in dairy desserts. Lwt. 2019;105:336–43.
72. Pursel VG, Solomon MB. Alteration of carcass composition in transgenic swine. Food Rev Intl. 1993;9(3):423–39.
73. Raes K, De Smet S, Demeyer D. Effect of dietary fatty acids on incorporation of long chain polyunsaturated fatty acids and conjugated linoleic acid in lamb, beef and pork meat: a review. Anim Feed Sci Technol. 2004;113(1–4):199–221.
74. Saeki K, Matsumoto K, Kinoshita M, Suzuki I, Tasaka Y, Kano K, et al. Functional expression of a Δ12 fatty acid desaturase gene from spinach in transgenic pigs. Proc Natl Acad Sci. 2004;101(17):6361–6.
75. Salter LJ, Mottram DS, Whitfield FB. Volatile compounds produced in maiilard reactions involving glycine, ribose and phospholipid. J Sci Food Agric. 1989;46(2):227–42.
76. Sañudo C, Enser ME, Campo MM, Nute GR, Marıa G, Sierra I, Wood JD. Fatty acid composition and sensory characteristics of lamb carcasses from Britain and Spain. Meat Sci. 2000;54(4):339–46.
77. Scerra M, Caparra P, Foti F, Galofaro V, Sinatra MC, Scerra V. Influence of ewe feeding systems on fatty acid composition of suckling lambs. Meat Sci. 2007;76(3):390–4.
78. Scollan N, Hocquette JF, Nuernberg K, Dannenberger D, Richardson I, Moloney A. Innovations in beef production systems that enhance the nutritional and health value of beef lipids and their relationship with meat quality. Meat Sci. 2006;74(1):17–33.
79. Simopoulos AP. The importance of the ratio of omega-6/omega-3 essential fatty acids. Biomed Pharmacother. 2002;56(8):365–79.
80. Sobczuk-Szul M, Wroński M, Wielgosz-Groth Z, Mochol M, Rzemieniewski A, Nogalski Z, Pogorzelska-Przybyłek P, Purwin C. The effect of slaughter season on the fatty acid profile in four types of fat deposits in crossbred beef bulls. Asian Australas J Anim Sci. 2013;26(2):275–81.
81. Solomon MB, Pursel VG, Campbell RG, Steeleae NC. Biotechnology for porcine products and its effect on meat products. Food Chem. 1997;59(4):499–504.
82. Tang M, Zheng X, Cheng W, Jin E, Chen H, Yang S, et al. Safety assessment of sFat-1 transgenic pigs by detecting their co-habitant microbe in intestinal tract. Transgenic Res. 2011;20:749–58.

83. Teerasak P, Tharawee W, Anawin C, Sirusa K, Chatchawan C, Preeyabhorn D. Developing a soft margarine with modified fatty acid profile having low *trans* fatty acids. Int J Food Sci Technol. 2023; https://doi.org/10.1111/ijfs.16692.
84. Tonial I, Aguia AC, Oliveira CC, Bonnafé EG, Visentainer JV, De Souza NE. Fatty acid and cholesterol content, chemical composition and sensory evaluation of horsemeat. S Afr J Anim Sci. 2009;39(4):328.
85. USDA Dietary Guidelines: https://www.dietaryguidelines.gov/.
86. USDA. FoodData Central, 2021. U.S. Department of Agriculture Agricultural Research Service. fdc.nal.usda.gov. Accessed 8 April 2021.
87. Van Wezemael L, Caputo V, Nayga RM Jr, Chryssochoidis G, Verbeke W. European consumer preferences for beef with nutrition and health claims: a multi-country investigation using discrete choice experiments. Food Policy. 2014;44:167–76.
88. Verbeke W. Consumer acceptance of functional foods: socio-demographic, cognitive and attitudinal determinants. Food Qual Prefer. 2005;16(1):45–57.
89. Verbeke W, Frewer LJ, Scholderer J, De Brabander HF. Why consumers behave as they do with respect to food safety and risk information. Anal Chim Acta. 2007;586(1–2):2–7.
90. Verma A, Banerjee R. Dietary fibre as functional ingredient in meat products: a novel approach for healthy living — a review. J Food Sci Technol. 2010;47(3):247–57.
91. Weiss J, Gibis M, Schuh V, Weiss HSJ, Gibis M, Schuh V, Salminen H. Advances in ingredient and processing systems for meat and meat products. Meat Sci. 2010;86:196–213.
92. Whitfield FB, Mottram DS. Volatiles from interactions of Maillard reactions and lipids. Crit Rev Food Sci Nutr. 1992;31(1–2):1–58.
93. Wood JD, Richardson RI, Nute GR, Fisher AV, Campo MM, Kasapidou E, et al. Effects of fatty acids on meat quality: a review. Meat Sci. 2004;66(1):21–32.
94. Woods VB, Fearon AM. Dietary sources of unsaturated fatty acids for animals and their transfer into meat, milk and eggs: a review. Livest Sci. 2009;126(1–3):1–20.
95. Yu K, Shu G. Fatty acid and transcriptome profiling of longissimus dorsi muscles between pig breeds differing in meat quality. Int J Biol Sci. 2013;9:108–18.
96. Lobb K, Chow CK. Fatty acid classification and nomenclature. In: CK Chow, editor, Fatty acids in foods and their health implications. CRC Press, Boca Raton; 2000. p. 1–16.
97. Jenkins TC. Lipid metabolism in the rumen. J Dairy Sci. 1993; 76:3851–3863. https://doi.org/10.3168/jds.S0022-0302(93)77727-9
98. Gunstone FD. Fatty acids—Nomenclature, structure, isolation and structure determination, biosynthesis and chemical synthesis. In: Fatty acid and lipid chemistry, Springer, Boston, MA; 1996. p. 1–34.
99. Gunstone FD. Fatty acid and lipid chemistry. Springer, Boston, MA; 2012.
100. Wallis JG, Watts JL, Browse J. Polyunsaturated fatty acid synthesis: What will they think of next? Trends Biochem Sci. 2002; 27:467–473. https://doi.org/10.1016/s0968-0004(02)02168-0

Meat Products Enriched with Micronutrients

Suman Bishnoi, Gauri Jairath, Monika Rani, Sanjay Yadav, and Pramod Kumar

1 Introduction

As per the recent report published by World Health Organization (WHO) in year 2020, about 3.1 billion population are on the edge of getting nutritional deficiencies due to the inaccessibility of nutritious diet. It is further reported that approximately 222 million people across 53 countries are facing severe food insecurity [1]. Thus, UN Food Systems Summit and Nutrition for Growth Summit was held in 2021 that emphasized on achieving zero hunger by 2030. Further, global nutrition research community addressed malnutrition comprehensively, defined malnutrition as "triple burden" in reference to their involvement as undernutrition, micronutrient deficiencies and overnutrition. Undernutrition results in stunting and wasting, micronutrient deficiencies result in abnormal growth and are often termed as hidden hunger and overnutrition results in overweight and obesity. Thus, these three nutritional challenges have advocated the need to develop such products that satisfy the hunger of consumers. In response to global challenges related to hunger, malnutrition, and micronutrient deficiencies, there is a growing interest in developing food products that not only satiate hunger but also contribute significantly to nutritional well-being [2]. In accordance with the nutritional needs of humans, dietary nutrients are divided into macro- and micronutrient categories. Vitamins and trace elements are subcategories of micronutrients. The B-complex vitamins are counted amongst the water-soluble vitamins, whereas the vitamins capable of getting soluble in fat include vitamins A, D, E and K. Trace elements fall into two categories: those that are not

S. Bishnoi (✉) · M. Rani · S. Yadav · P. Kumar
Department of Livestock Products Technology, Lala Lajpat Rai University of Veterinary and Animal Sciences, Hisar, India

G. Jairath
Department of Livestock Products Technology, ICAR-IVRI, Palampur, India

A. Kumar Pathera et al. (eds.), *Healthier Meat Products*,
https://doi.org/10.1007/978-3-031-78215-2_12

as affected by diet and liver absorption, such iodine and selenium, and those that are greatly influenced, like iron, copper, and zinc.

Human nourishment requires food containing necessary and vital nutrients for its existence and upright well-being. Nourishment is vital for physical, mental growth and appropriate biological growth of children. Every nutritious food item is composed of carbohydrates, lipids, fibre, minerals, proteins, vitamins, and water and they are categorised into macronutrients and micronutrients. Macronutrients (carbohydrates, lipids, and proteins), are main source of energy. Some macronutrients like water and fibre does not provide any energy. Vitamins and minerals are categorised as micronutrients as they are essential in very lesser amounts but provide assistance to the body in very significant ways. As multiple type of vitamins and minerals are present in lesser amount in our food so chances of their deficiency is greater. Mostly, except vitamin D, our body does not synthesize them and must be derivative from our diet.

All the vital processes that are carried out in the body require micronutrients. For regularly maintaining the metabolism of the body and proper working of body functions, micronutrients are an integral part of the body. Similarly, for effective immune response, the body needs optimal levels of micronutrients. These micronutrients can be made available effectively to address malnutrition or micronutrient-associated diseases by fortifying food products, especially those, whose production and demand is high.

In this context, the demand for meat is rising globally over the past 50 years that has triggered the meat production. The world now produces more than 340 million tonnes each year. Poultry and pig meats are most liked amongst the various varieties of meats that are available in the marketplaces [3]. As over a billion people globally, many of whom live in poor nations, can be employed and paid for their meat production, meat consumption can support sustainable growth [3]. Presently, the primary source of protein (28 g of protein per person per day) comprises livestock products, which account for 30% of total calories consumed [3]. Meat fulfils the nutritional requirement for human enriched with proteins, vitamins, and minerals, so meat consumption can assist in reducing dietetic insufficiencies and thus stimulate human health [4]. So, it is forecasted that global meat demand will continue to rise due to income and demographic projections [5] and has potential to ameliorate/reduce nutrition associated deficiencies. Thus, this chapter will focus on micronutrients, their importance, factors driving the need of fortification and the role of meat products in tackling micronutrient-associated deficiencies.

2 Micronutrients and Their Importance

Micronutrients is composed of two words 'micro' which mean 'very small' and 'nutrients' means 'a material required for a living thing's development and maintenance. So, micronutrients may be defined as those nutritional components which are required in trace amounts in the body for the overall performance of the body and deficiency of which can lead to serious malfunctioning in the body. According to UNICEF, micro-nutrients are the nutrients the body requires in small amounts, to produce hormones, enzymes, and other chemicals, for assisting in the regulation of development activity, progress and the working of the immune and reproductive systems.

The micronutrients iron, zinc, iodine, selenium, copper, vitamins A, E, C, D, B2, B6, and folate are known to be significant to the well-being of the public [4]. Deficiencies in iron, vitamin A, and iodine, prevalent globally, especially impact children and expecting women, with low- and middle-income countries facing a disproportionate burden. While severe deficiencies can lead to visible health issues, subtler effects include reduced energy, mental clarity, and overall capacity, affecting education and work productivity [6]. Without the micronutrients, the important physiological reactions in the body cannot be accomplished. As reported by WHO, 2023 [5], around the world in children and pregnant women, the most common, deficiencies are iron, folate, vitamins B12 and A which can result in serious health issues. For instance, anaemia, characterized by a diminished counts of red blood cells or haemoglobin concentration, leads to fatigue, weakness, breathlessness, and dizziness, impacting work, education, and community engagement. Anaemia affects about 40% of pregnant women and 42% of children under 5 worldwide. Severe iodine deficiency poses risks of brain damage, stillbirth, abortion, and congenital anomalies, while milder deficiencies may cause intellectual impairment. Universal salt iodization remains a key strategy, with 66% of households globally having admittance to iodized salt. Vitamin A deficiency, a major cause of preventable childhood blindness, increases susceptibility to severe infections. Breastfeeding and vitamin A supplementation are recommended for infants and children in affected areas. Further, over 2 billion individuals worldwide experience shortages in minerals and vitamins, mostly iron, zinc, vitamin A, and iodine, which can have serious health repercussions. In this context, nutrition education, a varied healthy diet, and strategic interventions like fortification and supplementation are effective preventive measures. Despite progress, persistent efforts are necessary to address ongoing challenges in combating micronutrient deficiencies. The increasing demand for nutritionally enhanced food products due to consumer awareness has also driven nutrition scientists to focus on nutritionally-enriched food products as discussed in the next Sect. 3.

3 Nutritionally Enhanced Food Products: Driving Factors for Rising Demands

Some of the simplest means to upsurge the nutritional value of food products is to add vitamins, minerals, or other nutrients that are naturally present or lacking in the raw materials or processing methods. Food is connected to our health and our wellness, food quality has a direct impact on our health. Wellness comes from a balanced diet to be included in our daily diet. The opportunities for health and wellness related to functional foods are growing at a faster rate. This may be due to the growing demand of consumers for healthy and tasty and organic food to adopt a well-balanced healthy lifestyle. There is a requirement of a balanced nutritional diet to match with an active lifestyle. The importance of physical activity, healthy diet and exercise is growing which is inciting the nutritionist and food technologist to develop more nutritious and well-proportioned food variants. The rising consumption of organic food products that are made from natural ingredients compared to inorganic food products is paving the doors for a healthy and functional food market. In current situations, due to rising health awareness, the demand for nutritionally enhanced food products is rising. Nowadays, people prefer those food products with high nutritional value and with various health benefits. These type of foods are helpful to maintain a healthy diet. Changing lifestyle and eating habits of the population leads to higher consumption of healthy and fresh foods.

Currently, individuals are concerned about health and fitness and they know its importance as well. So to meet the healthy lifestyle and fitness goals of the body, the demand for fortified foods is rising day by day. That's why the practice of nutritional enrichment of foods is also rising to augment the nourishing quality of food and to cater for the needs of nutritionally enhanced foods. Nowadays people are very selective of their food in respect of their nutritive value and also very conscious about what nutrition they are getting from what they are eating, whether it is nutritionally balanced or meets the needs of the body, etc. Nutritional gaps pose a great challenge that further shifts nutritionists towards nutritionally enhanced products as discussed in Sect. 3.1.

3.1 The Nutritional Gap in Modern Diets

In the age of fast food and pre-packaged meals, achieving optimal nutrition can be a challenge. Busy lifestyles often get in the way of preparing wholesome meals, leading many to rely on the convenience afforded by modern food options. The nutrition gap refers to the discrepancy between the nutrient levels that the typical person consuming a reasonable diet is primarily obtained from food and those nutrient levels that research has shown are required for the population's optimal health. To get rid of malnutrition, nutrient requirement must be fulfilled, but in many countries, nutritional requirement of susceptible crowds is not fulfilled the by the local diet, subsequently leading to "undernutrition" condition [6, 7]. Some of the common deficiencies are of calcium, potassium, dietary fiber, and vitamin D due to underutilisation is alarming for the general population. Similarly, for girls of adolescent age, reproductive women, and infants on mother's milk iron deficiency is raising a concern. Likewise, for pregnant ladies, folic acid, iron, and iodine deficiency a common problem. Research also directs that people from undernourished segment are more exposed to micronutrient insufficiency. Research also indicates that those who consume supplements are less prone to micronutrient deficit or anaemia than individuals who do not consume nutritive supplements [8]. Population-based

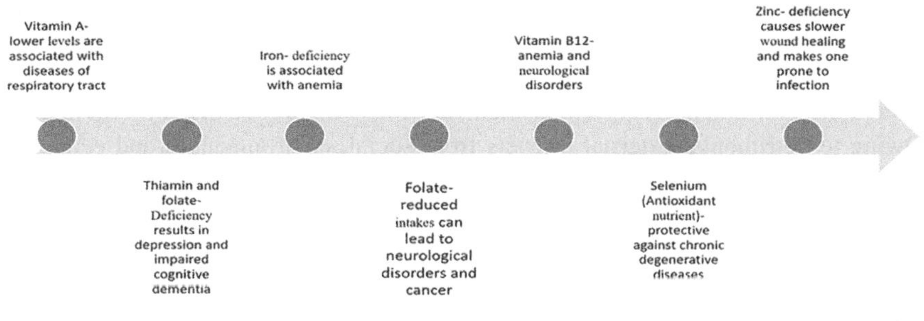

Fig. 1 Meat micronutrients and the implications of deficiencies

surveys indicated that the most common deficiencies which were observed were of iron, vitamin A, iodine, zinc, folate, and vitamin B12 (Fig. 1). It was observed that anaemia is a result of iron deficiency; additional impacts that have been noted include death, diminished productivity at work, and cognitive impairment [9]. Night blindness is a result of deficiency of vitamin A; other effects observed are like exposure to contaminations, and death. Similarly, iodine deficiency can cause growth and cognitive impairment, goitre, and death, often due to deficiency during pregnancy [10]. Likewise, zinc shortage leads to poor health, higher incidences of diarrhoea, and diminished cognitive and motor development in children [11]. While, vitamin B12 deficiency can lead to anaemia, developmental regression in infants, and adult depression, as well as it impacts adult cognitive performance [12] and folate deficiency in infants and young children leads to poorer brain development and anaemia and in adults can result in depression and induce neural tube abnormalities in foetuses while a woman is pregnant [12]. It is a well-recognized fact amid health and nutrition experts that nutritional requirement can be accomplished only through a healthy diet.

3.2 Need for Enrichment and Fortification of Various Micronutrients

The general health and well-being of society are greatly impacted by micronutrient deficiencies, which are also possible targets for replenishment. Diets in a number of small- and middle-income countries are allegedly lacking certain micronutrients, lack dietary variety, and have high levels of anti nutrients substances (ANFs) that hinder the absorption of iron, such as tannins and phytates during bodily functions [1]. Specially, in the developing countries where the vegetarian diet composed of rice, beans and maize is principally associated with low bioavailability of iron leading to its deficiency and hence the diseases associated with it. When 100 g of pig meat or even less liver is further supplemented with the vegetarian diet mentioned above, iron absorption improves by 3.6 times.

The need for enrichment of different food products including meat finds its place to supply certain nutrients for special categories or targeted populations such as children, pregnant women and old adults as they have specific needs during their altered life span and the requirement cannot be met with normal dietary food intakes. Owing to detrimental external impacts from social, environmental, and economic variables, human immunity, metabolism, and adaptive capacity must all be increased, and defensive processes must be stimulated with the development of functional supplemented with targeted nutrients [13]. The role that a nutritious diet has in boosting immunity and lowering illness has a long history of being known. Micronutrient balance is essential for immune system maintenance and prevention of many of the infections known to occurs due to lowered immunity and one such viral infection named Covid-19 could be prevented by proper nutritional balance [14].

In light of the available data, it is critical to develop a consensus statement and implement target-meeting tactics. It is necessary to expand the current iron, vitamin A, zinc, and universal salt iodization programs, particularly for high-risk populations. There must be widespread vitamin D supplementation, but extra research is desired to determine the appropriate dosage. Only high-risk populations should undergo screening for and supplementation with vitamin B12 deficiency. It is imperative that the availability of suitably fortified meals be addressed immediately.

4 Human Nutrition and Meat and Meat Products

In the upcoming times, the most challenging is that availability of food to the whole world with sustainability. And meat can be a big player in this. As meat is a vital source of nutrition for several people round the world, it is indispensable for optimum human development and growth [15]. Globally, micronutrient deficiencies are a burning issue. Protein which are found in meat have high biological value and engrossed with vital micronutrients like iron, selenium, zinc and vitamin B12. Vitamin A and folic acid are more commonly available in offal meats like liver. One significant thing is that meat protein is composed of all essential amino acids [16]. Two thirds of the daily requirement (DR) of vitamin B12 and 25% of the recommended dietary intakes of riboflavin, niacin, pantothenic acid, and vitamin B6 are satisfied by 100 g of red meat. Some vitamins are only found in meat like vitamin A and vitamin B_{12} and plant-derived provitamins cannot compensate it. Meat provides iron as heme iron which has more bioavailability than plant-derived iron. Likewise, bioavailability of folic from meat (especially, liver) and eggs is approximately 10-fold higher than from vegetables and play a pivotal role in human nourishment [17]. However, there are variation of micronutrient content based on meat type, feed, and processing. Different animal species (ruminant and monogastric) and breeds, as well as various methods of feeding and raising cattle, are known to cause variability in the nutritional composition of red meat. These factors have made it possible to modify the meat's nutritional value. Several studies have also

Table 1 Vitamin content range in red meats [17]

Vitamins	Beef	Lamb	Pork
Vitamin A (μg/100 g)	5.00–11.5	7.80–8.60	2.00–6.10
Vitamin E (mg/100 g)	0.03–1.10	0.08–1.20	0.01–0.86
B2 (mg/100 g)	0.09–0.80	0.11–0.25	0.05–1.23
B12 (μg/100 g)	0.40–3.10	0.60–2.50	0.30–1.10

demonstrated that the complex interplay between an animal's genetic makeup, its production environment, and their interactions determines the variance in nutrient composition found in red meat. Additionally, diet has been found to have the biggest influence on the composition of red meat. Environmental factors such as the production system, animal age, gender, and physical activity might affect the nutritional content of meat [17]. Consumption of meat provides high iron content, particularly in meats with high myoglobin content and myoglobin content of red meats is higher owing to the redder fibers as compared to other meat types. There exists a high variability in the content of different vitamins in different type of red meats, given in Table 1. However, processing affects the micronutrients composition in meat.

The processing starts right after the slaughter and can be divided into primary processing and secondary. Primary processing is accomplished during the conversion of muscles to meat and it has been ventured that a delay in the bleeding could lead to rise in heme iron in muscle tissue thereby in the meat and meat products. Conditions which result due to stress in animals like PSE (pale soft and exudative) have decreased amount of water-soluble micronutrients.

Secondary processing includes simply cutting of meat into wholesale and retail cuts or processing into the products with implication of cooking and restructuring. As variability exists in the severity of processing, similarly the impact on composition on micronutrients is also inconsistent like during cooking, there is loss because of degradation and drip loss. Also, during freezing and thawing alike losses results. Lipid oxidation also leads in loss of lipid-soluble vitamins (A and E) and some water-soluble vitamins comprising folic acid and thiamine. There may be reduction in bioavailability of certain trace minerals during restructuring due to inclusion of phosphates and non-digestible polysaccharides whereas addition of antioxidants can improve the stability of fat soluble micronutrients during processing. The heme iron concentration in processed and cooked meat decreases when it is cooked, simmered, or baked; on average, this loss occurs after 60 minutes at 95 °C, and after just a few minutes at 60 °C [18].

4.1 Rationale and Need for Micronutrient Enrichment in Meat Products

Meat being rich in vitamins and minerals seamlessly fulfils appetite although proportion of nutrients is not well balanced. So, to fulfill the demand of consumers seeking nutritionally enriched meat products, meat products are reformulated, for example; reduced fat percentage, less cholesterol, less sodium chloride and nitrites, as well as raising the beneficial components to improve human wellbeing. A diet with high cholesterol is linked with heart problems like CHD (coronary heart disease), so intake of cholesterol-rich food should be controlled. Meat and meat products come under this category. In lieu, meat and meat products fulfill the demand of individuals for superior proteins, potent iron, vitamin B12, zinc and selenium. So, the addition of some compounds which lower the cholesterol such as phytosterols and soy proteins, is the solution for the same [19]. Similarly, due to the harmful effects of consuming sodium on one's health (high blood pressure) [20], efforts are made to lower down the salt percentage in meat products [21]. Owing to the negative effects of phosphates in meat products on diabetes, obesity, and cardiovascular disease, phosphates are substituted with sodium citrate, carrageenan, or proteins derived from alternative sources [22].

Selenium deficiency in human metabolism is associated with a weakened immune system and greater vulnerability to certain chronic illnesses which are associated with a weakened immune system and greater vulnerability to certain chronic illnesses [23]. Customers frequently believe that meat and meat products are unhealthy. By introducing functional or reformed meat products, these impressions can be altered [24].

4.2 Meat Factor as an Effective Vehicle for Micronutrient Availability from Plant Sources

Animal source foods including meat serves as an excellent source of numerous micronutrients as zinc, iron, riboflavin, calcium, vitamin B-12, and vitamin A [3]. Additionally, it's speculated that foods derived from animals contain the meat factor, described as an active ingredient, or chemicals, found in the muscular tissue of animals, that improves the absorption of micronutrients from plant-based diets [25] thereby improving the nutritional status of individuals [2]. The two possible mechanisms of meat factors action are the production of low molecular weight compounds during protein breakdown and low pH conditions for longer duration resulting from the digestion of muscle foods, these favour the availability and absorption of the trace element absorption. The earliest documented proof of the meat factor was presented in 1968 by Layrisse et al. [21] who showed that adding fish or beef muscle increased the amount of non heme iron absorbed from meals based on maize. Initially, the contribution of meat, particularly red meat, and its constituent parts (such as heme iron), to improving the uptake of iron and zinc from meals has been studied extensively [4, 5]. Increasing meat demand, nutritional enrichment and consumers' demand suggest meat plays the role of carrier for micronutrient delivery in human diet to address malnutrition.

4.3 Role of Meat as a Micronutrient Carrier

Frequent consumption of meat and meat products can meet the demand of energy, protein and important micronutrients [26, 27]. In addition to providing high biological value proteins, beef is a recognised source of most B-vitamins, long chain n-3 fatty acids, and vital trace minerals (copper, iron, iodine, manganese, selenium, and zinc) [28]. Given that some micronutrients are only found in meat or because meat has a higher bioavailability of these elements than plant sources, meat, including liver, is a significant source of these vital nutrients. The inadequate intake of which can lead to several deficiency diseases in humans. As an account of the micronutrient profile meat, solely meat contains vitamin A and vitamin B12. Plant-based provitamins can hardly make up for these deficiencies. Because provitamin B12 is nonexistent and provitamin A, b-carotene, has a weak 1:12 conversion rate, it is necessary to consume substantial amounts to make up for the deficiencies in vitamin A and vitamin B12. The bioavailability of iron as heme iron and folic acid [about 10 times] is much higher from the meat sources as compared to their plant counterparts [29]. Consuming 100 grammes of pork and liver per day can supply as much as 50% of the recommended daily amount (RDA) for vitamins B12, B1, B2, B6, and A, zinc, iron, and selenium [29]. Meats are very much regarded bases of iron sources because of their heme iron content, which is higher in beef than in pig and chicken. Their around 25% absorption rate is unaffected by dietary variables, with the exception of calcium [30, 31]. Micronutrients such as iron and folic acid are especially important in expecting women as to prevent the risk of birth defects therefore, these are supplemented as nutraceuticals. Liver and meat are rich in a number of micronutrients: There are 1.8 mg of iron and 2.6 mg of zinc in 100 g of low-fat pork, compared to 360 mg of magnesium, 20 mg of iron, and 60 mg of selenium in 100 g of pig liver. Therefore, 100 grams of pork and liver per day can provide up to 50% of the RDA for zinc, iron, selenium, and vitamins B12, B1, B2, B6, and A [29]. Older adults are generally considered to be at risk of having deficiencies in vitamins and trace elements, particularly for iron, calcium, and the vitamins A, D, and E, as well as folate due to their lower energies and sedentary lifestyles. Because elevating levels of PUFA, CLA, EPA, HA, vitamin E, selenium, and other bioactive compounds in rabbit diets is so successful, regular eating of rabbit meat may provide consumers with these compounds [32]. B vitamins are vital micronutrients found in meat, fish and poultry that support immune system function, encourage cell growth, and are involved in the metabolism of carbohydrates. Vitamins B2, B3, B5, and B1 are necessary coenzymes for the metabolism of energy. Meat has a significant effect on vitamin availability. Specifically, it contributes between 4% and 56% of worldwide vitamin availability for vitamin A and the B vitamins (vitamin B9, folate and vitamin B12, cobalamin). The minerals that contributed the most were copper, zinc, iron, selenium, and phosphorus. However, meat contributed relatively little to the availability of other macronutrients, including fibre, starches, and vitamins C and E as micronutrients. Given that plant meals are typically the source of these nutrients, this outcome was anticipated [33].

Micronutrient deficiencies are increasingly prevalent in low- and middle-income nations, however, deficiencies in iron and iodine have also been recorded in high-income nations [12]. Meat provides a number of essential trace elements, the majority of which are in an organic, easily absorbed form [28]. Vitamin B12 is mostly found in animal foods (i.e., meat, milk, egg, fish, and shellfish) as compared to plant foods [34] that's why vitamin B12 deficiency is most likely to occur in vegetarians. Rising incidences of several chronic diseases such as cancer and cardiovascular diseases has led to detailed research into the foods associated with increased risk. Despite its nutritional abundance, meat consumption is linked to disease-promoting food. As it is well known that meat is extensively rich in vitamins and minerals. One hundred grams of red meat contributes approximately 25% of the daily requirement for riboflavin, niacin, vitamin B6, and pantothenic acid. A reduction in the amount of livestock products consumed by humans could be a potential result of measures intended to lower obesity rates in the general population and reduce greenhouse gas emissions. However, at the moment, consumption of some micronutrients is significantly influenced by meat products. Of the trace elements, it was difficult to alter the concentrations of copper (Cu), iron (Fe), and zinc (Zn) through food, while it was easy to alter the concentrations of iodine (I) and selenium (Se). Similar to this, riboflavin was resistant to dietary manipulation, outcomes of retinol, folate, and cobalamin supplementation were lower, and α-tocopherol concentrations were easily altered. The simplicity of manipulating the composition of the various goods varied: egg concentrations responded the best, followed by milk and muscle meats. However, a sizable amount of the daily recommended nutrient intake can come from livestock products with higher concentrations of micronutrients.

4.4 Strategies for Micronutrients Enrichment

Fortification and bio-fortification come into existence while micronutrients enrichment. Fortification has emphasis on aggregating the amount of one or more target micronutrients in the food and this is done mainly while the food is processed and also during value addition of food. Whereas, the bio-fortification biotechnological intrusions done at level of farm itself by means of modifications to plant breeding, biotechnology techniques, or agronomic methods to increase in the level of macro and micronutrients. Numerous techniques like genetic modification, transgenic and agronomic approach and plant growth-promoting agents have been discovered.

The target demographic group is a key factor in determining the most effective method of distributing micronutrients, be it through supplementation or fortification. Fortifying food can benefit a huge number of people in an economical way and helps rural populations easily meet their daily nutritional needs without the need for pharmaceutical supplements. Fortification of nutrients refers to the delivery of them through food for regular absorption and there is gradual increase in the level of the micronutrients, whereas supplementing of elements entails the oral administration of pharmaceutical compounds for quicker absorption and immediate action in the body. Numerous innovations and strategies are being developed; for instance, double and quadruple fortification is currently in use and offers the possibility of delivering multiple micronutrients [35]. Given that both the targeted micronutrient and the chosen food carrier affect the nutrients' bioavailability, they must work in unison. Vitamin A deficiency can be prevented by its supplementation in the subcutaneous fat of the livestock through dietary feeding. For instance, giving the Celta pigs chestnuts when they're nearing completion resulted in increase in retinol concentration in different tissues [36].

The four main elements that determine the utilitarian worth and ultimate presentation of functional materials or ingredients are their availability, sustainability, marketability, and acceptance. Successful administration of functional materials can be achieved through two main methods: direct and indirect administration, explained below.

(i) Direct administration: This method involves the direct supply of the functional ingredient in the diet of the consumers on daily basis, but this method is more difficult and taxing procedure to carry out as it involves quite some large regulatory clearances from food regulatory authorities, bioavailability of the products, availability of clinical data as well as the acceptance of the same by the consumers [37].
(ii) Indirect administration: This approach seeks to provide through livestock or cultivated animals that the buyer usually consumes. Both the delivery cost and the food standards clearance are lower with this, but the efficacy of uptake needs to be accessed before any beneficial claim can be made with technique. Different researches on fortification of meat and meat products have been conducted as briefed in next section [6].

5 Examples of Fortification in Meat

5.1 Vit-D Fortified Meat Products

The deficiency of micronutrients including the vitamins may lead to several deficiency diseases. One such vitamin deficiency commonly encountered is the insufficiency of vitamin D which is found in forms namely vitamin D_2 and vitamin D_3 of which D_3 is more effective in regulating and supplementing the vitamin D levels in the body. Low levels of vitamin D have a significant impact on public health and are a global concern. The course of severe Multisystem Inflammatory Syndrome (MIS) in children may be influenced by correcting aberrant levels of Vitamin D. Vitamin D levels may be useful in foreseeing unembellished types of MIS [38]. Both the necessary amount of dietary vitamin D and the necessary amount of sun exposure are often lacking in individuals. So, there remains a scope for improvement of dietary sources of vitamin D_3 [39]. As meat remains one of the important components of diets in many parts of the globe and specially the pork which indeed has vitamin D_3 and 25-hydroxyvitamin D_3, an enrichment of such meat will be helpful in increasing the levels of vitamin D_3 and thereby, reducing the risk of deficiency diseases. One of the ways of enriching vitamin D_3 in meat of different species of animals and poultry is by supplementation of same in the feeds or adjusting the animal husbandry practices [40]. Pork bio fortification could be used in addition to conventional fortification techniques to give consumers another dietary approach to help them meet recommended intake levels of vitamin D. The best way to increase the quantity of vitamin D in loin and plasma in pigs is to expose them to UVB sunlight throughout the concluding phase in addition to providing the maximum amount of EU-approved vitamin D-enriched feed thus decreasing the chances of hypovitaminosis D [41].

5.2 Selenium-Enriched Meat Products

Numerous hormones and enzymes contain selenium, a microelement that is involved in biological processes. The following are the roles that selenium plays in birds and other animals: boosts the production of macrophages, interferons, antibodies, and the immune system. Antioxidants that are particularly effective include selenium. Selenium's antioxidant qualities influence how glutathione peroxidase, one of the essential enzymes for the normal operation of the glutathione system, is activated. When this metal is lacking in the body, it affects the production of glutathione peroxidase in its active form. This leads to serious problems with the glutathione system as a whole [42, 43]. Depending on the severity of the deficit, an organism's selenium insufficiency might result in major metabolic abnormalities, particular illnesses, or physiological changes within the regulatory norm. Selenium deficiency is linked to around 75 distinct diseases and discomfort complaints. The majority of nations continue to consume low levels of selenium (20–40 µg/day) [8]. A population's selenium consumption can be increased in a number of ways, including by using selenium as a medication or dietary supplement, making bread that is enhanced with selenium, growing vegetables and greens that are high in selenium, making beverages that are enriched with selenium, and acquiring animal-derived products that are high in selenium. Addition of variable doses of selenium in mixed feeds and diets of poultry birds and deposition and delivery of the element in muscle tissues of breast have been recorded. Supplying broilers, baby geese and young ducks mixed feeds containing selenium in the studied doses resulted in a consistent increase in this microelement's concentration in the leg muscles by 13.0–85.7%, 57.4–61.7%, and 20.5–79.4%, and in the chest muscles by 21.7–106.7%, 35.1–40.0%, and 23.2–66.0%, respectively thereby enriching meat with micronutrient and ensuring the supply through these meat products. Consequently, intake of selenium-enriched meat [115 g of meat products/day] from chickens, baby geese, and ducklings within the suggested physiological norms fulfills the daily requirement of an adult for this microelement (70 µg) having high biological value and regarded as safe from the standpoint of food hygiene although, the kind of poultry, the type of muscle, and the amount of selenium present in mixed diets all have an impact on the muscular tissues' selenium deposits' value. With the information presented above, it is possible to classify selenium-enriched poultry meat as a dietary food product with a bio correctional impact and functional importance for humans.

5.3 Calcium Fortification in Meat Products

The milk and milk products like chocolate contain lactose which are cariogenic responsible for the dental carries and tooth decay. As well as to improve upon the nutritional status of such products there is scope to utilize meat in powder form to combat the low protein content of chocolates. Also, adding additional calcium and phosphorus can help to reduce the cariogenicity (anticaries activity). Consequently, a calcium-fortified, protein-rich composite chocolate was created by fortifying milk-based chocolate with chicken protein powder, calcium lactate, and animal fat in place of some of the cocoa butter [44]. 10% meat protein powder, 10% animal fat, and 1% calcium lactate made up the optimised composite chocolate had a 93% increase in both calcium and protein content ($p < 0.05$).

5.4 Iron-Fortified Products

Despite the fact that iron is one of the minerals that is most prevalent in the Earth's crust, iron insufficiency is the utmost common and pervasive nutritional problem worldwide. The iron deficiency results in anemia which is a disorder characterized by low blood hemoglobin (Hb) levels in the human body, or by decreased erythrocyte counts and abnormal morphological traits. As has been noted, poor socioeconomic groups that eat inexpensive fast food are more likely to be obese and iron deficient. These foods have more sugar and preservatives and less nutrients overall. The dietary Reference Values for iron were calculated by European Commission and it was estimated that a twenty percent coefficient of variance was used to estimate Population Reference Intake (PRIs). For infants (7–11 months), they are 11 mg/day; for children, 1–6 years old, they are 7 mg/day; and for boys, 12–17 years old, they are 11 mg/day. For girls between the ages of 12 and 17, the midpoint of the predicted dietary requirements for 97.5 percent of girls is the PRI of 13 mg/day [45]. World Health Organization/Food Agriculture Organization advocated supplementation, fortification, nutrition education, and dietary diversification as the four main measures to treat micronutrient deficiencies. Though non-haem iron absorption is greatly impacted by dietary intake patterns, meat provides the most effective and usually unaffected hem iron absorption when it comes to other dietary components [46]. Among the four tactics, food fortification is thought to be the most cost-effective and effective in targeting a particular portion of population. Many "food-based strategies" are available to reduce anemia, and they center on increasing the availability and intake of a varied, high-micronutrient diet. These approaches also seek to increase the nutrients' bioavailability.

5.5 *Meat Products Enriched with Bee Pollens*

By employing the fermentation method, meat products enhanced with bee products (pollens) can be produced that are safe, biologically complete, and provide useful qualities. As the honey bee pollens have therapeutic and preventive properties: it increases the immune system, improves blood formation, brain function, cardiovascular system. In this regard, there has been development of bee pollen-enriched dry beef jerks having concentration of 2%, designed to combat vitamin and mineral deficiencies. Vitamin and mineral enrichment of the product has been demonstrated through experimentation [47].

6 Consumer Perception and Micro-Nutrient Enriched Meat

Innovative food products ultimately fail or succeed based on public acceptance, thus it's critical to address customer concerns about these novel goods. Based on the diverse characteristics of processed meat products—including enhanced ones—consumers have varying opinions. Their key worries are on the food's capacity to provide nutrients, its general health advantages, and the necessity of the finished goods. As long as the products' taste, price, and shelf life are unaltered, people are willing to tolerate them [48, 49]. Subjective norms and perceived behavioural control are less important in predicting a consumer's desire to purchase than their attitude [15].

The final determination of whether to approve functional foods, processed meat, and meat products is based on the following characteristics:

6.1 *Attitude of Customers*

Customers' attitudes is the main determinant of whether processed and healthy meat products are accepted. Moderate influences on acceptance and consumption patterns come from eating habits and overall food selections. A number of elements, such as price, familiarity, natural content, health, convenience, and sensory appeal, influence the everyday food choices that consumers make. Additionally, the acceptance and perception of such meat products are influenced by the ease of processed food items [50].

6.2 Products' Purported Healthfulness

One of the key elements influencing whether or not enhanced, processed, and treated meat products are accepted is their purported healthfulness. Customers that are willing to purchase nutritious foods have a good outlook on the health benefits of functional foods that have been designed.

6.3 Consumption Rate

The frequency with which processed and enriched meat products are consumed will ultimately determine how well-liked these products are, as infrequent meat eaters won't influence the growth in consumption of these goods merely to consume the nutritious processed meat products.

6.4 Socio-Demographic Dynamics

Income is one of many sociodemographic characteristics that influence consumer demand; an increase in income is linked to a rise in the purchasing power of ready-to-eat goods, and meat products are no exception. Age is another factor; wherein older adults typically consume more ground meats. Ethnicity also affects a lot, when it comes to contemporary beef-based products, different ethnic consumers will have distinct needs. Further convenience drives the dynamics of consumers preferences. Ready-to-eat meat products are becoming more and more popular as a way to cater to today's time-pressed consumers. Gender is another factor that predicts socio-demographic dynamics as women are more discriminating about their eating habits and eat less processed meat than men [51].

6.5 Degree of Education

Customers with a third-level degree are less inclined to buy the enriched processed meats. Because the nutritional standards recommend limiting intake, more knowledgeable consumers may decide to reduce rather than consume a healthier but smaller quantity of the real product.

7 Challenges and Considerations

Although the fortification and enrichment of staple foods is thought to be a potential strategy to counteract the malnutrition caused by micronutrient deficiencies, the research and marketing of these products is costly, risky, and difficult. A product's success requires collaboration amongst several academic disciplines, including food technologists, epidemiologists, nutritionists, natural product chemists, and other interested parties. In addition, the judicial components, sensory analysis, and supermarket forecasting research are crucial. Additionally, since the body already has a reserve of certain nutrients and lacks the mechanisms to control the quantity of micronutrients at every stage, there are worries about overdosing on them. For example, elevated iron levels have also been associated with severe epidemic malaria because they promote an environment in the body that is favourable to the proliferation of pathogens. Furthermore, it has been proposed that elevated iron levels in the body have a role in metabolic diseases such as cancer, heart disease, and diabetes [52]. It has been noted that giving young children mixed micronutrient supplements for various elements results in fever and diarrhoea. Iron-fortified meals have also been related to iron excess in hemochromatosis patients [35]. Moreover, the commercial production of meats enhanced with additional nutrients necessitates the resolution of multiple regulatory barriers. One of them is an amendment to the set of requirements and specifications that approve the nutritional claims made on the front of the package. Healthier meat products cannot be produced in the absence of these regulatory changes because they cannot outperform meats without fortification in terms of benefits [37]. Priority one should be given to assessing how functional additives impact the physical, chemical, and sensory characteristics of meat products. Finally, the success or failure of the functional food is determined by the salient characteristics of the product, its commercial viability, and the type, scope, and governance of interdisciplinary collaboration.

8 Conclusion

A number of micronutrients, including folic acid, iron, zinc, selenium, potassium, and vitamin A, as well as a group of B vitamins including riboflavin, thiamin, and vitamins B6 and B12, are essential for human consumption. These items are derived from red meat and poultry meat, including organ meat. Plants either don't contain these nutrients at all or have such poor bioavailability that eating only plant-based foods could leave you deficient in them. Fortifying and enriching meat products with micronutrients in line with people's physiological needs would help counteract the growing number of deficiency disorders brought on by food scarcity and changing human lifestyles. The production and commercialization of enhanced meat products are costly, and sophisticated. A product's success requires collaboration between various disciplines, especially amongst product chemists, food technicians, epidemiologists, nutritionists, and others. Double and quadruple fortification are methods that can be used to include micronutrients both directly and indirectly. There are meat products that are enhanced with micronutrients including vitamins, iron, and selenium. A wide range of factors, including education, choices, money, and demographics, determine whether or not these products are acceptable. Customer satisfaction will ultimately depend on palatability and its effect on the proximate composition; therefore, more research is needed to evaluate quality attributes and the ways in which different methods of modifying the nutritional value of meat affect them. Advances in biotechnology and food engineering to make such customized products need to be focused on as it would address the concerns related to global nutritional sustainability and would make the populace micronutrient malnutrition-free. Further consumer demand and market trends are shifting in favour of scientific research and the manufacturing of such enriched and fortified functional meals as a result of rising levels of education, income, awareness, and affordability.

References

1. WHO. Micronutrients. 2023. https://www.who.int/health-topics/micronutrients#tab=tab.
2. Muller O. Malnutrition and health in developing countries. Can Med Assoc J. 2005;173(3):279–86.
3. Bonnet C, Bouamra-Mechemache Z, Réquillart V, Treich N. Viewpoint: regulating meat consumption to improve health, the environment and animal welfare. Food Policy. 2020;97:101847.
4. Ekweagwu E, Agwu AE, Madukwe E. The role of micronutrients in child health: a review of the literature. Afr J Biotechnol. 2008;7(21):1604–11.
5. Micronutrients 2023. https://www.who.int/health-topics/micronutrients#tab=tab).
6. Dewey KG. The challenge of meeting nutrient needs of infants and young children during the period of complementary feeding: an evolutionary perspective. J N. 2013;143:2050–4.
7. De Pee S. Special nutritious solutions to enhance complementary feeding. Matern Child Nutr. 2016;11:i–viii. https://doi.org/10.1111/mcn.12265.
8. Sobolev A, Gutyj B, Grynevych N, Bilkevych V, Mashkin Y. Enrichment of meat products with selenium by its introduction to mixed feed compounds for birds. Regul Mech Biosyst. 2017;8(3):417–22.
9. Petry N, Olofin I, Hurrell RF, et al. The proportion of anemia associated with iron deficiency in low, medium, and high human development index countries: a systematic analysis of national surveys. Nutrients. 2016;8:8. https://doi.org/10.3390/nu8110693.
10. Zimmermann MB. Iodine deficiency. Endocr Rev. 2009;30:376–408.
11. Black MM. Zinc deficiency and child development. Am J Clin Nutr. 1998;68:464S–9S. https://doi.org/10.1093/ajcn/68.2.
12. Black MM. Effects of vitamin B12 and folate deficiency on brain development in children. Food Nutr Bull. 2008;29:S126–S13113.
13. Borisova VL, Sazonova EA, Terentev SE, Vernigor VA, Nikonorova AA. Current state and prospects of enrichment of functional products based on poultry meat with targeted nutrients for the production of corrected nutrition for children and pregnant women. IOP Conf Ser Earth Environ Sci. 2022;979:012084.
14. Shavandi M, Sarani A. Effect of food and key micronutrients on Covid-19: a review. Heliyon. 2022;8(10):11216.

15. Higgs JD. The changing nature of red meat: 20 years of improving nutritional quality. Trends Food Sci Technol. 2000;11(3):85–95. https://doi.org/10.1016/S0924-2244(00)00055-8.
16. Williamson CS, Foster RK, Stanner SA, Buttriss JL. Red meat in the diet. Br Nutr Found Nutr Bullet. 2005;30:323–55.
17. Juárez M, Lam S, Bohrer BM, Dugan MER, Vahmani P, Aalhus J, et al. Enhancing the nutritional value of red meat through genetic and feeding strategies. Food Secur. 2021;10(4):872. https://www.mdpi.com/2304-8158/10/4/872
18. Czerwonka M, Tokarz A. Iron in red meat–friend or foe. Meat Sci. 2017;123:157–65. https://linkinghub.elsevier.com/retrieve/pii/S0309174016303096
19. Cohn JS, Kamili A, Wat E, Chung RWS, Tandy S. Reduction in intestinal cholesterol absorption by various food components: mechanisms and implications. Atheroscler Suppl. 2010;11(1):45–8.
20. Aburto NJ, Ziolkovska A, Hooper L, Elliott P, Cappuccio FP, Meerpohl JJ. Effect of lower sodium intake on health: systematic review and meta-analyses. Br Med J. 2013;346:1–20.
21. Inguglia ES, Zhang Z, Tiwari BK, Kerry JP, Burgess CM. Salt reduction strategies in processed meat products—a review. Trends Food Sci Technol. 2017;59:70–8.
22. Alvarado C, McKee S. Properties and safety of poultry meat. J Appl Poult Res. 2007;16(1):113–20.
23. Fisinin VI, Papazyan TT, Surai PF. Producing selenium-enriched eggs and meat to improve the selenium status of the general population. Crit Rev Biotechnol. 2009;29(1):18–28.
24. Grasso S, Brunton NP, Lyng JG, Lalor F, Monahan FJ. Healthy processed meat products—regulatory, reformulation and consumer challenges. Trends Food Sci. 2014;39(1):4–17.
25. Consalez F, Ahern M, Andersen P, Kjellevold M. The effect of the meat factor in animal-source foods on micronutrient absorption: a scoping review. Adv Nutr. 2022;13(6):2305–15. https://linkinghub.elsevier.com/retrieve/pii/S2161831323000789
26. Givens DI. The role of animal nutrition in improving the nutritive value of animal-derived foods in relation to chronic disease. Proc Nutr Soc. 2005;64:395–402.
27. McAfee AJ, McSorley EM, Cuskelly GJ, Moss BW, Wallace JMW, Bonham MP, Fearon AM. Red meat consumption: an overview of the risks and benefits. Meat Sci. 2010;84:1–13.
28. Rooke JA, Flockhart JF, Sparks NH. The potential for increasing the concentrations of micronutrients relevant to human nutrition in meat, milk and eggs. J Agric Sci. 2010;148(5):603–14.
29. Nohr D, Biesalski HK. 'Mealthy' food: meat as a healthy and valuable source of micronutrients. Animal. 2007;1(2):309–16. https://linkinghub.elsevier.com/retrieve/pii/S1751731107657796
30. Fewtrell M, Bronsky J, Campoy C, Domellöf M, Embleton N, Fidler Mis N, et al. Complementary feeding: a position paper by the European Society for Paediatric Gastroenterology, Hepatology, and Nutrition (ESPGHAN) Committee on nutrition. J Pediatr Gastroenterol Nutr. 2017;64(1):119–32.
31. de Pereira PMCC, dos Vicente AFRB. Meat nutritional composition and nutritive role in the human diet. Meat Sci. 2013;93(3):586–92.
32. Dalle Zotte A, Szendrő Z. The role of rabbit meat as functional food. Meat Sci [Internet]. 2011;88(3):319–31. https://linkinghub.elsevier.com/retrieve/pii/S030917401100074X
33. Smith NW, Fletcher AJ, Hill JP, McNabb WC. Modeling the contribution of meat to global nutrient availability. Front Nutr. 2022;9:766796. https://www.frontiersin.org/articles/10.3389/fnut.2022.766796/full
34. Truswell AS. Vitamin B12. Nutr Diet. 2007;64:S120–5.
35. Shubham K, Anukiruthika T, Dutta S, Kashyap AV, Moses JA, Anandharamakrishnan C. Iron deficiency anemia: a comprehensive review on iron absorption, bioavailability and emerging food fortification approaches. Trends Food Sci Technol. 2020;99:58–75.
36. Domínguez R, Martínez S, Gómez M, Carballo J, Franco I. Fatty acids, retinol and cholesterol composition in various fatty tissues of Celta pig breed: effect of the use of chestnuts in the finishing diet. J Food Compos Anal. 2015;37:104–11.
37. Hathwar SC, Rai AK, Modi VK, Narayan B. Characteristics and consumer acceptance of healthier meat and meat product formulations—a review. J Food Sci Technol. 2012;49(6):653–64. http://link.springer.com/10.1007/s13197-011-0476-z

38. Feketea G, Vlacha V, Bocsan IC, Vassilopoulou E, Stanciu LA, Zdrenghea M. Vitamin D in Corona Virus disease 2019 (COVID-19) related multisystem inflammatory syndrome in children (MIS-C). Front Immunol. 2021;12:648546.
39. Dunlop E, Kiely ME, James AP, Singh T, Pham NM, Black LJ. Vitamin D food fortification and biofortification increases Serum 25-Hydroxyvitamin D concentrations in adults and children: an updated and extended systematic review and meta-analysis of randomized controlled trials. J Nutr. 2021;151(9):2622–35.
40. Neill HR, Gill CIR, McDonald EJ, McRoberts WC, Pourshahidi LK. The future is bright: biofortification of common foods can improve vitamin D status. Crit Rev Food Sci Nutr. 2023;63(4):505–21.
41. Neill HR, Gill CIR, McDonald EJ, McMurray R, McRoberts WC, Loy R, et al. Improving vitamin D content in pork meat by UVB biofortification. Meat Sci. 2023;199:109115. https://linkinghub.elsevier.com/retrieve/pii/S0309174023000219
42. Santhosh Kumar B, Priyadarsini KI. Selenium nutrition: how important is it? Biomed Prev Nutr. 2014;4(2):333–41. https://linkinghub.elsevier.com/retrieve/pii/S2210523914000075
43. Hariv MI, Gutyj BV. Influence of the liposomal preparation Butaintervite on protein synthesis function in the livers of rats under the influence of carbon tetrachloride poisoning. Visnyk Dnipropetrovsk Univ Biol Med. 2016;7(2):123–6. http://medicine.dp.ua/index.php/med/article/view/290
44. Kaur M. Development of composite meat chocolate fortified with calcium and plant extracts. Food Biosci. 2021;42:101082.
45. Scientific opinion on dietary reference values for iron. EFSA, 2015.https://www.efsa.europa.eu/en/efsajournal/pub/4254.
46. Hunt JR. Dietary and physiological factors that affect the absorption and bioavailability of iron. Int J Vitam Nutr Res. 2005;75(6):375–84. https://econtent.hogrefe.com/doi/10.1024/0300-9831.75.6.375
47. Sukhov MA, Giro TM. Development of technology for meat products enriched with essential trace elements. IOP Conf Ser Earth Environ Sci. 2021;640(3):032032.
48. Shan LC, Regan A, Monahan FJ, Li C, Murrin C, Lalor F, et al. Consumer views on "healthier" processed meat. Br Food J. 2016;118(7):1712–30. https://www.emerald.com/insight/content/doi/10.1108/BFJ-11-2015-0447/full/html
49. Hung Y, Verbeke W, Dc Kok TM. Stakeholder and consumer reactions towards innovative processed meat products: insights from a qualitative study about nitrite reduction and phytochemical addition. Food Control. 2016;60:690–8. https://linkinghub.elsevier.com/retrieve/pii/S095671351530181X
50. Taste lovers versus nutrition fact seekers. How health consciousness and self-efficacy determine the way consumers choose food products. J Consum Behav. 2012; https://onlinelibrary.wiley.com/doi/abs/10.1002/cb.1390
51. Clonan A, Wilson P, Swift JA, Leibovici DG, Holdsworth M. Red and processed meat consumption and purchasing behaviours and attitudes: impacts for human health, animal welfare and environmental sustainability. Public Health Nutr. 2015;18(13):2446–56. https://www.cambridge.org/core/product/identifier/S1368980015000567/type/journal_article
52. Basuli D, Stevens RG, Torti FM, Torti SV. Epidemiological associations between iron and cardiovascular disease and diabetes. Front Pharmacol. 2014:5. https://www.frontiersin.org/articles/10.3389/fphar.2014.00117

Meat Products with Encapsulated Bioactive Ingredients

Neha Sharma and Ashish M. Mohite

1 Introduction

Since time immemorial, meat and meat products have been an important part of human food habits. Meat and meat products have not only been considered important for optimal human growth, but, also contributed significantly as a source of high quality proteins, energy and other essential nutrients. In the late 1990s and beginning of twenty-first century, as a result of new scientific insights, futuristic iconic terms like nutraceuticals, functional foods, immunity boosters, etc. were introduced in the food processing sector. This led to a change in consumer behaviour and people started showing greater interest in food with bioactive or functional components, resulting in increase in the demand for natural, organic and clean label meat products [1]. The bioactive compounds can be of plant, animal, or microbial origin, and are considered safer for human consumption than the artificial/synthetic formulations. There are many bioactive ingredients containing antioxidant, antimicrobial, flavoring, coloring, nutritional, and therapeutic properties, and they are widely used as preservatives and additives in the food industry [2]. The fruits and vegetables act as a major commercial source of the bioactive ingredients used in food products including meat products. These compounds exert their effects on the food systems and provide better quality and stability to the food products. The degradation of the natural bioactive compounds during processing and storage is the major hurdle faced by the manufacturers. These bioactive components behave and interact differently in food systems including meat and meat products [1]. The rapid degradation and undesirable reactions can be prevented while maintaining full functionality by stabilizing or immobilizing the bioactive prior to their addition in food products. The internal and external factors like oxidation reactions may lead to gradual degeneration of the bioactive compounds added to food systems, and may

N. Sharma (✉) · A. M. Mohite
Amity Institute of Food Technology, Amity University, Noida, Uttar Pradesh, India

A. Kumar Pathera et al. (eds.), *Healthier Meat Products*,
https://doi.org/10.1007/978-3-031-78215-2_13

even change their physicochemical and functional properties. The bioactive compounds once imbibed in food systems, may react with each other, thus reducing their bioavailability and ultimately may result in unacceptable colour or flavour.

Encapsulation is a technique which entraps the bioactive compounds within a carrier material, thus allowing these compounds to react in food systems and providing their targeted release during retail storage. This aspect of encapsulation technique helps in protecting bioactive substances like fungal metabolites, plant extracts, essential oils, etc. against external factors and degradation when used in food and agricultural areas. The encapsulation allows the biological integrity of the products and supports environmental conditions during storage, ensuring the viability of active ingredients for long periods [3]. It is a well-known fact that the micro-organisms are prone to abiotic and biotic factors. These factors are responsible for reducing the effectiveness of the micro-organisms and their metabolites, for example, when exposed to ultraviolet radiation and adverse temperatures; resulting in the loss of toxin integrity and spore viability [4]. In this context, encapsulation is a boon as it provides an alternative to these adversities. In the food and agricultural applications, encapsulation results in better biological integrity, increased efficiency, reduced losses by volatility, improved commercial viability, and increased formulation stability, among others [5–7].

2 Need and Importance of Encapsulated Meat Products

The term "food stability" describes how different environmental factors affect food as it is processed and stored, with the goal of maximising desirable food quality while achieving acceptable stability in nonequilibrium situations. Food stability is primarily influenced by chemical and physical variables. Physical factors include light, heat, humidity, pressure, noise etc. Chemical factors, such as oxidation. Encapsulation technique helps in improving the stability of the core material by protecting the sensory attributes and preventing the loss of nutrients [8]. Another factor which impacts the stability of the material is light, in which the exposure of the core material can be prevented by the process of encapsulation which act as a physical barrier [9].

The interaction between the components of a liquid encapsulation system, such an emulsion, is also necessary for physical stability. Elevating the concentration of salt in the emulsion, up to a certain point, enhances stability by reducing intermolecular spatial site resistance and providing protection against electrostatic repulsion between wall components. Emulsion droplet aggregation and decreased stability come from a decrease in the internal electrostatic repulsion between particles as the concentration of salt ions rises [10].

Foods containing fats and oils are susceptible to lipid oxidation, which can cause unfavourable flavour, colour changes, and a reduction in nutritional value. Hazardous materials might also be created. By efficiently preventing oxidation reactions and postponing the oxidation process, encapsulation treatments delivers a solution for lipid molecules. This improves the quality of the product and also helps in uplifting the shelflife of the products [11]. The wall material plays a significant role in lipid oxidation, the most preferred wall materials for lipid encapsulation includes carbohydrates, proteins and gums [12].

By delaying the oxidation fading of pigmented compounds, encapsulation treatment provides a way to increase their antioxidant activity. Improved oxidation stability can result from changes made to the interior properties of particles by the encapsulation process, including hydrophobicity, disulfide–thiol interactions, electrostatic forces, and hydrogen bonding [13]. Animal-derived pigments can also benefit from colour protection through encapsulation. The advantages of colour preservation are not just restricted to plant pigments. Haemoglobin, for instance, is more stable when encapsulated in chitosan and alginate microspheres, which lowers the rate of haemoglobin release and gives the product a consistent light brown hue [14]. Plant pigments are extensively employed in manufacturing and daily life, but because of their instability, they are vulnerable to oxidation and fading. Encapsulation technique isolates oxygen and performs antioxidant actions to give a solution. Because phospholipids and peptides have hydrogen bonds and ionic complexation, encapsulation can shield peptides from digestive secretions and enzymes. For instance, in experiments for simulated digestion and preservation of protein hydrolysates, liposomal encapsulation safeguards antioxidant peptides [15]. Protein aggregation or precipitation results from the addition of salt ions, which neutralises the charge on the protein molecules and reduces or eliminates electrostatic repulsion, which is essential for maintaining protein structure. Such precipitation and aggregation in the salt solution can be avoided by encapsulation treatment. In solutions with high salt content, the encapsulated protein did not dissociate or aggregate. Proteins in sauces, condiments, and meat items with a comparatively high salt content can be transported by this function [16].

3 Bioactive Compounds

Bioactive compounds are components which are present in trace amounts in various organisms and substances which helps in the enhancement of various biological phenomena [17], these are also referred as secondary metabolites [18]. These compounds are derived from various sources which includes plants, crustaceans and microorganisms [19]. Generally, bioactive compounds are subdivided into two categories which include lipophilic compounds (eg: essential oils, carotenoids, curcuminoids etc) and hydrophilic compounds (eg: polyphenols), the lipophilic compounds are extracted by non polar solvents and in hydrophilic compounds polar solvents are being used for extraction [20]. Some natural sources provide natural bioactive chemicals. These natural bioactive compounds fall into two categories: (a) those that have functional effects in food systems, like flavour enhancers, stabilisers, antioxidants, and antimicrobials; and (b) those that have functional effects on consumers, like antidiabetic, anti-inflammatory, antioxidant, and antihypertensive compounds. The use of naturally occurring bioactive compounds as food ingredients or medicinal agents is subject to certain constraints due to the impact of processing and storage conditions on their bioactivity [21]. These compounds undergo degradation when exposed to high temperatures, pressures, severe pH levels, light, and other circumstances [22]. Additional issues affecting their stability and bioactivity include chemical and physical changes. When it comes to meat and meat products, naturally occurring bioactive substances are the most favoured. Natural bioactive compounds are frequently chosen for use in meat products because of their capacity to prolong shelf life and their antibacterial qualities against both plant and human infections [23]. The primary natural sources of bioactive compounds used in meat products are plant extracts and essential oils.

3.1 Types of Bioactive Compounds Used in Meat Industry

There are three natural sources from which bioactive compounds can be extracted and further used in meat products for various purposes- Plant, animal and microbial.

3.1.1 Plant Sources

3.1.1.1 Plant Extracts

Plant extracts are biological materials that have undergone a process of washing, drying, grinding, and extraction using various solvents, including acetone, water, ethanol, and methanol, as well as a combination of these [24]. Plant extract is primarily composed of polyphenols such as caffeic acid, luteolin-7-0-rutinoside, gallic acid, quercetin, and epigallocatechin, which can reach concentrations of up to 85%. The minor components that remain are necessary for the compound's synergistic preservation capabilities [25].

3.1.1.2 Essential Oils

Essential oils are naturally occurring plant extracts that possess a strong scent due to the presence of volatile components such as fatty acids, terpenes, and polyphenols [26]. Oily liquids known as essential oils are produced by distilling stems and seeds. They have been utilised in MP and are a desirable substitute for synthetic antioxidants [27]. The primary Essential oils constituents that exhibit strong antibacterial activity are phenylpropanoids, terpenoids, and terpenes [28]. However, variables including pH, lipids, carbohydrates, proteins, water, salt, antioxidants, preservatives, and extrinsic variables like temperature, vacuum packaging, and microbes can affect the antibacterial action of bioactive compounds in essential oils

[25]. Essential oils tend to function more hydrophobically at pH levels of 5–7, where they exhibit optimal antibacterial efficacy, Essential oils are lipophilic, which allows them to readily penetrate bacterial cells and permeabilize or breach their membrane [27, 29]. pH is lowered as a result of the loss of K^+, Mg^{2+}, and Ca^{2+} ions that occurs when this occurs [28]. Essential oils depolarizes the mitochondrial and exterior membranes in eukaryotic cells, causing the membranes to become abnormally permeable [25]. This may eventually result in apoptosis, necrosis, and cell death due to the release of radicals, the disintegration of the proton pump, and the depletion of ATP, cytochrome c, ions, and proteins [30]. In conclusion, essential oils have shown encouraging antioxidant and antibacterial qualities, which makes them a desirable choice for application in the preservation and improvement of meat.

3.1.1.3 Proteins

Short amino acid sequences known as "bioactive peptides" are inert within the parent protein's structure but, when released, have a beneficial effect on several bodily systems. Many bioactive peptides and hydrolysates are currently thought to have health benefits, as well as the potential to replace synthetic conserving because of their antioxidant activity and ability to postpone oxidation reactions. The bioactive molecules exhibit potential as nutraceutical additives in meat products; nevertheless, their usage should be limited as certain peptides possess a stronger taste, similar to those released by the papain enzyme. In order to preserve lipids and proteins from oxidation, bioactive peptides can be added to emulsions that are water-in-oil-in-water or encapsulated [31].

3.1.1.4 Natural Colorants

To maintain the cured colour of processed meats, a variety of natural colourants derived from fruits and vegetables are utilised in place of sodium nitrite.

Furthermore, as customers want natural and organic ingredients free of artificial preservatives in health-conscious products, these natural colourants have consumer acceptance, antioxidant activity, and acceptable sensory flavour and colour [32]. Red beetroot pigment has been utilised to produce the colour of products that contain cured meat. Red beetroot alone was not as effective in maintaining colour as red beetroot plus sodium nitrite, which helped preserve the colour of the smoked sausage while it was being stored [33]. Paprika oleoresin has a higher pigment stability and is processed to retain its colour, flavour, and taste [34]. Therefore, as paprika can boost the redness (a*) values of meat products and lengthen their shelf life, it may be appropriate as a natural colouring ingredient to minimise the amount of nitrite added.

3.1.1.5 Natural Antioxidants

Plants are a constant source of valuable bioactive chemicals that humans can use to maintain and enhance the overall quality of meat and meat products, such as natural antioxidants. The delicate balance between oxidation and reduction is maintained by natural antioxidants. Antioxidants react with both radical and non-radical species to trigger defence mechanisms that protect both. Components both external and intracellular. The most plentiful source of antioxidants is the plant kingdom, and they are abundant in the spices (seeds), herbs, and essential oils that are employed in organoleptic uses for beef products. In addition, several fruits, vegetables, and leaves from trees are good sources of phytochemicals, which include antioxidants [35]. The annual plant of the Solanaceae family, Capsicum and Annuum, is known as paprika (Capsicum annuum var. Angulosum). It contains significant levels of antioxidants, including carotenoids, tocopherol, and vitamins C and E, which may help prevent major illnesses like cancer [36]. Furthermore, carotenoids—which are frequently utilised in a variety of foods—can be separated from the red, purple, and yellow coloured pigments in paprika.

Meat cuts are enhanced by the addition of garlic and onion. When 10% brine solution of onion and garlic was added to loin and belly cuts, the growth of Enterobacteriaceae was reduced and antioxidant activity matched that of sodium ascorbate [36]. Garlic and onion were added, and this decreased the amount of oxidative products, particularly hexanal and volatile chemicals. Two samples of powdered garlic (fresh and heated) in two different solvents (water and ethanol) were investigated. They found that the heated sample of garlic had less yield and water extract than the fresh onion, but the heated onion's methanol extract could be utilised as a natural antioxidant.

3.1.1.6 Antimicrobial

Protein breakdown by micro-organisms (MO) action during microbial deterioration in MP (meat products) can produce volatile organic compounds and biogenic amines. Consumers are prone to reject rotten and bitter flavours produced by biogenic amines such hypoxanthine and inosine [28]. It has been demonstrated that the bioactive chemicals in PE are useful in preventing the growth of foodborne pathogens, which can assist postpone the generation of biogenic amines and increase the shelf life of MP [37]. Numerous bioactive compounds (BC) included in plant extracts and essential oils have antibacterial qualities that can extend MP's shelf life. High concentrations of BC can also cause pro-oxidant effects in meat matrices. To guarantee their efficacy and prevent any negative side effects, it is crucial to thoroughly assess and optimise the dosages to be used [Gabriel].

3.1.1.7 Dietary Fibres

Dietary fibre includes polysaccharides, oligosaccharides, lignin and associated plant substances. It functions as a laxative agent, blood cholesterol and/or blood glucose attenuator. It has been found that addition of dietary fibre enhances the water-holding capacity and results in improved cooking properties of beef burgers. It also changes colour of meat depending upon the fibre type used as supplement. The redness of raw and cooked chicken hamburgers was increased when Grape antioxidant dietary fibre was added to the products. The addition of dietary fibre may also affect the textural properties of meat products like hardness (usually increases). Due to their adverse effect on textural properties when added in higher amounts, addition of small amount of dietary fibre is recommended in meat products [29].

3.1.2 Microbial Sources

3.1.2.1 Probiotics

Probiotics are live bacteria that help the host's gut microbiota balance and hence improve health. To efficiently function in the human body environment, they are formulated into live bacterial preparation using a variety of techniques and dosage forms. They are obtained from typical flora found in humans or animals, selected, and artificially reproduced. Probiotic viability must be maintained during the encapsulation procedure. Probiotic viability has been demonstrated to be increased by encapsulation treatments [38]. Probiotic encapsulation uses a variety of wall materials, including proteins (like whey, protein, and gelatin), plant gums (such gum arabic, carrageenan, pectin, and xanthan gum), and composite wall materials. Polysaccharides are also employed. Typical encapsulation techniques include spraying drying and other process.

3.1.2.2 Antibacterial

In food processing, bacteriostatic chemicals are essential for preventing bacterial development. They do, however, have certain drawbacks, including poor sensory flavour, limited solubility, quick oxidation, and noticeably low thermal stability. Thankfully, the process of encapsulating these bacteriostatic drugs improves their solubility in specific conditions and increases their chemical and thermal stability, providing a solution. Essential oils, plant extracts, bacteriocins, and other bacteriostatic substances are the main focus of encapsulation for bacteriostasis in food. Phenols, alcohols, terpenoids, and aldehydes included in essential oils have antibacterial properties that infiltrate and damage cell membranes, leading to leaking of cell contents and eventual cell death [39]. In line with the expanding market need for health and safety, essential oils are naturally extracted and have broad-spectrum inhibitory effects [40]. Natural plant components, aside from essential oils, include anti-inflammatory, antioxidant, and inhibitory properties that render them suitable for use in food bacteriostasis.

3.1.3 Animal Sources

3.1.3.1 Bioactive Peptides

Bioactive peptides are short-chain polymers of 2–20 amino acids linked together by peptide linkages. Usually, they remain inactive in food as they are pre-cursors of proteins. During various processes like fermentation, aging, enzymatic treatment, etc. these peptides get released in their active form. These peptides tend to affect various physiological functions in human body like anti-diabetic, antioxidant, immunomodulating, antihypertensive, anticancer, antithrombotic, antimicrobial, and probiotic [41]. These bioactive peptides are mostly bitter in taste. To avoid this sensory problem, hydrolysates of meat or gelatin are used as a source of these peptides in meat products as they are less bitter.

4 Techniques Used for Extraction of Bioactive Compounds

Usually, bioactive compounds can be categorised into two groups- those which are extracted more with non-polar solvents are grouped as lipophilic compounds, where as those which are more soluble in polar solvents are termed as hydrophilic compounds. Oleoresins, essential oils, terpenoids, etc. belong to first group, whereas, polyphenols are an example of hydrophilic compounds. Mostly, a broad spectrum of these compounds is found present in a single source, especially if the extract belongs to plant or animal source. Such crude extracts contain a mixture of bioactive compounds with varied biological activities, which in turn can be additive, synergistic, or antagonistic. For example, the hydrophobic botanical extract of turmeric contains a mixture of phytochemicals like carotenoids, xanthophylls, curcuminoids, alkaloids, fat soluble vitamins, pro-vitamins, etc. [42]

4.1 Conventional Techniques

Conventional extraction techniques majorly involve use of solid-liquid, or, liquid-liquid-liquid extraction methods.

4.1.1 Solid-Liquid Extraction Method

In this method, the source material like leaves, stem, etc. of a plant is mixed with a suitable solvent or solvent mixture. The required active compounds get solubilized in the solvent/solvent mixture.

4.1.2 Liquid-Liquid Extraction Method

This method of extraction includes use of two immiscible solvents for extraction of the bioactive compounds from the source material. Depending upon the polarity of the solvent, the targeted compounds are recovered from each fraction.

In both the methods, the BC are extracted from source material using different techniques like maceration, Soxhlet, percolation, etc. [2]. The conventional methods have their own advantages and disadvantages which are discussed below.

Advantages of conventional extraction methods:

(i) Low -cost of processing
(ii) Simple and easy to perform
(iii) Do not require use of highly sophisticated equipment

Disadvantages of conventional methods:

(i) Long processing times
(ii) Low -efficiency
(iii) Non-selectivity of target compounds
(iv) Use of high amount of solvents
(v) Formation of emulsions during extraction

4.2 *Modern or Emerging Techniques*

To overcome the above discussed disadvantages of conventional techniques, the use of new techniques have been emerged. Each of these techniques have its pros and cons. Few of these emerging extraction techniques have been discussed below.

4.2.1 Ultrasound-Assisted Extraction (USAE)

The ultrasonic methods are categorised as low intensity (100 kHz–10 MHz; <1 W/cm^2) and high -intensity (20 kHz–100 MHz; >1 W/cm^2) methods, based on the power levels and frequency used. USAE makes use of only high intensity methods as high-intensity methods have capability to alter the physical properties of sample. The ultrasonic waves need a medium for propagation, and disrupt materials (source cells) based on the cavitation phenomenon. This phenomenon helps in quick transfer of bioactive compounds from source to solvent [43].

4.2.2 Microwave-Assisted Extraction (MWAE)

It is another widely used technique, majorly for extraction of colorants, aromatic compounds and antioxidants from source material. Its principle is based upon propagation of EM (electromagnetic) waves (0.1–3 GHz) through source material for few seconds to few minutes. These MW are mildly ionizing radiations causing the generation of thermal energy due to vibration of polar molecules. The non-polar solvents with low dielectric constant are not affected by these radiation, resulting in no increase in temperature of the solvent. Due to increase in temperature and pressure, a series of changes happen in the source material like evaporation of moisture, swelling of cells of source material, dehydration of cellulosic material, etc. These changes result in disruption of material cells, causing faster transfer of BC from source to solvent [44].

4.2.3 Pulsed Electric Field-Assisted Extraction (PEFAE)

It is an effective non-thermal extraction method, which can be used alone or as a pre-treatment with other extractive techniques. It makes use of intermittent high-voltages pulses (0.5–50 kV) for very short intervals (microseconds to milliseconds). The application of PEF causes disruption of cell membranes of source material (electroporation), The formation of pores enhances the transfer rate of BC to the solvent. This technique is equally useful in extraction of both, hydrophilic as well as hydrophobic compounds [45].

4.2.4 Supercritical Fluid-Assisted Extraction

It a highly effective and efficient method for extraction of bioactive compounds. Supercritical fluids have a unique characteristic, ie, they show the properties of both liquids and gases under specific temperature and pressure ranges. This feature gives them unique density, viscosity, and solvation properties [44]. This combination of properties helps in their easy penetration, and, thus faster mass transfer rate of targeted active compounds.

Now-a-days, researchers are looking for the options to use these emerging extraction techniques in combination, for better efficiency and effectiveness. Their combined application may have synergistic effect and, can result in higher yields with better quality extracts. The pros and cons of these techniques are tabulated below (Table 1):

Table 1 Advantages and drawbacks of emerging extraction technologies

S.no.	Technique	Benefits	Constraints
1	Ultrasound-assisted extraction	Decrease energy, time and solvent consumption Increase solvent penetration into plant material (mixing effect) and extraction yields Easy to use and low equipment cost Facilitate mass transfer Compatibility with GRAS solvents	Can induce oxidation pyrolysis Promote free radical formation High ultrasound waves has deleterious effects on bioactive compounds Increase in "temperature by cavitation" Low selective
2	Microwave-assisted extraction	Decrease time and solvent consumption Increase extraction yields Cost-effective equipment Easily to scale up	High energy consumption Excessive temperature (bioactive compound degradation) Oxidation reactions Low selective (large number of compounds extracted)
3	Pulsed electric field-assisted extraction	Decrease energy, time and solvent consumption Very low changes in temperature Minimize degradation of thermolabile Bioactive compounds Easily to scale up High selectivity	Very expensive equipment Need proper solvent and electrical conductivity
4	Supercritical fluid-assisted extraction	Decrease extraction time No use toxic solvents Extracts are pure, and present high quality Effective (low viscosity and high diffusivity) Increase extraction yields Continual process Recycling supercritical fluid Preserve thermolabile bioactive compounds Tunable supercritical fluid (solvent) density	High energy consumption Very expensive and complex equipment Need co-solvent to ensure the correct extraction of polar compounds Scale up not feasible

Information was obtained from previous studies [42]

5 Encapsulation Technology

The process of entrapping one substance within another, resulting in particles with dimensions of a few nanometers to a few millimetres, is known as encapsulation. The encapsulation process involves two materials ie; an active agent which is being encapsulated, is referred as core material and a coating material is used for encapsulation is referred as wall material [46]. Wall material and core material, properties – Encapsulation is a two-step procedure that begins with emulsifying the core material (such as the "lipid-essence" system) with a wall material solution (such as a polysaccharide or protein solution) at a specific density. The emulsion is then dried or condensed. Two immiscible liquids, often water and oil, combine to produce an emulsion. The dispersion of one liquid into the other occurs as tiny, spherical droplets due to surfactant activity and mechanical shear force [47].

The material forming the surrounding liquid is referred to as the continuous or external phase, and the droplets that develop in the emulsion are referred to as the dispersed, discontinuous, or internal phase. There are two categories of emulsions: single, or conventional, and multiple, or double (Freire [48]). Water-in-oil-in-water (W1/O/W2) and oil-in-water-in-oil (O1/W/O2) are the two types of multiple or double emulsions (DE), whereas simple emulsions are those with an oil-in-water (O/W) and water-in-oil (W/O) composition [49] whereby smaller droplets of a different phase are contained in the scattered droplets.

There are two steps in the development of a DE. The process begins with the creation of the internal emulsion (W1/O), which is then spread in the external aqueous phase (W2) in a subsequent stage [50]. Applications for food-grade double emulsions are numerous, but one of the most popular uses is the encapsulating of bioactive substances, such as phytochemicals like quercetin, iron, peptides, etc.

5.1 Micro Encapsulation

In micro encapsulation, the core material will be of micro sized which are protected from external conditions by wall material. Here the micro emulsions are used which are characterized by the size of droplet. As microemulsions are thermodynamically stable, they are formed by selfassembly [51]. microencapsulation can also be defined as the process of packing solids, liquids, or gases with thin polymeric coverings to create tiny particles known as microcapsules. The food industry's use of microencapsulated substances may shield key components from the impacts of the environment [52]. The microcapsules' size ranges from 1 to 1000 μm, contingent on the microencapsulation methods employed, as well as the processing parameters, wall and core materials, etc. Microcapsules smaller than 100 μm are often preferred for food applications to prevent any negative effects on the food product's mouthfeel [53]. Microencapsulation is a commonly used technique in the pharmaceutical, agricultural, medical, and culinary industries. It is particularly useful for encapsulating essential oils, colorings, flavourings, sweeteners, and microbes, among other things.

The four processes of microencapsulation are, development of the core and encapsulants, incooperation and solidification [54]. Grinding, melting, or milling reduces the core material's particle size and increases its solubility. To improve stability, a suitable emulsifier or surfactant is added to soluble core materials; porous solids or matrices are utilised for gaseous materials [55]. Nevertheless, heat-sensitive materials may degrade, diffuse, or leak when solidified or exposed to high temperatures, which would reduce the preservation of the core material. Freeze drying, coacervation, spray chilling, or cross-linking are effective encapsulation strategies to reduce the loss of core components [42]. This can be accomplished by using physical or chemical treatments, including heating or cooling, or a combination of both. The size of encapsulating has no direct effect on encapsulation efficiency, but it affects the controlled release of active ingredients in addition to physical and chemical properties of encapsulants, as smaller sizes result in larger contact areas for enzymes or releasing media and thus the release of larger amounts as compared to encapsulates with larger sizes [56]. The shell or wall material should preferably be insoluble, non-reactive to active ingredients, have excellent film-forming, and have the desired protective properties against various environmental conditions [42]. The encapsulation method, properties of core and encapsulant materials, and load quantity result in the different microstructures of encapsulates. The microstructure of capsules has an impact on the efficiency and retention of the core material and the encapsulation efficiency and regulates the release of active ingredients. Nonetheless, there are two primary obstacles to the microencapsulation of oils and fats. Initially, powdered microcapsule materials have the potential to aggregate as they store. Reducing the amount of wall material adsorption or include anticoagulants in the formulation are two ways to deal with this problem. The high cost of equipment and energy usage related to this technique presents another challenge. Thus, in order to make this technology more widely available and long-lasting, more affordable and energy-efficient encapsulating equipment must be developed.

5.2 Nanoencapsulation

The use of films, layers, and coatings to encapsulate on a nanoscale is referred to as nanoencapsulation. The encapsulation layer, which forms a protective coating over food or flavour molecules/ingredients, is nanoscale in size.

The advantage of nanoencapsulation over microencapsulation arises from its smaller size; the smaller the capsules, the more bioavailable they are and the more easily their release may be regulated. Particle sizes of nano-capsules typically vary from 1 to 100 nm. In addition to the food industry, the packaging sector uses nanoencapsulation to package meat and fruits to prolong their shelf life while maintaining their nutritional value [56]. The methods used to encapsulate these compounds are categorised according to the amount of power required, including combination treatment and top-down and bottom-down methods [57]. The top-down encapsulation method uses a lot of power because it uses devices like homogenizers, spray-drying, ultra-sonication, and many more, whereas the bottom-down method uses less energy and uses techniques like precipitation, micro-emulsification, conjugation, atom exchange, etc. The selection of a certain nanoencapsulation technique is contingent upon several key aspects, including the delivery motive, release rate, stability and solubility of the nano-carrier, and production costs.

5.3 *Methods for Encapsulation*

5.3.1 Spray Drying

The process involves first dissolving and emulsifying the dispersion of the core component in the carrier material solutions, then spraying the mixture into a heated chamber [58]. The wall material covers the core when the solvent evaporates, resulting in the production of polynuclear or matrix-type microcapsules. This method therefore involves atomizing an emulsion or solution of core-carrier chemicals and then recovering the dehydrated microcapsules thereafter [59]. In nanoencapsulation, a slight change has been made in the case of nano-spray, involving the application of a specific nozzle that creates nano-droplets by continuously supplying drying gas from a laminar flow. Additionally, vibrating mesh with holes of 4.0, 5.5, 7.0 m, Encyclopaedia 2021, and 1652 size is used to the nano-droplets. The concentration of the fluid, the temperature of the drying gas, the spray velocity, and the size of the droplets all affect how big the fragments are in the spray-dried sample [19].

5.3.2 Coacervation Technique

The fundamental mechanism of capsule wall development is called "coacervation." Coacervation is a colloidal phenomena that encapsulates liquids and solids under constant agitation. The process known as "coacervation" entails changing the physicochemical characteristics of the medium, such as its pH, polarity, temperature, and ionic strength, to deposit polymer around the core.

The process of coalescence is inexpensive, straightforward, and does not call for high temperatures or organic solvents.

5.3.3 Extrusion

This method involves dropping droplets of an aqueous polymer solution into a gelling bath solution. The most often used polymer with calcium chloride gelling bath solution is sodium alginate (0.6–3%). A pipette, syringe, vibrating nozzle, spraying nozzle, jet cutter, or vibrating disc can all be used as the dropping tool. For the encapsulation of volatiles and unstable flavours, this technique is frequently utilised. Co-extrusion is used for highly concentrated and viscous cores or encapsulant incorporation and particle fabrication, resulting in the formation of matrix microstructures, as opposed to conventional extrusion, which typically uses force to force a mixture of core and encapsulants through the nozzle. It involves the glassy carbohydrate matrix encapsulating volatile and unstable flavours. This technique confers a very long shelf life to flavour chemicals that are often susceptible to oxidation, like citrus oils [60].

5.3.4 Fluidized Bed Coating

Coating is applied onto a fluidized bed of solid particles in this technology, which is a modified form of the spray drying method [61]. It is mostly employed in the pharmaceutical sector. This method uses porous solid material to absorb and encapsulate liquid and solid core components. It involves three steps: first, suspending the solid materials in the air; second, encasing the solid particles with a liquid material by spraying; and third, solidifying the shell by cooling or solvent vaporisation. Hydrogen bonds or liquid bridges, as well as electrostatic, hydrophobic, and hydrophilic interactions, hold the core material to the encapsulants. Surface tension and viscosity are crucial encapsulant solution parameters. These three steps are carried out once more till the wall reaches the required thickness [55].

5.3.5 Ionic Gelation

It is the process of combining an anionic salt with a charged biopolymer to produce complexes. Encapsulation efficiency is not significantly affected by either an internal or external gelation process. Using extrusion, a syringe needle, or electrospinning, core material and charged biopolymer drops are dropped into an anionic salt solution to generate the exterior gelation. When internal gelation occurs, the encapsulants, anionic salt solutions, and core components are combined and continually combined with an emulsion to achieve the required encapsulate size [42].

5.3.6 Micelle and Liposome-Based Technique

Liposomes are small membrane vesicles based on lipids that are created through phospholipid hydrolysis. They are becoming more and more popular in the production of functional foods, pharmaceuticals, cosmetics, and nutraceuticals for the transport and sustained release of bioactive ingredients [60]. Liposomes contain phospholipids in large quantities. The most widely utilised phospholipids are semi-synthetic or naturally occurring, like soy, egg, and marine lecithin. When encapsulating bioactive substances that are chemically unstable or have poor water solubility, liposome delivery systems are used. It offers increased bioavailability and encapsulation efficiency for the active components. The high cost, inadequate chemical and physical qualities, irregular particle size, increased susceptibility to lipid oxidation, and need for post-processing make this method unsuitable for large-scale use.

5.3.7 Ultrasonication

A cluster of sound waves above the human hearing threshold (>16 kHz) is referred to as "ultrasound." Essentially, it is separated into two streams: low- and high-intensity waved. Low-intensity waves are typically used for sonography, or detection, while high-intensity waves are used to modify molecules, including their size reduction, develop emulsification, and are widely used in the food industry [61]. Ultrasonic components include an electric generator, a piezoelectric transducer that converts electrical energy into sound, and a titanium horn sound emitter. Form for transferring ultrasonic waves into the medium or sample [62]. Numerous nano-delivery methods utilising lipid and surfactant have been developed using molecules as a wall substance to create nano-emulsions,

5.3.8 High Pressure Homogenization

The process of creating nano-fragments of a homogenous size in a fluid at a certain high pressure is known as high-pressure homogenization. It has a pressure that is 10–15 times higher (100–400 MPa) than a traditional homogenizer. The study indicates that the milk and milk product industries heavily use this technique to improve the texture, taste, and flavour of their products as well as their shelf life and antimicrobial quality (in terms of activating Salmonella spp., Listeria monocytogenes, Staphylococcus aureus, and Escherichia coli). At 200 and 300 MPa homogenization pressure and 30 and 40 C inlet temperatures, respectively, a decrease of around 3 and 4 log cycles was found. Because of its remarkable efficacy in rendering enzymes and microbiological activity inactive, it serves as a stand-in for heat processing [63].

6 Choice of Matrices and Delivery Systems for Encapsulation

The delivery system is a type of reaction system that fully controls how compounds are distributed in organisms in terms of time, space, and dosage. Its goal is to guarantee the accurate supply of the appropriate quantity of material at the appropriate time to the appropriate location, boosting efficacy, cutting expenses, and increasing substance utilisation efficiency. Delivery via encapsulation has specific demands regarding the kind of wall material. The majority of food delivery processes take place in the stomach's and intestines' acidic environment, where they come into contact with digestive enzymes like lipase, amylase, and protease in the digestive tract. As a result, it's critical to choose wall materials resistant to acid that can tolerate digestive enzymes and permit the embedded ingredient to be released and digested gradually within the gastrointestinal tract. Non-starch Because they can withstand digestive enzymes and allow for delayed release in the gut, polysaccharides—including chondroitin sulphate, pectin, chitosan, guar gum, and sodium alginate—are frequently utilised as wall materials for this purpose [64] The hydrophilic/lipophilic, polar/nonpolar solubility, and other characteristics of the core material can also be altered by encapsulation, which can have an impact on the release site. For instance, enzymes have less of an impact on the resulting microcapsules when xanthan gum and starch are combined as the wall material. As a result, the wall components' deterioration brought on by digestive enzymes is slowed down. Other insoluble compounds can be encapsulated and released into the gastrointestinal tract by using the hydrophilicity and solubility of starch and xanthan gum. This illustrates how targeted delivery vehicles for bioactive compounds may be used.

7 Application of Encapsulated Bioactive Ingredients in Meat Industry

In addition to helping to reduce meat deterioration by initially lowering the viable microbiological population, bioactive compounds added to MP can also stop the development of oxidised chemicals that arise during processing, heat treatment, smoking, fermenting, or curing [65]. But because BC are unstable, very susceptible to environmental changes, and have poor stability, their application is limited. If not employed in the right amounts, these substances can also have a detrimental effect on the food's sensory qualities [66]. Despite these drawbacks, the fact that BC is generally safe to use in food products is one of its main advantages. Addition of moringa leaf powder in chicken burger improved the flavour, colour, tenderness and juiciness of the meat. In ground meat, nettle extracts were added, but the addition of half the quantity of nettle has reduced the lipid autooxidation in the product even though there was no much change in the colour of the product. A study was done on the shelf life of the pork burgers, in which one sample was provided with 0.5% walnut leaf powder and other with 0.1% BHT, both were stored under refrigerated condition for 15 days, the one with walnut leaf powder had better properties when compared to BHT.

Extending the shelf-life of meat products is a significant usage of encapsulated bioactive compounds. In order to stop minced meat from deteriorating, the authors of a fairly recent study suggested encapsulating garlic essential oil using the spray drying process and employing gum Arabic and maltodextrin as wall materials [27]. In this instance, the 20% concentration of microencapsulated essential oil demonstrated potent antibacterial activity, extending the shelf life of the minced beef. In a similar vein, casein-maltodextrin thyme essential oil that had been spray-dried shown potent antibacterial and antioxidant properties [67]. Furthermore, these authors draw the conclusion that spray drying essential oil encapsulation with casein and maltodextrin as wall materials worked well to create capsules with high encapsulation efficiency, superior thermal stability, and consistent. In a another study, the authors increase the shelf life of beef burgers by using extracts from prickly pears that have been encapsulated. In this study, the addition of prickly pear extracts to burgers had a noteworthy antibacterial impact; among these, the usage of encapsulated extract exhibited higher activity than unencapsulated extract. Furthermore, the extract's encapsulation revealed more appealing texture and colour characteristics [27]. Promising outcomes were also observed when pitaya peel extract was encapsulated using a spray-drying process and maltodextrin was used as the wall material. Using these microcapsules prevented protein oxidation and showed a shielding effect against changes brought about by the high-pressure process treatment.

The use of encapsulated oils to swap out animal fat, which is heavy in cholesterol and saturated fatty acids, with beneficial marine or vegetable oils has been extensively researched in the meat business. In this way, a functional chicken sausage (enriched with omega-3) was developed using encapsulated flaxseed oil. According to the study, the inclusion of encapsulated oil significantly affects the sausages' physical qualities, primarily their texture and rheological metrics, while also improving their nutritional profile when compared to control samples [27]. Research was done to improve the physicochemical and nutritional qualities of deer pâté by substituting microencapsulated healthy oils (tigernut, chia, and sesame) for a portion of the animal fat (50%). The nutritional qualities significantly improved, according to the authors (increased mono-and/or polyunsaturated fatty acids, depending on the formulation, and decreased total fat, cholesterol, and saturated fatty acids). On the other hand, a decrease in sensory acceptability and an increase in lipid oxidation were the results of the reformulation using high PUFA oils (chia and linseed). Tigernut oil microcapsules, on the other hand, showed the same acceptability to control pâtés and did not encourage oxidation, proving that this oil can be reformulated [66]. The authors of a relatively recent study suggested using various encapsulated fish oils in the chicken nugget formulation. The encapsulation process significantly enhanced the fatty acid stability (protective action against lipid and protein oxidation), and a significant fortification of EPA and DHA fatty acids was noted, which had positive effects on human health.

Furthermore, these authors emphasised that the use of encapsulated fish oil did not affect sensory quality, whereas the addition of unencapsulated fish oil in chicken nuggets resulted in a decrease in sensory acceptance [68]. This finding demonstrated the high potential for using encapsulated fish oil to produce healthy meat products. Similar outcomes were seen in cooked and dry-cured sausages, where mono-and multilayered fish oil microcapsules had no effect on the physicochemical. Furthermore, fish oil encapsulation shields against oxidation of proteins and lipids [69]. However, other writers who produced long-chain omega-3 enriched burgers using encapsulated fish oil had different outcomes [70, 71]. In this instance, the addition of microparticles to the burgers reduced their oxidative stability and enhanced the release of volatiles derived from lipids, which had a detrimental effect on the sensory profile and general appeal of the burgers, even though the nutritional quality had significantly improved. Using a combination of oregano (Origanum vulgare) in the form of oleoresin or EO and vacuum packaging, Vergara et al. [72] assessed the shelf life of lamb burgers as well as the quality attributes of pH and colour coordinates, including LA and MO growth (total viable count, Enterobacteriaceae, Pseudomonas Spp., and lactic acid bacteria [LAB]). They concluded that this approach kept the meat for 14 days with a better colour than the control, all at an acceptable pH of 5.6. Because it contains phenolic components like luteolin, quercetin, naringenin, and pinocembrin, oregano can delay rancidity and extend its shelf life. Another example is the research done by Cozar [72, 73] on the combined effects of vacuum packaging and spice powder (rosemary, thyme, sage and garlic at a concentration of 0.1% w/w) with lamb burgers. The researchers looked at colour coordinates, MO counts (including total viable count,

Pseudomonas spp., Enterobacteriaceae and LAB), and found that the effects of rosemary (*Salvia rosmarinus*), thyme (*Thymus vulgaris*), and sage (*Salvia officinalis*) were more successful in delaying rancidity (Table 2).

Table 2 Few examples of application of encapsulated plant-based active ingredients in meat industry

S.no.	Extract	Meat product	Effect on product
1	Rosemary extract added @ 800–16,000 ppm	Beef product	Reduces primary and secondary lipid oxidation; inhibit microbial growth at refrigerated temperature; minimal changes in colour parameters
2.	Orange essential oil and cactus acid fruit extract @ 0–5%	Emulsified meat system	Increases antioxidant activity; reduces lipid oxidation during storage; increases fat content (due to addition of orange essential oil), increases the phenolic content
3.	Radish, hibiscus and beetroot extracts @ 0.4–7.29 g/kg	Cooked ham	Enhances colour (as colourant)
4.	Lupulon–xanthohumol nanoliposome @ 50–200 ppm	Cooked beef sausage	Prevents lipid oxidation; inhibits the growth of microorganisms (total viable counts and molds/yeast) when added with nitrite under refrigerated conditions; inhibit the growth of *Clostridium perfringers* and coliforms when added with nitrate under refrigerated conditions; acts as colour fixative and maintains red colour
5.	Allium sativum essential oil @ 0.10%	Minced meat	Act as antimicrobial agent; essential oil microcapsules showed inhibitory effect against microbial growth (total aerobic mesophilic flora, sulphite-reducing anaerobes and *E. coli*)
6.	Thyme essential oil @ 1%	Hamburger-like meat products	Inhibits the growth of thermotolerant coliforms and E. coli (antimicrobial agent)
7.	Prickly pear fruit extract @ 5%	Beef burger patties	Encapsulants showed lower values of mesophilic bacteria, Enterobacteriaceae and Pseudomonas spp.; help in maintenance of colour (redness) and texture; Maintain the pH values during storage (buffering action)
8.	Olive leaves extract 100 mg oleuropein/kg	Meat systems (with healthy oil mixture)	Shows high antioxidant activity with higher oxidative stability of product; improves binding properties and texture (binding and texturizing agent)
9.	Laurus nobilis leaf extract 1000–1500 ppm	Minced beef	Inhibits oxidative degradation; antimicrobial agent as controls total viable count and psychotropic count; inhibits the growth of *Staphylococcus aureus* and *E. coli*; nano encapsulants reduce lipolytic and non-protein nitrogen breakdown; reduces sensory properties minutely
10.	Quinoa peptide-loaded nanoliposomes 3–5 mg/mL	Burger	Reduces primary and secondary lipid oxidation; Reduces the total bacterial count and growth of *S. aureus*, mold, and yeast; Reduce proteolytic activity derived from enzyme and/or microbial spoilage

8 Conclusion

In conclusion, the rise of deadly diseases underscores the importance of strengthening the human immune system, which can be achieved through bioactive compounds extracted from natural sources like plants and micro-organisms. Researchers are thus motivated to find conventional and sustainable methods for extracting these compounds. Encapsulation of bioactive compounds significantly enhances their stability against environmental factors such as pH variations, high temperatures, light, and oxidative conditions. Additionally, encapsulation proves effective in controlling the release of phytochemicals in meat products, thereby improving their quality, nutritional properties, and resistance to microbial contamination and oxidation. Various encapsulation methods have shown that this technology not only enhances the nutritional content of meat products but also improves their texture and overall quality.

References

1. Stajić S, Vasilev D. Encapsulation of meat product ingredients and influence on product quality. In: Encapsulation in food processing and fermentation. 1st ed. CRC Press; 2022. p. 26. ISBN 978042932491.
2. Lorenzo JM, Pateiro M, Domínguez R, Barba FJ, Putnik P, Kovačević DB, Shpigelman A, Granato D, Franco D. Berries extracts as natural antioxidants in meat products: a review. Food Res Int. 2018;106:1095–104. https://doi.org/10.1016/j.foodres.2017.12.005.
3. De Oliveira JL, Fraceto LF, Bravo A, Polanczyk RA. Encapsulation strategies for *Bacillus thuringiensis*: from now to the future. J Agric Food Chem. 2021;69:4564–77.
4. Vassilev N, Vassileva M, Martos V, Garcia del Moral LF, Kowalska J, Tylkowski B, Malusá E. Formulation of microbial inoculants by encapsulation in natural polysaccharides: focus on beneficial properties of carrier additives and derivatives. Front Plant Sci. 2020;11:270.
5. Bae M, Lewis A, Liu S, Arcot Y, Lin Y-T, Bernal JS, Cisneros-Zevallos L, Akbulut M. Novel biopesticides based on nanoencapsulation of azadirachtin with whey protein to control fall armyworm. J Agric Food Chem. 2022;70:7900–10.
6. Kadmiri IM, El Mernissi N, Azaroual SE, Mekhzoum MEM, Qaiss AEK, Bouhfid R. Bioformulation of microbial fertilizer based on clay and alginate encapsulation. Curr Microbiol. 2021;78:86–94.
7. Riseh RS, Skorik YA, Thakur VK, Pour MM, Tamanadar E, Noghabi SS. Encapsulation of plant biocontrol bacteria with alginate as a main polymer material. Int J Mol Sci. 2021;22:11165.
8. Meng R, Wu Z, Xie QT, Cheng JS, Zhang B. Preparation and characterization of zein/carboxymethyl dextrin nanoparticles to encapsulate curcumin: physicochemical stability, antioxidant activity and controlled release properties. Food Chem. 2021;340:127893. https://doi.org/10.1016/j.foodchem.2020.127893.
9. Chen Y, Tai K, Ma P, Su J, Dong W, Gao Y, Mao L, Liu J, Yuan F. Novel γ-cyclodextrin-metal–organic frameworks for encapsulation of curcumin with improved loading capacity, physicochemical stability and controlled release properties. Food Chem. 2021;347:128978. https://doi.org/10.1016/j.foodchem.2020.128978.
10. Tang XM, Liu PD, Chen ZJ, Li XY, Huang R, Liu GD, et al. Encapsulation of a desmodium intortum protein isolate pickering emulsion of β-carotene: stability, bioaccesibility and cytotoxicity. Food Secur. 2022;11(7):936. https://doi.org/10.3390/foods11070936.

11. Le Priol L, Gmur J, Dagmey A, Morandat S, Kirat KE, Saleh K, Nesterenko A. Oxidative stability of encapsulated sunflower oil: effect of protein- polysaccharide mixtures and long-term storage. J Food Meas Charact. 2022;16(2):1483–93. https://doi.org/10.1007/s11694-021-01254-5.
12. Krstonošić VS, Kalić MD, Dapčević-Hadnađev TR, Lončarević IS, Hadnađev MS. Physico-chemical characterization of protein stabilized oil- in-water emulsions. Colloids Surf A Physicochem Eng Asp. 2020;602:125045. https://doi.org/10.1016/j.colsurfa.2020.125045.
13. Chen B, Liu X, Zhou K, Xie Y, Wang Y, Zhou H, Bai Y, Xu B. Differentiating the effects of hydrophobic interaction and disulfide bond on the myofibrillar protein emulsion gels at the high temperature and the protein interfacial properties. Food Chem. 2023;412:135472. https://doi.org/10.1016/j.foodchem.2023.135472.
14. Silva CM, Ribeiro AJ, Figueiredo M, Ferreira D, Veiga F. Microencapsulation of hemoglobin in chitosan-coated alginate microspheres prepared by emulsification/internal gelation. AAPS J. 2005;7(4):88. https://doi.org/10.1208/aapsj070488.
15. Xu J, Jiang S, Liu L, Zhao Y, Zeng M. Encapsulation of oyster protein hydrolysates in nanoliposomes: vesicle characteristics, storage stability, in vitro release, and gastrointestinal digestion. J Food Sci. 2021;86(3):960–8. https://doi.org/10.1111/1750-3841.15606.
16. Zou L, Zhang Z, Zhang R, Liu W, Liu C, Xiao H, McClements DJ. Encapsulation of protein nanoparticles within alginate microparticles: impact of ph and ionic strength on functional performance. J Food Eng. 2016;178:81–9. https://doi.org/10.1016/j.jfoodeng.2016.01.010.
17. Filipsky T, Riha M, Macakova K, Anzenbacherova E, Karlickova J, Mladenka P. Antioxidant effects of coumarins include direct radical scavenging, metal chelation and inhibition of ROS-producing enzymes. Curr Top Med Chem. 2015;15(5):415–31. https://doi.org/10.2174/1568026615666150206152233.
18. Noore S, Rastogi NK, O'Donnell C, Tiwari B. Novel bioactive extraction and nano-encapsulation. Encyclopedia. 2021;1:632–64. https://doi.org/10.3390/encyclopedia1030052.
19. Bor T, Aljaloud SO, Gyawali R, Ibrahim SA. Antimicrobials from herbs, spices, and plants. Elsevier Inc; 2016. https://doi.org/10.1016/B978-0-12-802972-5.00026-3.
20. Domínguez R, Pateiro M, Munekata PES, McClements DJ, Lorenzo JM. Encapsulation of bioactive phytochemicals in plant-based matrices and application as additives in meat and meatproducts. Molecules. 2021;26:3984. https://doi.org/10.3390/molecules26133984.
21. Zhao L, Temelli F, Curtis JM, Chen L. Encapsulation of lutein in liposomes using supercritical carbon dioxide. Food Res Int. 2017;100:168–79. https://doi.org/10.1016/j.foodres.2017.06.055.
22. Cheng YS, Lu PM, Huang CY, Wu JJ. Encapsulation of lycopene with lecithin and α-tocopherol by supercritical antisolvent process for stability enhancement. J Supercrit Fluids. 2017;130:246–52. https://doi.org/10.1016/j.supflu.2016.12.021.
23. Olvera-Aguirre G, Piñeiro-Vázquez ÁT, García JRS, Zárate AS, Ochoa-Flores AA, Campos MRS, et al. Using plant-based compounds as preservatives for meat products: a review. Heliyon. 2023;9:e17071.
24. Olvera-aguirre G, Mendoza-taco MM, Moo-huchin VM, Roque-jim A, Armando G. Effect of extraction type on bioactive compounds and antioxidant activity of moringa oleifera lam, leaves. Agriculture. 2022;12:10–3390.
25. Perricone M, Arace E, Corbo MR, Sinigaglia M, Bevilacqua A. Bioactivity of essential oils: a review on their interaction with food components. Front Microbiol. 2015;6:1–7. https://doi.org/10.3389/fmicb.2015.00076.
26. da Silva BD, Bernardes PC, Pinheiro PF, Fantuzzi E, Roberto CD. Chemical composition, extraction sources and action mechanisms of essential oils: natural preservative and limitations of use in meat products. Meat Sci. 2021;176:108463. https://doi.org/10.1016/j.meatsci.2021.108463.
27. Radünz M, dos Santos Hackbart HC, Camargo TM, Nunes CFP, de Barros FAP, Dal Magro J, Filho PJS, Gandra EA, Radünz AL, da Rosa Zavareze E. Antimicrobial potential of spray drying encapsulated thyme (Thymus vulgaris) essential oil on the conservation of hamburger-like meat products. Int J Food Microbiol. 2020;330:108696. https://doi.org/10.1016/j.ijfoodmicro.2020.108696.

28. Hao R, Roy K, Pan J, Shah BR, Mraz J. Critical review on the use of essential oils against spoilage in chilled stored fish: a quantitative meta-analyses: essential oils and seafood. Trends Food Sci Technol. 2021;111:175–90. https://doi.org/10.1016/j.tifs.2021.02.054.-76.
29. Pogorzelska-Nowicka E, Atanasov AG, Horbańczuk J, Wierzbicka A. Bioactive compounds in functional meat products. Molecules. 2018;23(2):307. https://doi.org/10.3390/molecules23020307. PMID: 29385097; PMCID: PMC6017222
30. Srinivasan K. Spices and flavoring crops: uses and health effects. 1st ed. Elsevier Ltd; 2015. https://doi.org/10.1016/B978-0-12-384947-2.00645-0.
31. Lafarga T, Hayes M. Bioactive protein hydrolysates in the functional food ingredient industry: overcoming current challenges. Food Rev Int. 2017;33(3):217–46.
32. Bázan-Lugo E, Garcia-Matinez I, Alfaro-Rodriguez RH, Totosaus A. Color compensation in nitrite-reduced meat batter incorporation paprika or tomato paste. J Sci Food Agr. 2012;92:1627–32. https://doi.org/10.1002/jsfa.4748.
33. Jeong HJ, Lee HC, Chin KB. Effect of red beet on quality and color stability of low-fat sausages during refriger ated storage. Korean J Food Sci An. 2010;30:1014–23. https://doi.org/10.5851/kosfa.2010.30.6.1014.
34. Lee HC, Chin KB. Effectofmungbeanflourlevelin combination with microbial transglutaminase on physico chemical and textural properties of low-salt pork model sau sages. Korean J Food Sci An. 2013;33:221–8. https://doi.org/10.5851/kosfa.2013.33.2.221.
35. Jiang J, Xiong YL. Natural antioxidants as food and feed additives to promote health benefits and quality of meat products: a review. Meat Sci. 2016;120:107–17.
36. Daood HG, Kapitány J, Biacs P, Albrecht K. Drying temperature, endogenous antioxidants and capsaicinoids affect carotenoid stability in paprika (red pepper spice). J Sci Food Agr. 2006;86:2450–7. https://doi.org/10.1002/jsfa.2639.
37. Park SY, Yoo SS, Shim JH, Chin KB. Physicochemical properties, and antioxidant and antimicro bial effects of garlic and onion powder in fresh pork belly and loin during refrigerated storage. J Food Sci. 2008;73(C577):C584. https://doi.org/10.1111/j.1750-3841.2008.00896.x.
38. Azeem M, Saeed F, Afzaal M, Ateeq H, Ahmad A, Liaqat A, Busquets R, Lorenzo JM, Asif Shah M. Encapsulation of probiotics in solid lipid micro particle for improved viability and stability under stressed conditions. Int J Food Prop. 2023;26(1):1612–23. https://doi.org/10.1080/10942912.2023.2223776.
39. Teixeira RF, Balbinot Filho CA, Borges CD. Essential oils as natural antimicrobials for application in edible coatings for minimally processed apple and melon: a review on antimicrobial activity and characteristics of food models. Food Packag Shelf Life. 2022;31:100781. https://doi.org/10.1016/j.fpsl.2021.100781.
40. Zuidam NJ, Nedovic V. Encapsulation technologies for active food ingredients and food processing. New York: Springer; 2010. https://doi.org/10.1007/978-1-4419-1008-0.
41. Zhang W, Xiao S, Samaraweera H, Lee EJ, Ahn DU. Improving functional value of meat products. Meat Sci. 2010;86:15–31.
42. Domínguez R, Pateiro M, Munekata PES, McClements DJ, Lorenzo JM. Encapsulation of bioactive phytochemicals in plant-based matrices and application as additives in meat and meat products. Molecules. 2021;26:3984. https://doi.org/10.3390/MOLECULES26133984.
43. Chemat F, Rombaut N, Sicaire AG, Meullemiestre A, Fabiano-Tixier AS, Abert-Vian M. Ultrasound assisted extraction of food and natural products. Mechanisms, techniques, combinations, protocols and applications. A review. Ultrason Sonochem. 2017;34:540–60.
44. Ameer K, Shahbaz HM, Kwon J-HH. Green extraction methods for polyphenols from plant matrices and their byproducts: a review. Compr Rev Food Sci Food Saf. 2017;16:295–315. [CrossRef]
45. Alirezalu K, Pateiro M, Yaghoubi M, Alirezalu A, Peighambardoust SH, Lorenzo JM. Phytochemical constituents, advanced extraction technologies and techno-functional properties of selected Mediterranean plants for use in meat products. A comprehensive review. Trends Food Sci Technol. 2020;100:292–306.
46. Xu Y, Yan X, Zheng H, Li J, Wu X, Xu J, et al. The application of encapsulation technology in the food industry: classifications, recent advances, and perspectives. Food Chem X. 2024;21:101240.

47. Kale S, Deore SL. Emulsion micro emulsion and nano emulsion: a review. Sys Rev Pharm. 2016;8:39–47.
48. Freire M, Bou R, Cofrades S, Jiménez-Colmenero F. Technological characteristics of cold-set gelled double emulsion enriched with n-3 fatty acids: Effect of hydroxytyrosol addition and chilling storage. Food Res Int. 2017;100(2):298-305. https://doi.org/10.1016/j.foodres.2017.08.047. Epub 2017 Aug 19.
49. Freire M, Cofrades S, Pérez-Jiménez J, Gómez-Estaca J, Jiménez-Colmenero F, Bou R. Emulsion gels containing n-3 fatty acids and condensed tannins designed as functional fat replacers. Food Res Int. 2018;113:465–73.
50. Kumar S, Kumar A. Nanoemulsions: techniques for the preparation and the recent advances in their food applications. Innovative Food Sci Emerg Technol. 2022;76:102914.
51. Espinoza-Leandro YK, Olivera-Montenegro L, Paredes-Concepción P. Meat, dairy and vegetable emulsions: recent innovations in the development of functional, healthy and more stable foods. Sci Agropecu. 2023;14(2):201–12.
52. Gharsallaoui A, Roudaut G, Chambin O, Voilley A, Saurel R. Applications of spray-drying in microencapsulation of food ingredients; an overview. Food Res Int. 2007;40:1107–21. https://doi.org/10.1016/j.foodres.2007.07.004.
53. Cittadini A, Munekata PES, Pateiro M, Sarriés MV, Domínguez R, Lorenzo JM. Encapsulation techniques to increase lipid stability. In: Lorenzo JM, Munekata PES, Barba FJ, Pateiro M, Domínguez R, editors. Food lipids sources, health implications, and future trends. Oxford: Academic Press; 2022. p. 1–46. ISBN 9780128233719.
54. Gómez B, Barba FJ, Domínguez R, Putnik P, Bursać Kovačević D, Pateiro M, Toldrá F, Lorenzo JM. Microencapsulation of antioxidant compounds through innovative technologies and its specific application in meat processing. Trends Food Sci Technol. 2018;82:135–47. https://doi.org/10.1016/j.tifs.2018.10.006.
55. Benelli L, Oliveira WP. Fluidized bed coating of inert cores with a lipid-based system loaded with a polyphenol-rich Rosmarinus officinalis extract. Food Bioprod Process. 2019;114:216–26. https://doi.org/10.1016/j.fbp.2019.01.004.
56. Yun P, Devahastin S, Chiewchan N. Microstructures of encapsulates and their relations with encapsulation efficiency and controlled release of bioactive constituents: a review. Compr Rev Food Sci Food Saf. 2021;20:1768–99. https://doi.org/10.1111/1541-4337.12701.
57. Liu T, Liu L, Gong X, Chi F, Ma Z. Fabrication and comparison of active films from chitosan incorporating different spice extracts for shelf life extension of refrigerated pork. LWT. 2021;135:110181.
58. Seyedabadi MM, Rostami H, Jafari SM, Fathi M. Development and characterization of chitosan-coated nano-liposomes for encapsulation of caffeine. Food Biosci. 2021;40:100857.
59. Mehta N, Kumar P, Verma AK, Umaraw P, Kumar Y, Malav OP, et al. Microencapsulation as a noble technique for the application of bioactive compounds in the food industry: a comprehensive review. Appl Sci. 2022;12(3):1424.
60. Schäfer C, Meyer SP, Osswald TA. A novel extrusion process for the production of polymer micropellets. Polym Eng Sci. 2018;58:2264–75. https://doi.org/10.1002/pen.24847.
61. Petraitytė S, Šipailienė A. Enhancing encapsulation efficiency of alginate capsules containing lactic acid bacteria by using different divalent cross-linkers sources. LWT. 2019;110:307–15. https://doi.org/10.1016/j.lwt.2019.01.065.
62. Lorimer JP, Mason TJ. Sonochemistry. Part 1—the physical aspects. Chem Soc Rev. 1987;16:239–74.
63. Bermudez-Aguirre D, Mobbs T, Barbosa-Cánovas GV. Ultrasound applications in food processing. New York: Springer; 2011. p. 65–105.
64. Georget E, Miller B, Aganovic K, Callanan M, Heinz V, Mathys A. Bacterial spore inactivation by ultra-high pressure homogenization. Innov Food Sci Emerg Technol. 2014;26:116–23.
65. Xiao J, Tian W, Abdullah, Wang H, Chen M, Huang Q, Zhang M, Lu M, Song M, Cao Y. Updated design strategies for oral delivery systems: maximized bioefficacy of dietary bioactive compounds achieved by inducing proper digestive fate and sensory attributes. Crit Rev Food Sci Nutr. 2024;64(3):817–36. https://doi.org/10.1080/10408398.2022.2109583.

66. Parafati L, Palmeri R, Trippa D, Restuccia C, Fallico B. Quality maintenance of beef burger patties by direct addiction or encapsulation of a prickly pear fruit extract. Front Microbiol. 2019;10:1760. https://doi.org/10.3389/fmicb.2019.01760.
67. Najjaa H, Chekki R, Elfalleh W, Tlili H, Jaballah S, Bouzouita N. Freeze-dried, oven-dried, and microencapsulation of essential oil from allium Sativum as potential preservative agents of minced meat. Food Sci Nutr. 2020;8:1995–2003. https://doi.org/10.1002/fsn3.1487.
68. Bolger Z, Brunton NP, Monahan FJ. Impact of inclusion of flaxseed oil (pre-emulsified or encapsulated) on the physical characteristics of chicken sausages. J Food Eng. 2018;230:39–48. https://doi.org/10.1016/j.jfoodeng.2018.02.026.228.
69. Vargas-Ramella M, Pateiro M, Barba FJ, Franco D, Campagnol PCB, Munekata PES, Tomasevic I, Domínguez R, Lorenzo JM. Microencapsulation of healthier oils to enhance the physicochemical and nutritional properties of deer Pâté. LWT. 2020;125:109223. https://doi.org/10.1016/j.lwt.2020.109223.
70. Pourashouri P, Shabanpour B, Heydari S, Raeisi S. Encapsulation of fish oil by carrageenan and gum tragacanth as wall materials and its application to the enrichment of chicken nuggets. LWT. 2021;137:110334. https://doi.org/10.1016/j.lwt.2020.110334.
71. Pérez-Palacios T, Ruiz-Carrascal J, Jiménez-Martín E, Solomando JC, Antequera T. Improving the lipid profile of ready to-cook meat products by addition of omega-3 microcapsules: effect on oxidation and sensory analysis. J Sci Food Agric. 2018;98:5302–12. https://doi.org/10.1002/jsfa.9069.
72. Vergara H, Cózar A, Rubio N. Effect of adding of different forms of oregano (Origanum vulgare) on lamb meat burgers quality during the storage time. CyTA J Food. 2020;18:535–42. https://doi.org/10.1080/19476337.2020.1794981.
73. Cózar A, Vergara H. Lamb burgers made with low and high value cuts: effect of the spice added and the packaging method on shelf life. CyTA J Food. 2018;16:1115–24. https://doi.org/10.1080/19476337.2018.1537303.

Meat Products with Soy

Sweta Joshi, Irfan Khan, Akanksha Soni, Tohfa Siddiqui, Mohammad Umar Khan, Sadaf Ahmad, and Nazia Tabassum

1 Introduction

Over a decade, the demand for plant-based meat alternatives is on rise. The high protein content and less fat content in soy provides an edge to its rising demand. The availability of high biological value proteins is becoming increasingly scarce due to the world's population growth and the resulting consequences on the consumption of natural resources. Other factors that have encouraged the development of plant-based meat substitutes include concerns about human health and animal welfare. In

S. Joshi (✉)
Department of Food Technology, School of Interdisciplinary Sciences and Technology, Jamia Hamdard, New Delhi, India
e-mail: swetajoshi@jamiahamdard.ac.in

I. Khan
Department of Bioengineering, Integral University, Lucknow, India

GVICAA-A FoodTec Research & Development Consultancy, Aonla, Bareilly, UP, India

A. Soni
GVICAA-A FoodTec Research & Development Consultancy, Aonla, Bareilly, UP, India

T. Siddiqui
Galgotias College of Pharmacy, Greater Noida, India

M. U. Khan
Dehlvi Natural, New Delhi, India

S. Ahmad
Centre for Food Science and Technology, Faculty of Engineering and Technology, Aligarh Muslim University, Aligarh, UP, India

N. Tabassum
Department of Post-Harvest Engineering and Technology, Faculty of Agricultural Sciences, Aligarh Muslim University, Aligarh, India

A. Kumar Pathera et al. (eds.), *Healthier Meat Products*,
https://doi.org/10.1007/978-3-031-78215-2_14

response to the rising demand from consumers, the market for plant-based meat substitutes is rising quickly. Owing to its superior gelation qualities and ability to form fibrous structures, soy protein has been effectively used to prepare meat substitutes and has gained widespread recognition as an alternative to animal protein [1].

The plant-based proteins are known as "incomplete protein packages" as they often lacking one or two of the essential amino acids. In comparison to other plant-based proteins like pea protein, barley protein and whey protein, soy protein possesses balanced essential amino acid profile except Sulphur containing amino acids. Plant-based protein has reduced digestibility resulting in weak muscle formation whereas animal protein has high digestibility potential [2].

Plant-based protein diets having more health-promoting advantages than animal protein diets, even though it is not yet proven that they can completely replace meat products. To create a system of food production that is sustainable, plant-based proteins still have the potential to replace animal products. Another reason for the rise in the demand of plant-based protein is due to the correlation of meat consumption and development of chronic diseases like cardio vascular ailments [3, 4].

In India approximately 50% of goat meat is mostly derived from adult and spent animals resulting in meat that is chewy and fibrous with low juiciness, palatability, and reduced consumer acceptance. Animal meat being expensive has increased the need of alternative protein sources without affecting the nutritional and organoleptic properties of meat products. Due to objectionable flavor, beany note and bitter properties of soy, its incorporation into products is challengeable [5]. Many researchers are working to find the best ways to mask or eliminate the off-flavor attribute of soy. In addition to soy protein, wheat gluten (WG) needs be included as a binder to stabilize the fiber structure to successfully formulate a soy protein meat substitute. Other plant proteins, such as pea protein, have also shown promising results in the development of meat alternatives. Non-protein ingredients are crucial for creating soy protein meat substitutes since they improve flavor and improve the nutritional profile in addition to helping to create the structure [1].

In a time when people are becoming more aware of sustainability and health, need for novel and eco-friendly food solutions has become crucial. The purpose of this chapter is to explain the purpose of adding soy to conventional meat dishes. There is a paradigm shift in the modern food consumption landscape due to the growing desire for alternatives that support the health of the planet as well as individual well-being. Considering this, the chapter aims to fulfill a vital need by investigating the potential role of soy in transforming traditional meat products. A flexible and high-protein legume, soy appears to be a ray of light for those seeking more sustainable and healthful food options.

2 The Rise in Plant-Based Protein Sources

The Protein Digestibility Corrected Amino Acid Score (PDCAAS) is used for the evaluation of protein quality which is dependent on its essential amino acid content and true fecal protein digestibility. Owing to their amphiphilic nature proteins are utilized as surface-active agents in the food industry. Though animals are known to be good sources of proteins but they impact environment due to greenhouse gas emissions during meat production and its consumption increases the risk of chronic diseases [4].

The environmental issues raised by livestock agriculture sector are the other factors contributing to the growing popularity of plant-based proteins. The use of natural habitats as food sources for millions of animals raised for meat and the amount of manure these animals create are the risks linked with the livestock industry [6]. The world's levels of greenhouse gases have risen in part due to the gases and fumes released by animal manure. The destruction of natural ecosystems and flora to make room for new meat farms has been a major topic of discussion throughout the world [7]. Conventional livestock production is frequently associated with substantial water use, greenhouse gas emissions, and deforestation. The production of 200 g of beef is estimated to require 792 L of drinking water, 4 kg of feed grains, 6.6 m^2 of deforestation, and 50 kg of CO2 emissions into the atmosphere [8]. This has led to the gaining popularity of alternate proteins sources like legumes, oilseeds, cereal & pseudocereals, algae, insects, and edible fungus protein sources in food Industry.

As the chapter focuses on the application of soy in meat products, the characteristics of soy has been discussed. For addressing the growing concerns about the environmental effects of conventional meat production, soy is a potent trigger. Soy offers a convincing alternative in light of the growing need for ecological products. As a leguminous crop, soy production often uses less water and land than traditional livestock farming, making it a more resource- and environmentally-conscious method of producing meat [4, 9]. Due to lower agricultural footprint more soy may be produced in each area, which lessens the need for major land clearing and allays fears about deforestation. This effectiveness helps to protect biodiversity and natural environments. Water Conservation: Growing soybeans usually uses less water than rearing cattle. Soy is a more water-efficient option in areas where water scarcity is becoming an increasing concern due to its much less water footprint [10]. This perspective of the soy production process encourages a more sustainable approach to agriculture and is in line with the worldwide urge to protect water resources. The methane from ruminant animals is one of the main greenhouse gas emissions caused by livestock operations. Conversely, soy farming results in lower emissions, making it an eco-friendlier choice. The total carbon footprint can be decreased by adding soy to meat products, aiding in the fight against climate change [10]. Selecting soy over conventional meat sources contributes to the preservation of natural habitats. The preservation is crucial for preserving biodiversity, sustaining wildlife habitats, and preserving the delicate balance of ecosystems [11].

Owing to its nutritional & functional properties, soy has emerged as a transformative and versatile ingredient, playing a pivotal role in reshaping traditional meat products. By choosing soy, not only can consumers enjoy a sustainable and nutritious alternative, but they also actively contribute to the global movement towards a more environmentally friendly and resilient food system.

2.1 Soy as a Protein Source

The soybeans possess 35–40% protein, 20% lipids, 9% dietary fibre and 8.5% moisture on dry weight basis [12]. Soy protein contains globulins (salt soluble) and albumins (water soluble). Globulin is the main type of protein found in soybeans. Two major storage proteins, glycinin (11S) and β-conglycinin (βCG, 7S), which make up 30% and 40% of the soy protein, respectively, can be separated using ultracentrifugation treatment [13]. The largest producer of soy, US contributes to 45% of world population followed by Brazil (20%) and China (12%) [14]. Soybeans undergo several processes such as dehulling, flaking, and defatting for the extraction of soy protein. Soy flour, concentrates, and isolates are the three high-protein commercial products that can be made from it. With a protein level greater than 90% and enhanced digestion, soy protein isolate (SPI) is the most refined type among them [15]. The SPI possesses water-insoluble globulins which results in enhanced swelling property [16].

Soy possesses an excellent source of complete protein profile that includes all essential amino acids necessary for human health. The amino acids included in soy enhances immune system and general muscle health by aiding in tissue growth and repair. Soy has wealth of other nutrients, such as fiber, vitamins, minerals, isoflavones, and other phytochemicals [17]. Producers of meat products can improve the nutritional profile of their goods by adding soy, giving consumers a healthy and well-balanced source of protein [1, 12].

The processesed meat products have high saturated fat content. Their consumption results in cardiovascular diseases. The incorporation of soy into meat products improves nutrition, intake of saturated fats thus lowering cholesterol levels and alleviates major health concerns. The proteins and isoflavones in soy provide protection to heart and reduces related ailments [18]. In addition to its nutritional content, protein and fibre portion of soy leads to the feelings of satiety and assist in weight management. It is a nutrient-dense plant- based alternative for those suffering from obesity [11, 19].

Soy protein can be played with to reproduce the texture of various meat products [20]. Through various advanced food processing techniques, producers can produce soy-based meat products that are as moist, tender, and delicious as conventional meats. This enables soy to be used by food technologists aiming to produce soy-based meat analogues that provide a genuinely satisfying eating experience [21] with a variety of flavors, from salty and umami to spicy or smoky. With proper selection and combination of spices & condiments, researchers can produce meat alternatives that do not just mimic the flavor of traditional meats [22]. The adaptability of soybeans extends to diverse culinary traditions. Whether used in American burgers, Asian stir-fries, or European stews, soy-based meat substitutes fit seamlessly into countless global cuisines. This makes soybeans a useful ingredient for producing diverse flavors and preferences, transcending cultural boundaries and pleasing to a broad consumer base [11]. Food scientists can experiment with soy ingredients to create new textures, flavors, and shapes [23].

Soy-based alternatives can provide a sustainable and cost-effective alternative to traditional meat sources. Soybeans provide a cost-effective source of protein. The efficiency of growing soybeans, with its relatively high protein yield per acre, makes it an cost effective option to meet the growing global demand for protein without compromising protein quality Thus, soybeans are becoming a key player in addressing global food security challenges and meeting diverse consumer needs [24].

3 Nutritional Advantages of Soy

For consumers who are concerned about their health, adding soy into meat products offers a variety of nutritional benefits. The nutritional content of soy-based meat analogues has been summarized in Table 1.

3.1 High-Quality Protein Content

Because soy is known to have a high protein content, it is a valuable and high-quality source of protein for people looking for nutrient-dense food options. The idea that soy is a superior source of protein comes from several sources. Soy is categorized as a complete protein since it has every necessary amino acid that the human body needs. The amino acids classified as essential are those that the body is

Table 1 Nutritional content of soy-based meat analogues

S.No.	Soy-based meat analogue	Protein content	Nutritive value	References
1.	Soy-Based Burgers (Soy Flour/Soy Protein Isolate/Concentrate)	20–25 gm	Contains fibre, iron & potassium	[11]
2.	Soy-Sausage frankfurter mortadella	20 g/100 g 17.5 g/100 g	Possess palmitic acid, oleic acid and linoleic acid.	[25]
3.	Soy Chicken Nuggets (Soy Flour/Soy Protein Concentrate)	50–70%	Balanced profile of amino acids	[26]
4.	Soy Meatballs (Wheat Protein Concentrate/Soy Protein Isolate)	22 gm	Cooked meatballs were inhibitory toward oxidation	[27]

unable to synthesize on its own and need to be received through diet [4]. Soy is comparable to animal-based proteins because it has a complete amino acid profile, which provides a wide range of building blocks needed for different physiological processes. The body can effectively absorb and use the nutrients found in soy proteins because they are highly bioavailable. Because of its bioavailability, people who include soy in their diets can benefit most from the necessary amino acids found in soy protein [4]. This trait is especially crucial for maintaining overall metabolic processes, muscular growth, and tissue healing. Although the amount of protein in soy products varies, soybeans are inherently high in protein. Because of its high protein content, soy is a practical and efficient way to get the recommended daily intake of protein [4]. Soy can provide a significant protein increase to people's diets, regardless of whether it is ingested in the form of soy-based meat analogues, tofu, or soy protein isolates. The protein content of soy is flexible and can be used in a variety of culinary contexts. Because of its adaptability, a large variety of soy-based goods can be made to suit a variety of tastes and preferences. The adaptability of soy protein allows for the creation of meat substitutes that complement a variety of dietary preferences without sacrificing the quality of the protein, such as plant-based sausages and burgers [28]. Soy's high-quality protein content becomes especially advantageous for people who adhere to dietary patterns, such as vegetarianism, veganism, or reducing their reliance on animal products. It offers a plant-based substitute that can provide protein requirements without compromising vital amino acids, promoting general health and wellbeing. One of the main factors that highlights soy's nutritional importance is its high-quality protein content [29]. Soy's protein profile makes it a valuable and long-term food source of meat protein substitute regardless of aspects for evaluation such as completeness of amino acids, bioavailability, protein richness, variety, or alignment with dietary goals.

3.2 Essential Amino Acids

The essential amino acid makeup of soy is one of its unique qualities that adds to its nutritional worth. The building blocks of proteins are amino acids, and essential amino acids are those that the body cannot manufacture on its own and must obtain through nutrition [30]. The fact that soy includes each of the nine essential amino acids needed by the human body makes it a complete protein source. These essential amino acids include valine, phenylalanine, histidine, isoleucine, leucine, lysine, methionine, phenylalanine, and threonine. Since soy contains every essential amino acid, it differs from many other plant-based protein sources that may be lacking in one or more of these vital amino acids [31].

The essential amino acid content of soy is essential for encouraging the body to produce proteins. Protein synthesis is the process by which the body makes new tissues and repairs old ones, including muscles, organs, and enzymes. People who include soy in their diets gets all the nutrients needed for efficient protein synthesis, which supports overall physiological processes and maintenance. This is because soy contains the complete amino acid profile. To be more precise, leucine is needed to make muscle proteins. To be more precise, leucine is needed to make muscle proteins. Because of its leucine content, soy is a great dietary choice for anyone looking to maintain or increase their muscle mass as well as for anyone who exercises or participates in physical activity. Its well-balanced amino acid composition explains why soy is a great source of protein for muscle growth. The essential amino acid ratios of soy closely resemble the optimal ratios required for human nutrition utilization [32]. This balance ensures that the body can utilize and absorb the amino acids found in soy protein in an efficient manner, promoting optimal health and well-being. Because of its well-balanced amino acid composition, soy is an effective source of protein. As an alternative to animal-derived proteins, soy stands out as a plant-based choice that closely resembles the completeness of amino acids found in animal proteins [32]. This completeness is necessary to meet the protein and amino acid needs of vegans, vegetarians, and other people who may prioritize plant-based nutrition. Because soy contains all the essential amino acids, it is a complete and highly nutritious source of protein that supports a variety of physiological activities. This makes it a flexible and inclusive option for people with a range of dietary preferences and objectives [33].

3.3 Rich in Fiber

In addition to its protein content, Soy is also a good source of dietary fiber, which promotes a healthy digestive system and overall wellbeing. Soybeans naturally contain both soluble and insoluble fiber. Soluble fiber dissolves in water to form a gel-like substance, while insoluble fiber does not dissolve during digestion. This combination of fiber types affects the total amount of fiber in products made from soy. The amount of fiber in different soy products varies; some retain more fiber than derivatives that have undergone extensive processing [34] .

Eating dietary fiber is essential for maintaining the health of the digestive system. Insoluble fiber found in the outer layers of soybeans gives stool more volume and promotes regular bowel movements to prevent constipation. Soluble fiber, on the other hand, gels in the digestive system to aid in nutrient absorption and heighten feelings of fullness. The combined advantages of these two types of fiber supports the digestive system's overall health. Because soy contains a lot of fiber, it makes people feel satisfied or full after eating [35]. This satiating effect may be especially beneficial for those who are attempting to control their weight. By promoting a feeling of fullness, soy-based products can potentially support weight management efforts by helping to regulate calorie intake. Soluble fiber and blood sugar regulation are interlinked [36]. It can lessen the rate at which sugar is absorbed, preventing sudden spikes in blood sugar levels following meals. For those who already have diabetes or are at risk of getting the condition, including soy products with their inherent fiber content in the diet may be a wise choice for blood sugar management. Fiber is known to enhance cardiovascular health. It can aid in lowering cholesterol levels by binding to cholesterol and promoting its excretion from the body. Consuming soy, a diet high in fiber, may help people maintain heart health and lower their risk of cardiovascular diseases. In the digestive system, fiber helps to improve nutritional absorption. Although fiber is not digested itself, its presence can affect how other nutrients are absorbed, thereby maximising the body's utilisation of the wide range of vitamins and minerals included in soy. Soy's high fiber content makes it beneficial for health in addition to providing protein. Both soluble and insoluble fiber support healthy digestion, control of weight, control of blood sugar, and cardiovascular health. Because of its diverse range of benefits, soy is a very nutritious option for anyone looking to eat a diet that supports overall health [37].

3.4 Low in Saturated Fats

Soy is a heart-healthy substitute for several traditional meat products because of its low saturated fat level, which is one of its noteworthy health benefits. There can be many reasons why soy's low saturated fat content is related to improvement in eating habits and general health.

Soy-based products such as tempeh, tofu, and soy protein isolates have a naturally low saturated fat content. A high consumption of saturated fats has been linked to an increased risk of cardiovascular diseases because it can raise LDL (low-density lipoprotein), or "bad" cholesterol [38]. People who choose to get their protein from soy can make a heart-healthy dietary choice. It is reported that elevated blood vessel plaque and a higher risk of heart-related complications can result from high LDL cholesterol levels [39]. Soy is a heart-healthy food substitute that is cholesterol-friendly and conforms with dietary guidelines because it contains less saturated fat than other foods. Saturated fats are heavy in calories and can make you gain weight if you eat too much of them. Because soy has a low saturated fat content, it can be included in a weight-management plan. By selecting soy-based substitutes that enable them to maintain a balance between calorie intake and expenditure, people can support overall weight control and wellbeing [40].

Plant-based diets are typically associated with lower intakes of saturated fat when compared to diets high in animal products. Soy can be incorporated into plant-based meals to provide a source of protein that complements a range of recipes while allowing people to stick to a plant-centric fat intake strategy. Soy's low content of saturated fat does not diminish its culinary diversity. Because of its ability to absorb and enhance flavors, soy is a versatile ingredient that works well in a wide range of dishes, including savoury foods, sweets, and drinks. Owing to its versatility, people can choose carefully how much saturated fat they eat and still enjoy a wide range of soy-based options. Choosing soy over certain animal products high in saturated fat contributes to a balanced approach to fat intake, enhancing overall health and adhering to guidelines for cardiovascular health [41].

3.5 *Cholesterol-Free*

Soy's cholesterol-free status stands out as a major health benefit, especially for those who value heart health and lowering cholesterol levels. Unlike animal-based products, soy is naturally free of dietary cholesterol, making it a heart-healthy and wise dietary choice. An increased risk of cardiovascular diseases is associated with a high dietary cholesterol intake. A health claim linking soy protein to a lower risk of cardiovascular disease (CVD) has been approved by the regulatory body because to the well-established hypocholesterolemic impact of soy. Soybeans also possess fiber, lecithins, saponins, and isoflavones known to enhance cardiovascular health [39].

3.6 *Isoflavones and Antioxidants*

Isoflavones are bioactive substances found in soy that possess antioxidant qualities. By assisting the body in combating free radicals, antioxidants help lower inflammation and oxidative stress. Soy's isoflavone content provides another level of possible health benefits, such as support for the cardiovascular system and hormone balance. Soy's reputation as a healthful and complete food comes partly from its nutritional benefits. When it comes to heart health, protein requirements, or essential nutrients, soy is a flexible ingredient that can be used to satisfy the dietary preferences and health objectives of a diverse range of consumers [42].

In addition to having a high protein content, soy is unique in that it has bioactive ingredients, such as antioxidants and isoflavones, which have extra health benefits. Isoflavones are phytoestrogens with antioxidant properties that help the body scavenge free radicals. Oxidative stress, inflammation, and the aging process are linked to free radicals. Numerous metabolic processes and environmental exposures produce them. Because soy contains isoflavones, it can improve cardiovascular health and hormonal balance. Soy has been linked to enhanced arterial function and cholesterol levels, which supports its beneficial effects on the cardiovascular system. Furthermore, soy contains a variety of advantageous substances that can help reduce oxidative stress in addition to isoflavones [43]. Antioxidants are essential for decreasing inflammation and shielding cells from deterioration. With soy's high isoflavone and antioxidant content, incorporating it into the diet offers a comprehensive strategy for enhancing general health and wellbeing. People can take advantage of these bioactive chemicals' potential health benefits by consuming soy in the form of whole foods or soy-based products, which makes soy a great complement to a balanced and health-conscious diet [44].

3.7 Processing Techniques

3.7.1 Extrusion in Soy-Based Meat Products

Extrusion technology is the crucial cooking method where gelatinization of starch takes place at high temperature (55–79 °C for rice starch, 72–74 °C for corn starch) along with inactivation of anti-nutrient components (Phytic acids, lectins, trypsin inhibitor, tannins, polyphenols, saponins etc.) present in the food system. Extrusion technology helps in the development of texturized plant-based products such as soy chunks, soy nuggets which can be claimed as plant-based products due to similar texture to that of meat. The extrusion process involves the use of an extruder, a specialized machine that subjects a mixture of soy proteins, water, and other ingredients to a combination of high temperature and pressure [45]. During extrusion process, the material mixture or blend of soy products forcefully passed through die at standard temperature and pressure. On passing through main section of extruder, the material mixture experiences intense mechanical shear and temperature. The combination of pressure, heat, and shear forces leads to the denaturation and reformation of proteins, resulting in the creation of a semi-solid, fibrous structure [46]. The extrusion process is particularly effective in aligning the proteins in the soy mixture, creating a texture that closely resembles the fibrous composition of muscle tissue in traditional meats. This protein alignment is essential to obtaining the chewiness and toughness that come with eating meat. Additionally, extrusion helps to eliminate the raw, beany flavor that is frequently connected to soy [47]. Unwanted flavors are reduced by the heat produced during extrusion, creating a neutral base that easily takes up flavors introduced during later processing stages. Extrusion's adaptability allows for the creation of a wide range of meat substitutes made from soy, such as sausages, nuggets, and burgers.

3.7.2 Texturization in Soy-Based Meat Products

Texturization of proteins is the process of constructing a three-dimensional structure to either fully or partially replace lipids (fat substitutes) or other proteins (meat analogues) in conventional diets. Additionally, texturization can be used to restructure meat and byproducts of meat processing. Depending on the type of meat, a high percentage of meat—up to 70%—can be used in the feed; for example, the extruder may receive 70% meat and 30% soy protein.

In a wet extrusion process, roughly thirty percent of vegetable material—mostly proteins, though small amounts of fiber and carbs can also be beneficial—is needed instead of pure meat material.

Concurrent with extrusion, texturization is a crucial stage in the production of soy-based meat substitutes. Its goal is to produce a texture that closely resembles traditional meats [48].

Texturized soy protein (TSP) or texturized vegetable protein (TVP) is the term used to describe the fibrous and chewy qualities associated with meat and meat products that are added to the soy protein mixture after it has been processed in the extruder portion. Many texturizing methods are available to the food industry, such as freezing, high pressure, chemical or enzymatic texturization, extrusion cooking, and spinning (dry, wet, and spinneretless).

Extruded soy protein is modified during the texturization process to replicate the structure that resembles the fibrous makeup of animal muscle. The use of mechanical shear and heat generation during the extrusion process causes a variety of physical and chemical changes that are necessary for the best possible gelatinization [49].

Depending on the result that is intended, different texturization procedures may be used. For example, to create the desired structure while texturizing soy-based burgers or sausages, shaping and molding may be necessary. The sensory sensation of biting into a soy-based product should be very similar to biting into a regular meat product, hence this phase is critical for consumer acceptability.

Texturization improves the overall palatability of soy-based meat substitutes while still maintaining their structural integrity. These goods are a great alternative to a variety of traditional meat products because of their fibrous texture, which is formed during texturization and makes them versatile in culinary applications like grilling and sautéing [46].

A crucial step in creating soy-based meat products is texturization, which guarantees that the finished goods are gratifying and really flesh-like while still being nutritionally sound. Table 2 lists the ingredients used in the extrusion processing of soy-based meat substitutes.

Table 2 Common ingredients used in extrusion processing of soy-based meat products

Ingredient	Role of ingredient	References
Soy Proteins/Wheat Gluten	Main protein source, crucial for anisotropic meat-like structure/acts as a binder to stabilize the fiber structure	[50]
Wheat Gluten	Enhances texture and provides elasticity in some products	[51]
Legume Flours	Adds nutritional value and complements soy protein	[52]
Flavor Enhancers	Enhances taste, often includes yeast extracts or umami-rich compounds	[53]
Texturizing Agents	Alters the texture to mimic the fibrous nature of meat	[54]
Seasonings and Spices	Contribute to the overall flavor profile of the end product	[4]
Preservatives	Extends shelf life by inhibiting microbial and oxidative reactions.	[55]

3.8 Challenges and Considerations in Soy-Based Meat Products

Although extrusion technology provides several advantages over conventional cooking methods but it has some limitations on which people are working to optimize and taking those as technical and non-technical challenges in the forthcoming extrusion improvement. Understanding and addressing these aspects is essential for the continued improvement and acceptance of soy-based alternatives to traditional meats [11].

3.8.1 Allergen Concerns in Soy-Based Meat Products

Allergen concerns related to soy-based meat products revolve around the fact that soy is one of the eight major allergens responsible for a significant portion of food allergies. Individuals with soy allergies may experience adverse reactions when exposed to soy proteins or other soy-derived components [56]. Soy allergies are relatively common, especially in children while many individuals may outgrow soy allergies, some may carry this allergy into adulthood. Symptoms of soy allergies can range from mild, such as hives and digestive discomfort, to severe, including anaphylaxis, a life-threatening allergic reaction. Hence it is mandatory to print the complete information on the food label to aware the consumer and select the food according to their preference or choice [57]. The food labeling helps individuals to identify soy allergies and avoid products that could trigger an allergic reaction [58]. This transparency is crucial for consumer safety and regulatory compliance. For those who are allergic to soy, cross-contamination during manufacturing processes is a serious worry.

Strict production guidelines and cleaning procedures are necessary to reduce the chance of cross-contamination and safeguard customers who are allergic to soy. Variability in the allergenic potential is introduced using different forms of soy in different soy-based products. For instance, whereas soy protein isolates are frequently utilized, other soy derivatives may also be present [59]. Aware of how common soy allergies are, producers are working harder to create substitutes that are free of allergens. The goal of these substitutes is to give those who are allergic to soy access to plant-based protein choices without running the risk of negative reactions.

Soy-free meat alternatives are investigating the use of plant-based protein sources such mycoprotein (a protein derived from fungi), rice protein, and pea protein as replacements [21]. Raising awareness can help manage soy allergies more effectively and lessen the chance of unintentional exposures. Given the rising acceptance of plant-based diets and the emergence of soy-based substitutes in a wide range of food items, this knowledge is particularly crucial [57]. A more secure and welcoming food environment for people with soy allergies is facilitated by open communication and increased consumer and industry understanding [60]. The term "cross-contamination" describes the unintentional addition of soy proteins to soy-free meat products [61].

Strict cleaning procedures and allergy control techniques are necessary to avoid cross-contamination when food products are being produced and packaged [62].

Between production cycles, manufacturers need to follow stringent cleaning protocols to make sure that all traces of soy are eliminated from tools and surfaces. To stop allergen carryover, this entails disassembling and cleaning conveyors, utensils, and equipment.

Using color-coding schemes for tools and equipment can assist avoid cross-contamination when cleaning and producing goods [63]. Diverse computer-aided techniques are utilized to differentiate between meat products that are based on soy and those that are not, hence decreasing the probability of unintentional cross-contamination. The robustness and dependability of the applied cleaning methods are guaranteed by validation procedures. This additional layer of assurance in allergy control is provided by these testing and validation procedures [64].

Manufacturers may improve consumer safety when it comes to soy-based meat products by addressing cross-contamination issues with a mix of strict cleaning procedures, employee training, and clear labeling. This is especially important for consumers who have soy allergies.

This all-encompassing strategy guarantees that soy-based goods satisfy strict quality requirements and allergen control measures during the whole production process [65].

4 Consumer Trends and Market Growth in Soy-Based Meat Products

Consumer trends and market expansion in the soy-based meat product space show a dynamic environment driven by changing dietary habits, heightened consciousness of sustainability, and an increasing desire for plant-based substitutes [66]. The soy food market was projected to be valued USD 40.8 billion in 2022.

The market for soy food is anticipated to develop at a compound annual growth rate (CAGR) of 4.98% between 2023 and 2032, from USD 42.8 billion in 2023 to USD 63.1 billion in 2032.

Health and nutrition awareness, the growing demand for plant-based proteins, lactose intolerance and dairy allergies, functional qualities and versatility, the growing demand for non-GMO and organic foods, economic factors, government support and promotion, sustainable agriculture, and environmental concerns are the main market drivers boosting market growth [67, 68].

4.1 Rising Demand for Plant-Based Meat Alternatives

Consumer preferences for dietary options have been changed drastically for many years and consumers are looking for healthier food options in the market such as plant-based meat products over meat and meat products. This behavioral change in the consumer choices is influenced by their health concerns, religious endorsement, environment, and sustainability and to oppose any activity regarding animal cruelty. In the current market, consumers are looking for foods with healthy protein in spite animal protein option. Among the plant food categories, the soy protein is the healthy option both in respect of quality and quantity as an alternative of meat protein, in the form of soy chunks, extruded soy products, soy nuggets etc. [69]. The factors contributing to the rising demand for plant-based foods is summarized in Table 3.

4.2 Consumer Likings towards Functional Food Product

The increasing problems of non-communicable diseases has been influenced the consumer preferences towards their food choices. The consumers are now preferring healthier food options that are low in saturated fats, low in salt, rich in unsaturated fatty acids such as MUFA & PUFA for cardiovascular health and fiber rich foods to get rid of constipation problems that might increase the risk of colon cancer. The soy-based food products are the healthier protein rich foods for the vegan and vegetarian population at the place of animal food options. Soybean being rich in all the amino acids provides balance nutrition to the consumers that ultimately take part in muscle building and various metabolic pathways that synthesize important co-factors, vitamins and other essential molecules. The demand of soy based meat products among the consumers is increasing as they are becoming conscious for their health [75].

Table 3 Rising demand for plant-based foods and their expected reasons

S.No.	Factors driving demand for plant-based food	Type of food	References
1	Health and Wellness Focus	Plant-based burgers	[70]
2	Flexitarian and Vegan Lifestyle	Almond milk	[71]
3	EnvironmentalSustainability Considerations	Quinoa	[72]
4	Innovation in Product Offerings	Soy-basedmeat alternatives	[11]
5	Increased Availability in Retail and Food Services	Chickpea pasta	[73]
6	Investments and Partnerships	Plant-based protein bars	[74]

4.3 Technological Advances in Soy-Based Meat Product Production

Technological advancement improves the desirable quality of soy-based meat product and increase the consumer appeal by mimicking the soy-based meat products like meat products. Particularly texture profile viz., hardness, cohesiveness, springiness, gumminess, chewiness etc. [76].

4.3.1 Emulsification

Emulsification plays a crucial role in the production of soy-based meat products, contributing to the development of desirable textures, structures, and mouthfeel that closely mimic traditional meat. Emulsification involves the creation of a stable emulsion, a finely dispersed mixture of fat, water and protein which is essential for achieving the desired characteristics in certain plant-based products. Emulsification is the result of optimum entrapment of fat molecules inside the protein matrix in a sequential event while adding the different ingredients in meat mixture such as salt, spices, condiments, sodium nitrite, sodium phosphate etc. [28]. In soy-based meat products, soy proteins are often used as the primary ingredient to form the protein matrix. Oil and protein, along with water, form the basis for emulsification. The soy protein, fats, and water are subjected to intense mechanical mixing for optimum emulsification at low temperature condition (around 8–10 °C). This process breaks down the fat into smaller droplets dispersed into the water and protein, creating a stable emulsion [11].

Emulsification process is highly sensitive towards temperature i.e. it is important to bring the temperature in desirable range for optimum emulsification. Temperature not only regulates the emulsification process but also prevent over heating of proteins that may affect the functional properties of meat product [77]. Emulsifiers, such as lecithin derived from soy or other sources, are often added to soy-based meat formulations. These emulsifiers stabilize the fat droplets in water phase, preventing separation during processing. Homogenization is a process that further refines the emulsion by reducing the size of fat droplets [78]. The smaller droplet size enhances the smoothness and creaminess of the emulsion, contributing to a better mouth feel and palatability of final product. The emulsion of meat can also be influence by its pH value that ultimately affects the quality of final product. The optimization of desired pH value controls the solubility and gelation properties of soy proteins influencing the overall structure and stability of the emulsion [79] for instance, in the production of sausages and hot dogs, emulsification results in a fine and uniform protein matrix, giving these products their characteristic bite and juiciness.

The advanced emulsification techniques bring the remarkable changes in the quality of soy-based meat products in comparison to the conventional emulsification techniques. Soon better and sophisticated techniques will be available that could even reduce the overall cost of production with influential product quality.

4.3.2 Soy-Based Meat Binders and Extenders

Soy-based meat binders and extenders are essential components in the formulation of plant-based meat products, playing a key role in improving texture, moisture retention, and overall product integrity. These ingredients are particularly important in creating plant-based alternatives that closely resemble the sensory qualities of traditional meat [80]. Binders are ingredients that enhance the cohesiveness among the different components of products and bring it in a single food matrix. Soy protein isolates are commonly used as binders in variety of meat products and other meat alternatives. On hydration, soy protein isolates form a gel-like structure, creating a network that binds other ingredients together that mimic the fibrous and cohesive characteristics of traditional meat [28]. Soy-based binders trap the water molecule within the protein matrix and results in a juicier and more tender product with overall acceptability. Soy-based binders contribute to the cooking stability of plant-based products i.e. maintain its structural integrity and form during cooking processes such as grilling, baking, or frying [81]. Manufacturers can tailor the concentration and type of binders to achieve specific textural characteristics based on the desired product, such as burgers, sausages, nuggets, and other meat analogues. Soy based binders being cost effectively contribute to the economic viability of producing soy-based meat products on a larger scale, making plant-based alternatives more accessible to a broader consumer base. Soy-based meat binders and extenders are integral components in the development of high-quality plant-based alternatives [21]. They contribute to the overall sensory experience, texture, and nutritional content of soy-based meat products, making them more appealing and acceptable to consumers seeking sustainable and flavorful alternatives to traditional meat.

4.3.3 3D Food Printing Technology

In the realm of soy-based meat product production, technological advances such as 3D food printing have revolutionized the food industry. The 3D food printing technology, allows for meticulous layer-by-layer construction of soy-based alternatives, offering unprecedented customization and personalization [82]. This technology enables the precise and accurate replication of textures found in traditional meats, ensuring a mouthfeel that closely resemble the traditional meat. Beyond its culinary applications, 3D food printing reduces food waste by optimizing ingredient use and supports sustainable practices [83]. The technology also serves as a valuable tool in research and development, accelerating the innovation process in creating novel soy-based formulations. The potential of 3D food printing technology in soy-based meat production holds promise for producing visually appealing, architecturally interesting, and sustainable plant-based alternatives.

5 Conclusion

The consumer interest in soy-based meat products is predominantly a result of the adverse effect of animal meat product on human health and the environment. Recent research endeavors pertaining to meat analogues have demonstrated that a meat-like texture may be created by the utilization of plant-based proteins such as soy proteins and process including mixing, shearing and extrusion. Extrusion, a high temperature short time process that involves application of high pressure can be used to create soy-based meat analogues. Extrusion enhances the digestibility of soy protein by altering its structure. The adverse effects of soy protein consumption can be minimized by denaturing the soy anti-nutritional factors like lecithin, trypsin inhibitor, phytates etc. during extrusion. The 3D food printing technology also involved to shape the food structure and customization of its ingredient according to personalized nutrition that may be proved beneficial to produce 3-dimensional soy-based meat products in association with extrusion technology. The chapter not only serves as a comprehensive guide to understanding the nuances of this burgeoning field but also emphasizes the pivotal role soy plays in shaping the future of sustainable, nutritious, and appealing plant-based alternatives.

References

1. Zhang T, Dou W, Zhang X, Zhao Y, Zhang Y, Jiang L, et al. The development history and recent updates on soy protein-based meat alternatives. Trends Food Sci Technol. 2021;109:702–10.
2. van Vliet S, Burd NA, van Loon LJ. The skeletal muscle anabolic response to plant-versus animal-based protein consumption. J Nutr. 2015;145(9):1981–91.
3. Richter CK, Skulas-Ray AC, Champagne CM, Kris-Etherton PM. Plant protein and animal proteins: do they differentially affect cardiovascular disease risk? Adv Nutr. 2015;6(6):712–28.
4. Qin P, Wang T, Luo Y. A review on plant-based proteins from soybean: health benefits and soy product development. J Agric Food Res. 2022;7:100265.
5. Das AK, Anjaneyulu ASR, Kondaiah N. Development of reduced Beany Flavor full-fat soy paste for comminuted meat products. J Food Sci [Internet]. 2006 [cited 2024 Jul 2];71(5). Available from: https://ift.onlinelibrary.wiley.com/doi/10.1111/j.1750-3841.2006.00045.x
6. Sha L, Xiong YL. Plant protein-based alternatives of reconstructed meat: science, technology, and challenges. Trends Food Sci Technol. 2020;102:51–61.
7. Harwatt H. Including animal to plant protein shifts in climate change mitigation policy: a proposed three-step strategy. Clim Pol. 2019;19(5):533–41.
8. Romão B, Botelho RBA, Torres ML, da Maynard DC, de Holanda MEM, Borges VRP, et al. Nutritional profile of commercialized plant-based meat: an integrative review with a systematic approach. Food Secur. 2023;12(3):448.
9. Dreoni I, Matthews Z, Schaafsma M. The impacts of soy production on multi-dimensional well-being and ecosystem services: a systematic review. J Clean Prod. 2022;335:130182.

10. Escobar N, Tizado EJ, Zu Ermgassen EKHJ, Löfgren P, Börner J, Godar J. Spatially-explicit footprints of agricultural commodities: mapping carbon emissions embodied in Brazil's soy exports. Glob Environ Change. 2020 May;62:102067.
11. Messina M, Sievenpiper JL, Williamson P, Kiel J, Erdman JW. Perspective: soy-based meat and dairy alternatives, despite classification as ultra-processed foods, deliver high-quality nutrition on par with unprocessed or minimally processed animal-based counterparts. Adv Nutr. 2022;13(3):726–38.
12. Michelfelder AJ. Soy: a complete source of protein. Am Fam Physician. 2009;79(1):43–7.
13. Sui X. The development history and recent updates on soy protein-based meat alternatives. J Am Oil Chem Soc Wiley. 2021;109:188.
14. Cheng MH, Rosentrater KA. Profitability analysis of soybean oil processes. Bioengineering. 2017;4(4):83.
15. Kutzli I, Weiss J, Gibis M. Glycation of plant proteins via maillard reaction: reaction chemistry, technofunctional properties, and potential food application. Food Secur. 2021;10(2):376.
16. Sze-Tao KWC, Sathe SK. Functional properties and in vitro digestibility of almond (Prunus dulcis L.) protein isolate. Food Chem. 2000;69(2):153–60.
17. Kumar A, Sharma A, Upadhyaya C. Vegetable oil: nutritional and industrial perspective. Curr Genomics. 2016;17(3):230–40.
18. Sacks FM, Lichtenstein A, Van Horn L, Harris W, Kris-Etherton P, Winston M. Soy protein, isoflavones, and cardiovascular health: an American Heart Association Science Advisory for professionals from the Nutrition Committee. Circulation. 2006;113(7):1034–44.
19. Zampelas A. The effects of soy and its components on risk factors and end points of cardiovascular diseases. Nutrients. 2019;11(11):2621.
20. Ismail I, Hwang YH, Joo ST. Meat analog as future food: a review. J Anim Sci Technol. 2020;62(2):111–20.
21. Arora S, Kataria P, Nautiyal M, Tuteja I, Sharma V, Ahmad F, et al. Comprehensive review on the role of plant protein as a possible meat analogue: framing the future of meat. ACS Omega. 2023;8(26):23305–19.
22. Wu YFG, Cadwallader KR. Characterization of the aroma of a Meatlike process flavoring from soybean-based enzyme-hydrolyzed vegetable protein. J Agric Food Chem. 2002;50(10):2900–7.
23. Astawan M, Prayudani AP. The overview of food technology to process soy protein isolate and its application toward food industry. World Nutr J. 2020;4(S1):12–7.
24. Calicioglu O, Flammini A, Bracco S, Bellù L, Sims R. The future challenges of food and agriculture: an integrated analysis of trends and solutions. Sustain For. 2019;11(1):222.
25. Monge R, Arias ML, Alfaro T, Jimenez M. Nutritional and microbiological profile of soybean sausages available in Costa Rica. Arch Latinoam Nutr. 2000;50(2):142–7.
26. Yeater M, Casco G, Miller RK, Alvarado CZ. Comparative evaluation of texture wheat ingredients and soy proteins in the quality and acceptability of emulsified chicken nuggets. Poult Sci. 2017;96(12):4430–8.
27. Ulu H. Effect of wheat flour, whey protein concentrate and soya protein isolate on oxidative processes and textural properties of cooked meatballs. Food Chem. 2004;87(4):523–9.
28. Kyriakopoulou K, Keppler JK, van Der Goot AJ. Functionality of ingredients and additives in plant-based meat analogues. Food Secur. 2021;10(3):600.
29. Bryant CJ. Plant-based animal product alternatives are healthier and more environmentally sustainable than animal products. Fut Foods. 2022;6:100174.
30. Lopez MJ, Mohiuddin SS. Biochemistry, essential amino acids. In: StatPearls [Internet] [Internet]. StatPearls Publishing; 2024 [cited 2024 Jul 3]. Available from: https://www.ncbi.nlm.nih.gov/books/NBK557845/?report=classic.
31. Mai K, Xue M, He G, Xie SQ, Kaushik SJ. Protein and amino acids. In: Fish nutrition [Internet]. Elsevier; 2022 [cited 2024 Jul 3]. p. 181–302. Available from: https://www.sciencedirect.com/science/article/pii/B9780128195871000124.

32. Lynch HM, Buman MP, Dickinson JM, Ransdell LB, Johnston CS, Wharton CM. No significant differences in muscle growth and strength development when consuming soy and whey protein supplements matched for leucine following a 12 week resistance training program in men and women: a randomized trial. Int J Environ Res Public Health. 2020;17(11):3871.
33. Hertzler SR, Lieblein-Boff JC, Weiler M, Allgeier C. Plant proteins: assessing their nutritional quality and effects on health and physical function. Nutrients. 2020;12(12):3704.
34. Huang L, Cai Y, Liu T, Zhao X, Chen B, Long Z, et al. Stability of emulsion stabilized by low-concentration soybean protein isolate: effects of insoluble soybean fiber. Food Hydrocoll. 2019;97:105232.
35. Szymandera-Buszka K, Waszkowiak K, Kaczmarek A, Zaremba A. Wheat dietary fibre and soy protein as new carriers of iodine compounds for food fortification–the effect of storage conditions on the stability of potassium iodide and potassium iodate. LWT. 2021;137:110424.
36. Chang JH, Kim MS, Kim TW, Lee SS. Effects of soybean supplementation on blood glucose, plasma lipid levels, and erythrocyte antioxidant enzyme activity in type 2 diabetes mellitus patients. Nutr Res Pract. 2008;2(3):152–7.
37. Agyekum AK, Nyachoti CM. Nutritional and metabolic consequences of feeding high-fiber diets to swine: a review. Engineering. 2017;3(5):716–25.
38. Messina M. Soy and health update: evaluation of the clinical and epidemiologic literature. Nutrients. 2016;8(12):754.
39. Ramdath DD, Padhi EM, Sarfaraz S, Renwick S, Duncan AM. Beyond the cholesterol-lowering effect of soy protein: a review of the effects of dietary soy and its constituents on risk factors for cardiovascular disease. Nutrients. 2017;9(4):324.
40. Eilat-Adar S, Sinai T, Yosefy C, Henkin Y. Nutritional recommendations for cardiovascular disease prevention. Nutrients. 2013;5(9):3646–83.
41. Jenkins DJA, Blanco Mejia S, Chiavaroli L, Viguiliouk E, Li SS, Kendall CWC, et al. Cumulative meta-analysis of the soy effect over time. J Am Heart Assoc. 2019;8(13):e012458.
42. Yoon GA, Park S. Antioxidant action of soy isoflavones on oxidative stress and antioxidant enzyme activities in exercised rats. Nutr Res Pract. 2014;8(6):618–24.
43. Rizzo G. The antioxidant role of soy and soy foods in human health. Antioxidants. 2020;9(7):635.
44. Samaranayaka AG, Li-Chan EC. Food-derived peptidic antioxidants: a review of their production, assessment, and potential applications. J Funct Foods. 2011;3(4):229–54.
45. Extrusion Techniques for Meat Analogues. Extrusion Techniques for Meat Analogues. Cereal Foods World. 2020;65:4.
46. Zahari I, Östbring K, Purhagen JK, Rayner M. Plant-based meat analogues from alternative protein: a systematic literature review. Food Secur. 2022;11:2870.
47. Wittek P, Zeiler N, Karbstein HP, Emin MA. High moisture extrusion of soy protein: investigations on the formation of anisotropic product structure. Food Secur. 2021;10(1):102.
48. Samard S, Maung TT, Gu BY, Kim MH, Ryu GH. Influences of extrusion parameters on physicochemical properties of textured vegetable proteins and its meatless burger patty. Food Sci Biotechnol. 2021;30(3):395–403.
49. Maningat CC, Jeradechachai T, Buttshaw MR. Textured wheat and pea proteins for meat alternative applications. Cereal Chem. 2022;99:37–66.
50. Sui X. The development history and recent updates on soy protein-based meat alternatives. Trends Food Sci Technol. 2021;109:702.
51. Pietsch VL, Emin MA, Schuchmann HP. Process conditions influencing wheat gluten polymerization during high moisture extrusion of meat analog products. J Food Eng. 2017;198:28–35.
52. Subuola F, Widodo Y, Kehinde T. Processing and utilization of legumes in the tropics. In: Trends in vital food and control engineering; 2012.
53. Shazer WH, Sindelar JJ, Jimenez-Maroto L, Rankin S, Sato T. Reduction of sodium in processed meats using soy sauce and natural flavor enhancer. Meat Sci. 2014;96(1):456.
54. Ozturk OK, Hamaker BR. Texturization of plant protein-based meat alternatives: processing, base proteins, and other constructional ingredients. Fut Foods. 2023;8:100248.

55. Kohli D, Kumar S, Upadhyay S, Mishra R. Preservation and processing of soymilk: a review. Int J Food Sci Nutr. 2017;2(6):66–70.
56. Savage JH, Kaeding AJ, Matsui EC, Wood RA. The natural history of soy allergy. J Allergy Clin Immunol. 2010;125(3):683–6.
57. Katz Y, Gutierrez-Castrellon P, González MG, Rivas R, Lee BW, Alarcon P. A comprehensive review of sensitization and allergy to soy-based products. Clin Rev Allergy Immunol. 2014;46:272–81.
58. Chang F, Eng L, Chang C. Food Allergy Labeling Laws: international guidelines for residents and travelers. Clin Rev Allergy Immunol. 2023;65:148–65.
59. Meinlschmidt P, Sussmann D, Schweiggert-Weisz U, Eisner P. Enzymatic treatment of soy protein isolates: effects on the potential allergenicity, technofunctionality, and sensory properties. Food Sci Nutr. 2016;4(1):11–23.
60. Blom WM, van Dijk LM, Michelsen-Huisman A, Houben GF, Knulst AC, Linders YFM, et al. Allergen labelling: current practice and improvement from a communication perspective. Clin Exp Allergy. 2021;51(4):574–84.
61. Hadi J, Brightwell G. Safety of alternative proteins: technological, environmental and regulatory aspects of cultured meat, plant-based meat, insect protein and single-cell protein. Food Secur. 2021;10(6):1226.
62. Crépon K, Duyme F. Assessment and management of the risk of unintended contamination of wheat by soy allergens in silos. Cereal Chem. 2022;99(6):1196–204.
63. Starovoytova D. Universal design to limit food cross-contamination: incased set of kitchen utensils with five color-coded food chopping boards and knifes. Innov Syst Des Eng. 2019;10:43.
64. Zwietering MH, Stewart CM, Whiting RC. Validation of control measures in a food chain using the FSO concept. Food Control. 2010;21(12 SUPPL):1716–22.
65. Martínez-Pineda M, Yagüe-Ruiz C. The risk of undeclared allergens on food labels for pediatric patients in the European Union. Nutrients. 2022;14(8):1571.
66. Safdar B, Zhou H, Li H, Cao J, Zhang T, Ying Z, et al. Prospects for plant-based meat: current standing, consumer perceptions, and shifting trends. Food Secur. 2022;11:3770.
67. Newswire PR. Soy protein market- global opportunity analysis and industry forecast (2017–2022). LON-REPORTBUYER, 2017.
68. Everstine K. Food authenticity: 2020 in review. Food Safety Tech. 2021.
69. Profeta A, Baune MC, Smetana S, Broucke K, Van Royen G, Weiss J, et al. Consumer preferences for meat blended with plant proteins – empirical findings from Belgium. Fut Foods. 2021;4:100088.
70. De Marchi M, Costa A, Pozza M, Goi A, Manuelian CL. Detailed characterization of plant-based burgers. Sci Rep. 2021;11(1):2049.
71. Torna E, Rivero-Mendoza D, Dahl W. Plant-based milks: almond. EDIS. 2020;2020(5):12.
72. Gamboa C, Bojacá CR, Schrevens E, Maertens M. Sustainability of smallholder quinoa production in the Peruvian Andes. J Clean Prod. 2020;264:264.
73. Saget S, Costa M, Barilli E, Wilton de Vasconcelos M, Santos CS, Styles D, et al. Substituting wheat with chickpea flour in pasta production delivers more nutrition at a lower environmental cost. Sustain Prod Consum. 2020;24:26–38.
74. Pang M, Trier C, Alexon C, Johnston CS. Daily ingestion of protein bars (with or without added fiber) increased energy intake and body fat mass after one week in healthy adults: a crossover trial. J Funct Foods. 2023;104:104.
75. Hertzler SR, Lieblein-Boff JC, Weiler M, Allgeier C. Plant proteins: assessing their nutritional quality and effects on health and physical function. Nutrients. 2020;12:1–27.
76. Andreani G, Sogari G, Marti A, Froldi F, Dagevos H, Martini D. Plant-based meat alternatives: technological, nutritional, environmental, market, and social challenges and opportunities. Nutrients. 2023;15:452.
77. Silva TJ, Barrera-Arellano D, Ribeiro APB. Margarines: historical approach, technological aspects, nutritional profile, and global trends. Food Res Int. 2021;147:110486.

78. Deng L. Current progress in the utilization of soy-based emulsifiers in food applications—a review. Food Secur. 2021;10:1354.
79. Chen Y, Li L, Zhao X, Zeng X, Xu X. How environmental stresses affect the physical stability of oil in water emulsion prepared using pH-shifted myofibrillar protein? LWT. 2023;186:115200.
80. Ahmad M, Qureshi S, Akbar MH, Siddiqui SA, Gani A, Mushtaq M, et al. Plant-based meat alternatives: compositional analysis, current development and challenges. Appl Food Res. 2022;2:100154.
81. Kołodziejczak K, Onopiuk A, Szpicer A, Poltorak A. Meat analogues in the perspective of recent scientific research: a review. Food Secur. 2022;11:105.
82. Dong H, Wang P, Yang Z, Xu X. 3D printing based on meat materials: challenges and opportunities. Curr Res Food Sci. 2023;6:100423.
83. Pereira T, Barroso S, Gil MM. Food texture design by 3d printing: a review. Food Secur. 2021;10:320.

Meat Products with Low Cholesterol

Riya Barthwal, Shikha Jain, Poonam Maurya, and Meghana K. P.

1 Introduction

Meat has been a staple in human diets for centuries, providing essential nutrients and contributing to diverse culinary traditions. The types of meat consumed vary widely, encompassing beef, pork, poultry, lamb, etc., all of which are rich source of essential nutrients, including protein, fat, minerals (Fe, Zn, P, Se) and vitamins (B3, B6, B12) [1]. Proteins found in meat have high biological value and considered complete proteins as they comprise of almost all essential amino acids (leucine, isoleucine, valine, lysine, methionine, tryptophan) required by the human body. Apart from proteins, meat and its products typically contain varying levels of cholesterol. Organ meats, such as liver, are inherently high in cholesterol, which is an essential part of cell membranes and is involved in production of calciferol, hormones and bile acids that helps in the digestion of fats. There are two main types of cholesterol, namely, low-density lipoprotein (LDL) and high-density lipoprotein (HDL) cholesterol. LDL, often referred to as "bad" cholesterol, is lipoprotein of blood plasma composed of moderate proportion of protein and high cholesterol which carries cholesterol from the liver throughout the body cells. If too much is delivered, it can build up in the walls of the arteries, forming plaque and potentially leading to atherosclerosis and other cardiovascular issues. On the other hand, HDL

R. Barthwal
Department of Food Science and Technology, Graphic Era (Deemed to be University), Dehradun, Uttarakhand, India

S. Jain (✉) · P. Maurya
Division of Fruits and Horticultural Technology, ICAR-Indian Agricultural Research Institute, New Delhi, India

Meghana K. P.
Department of Food Science and Technology, G. B. Pant University of Agriculture and Technology, Pantnagar, Uttarakhand, India

A. Kumar Pathera et al. (eds.), *Healthier Meat Products*,
https://doi.org/10.1007/978-3-031-78215-2_15

cholesterol, often referred to as "good" cholesterol, is composed of a high proportion of protein and small portion of cholesterol. It helps to remove LDL cholesterol from the bloodstream by scavenging LDL and supplying it to liver, thus reducing blood cholesterol levels. In the modern era, public health recommendations and dietary guidelines often emphasize the importance of managing cholesterol intake for overall health. This guidance influences consumer choices and contributes to the demand for meat and meat products that align with these recommendations and prioritize health. In the modern era, there has been a surge in consumer preferences for meat and meat products that prioritize health and feature lower fat and cholesterol levels and improved fatty acid profiles [2]. As people's perceptions of meat consumption continue to evolve and their awareness of nutrition and health grows, the meat consumption market is undergoing a notable transformation.

2 Relation Between Fat and Cholesterol Content of Meat

Depending on the animal's species, age and the portion of the carcass utilized, the fat content in meat and meat products varies significantly around 3–25 g/100 g of food [3, 4]. Meat fat comprises mostly saturated and monounsaturated fatty acids. Poultry and pork exhibit a slightly higher proportion of unsaturated fatty acids (approximately 10–15% of total fatty acids) compared to beef and lamb. Additionally, they contain a significant quantity of polyunsaturated fatty acids (PUFAs). Besides fatty acids, cholesterol is a nutritionally important component of meat, which varies between about 30 and 120 mg/100 g of food [3]. Cholesterol in meat is predominantly found in the fatty portions, particularly in the muscle cell membranes and connective tissues. Structurally, cholesterol is a complex organic sterol molecule characterized by a four ringed hydrocarbon structure often referred to as a steroid nucleus. This nucleus includes three six-membered rings and one five-membered ring, all fused together. Attached to the steroid nucleus is a hydrocarbon tail, which is a hydrophobic chain of carbon atoms. Additionally, a hydroxyl (-OH) group is present at one end of the molecule, rendering it amphipathic with both hydrophobic and hydrophilic properties making it an essential component of cell membranes [5, 6]. In meat, it exists in both free and esterified forms, with esterified cholesterol being associated with fat globules. The concentration of cholesterol varies among different cuts of meat, and its distribution within the tissue influences the overall cholesterol content.

The interconnection between meat's fat and cholesterol content is closely intertwined while they are distinct components, their concentrations in meat are correlated. High fat, especially high saturated fatty acid, rich food intake is well known to raise total and LDL cholesterol. Literature data have provided several evidences of positive correlation between fat and cholesterol content. A favourable correlation between cholesterol and saturated and monounsaturated fatty acids, and negative correlation between cholesterol content and polyunsaturated fatty acids have been proved by various researchers. It was revealed that 60–80% of total cholesterol in cattle muscle tissue was found in cell membranes, with 20–40% in the cytoplasm. This suggested that the cholesterol content in meat was likely to rise with increased fat [7]. Higher intake of fatty acids leads to increased cholesterol and lipoprotein secretion by the liver of animals [8]. Table 1 enlists the fat and cholesterol percentages of various meat parts of different animals.

3 Technologies for Developing Low Cholesterol Meat and Meat Products

Various efforts have been made till now to reduce the cholesterol levels in meat and its products. One such attempt in earlier days was to alter cholesterol levels in animal body by feeding them a diet rich in unsaturated fatty acids. Replacing corn and soybean meal of broiler chicks with full-fat flax seed meal resulted in reduced total carcass fat and tissue cholesterol [13]. In further work, it was determined that flax seed rich diet fed to hens resulted in dark meat with a significantly lower cholesterol content (72 mg per 100 g) than the control diet (100 mg per 100 g) [14]. Even after 20 years, studies are being done on using flaxseed as a substitute in animal diets to reduce cholesterol. For broiler feed, rice bran was substituted with flaxseed which resulted in notable increase in the elasticity of meat texture as well as reduction in cholesterol content in a dose dependent manner. Feeding with 10% flaxseed demonstrated the most favorable outcome, reducing the cholesterol content of broiler meat to 73.54 mg/100 g [15]. Another attempt to reduce cholesterol levels was to select genetically rich and modified breeds of different animals for meat production which already have low cholesterol [2]. These dietary attempts can only be done before the animals are slaughtered for meat, so with increasing demands of healthier meat, studies are also being done to reduce the cholesterol content in raw meat as well as processed meat.

3.1 Technologies for Raw and Processed Meat

In the last few years, significant efforts have been made to scale down the fat content of animal carcasses in response to consumer preferences for leaner meat. Various strategies have been employed, including the selection of specific genetic lines and breeds, changes in animal feeding practices, such as the use of probiotics and antibiotics, and interventions in animal metabolism, such as anabolic implants, beta-agonists, and growth hormones. Molecular genetics techniques have played a crucial role in achieving a substantial reduction in different types of fat deposits in carcasses [16]. These fat deposits include intramuscular fat (found among muscle fibers), intermuscular fat (deposited between muscles), subcutaneous fat (under the skin) and visceral fat (in body cavities). Among these, intramuscular fat is

Table 1 Fat and cholesterol percentage of various meat parts of different animals

S.No.	Meat type	Fat (%)	Cholesterol (mg/100 g)	References
A	**Beef**			
1	Frontal thigh muscles	2.2	49.0	[9]
2	Shoulder	2.9	56.7	[9]
3	Posterior thigh muscle	3.2	49.5	[9]
4	Fillet (small end)	3.7	50.8	[9]
5	Posterior thigh muscle	3.9	53.5	[9]
6	Back muscle	6.3	48.5	[9]
7	Entrecote (Rib eye)	11.3	46.7	[9]
8	Belly (middle part)	13.9	66.5	[9]
9	Belly (front part)	29.2	68.3	[9]
10	Heart	6.0	150.0	[10]
B	**Lamb**			
1	Posterior thigh muscle	2.0	50	[11]
2	Anterior thigh muscle	2.6	50	[11]
3	Posterior thigh muscle	2.1	49	[11]
4	Gluteus (buttock)	2.2	50	[11]
C	**Pork**			
1	Round	1.6	51.3	[9]
2	Fillet	1.6	54.9	[9]
3	Thigh muscle	1.7	48.6	[9]
4	Back muscle	3.3	45.3	[9]
5	Posterior thigh muscle	3.5	49.9	[9]
6	Chops	7	53.6	[9]
7	Neck	11.9	62.2	[9]
8	Belly	27.1	59.0	[9]
9	Tongue	11.3	86.7	[10]
10	Stomach	5.79	141.2	[10]
11	Heart	3.6	150.0	[10]
C	**Poultry**			
1	Skinless breast	0.7	43.4	[9]
2	Breast with skin	6.2	61.4	[9]
3	Skinless upper leg	6.5	84.0	[9]
4	Skinned wings	11.5	95.3	[9]
5	Skinned upper leg	15.1	84.6	[9]
6	Heart	5.8	170.0	[10]
D	Turkey			
1	Skinless breast (average)	1	44	[9]
2	Skinned breast roll	5.3	50.8	[9]
3	Skinned lower leg	5.7	71.9	[9]
4	Skinned upper leg	7.2	60.1	[9]
5	Skinned wings	8.7	64.9	[9]
E	Ostrich			
1	Calf muscle (Common ostrich)	1.3	65.5	[12]
2	Calf muscle (Somali ostrich)	1.5	63.4	[12]

particularly important from a dietary perspective because it cannot be easily trimmed before or during consumption. Other types of adipose tissue pose more of an economic concern. While they can be cut off and eliminated from carcasses or snipped from the external surfaces of muscles and excluded from the food chain, they are often degraded into by-products of very little value. Overall, these various tools and techniques contribute to meeting consumer demand for leaner meat while addressing both nutritional and economic considerations in the livestock industry. The cholesterol levels found in meat and its derived products can be influenced by various factors such as the type of meat, specific cut, and cooking methods employed (e.g., broiling, pan-frying, boiling). While there may be variations attributed to these factors, the standard cholesterol concentration typically falls within the range of 75 to 95 mg per 100 g of meat. Notably, organ meats like kidney, heart, and liver deviate from this general pattern, displaying notably higher cholesterol contents ranging from 300 to 375 mg per 100 g of meat [17].

Another technology used to prepare meat products with low cholesterol levels is to replace fat in the raw material with plant-based substitutes, such as vegetable oils or proteins. This is due to the genetically lower cholesterol levels present in plant tissues when compared to animal tissues. Table 2 enlists the effects of diet substitution and supplementation on the cholesterol quality of different meats. Plant based meat and meat analogues are also gaining popularity with significant advancements as low cholesterol or cholesterol free meat. Innovations in food science and technology have allowed for the formulation of plant-based alternatives that replicate the taste, texture, and nutritional profile of traditional meat products, all while maintaining reduced cholesterol levels. High-pressure processing, extrusion technology, and precision blending techniques are employed to create plant-based meat products with optimal cholesterol content. Meat alternatives derived from soy not only boast elevated protein levels and nutritional values comparable to traditional meat but also exhibit minimal to no cholesterol and fat content [18]. Furthermore, incorporation of microorganisms with probiotics [19] and utilization of plant extracts and compounds [20] have also been employed to decrease the cholesterol content in poultry meat. These strategies not only help minimize the use of additives and ingredients linked to health issues but also boost the overall nutritional value of meat products. Prebiotics, probiotics, and phytochemical extract supplementation not only improve the meat quality but also enhance the fat profile, preventing spoilage and extending the shelf life of fermented meat products. Probiotic microorganisms commonly incorporated into feeds comprise *Bacillus subtilis, Bifidobacterium rohita, Clostridium butyricum, Escherichia coli, Lactobacillus acidophilus, Lactobacillus casei, Lactobacillus coagulans, Lactobacillus lactis, Lactobacillus rhamnosus, Lactobacillus rohita* and *Streptococcus* [21]. The inclusion of Lactic Acid Bacteria (LAB) in broiler chicken feeds significantly contributes to lowering cholesterol levels in poultry meat. Numerous studies have demonstrated the efficacy of probiotic *Bacillus coagulans* in reducing both cholesterol and malondialdehyde content [22, 23]. Another research on prebiotics indicated beneficial effects on broilers,

Table 2 Effects of diet substitution and supplementation on the quality of different meat

Ingredients	Amount (%)	Duration of experiment	Animal	Effect on cholesterol	Reference
Ginger and cinnamon oils	1–2	50 days	Japanese Quails	Decrease	[26]
Moringa powder	0.25–0.75	42 days	Broiler chickens	Decrease	[27]
Guar (*Cyamopsis tetragonoloba*)	15–17	99 days	Pigs	Decrease	[28]
Safflower seed	2.5–10	42 days	Broiler chickens	Decrease	[29]
Flax seed	10		Broiler chickens	Decrease	[15]
Sacha inchi oil with herbal plants (B*idens pilosa, Angelica sinensis* and *Ramulus cinnamomi*)	2	70 days	Pigs	Decrease	[30]
Freshwater *Cladophora glomeratamacroalgal* biomass	4 and 8	70 days	Rabbits	No effect	[31]
Hybrid Rye	20–40	101 days	Pigs	No significant effect	[32]
Sacha inchi (*Plukenetia volubilis* L.) oil	2	70 days	Broiler chickens	Decrease	[33]
Rosemary and Ginger Essential Oils	0.5	49 days	Rabbits	Decrease	[34]
Alfalfa powder	22	60 days	Tibetan sheep	Decrease	[35]
Guanidinoacetic acid (GAA)	6–12	35 days	Broiler chickens	Decrease	[36]
Mucuna leaf meal	0.5–1.5	42 days	Broiler chicken	Decrease	[37]
Amaranth oilcake	20	60 days	Rabbits	Decrease	[38]
Moringa oleifera leaves	1.5	42 days	Rabbits	Decrease	[39]
Tetrapleura tetraptera	0.5–1.5	56 days	Rabbit bucks	Decrease	[40]
Olive cake	15	150 days	Limousin bulls	Decrease	[41]
Yellow lupin	68	16 weeks	Geese	No significant effect	[42]
Copper	0.2	40 days	Chickens	Decrease	[43]

including increased serum protein and albumin levels, along with a reduction in total cholesterol and triglyceride levels in the blood [24]. Moreover, augmentation with *Bacillus subtilis* probiotic has been proven to decrease cholesterol and fat content of broiler chickens breast and thigh meat along with enhanced meat quality [25].

3.2 Fat/Cholesterol Replacers

Fat/cholesterol replacers are excellent substitutes for dietary fat and cholesterol, offering the dual benefits of lowering the overall fat and calorie content while emulating the physical as well as chemical properties of dietary fat. Various cholesterol replacers are utilized in meat in order to make meat healthy and reduce the amount of cholesterol. Modified starch and maltodextrins exhibit promising capabilities for enhancing the overall quality of reduced-cholesterol foods. Modified starch granules directly influence the structure and sensorial characteristics of foods by acting similarly to fat globules [44]. A significant 45.7% decrease in cholesterol contents was observed when 2% each of citrus fibre and soy protein concentrate was incorporated as fat replacers in sausages [45]. In one study, the substitution of pork back fat with vegetable oil and rice bran fiber in frankfurters resulted in reduced cholesterol and trans-fat levels [46]. Animal fat of beef burgers when partially replaced with pineapple peel, pomace and canola oil which resulted in reduction in cholesterol content significantly [47]. Additionally, the incorporation of alternative sources like olive oil, chia oil, or rice bran fiber as partial or total fat replacement has been studied to reduce cholesterol levels and improve overall product quality. In one research, canola-based pickering emulsions was used as a fat replacer in meat systems, replacing 25%, 50%, 75%, and 100% of animal fat content with pickering emulsion. It was discovered that canola protein based emulsions when used as complete fat replacers in meat formulations significantly boosted nutritional aspects. This included increased protein content, enhanced monounsaturated and polyunsaturated fatty acid composition, and a decrease in saturated fatty acid content compared to the control with 100% animal fat. While significant color differences were observed only in samples with 100% fat replacement, meat systems containing fat replacement up to 50% exhibited texture parameters closest to the control [48]. Replacing 100% pork back fat in pate with olive oil increased α-tocopherol content and decreased cholesterol content [49]. Strategies involving beeswax oleogels and emulsion gels, especially those prepared with health-conscious lipid mixtures, offer innovative approaches to replace pork fat in meat formulations. Beeswax oleogels and emulsion gels, formulated using blend of olive oil and chia oil also serve as alternatives to pork fat in the preparation of a dry fermented meat product [50]. Replacing animal fat with ground poppy seed paste in koefte meat burger production significantly reduced cholesterol and saturated fatty acid content and increased PUFAs in the meat [51]. In one research, bigels (olive oil oleogels) were combined with κ-carrageenan and gelatin hydrogels, as fat replacers in semi-dry sausages. Substituting 50% of pork backfat with these bigels in sausages resulted in significant reduction in fat, saturated fatty acids and 6% cholesterol content. This finding highlights the potential of bigels as a promising fat alternative, offering the ability to produce semi-dry meat products with lower fat content and an improved nutritional profile, making them a viable and appealing option for health-conscious consumers [52]. Apart from this, water-in-oil-in-water double emulsions are also emerging as fat replacers for improving the lipid profile of meat products. Various studies have shown that application of double emulsions in meat systems replacing animal fat formulated with oils like olive, perilla, or chia is technologically feasible not only protects against lipid oxidation but also optimizes the presence of bioactive compounds in foods [53]. Double emulsions with a healthy oil blend as the lipid phase and an olive leaf extract (OLE) encapsulated in the internal aqueous phase are used as fat replacers in meat systems. The compound oleuropein present in olives was retained upto 69% during encapsulation, which showed cholesterol lowering effects [54]. These advancements in fat and cholesterol replacers contribute to the development of healthier meat options, aligning with the growing demand for nutritious and balanced food choices.

4 Effect of Processing on Cholesterol Content of Meat

Cholesterol content of processed meat vary from that of raw meat in qualitative as well as quantitative manner. During high-temperature processing, some cholesterol undergoes oxidation and gets converted to cholesterol oxidation products causing a subside in the cholesterol content in raw meat. A number of researchers examined how heat treatment affects the production of cholesterol oxides in raw and processed meat products. These cholesterol oxides are regarded harmful than cholesterol, contributing to the formation of arteriosclerotic plaque, functioning as cytotoxic, mutagenic and carcinogenc agents. Various researchers have observed variations in cholesterol and total lipid contents in different meat products based on cooking and processing methods. Three distinct cooking techniques to prepare different meat including sausages, ham and bacon, were used to explore the impact of cooking on both overall cholesterol levels and the generation of cholesterol oxidation products. The generation of cholesterol oxidation products was notably elevated in processed meat when subjected to microwaving and oven grilling, surpassing the levels observed with alternative cooking techniques. The cholesterol content in processed meat ranged from 76 mg/100 g to 201.70 mg/100 g, with microwaved ham exhibiting the lowest cholesterol levels among the various processed meat products. A total of six cholesterol oxides were identified in the processed meat, with 7β-hydroxycholesterol and α-epoxides identified as the primary oxidation products. Furthermore, refrigerated storage substantially amplified the cholesterol oxidation in processed meat [55]. Another researcher explored the impact of boiling and frying on the oxidation of cholesterol in chicken sausages, minced beef and fish fillets. Heat treatment led to elevated generation of cholesterol oxide and a simultaneous reduction in total cholesterol content. Frying resulted in notably higher oxide formation compared to boiling. When comparing minced beef to chicken sausages, minced beef exhibited a higher oxide level. However, fish fillets, whether raw or boiled, exhibited notably lower total cholesterol oxide content [56]. Similarly, another researcher investigated the effect of frying on the cholesterol contents of lamb and beef on a wet and dry basis. When examining the cholesterol content on a wet basis, the values represented the total weight of the meat product. Consequently, alterations in cholesterol were inherently linked to changes in the total weight of the meat. Conversely, on a dry basis, the cholesterol was expressed per gram of non-fat dry solids. This approach ensured that the total weight of the meat product remained perpetual even as the cholesterol content underwent variations. On wet basis, cholesterol and fat levels raised during lamb frying because of greater evaporation of water during the frying process, whereas on a dry basis cholesterol and fat decreased. In frying of beef, there was a notable increase in cholesterol and fat content on a wet basis, whereas a simultaneous decrease was observed on a dry basis [57]. In another study the impact of grilling, frying and roasting was studied on the cholesterol content of Muscovy drake breast meat. The cholesterol content was highest in deep-fried meat fillets, followed by roasted fillets, while grilling resulted in the lowest cholesterol content [58].

Processing methods such as curing, smoking, and adding preservatives influence the cholesterol content of meat. For instance, processed meats like bacon, sausages, and hot dogs often undergo curing and smoking processes, which can contribute to higher levels of cholesterol and saturated fats. Additionally, certain processed meats may incorporate animal fats and other ingredients that elevate their cholesterol content. On the other hand, lean and minimally processed cuts of meat, especially when prepared through healthier cooking methods like grilling or baking, tend to retain a lower cholesterol profile. The influence of canning, thermal processing and storage was investigated on the cholesterol and fatty acid content of goshtaba, a traditional low-fat meat product from J & K. There was a significant decrease in cholesterol content and two oxides were also generated in all goshtaba products. Thus, it can be seen that processing methods have great impact on the quality and quantity of cholesterol content in meat products [59].

5 Health Benefits of Low Cholesterol Meat

Meat low in cholesterol offers several health benefits, especially when incorporated into a balanced diet. Though cholesterol is a crucial component in the body, its excess in the bloodstream over the long term may cause a number of health problems, such as high blood pressure and cardiac diseases. Also, the heme-iron present in meat facilitates the oxidation process of fats and cholesterol, producing harmful free radicals that can adversely impact health. Opting for lean meats and low cholesterol meats assists in managing saturated fat intake within recommended limits, consequently lowering the risk of chronic diseases. Further, low cholesterol meat aid in blood sugar level regulation, making it beneficial for diabetic people. Consuming low-cholesterol meats has also been linked to mitigate digestive and respiratory cancers. Prospective studies conducted in populations with low cholesterol meat intake have demonstrated converse relations with oesophagal, liver, lung, colon and anal cancers.

Poultry lean fat is considered advantageous due to its composition, which contain substantial quantities of monounsaturated fatty acids as it consists only of a third of saturated fatty acids. In contrast to red meat, poultry fat also contains noteworthy levels of polyunsaturated fats, particularly omega-6, n-6 linoleic acid and arachidonic acid [60]. On the contrary, a systematic study was conducted regarding the effect of consuming red meat and its possibility for dyslipidaemia and inflammation. It was revealed that red meat consumption may lead to an elevation in serum lipid concentrations, particularly triglycerides. However, its impact on total cholesterol, C-reactive protein, low and high density lipoprotein cholesterol appears to be minimal [61]. Consequently, taking into account the influence of red meat on blood lipids, a discouraging stance is adopted regarding red meat consumption, particularly for individuals with higher triglyceride concentrations. In various other meats, camel meat is considered a healthier option due to its lower fat content and reduced cholesterol levels in comparison to other meat sources as it exhibits relatively high levels of polyunsaturated fatty acids compared to beef. Moreover, camel meat is employed for therapeutic purposes, addressing conditions such as hyperacidity, hypertension, pneumonia and respiratory ailments, and is even recognized for its aphrodisiac properties [62]. Likewise Meat from ruminant animals is abundant in saturated fatty acids (SFA), constituting 30–40% of the total fatty acids. This high SFA content results from the microbial hydrogenation of feed-derived unsaturated fatty acids (UFA) in the rumen. Consequently, ruminant meat significantly contributes to the overall intake of saturated fats in humans [63]. A study examining the connection between meat consumption and insulin resistance (IR) in 292 non-diabetic women was conducted. The results revealed that women with moderate or high meat intake exhibited significantly elevated homeostasis model assessment (HOMA) levels compared to their peers. Increased insulin resistance in non-diabetic women was associated with moderate and high meat intakes, but it was noted that variations in body fat also played a significant role in this relationship [64]. An inaugural meta-analysis was conducted investigating the correlation between dietary cholesterol and breast cancer. In the dose-response analysis, a non-linear relationship was identified between dietary cholesterol and breast cancer, however, the correlation became statistically significant when cholesterol consumption exceeded 370 mg/day [65]. A study reported a significant decrease in bladder cancer with limited consumption of white meat. This may be because cooking releases few mutagens and reduces the amount of heme iron and saturated fats, which can cause oxidative stress and DNA damage [66]. Thus, offering vital proteins, vitamins, and minerals without substantially impacting cholesterol levels low cholesterol meats can be integrated into a well-rounded diet as part of a broader approach to enhance heart health and overall well-being. Additionally, choosing lean cuts, trimming visible fat, and incorporating a variety of cooking techniques can contribute to a balanced and nutritious approach to meat consumption. However, addressing cholesterol levels, it is crucial to take into account overall dietary patterns and lifestyle factors, emphasizing a holistic perspective rather than concentrating solely on specific food items.

6 Low Cholesterol Meat Products Available in Market

When it comes to low cholesterol meats, choosing lean cuts is generally recommended. Lean cuts of pork (tenderloin, loin, sirloin, chop), skinless, lean, and ground chicken and turkey breast, fish (Salmon, Mackerel, Tuna, Trout,Herring) and lean and extra lean cuts of beef (top sirloin steak, bottom-round steak, top-round steak, sirloin tip steak, eye of round steak) are distinguishable by having less visible fat and cholesterol as compared to other fatty cuts. There are many food industries that are processing and marketing these low cholesterol meat products due to their high among the consumers. Perdue Farms, Bell and Evans, Applegate Farms, Laura's Lean Beef, and Tyson Foods are some of the companies known for their commitment to produce high-quality meat products, offering a range of lean and skinless chicken, lean pork and beef cut options.

Besides the traditional meats, rabbit meat is grasping the low cholesterol meat market as a lean and nutritious protein source. Rabbit meat is naturally lower in fat, cholesterol (48–60 mg/100 g) and sodium, and rich in PUFAs compared to traditional meats like beef and pork. Also, rabbits have a short gestation period of about 30 days and a high feed conversion capacity, which makes them one of the best sellers [67]. Amid the ongoing military conflict in Ukraine, the vulnerabilities in the food supply have led to rabbit meat gaining prominence as a potential food source to address food security issues. The country is grappling with a fragile food supply chain due to the conflict, resulting in shortages that impact both individuals and the wider population. Rabbit meat is increasingly recognized as a practical solution, offering a reliable and sustainable food option to alleviate the challenges posed by the current food security situation in Ukraine [68]. Another low-cholesterol meat emerging in market is horse meat. Though less common in Western diets, it has been consumed in certain cultures for centuries and is recognized for its unique nutritional profile. It is generally lean and contains lower levels of saturated fat and cholesterol (40.5 mg/100 g meat) compared to other traditional meats [69]. The consumption of venison (fallow deer meat) has also increased in recent years because of its low cholesterol and fat content. Raising deer has become increasingly appealing to consumers seeking alternatives to conventional meat sources as it offers a unique flavor profile is lean. As consumers become more health-conscious and environmentally aware, the interest in alternative meats like venison continues to rise [70].

Apart from the real meat available on the market, increasing demand for low cholesterol meats has led to arise in plant-based meat prepared from ingredients like soy, peas, wheat, mushrooms and other plant sources, which not only deliver a cholesterol-free option but also typically contain lower levels of saturated fats compared to traditional animal meats. Additionally, these products often showcase a reduced environmental footprint, aligning with both health-conscious and eco-friendly consumer preferences. As the demand for plant-based alternatives continues to rise, the development of low-cholesterol plant-based meats represents a positive step towards promoting heart-healthy choices and fostering sustainability in the food industry. Beyond Meats, Impossible Foods, Lightlife, Morning Star Farms, Gardein, Quorn, etc. are some of the notable internationally recognized companies known for producing low-cholesterol plant-based meat alternatives.

7 Future Trends and Potential Advancements

As we look toward the future, the consumption of low-cholesterol meats is likely to undergo significant advancements driven by technological innovations, increasing consumer preferences, and a growing emphasis on health and sustainability. Anticipating a growing demand for heart-friendly options, innovators in the food industry are likely to invest in advanced technologies and formulations for producing low-cholesterol meat alternatives. Lab-grown meats or cultivated meats have the potential to be engineered with lower cholesterol content, providing a healthier alternative for consumers concerned about their cardiovascular health. The integration of health monitoring and wearable technology may influence dietary choices. As individuals gain more insights into their health in real-time, there may be increased demand for meats that align with specific health goals, such as maintaining optimal cholesterol levels. Further advances in genetic engineering techniques may help to modify the genetic makeup of animals to produce meat with lower cholesterol levels. Incorporating functional ingredients like sterols into animal feed or meat processing also holds potential to help reduce cholesterol levels in meat. Moreover, as the demand for plant-based options continues to rise, incorporating them into one's diet can contribute not only to lower cholesterol levels but also to a more environmentally friendly and ethically conscious approach to food choices, fostering a positive impact on personal well-being and the broader ecosystem. As environmental and ethical concerns continue to shape consumer choices, the production of low-cholesterol meats may align with sustainable and ethical practices. This could involve a focus on reducing the environmental impact of meat production, including lower greenhouse gas emissions and responsible land use. In summary, the future of low-cholesterol meats will be shaped by fusion of technological innovations, ethical considerations and advancing preferences of consumers. As the demand for healthier and more sustainable food options continues to rise, the meat industry is expected to respond with novel approaches to meat production and consumption.

8 Conclusion

Meat serves as a vital component in human diets, offering essential nutrients like proteins, fats, vitamins, and minerals that contribute to overall health and well-being. However, excessive consumption of meat, particularly red and processed meats, can contribute to elevated cholesterol levels, increasing the risk of cardiovascular diseases. The utilization of low-cholesterol meats presents a promising avenue for addressing public health concerns related to cardiovascular well-being along with other health issues. Ongoing advancements in the field, ranging from selective breeding techniques and targeted dietary interventions to the rise of marketing of lean cuts of meat, reflect a collective effort to offer consumers healthier options. As awareness of dietary impacts on cardiovascular health grows, low-cholesterol meats not only cater to individuals seeking heart-friendly choices but also contribute to a more comprehensive approach to balanced nutrition. As sustainability concerns gain prominence, a convergence of health and environmental consciousness is expected to drive the evolution of low-cholesterol meat options, offering consumers a diverse and nutritionally rich array of choices to meet their dietary preferences and health goals. By embracing these innovations, individuals can enjoy the savoury experience of meat consumption while making informed decisions that align with their health-conscious lifestyles, fostering a positive impact on both personal well-being and broader public health initiatives.

References

1. Biesalski HK. Meat as a component of a healthy diet–are there any risks or benefits if meat is avoided in the diet? Meat Sci. 2005;70(3):509–24. https://doi.org/10.1016/j.meatsci.2004.07.017.
2. Clarke AD. Reduction of cholesterol levels in meat, poultry and fish products. In: Production and processing of healthy meat, poultry and fish products. Boston: Springer US; 1997. p. 101–17.
3. Ovaskainen ML, Reinivuo H, Korhonen T. Food table. Information on nutrient composition (in Finnish). Keuruu: Publishing Company Otava; 2001.
4. Valsta LM, Tapanainen H, Männistö S. Meat fats in nutrition. Meat Sci. 2005;70(3):525–30. https://doi.org/10.1016/j.meatsci.2004.12.016.
5. Bragagnolo N. Cholesterol and cholesterol oxides in meat and meat products. In: Handbook of muscle foods analysis. CRC Press; 2008. p. 207–40.
6. Shieh HS, Hoard LG, Nordman CE. The structure of cholesterol. Acta Crystallograph Sect B Struct Crystallograph Cryst Chem. 1981;37(8):1538–43. https://doi.org/10.1107/S0567740881006523.
7. Hoelscher LM, Savell JW, Smith SB, Cross HR. Subcellular distribution of cholesterol within muscle and adipose tissues of beef loin steaks. J Food Sci. 1988;53(3):718–22. https://doi.org/10.1111/j.1365-2621.1988.tb08939.x.
8. Zhou Y, Chen Z, Lin Q, Yang Y, Hang Y, Zhou X, Wu C, Xie Z. Nuciferine reduced fat deposition by controlling triglyceride and cholesterol concentration in broiler chickens. Poult Sci. 2020;99(12):7101–8. https://doi.org/10.1016/j.psj.2020.09.013.
9. Honikel KO, Arneth W. Cholesteringehalt in fleisch und eiern. Fleischwirtschaft (Frankfurt). 1996;76(12):1244–53.
10. Scherz H, Senser F. Food composition and nutrition tables 1989/1990. Stuttgart: Scientific Publishing Company; 1989. p. 1028.
11. Garcia PT, Casal JJ, Margaria CA, Pensel NA. Cholesterol content in different meats. In: Annual international congress of meat science and technology, vol. 41; 1995. p. 54–5.
12. Horbanczuk J, Sales J, Celeda T, Konecka A, Zieba G, Kawka P. Cholesterol content and fatty acid composition of ostrich meat as influenced by subspecies. Meat Sci. 1998;50(3):385–8. https://doi.org/10.1016/S0309-1740(98)00028-X.

13. Ajuyah AO, Lee KH, Hardin RT, Sim JS. Changes in the yield and in the fatty acid composition of whole carcass and selected meat portions of broiler chickens fed full-fat oil seeds. Poult Sci. 1991;70(11):2304–14. https://doi.org/10.3382/ps.0702304.
14. Ajuyah AO, Hardin RT, Cheung K, Sim JS. Yield, lipid, cholesterol and fatty acid composition of spent hens fed full-fat oil seeds and fish meal diets. J Food Sci. 1992;57(2):338–41. https://doi.org/10.1111/j.1365-2621.1992.tb05489.x.
15. Sinambela AF, Susilo A, Sjofjan O. Effect of rice bran substitution with flaxseed (*Linum usitatissimum*) flour in feed on pH, water holding capacity, texture, and cholesterol of broiler meat. In: AIP conference proceedings, 2023 Jun 5, vol. 2628, No. 1. AIP Publishing; 2023. https://doi.org/10.1063/5.0144251.
16. Zerehdaran SA, Vereijken AJ, Van Arendonk JA, Van der Waaijt EH. Estimation of genetic parameters for fat deposition and carcass traits in broilers. Poult Sci. 2004;83(4):521–5.
17. Biel W, Czerniawska-Piątkowska E, Kowalczyk A. Offal chemical composition from veal, beef, and lamb maintained in organic production systems. Animals. 2019;9(8):489.
18. Sun C, Ge J, He J, Gan R, Fang Y. Processing, quality, safety, and acceptance of meat analogue products. Engineering. 2021;7(5):674–8. https://doi.org/10.1016/j.eng.2020.10.011.
19. Bis-Souza CV, Pateiro M, Domínguez R, Penna AL, Lorenzo JM, Barretto AC. Impact of fructooligosaccharides and probiotic strains on the quality parameters of low-fat Spanish Salchichón. Meat Sci. 2020;159:107936. https://doi.org/10.1016/j.meatsci.2019.107936.
20. Pérez-Burillo S, Mehta T, Pastoriza S, Kramer DL, Paliy O, Rufián-Henares JÁ. Potential probiotic salami with dietary fiber modulates antioxidant capacity, short chain fatty acid production and gut microbiota community structure. Lwt. 2019;105:355–62. https://doi.org/10.1016/j.lwt.2019.02.006.
21. Hassan N, Mostafa I, Elhady MA, Ibrahim MA, Amer H. Effects of probiotic feed additives (biosol and Zemos) on growth and related genes in broiler chickens. Ital J Anim Sci. 2022;21(1):62–73. https://doi.org/10.1080/1828051X.2021.2016509.
22. Sarkar G, Mondal S, Bhattacharya D, Ponraj P, Sawhney S, Bala P, Chakraborty D, Sunder J, De AK. Effect of a multi-strain probiotic on growth performance, lipid panel, antioxidant profile, and immune response in Andaman Local Piglets at weaning. Fermentation. 2023;9(11):970. https://doi.org/10.3390/fermentation9110970.
23. Majeed M, Majeed S, Nagabhushanam K, Arumugam S, Beede K, Ali F. Evaluation of the in vitro cholesterol-lowering activity of the probiotic strain *Bacillus coagulans* MTCC 5856. Int J Food Sci Technol. 2019;54(1):212–20. https://doi.org/10.1111/ijfs.13926.
24. Yazhini P, Visha P, Selvaraj P, Vasanthakumar P, Chandran V. Dietary encapsulated probiotic effect on broiler serum biochemical parameters. Veterin world. 2018;11(9):1344. https://doi.org/10.14202/vetworld.2018.1344-1348.
25. Tang X, Liu X, Liu H. Effects of dietary probiotic (*Bacillus subtilis*) supplementation on carcass traits, meat quality, amino acid, and fatty acid profile of broiler chickens. Front Veterin Sci. 2021;8:767802. https://doi.org/10.3389/fvets.2021.767802.
26. Abd El-Hack ME, AboElMaati MF, Abusudah WF, Awlya OF, Almohmadi NH, Fouad W, Mohamed HS, Youssef IM, Al-Gabri NA, Othman SI, Allam AA. Consequences of dietary cinnamon and ginger oils supplementation on blood biochemical parameters, oxidative status, and tissue histomorphology of growing Japanese quails. Poult Sci. 2024;103(2):103314. https://doi.org/10.1016/j.psj.2023.103314.
27. Majidi KA, Alshelmani MI, Imdakim MM. Effect of diet supplemented with different levels of moringa powder on growth performance, carcass characteristics, meat quality, hematological parameters, serum lipids, and economic efficiency of broiler chickens. Arch Razi Instit. 2023;78(5):1647–56. https://doi.org/10.32592/ARI.2023.78.5.1647.
28. Karpiesiuk K, Kozera W, Daszkiewicz T, Lipiński K, Kaliniewicz J, Okorski A, Pszczółkowska A, Żak G, Matusevičius P. The effect of dietary supplementation with GUAR () meal protein on the quality and chemical composition of pig carcasses. Ann Anim Sci. 2023;23(4):1095–104. https://doi.org/10.2478/aoas-2023-0020.

29. Rathaur A, Rai DC, Agarwal A, Tripathi AD. Effect of dietary supplementation of safflower (Carthamus tinctorius L.) seed on the growth performance, blood lipid and meat quality of broiler chickens. Indian J Anim Res. 2023;57(6):742–8.
30. Oanh NC, Thu CT, Don NV, Hornick JL. Growth performance and pork quality of finishing pigs fed diet supplemented with *Sacha Inchi* oil and herbal plants. Trop Anim Sci J. 2023;46(4):471–7. https://doi.org/10.5398/tasj.2023.46.4.471.
31. Nutautaitė M, Racevičiūtė-Stupelienė A, Pockevičius A, Vilienė V. Sensory evaluation of rabbit meat from individuals fed functional and more sustainable diets enriched with freshwater *Cladophora glomerata* macroalgal biomass. Animals. 2023;13(13):2179. https://doi.org/10.3390/ani13132179.
32. Lisiak D, Janiszewski P, Borzuta K, Schwarz T, Grześkowiak E, Siekierko U, Lasek J. The effect of feeding fattener pigs with hybrid rye on selected growth and carcass traits and on meat quality characteristics. Ann Anim Sci. 2023;23(2):529–36. https://doi.org/10.2478/aoas-2022-0077.
33. Cong ON, Viet DN, Kim DP, Hornick JL. Effects of dietary sacha inchi (*Plukenetia volubilis* L.) oil and medicinal plant powder supplementation on growth performance, carcass traits, and breast meat quality of colored broiler chickens raised in Vietnam. Trop Anim Health Prod. 2022;54(2):87. https://doi.org/10.1007/s11250-021-02994-8.
34. Elazab MA, Khalifah AM, Elokil AA, Elkomy AE, Rabie MM, Mansour AT, Morshedy SA. Effect of dietary rosemary and ginger essential oils on the growth performance, feed utilization, meat nutritive value, blood biochemicals, and redox status of growing NZW rabbits. Animals. 2022;12(3):375. https://doi.org/10.3390/ani12030375.
35. Su Y, Sun X, Zhao S, Hu M, Li D, Qi S, Jiao X, Sun Y, Wang C, Zhu X, Li Z. Dietary alfalfa powder supplementation improves growth and development, body health, and meat quality of Tibetan sheep. Food Chem. 2022;396:133709. https://doi.org/10.1016/j.foodchem.2022.133709.
36. Rahmawati D, Hanim C. Supplementation of guanidinoacetic acid in feed with different levels of protein on intestinal histomorphology, serum biochemistry, and meat quality of broiler. J Indo Trop Anim Agric. 2022;47(4):265.
37. Oloruntola OD, Ayodele SO, Omoniyi IS, Adeyeye SA, Adegbeye MJ. The effect of dietary supplementation of mucuna leaf meal on the growth performance, blood parameters, and carcass quality of broiler. Acta Scientiarum Anim Sci. 2022;44:e55362. https://doi.org/10.4025/actascianimsci.v44i1.55362.
38. Shevchik RS, Duda YV, Gavrilina OG, Kuneva LV, Samoyluk HV. Impact of Amaranthus hypochondriacus in nutrition for rabbits on meat quality. J Hellenic Veterin Med Soc. 2021;72(1):2713–22. https://doi.org/10.12681/jhvms.26756.
39. Selim S, Seleiman MF, Hassan MM, Saleh AA, Mousa MA. Impact of dietary supplementation with Moringa oleifera leaves on performance, meat characteristics, oxidative stability, and fatty acid profile in growing rabbits. Animals. 2021;11(2):248. https://doi.org/10.3390/ani11020248.
40. Ingweye JN, Lamidi AA, Uzodinma DO. Chemical composition, cholesterol profile and sensory quality of meat from rabbits Fed Aidan (Tetrapteratetrapleura) as dietary additive. Energy (kcal ME/kg). 2021;2(2,750):2–767. https://doi.org/10.9790/2380-1404011419.
41. Chiofalo V, Liotta L, Lo Presti V, Gresta F, Di Rosa AR, Chiofalo B. Effect of dietary olive cake supplementation on performance, carcass characteristics, and meat quality of beef cattle. Animals. 2020;10(7):1176. https://doi.org/10.3390/ani10071176.
42. Biesek J, Kuźniacka J, Banaszak M, Maiorano G, Grabowicz M, Adamski M. The effect of various protein sources in goose diets on meat quality, fatty acid composition, and cholesterol and collagen content in breast muscles. Poult Sci. 2020;99(11):6278–86. https://doi.org/10.1016/j.psj.2020.08.074.
43. Skrivan M, Skrivanova V, Marounek M, Tumova E, Wolf J. Influence of dietary fat source and copper supplementation on broiler performance, fatty acid profile of meat and depot fat, and on cholesterol content in meat. Br Poult Sci. 2000;41(5):608–14. https://doi.org/10.1080/713654978.

44. Chen Y, She Y, Zhang R, Wang J, Zhang X, Gou X. Use of starch-based fat replacers in foods as a strategy to reduce dietary intake of fat and risk of metabolic diseases. Food Sci Nutr. 2020;8(1):16–22. https://doi.org/10.1002/fsn3.1303.
45. Cengiz E, Gokoglu N. Changes in energy and cholesterol contents of frankfurter-type sausages with fat reduction and fat replacer addition. Food Chem. 2005;91(3):443–7. https://doi.org/10.1016/j.foodchem.2004.06.025.
46. Choi YS, Choi JH, Han DJ, Kim HY, Lee MA, Jeong JY, Chung HJ, Kim CJ. Effects of replacing pork back fat with vegetable oils and rice bran fiber on the quality of reduced-fat frankfurters. Meat Sci. 2010;84(3):557–63. https://doi.org/10.1016/j.meatsci.2009.10.012.
47. Selani MM, Shirado GA, Margiotta GB, Rasera ML, Marabesi AC, Piedade SM, Contreras-Castillo CJ, Canniatti-Brazaca SG. Pineapple by-product and canola oil as partial fat replacers in low-fat beef burger: effects on oxidative stability, cholesterol content and fatty acid profile. Meat Sci. 2016;115:9–15. https://doi.org/10.1016/j.meatsci.2016.01.002.
48. Rezaee M, Aider M. Study of the effect of canola proteins-xanthan based Pickering emulsion as animal fat replacer in a food matrix produced from mechanically separated meat. Meat Sci. 2023;204:109283. https://doi.org/10.1016/j.meatsci.2023.109283.
49. Domínguez R, Agregán R, Gonçalves AA, Lorenzo Rodriguez JM. Effect of fat replacement by olive oil on the physico-chemical properties, fatty acids, cholesterol and tocopherol content of pâté. Grasas Aceites. 2016;67(2):e133. https://doi.org/10.3989/gya.0629152.
50. Pintado T, Cofrades S. Quality characteristics of healthy dry fermented sausages formulated with a mixture of olive and chia oil structured in oleogel or emulsion gel as animal fat replacer. Food Secur. 2020;9(6):830. https://doi.org/10.3390/foods9060830.
51. Gök V, Akkaya L, Obuz E, Bulut S. Effect of ground poppy seed as a fat replacer on meat burgers. Meat Sci. 2011;89(4):400–4. https://doi.org/10.1016/j.meatsci.2011.04.032.
52. Zampouni K, Filippou A, Papadimitriou K, Katsanidis E. Evaluation of bigel systems as potential substitutes to partially replace pork backfat in semi-dry sausages. Meat Sci. 2024;208:109392. https://doi.org/10.1016/j.meatsci.2023.109392.
53. Lamba H, Sathish K, Sabikhi L. Double emulsions: emerging delivery system for plant bioactives. Food Bioprocess Technol. 2015;8:709–28. https://doi.org/10.1007/s11947-014-1468-6.
54. Robert P, Zamorano M, González E, Silva-Weiss A, Cofrades S, Giménez B. Double emulsions with olive leaves extract as fat replacers in meat systems with high oxidative stability. Food Res Int. 2019;120:904–12.
55. Khan MI, Min JS, Lee SO, Yim DG, Seol KH, Lee M, Jo C. Cooking, storage, and reheating effect on the formation of cholesterol oxidation products in processed meat products. Lipids Health Dis. 2015;14:1–9. https://doi.org/10.1186/s12944-015-0091-5.
56. Hashari SZ, Rahim AA, Meng GY, Ramiah SK. Quantification of cooking method effect on COP content in meat types using triple quadrupole GC-MS/MS. Molecules. 2020;25(21):4978. https://doi.org/10.3390/molecules25214978.
57. Keklik NM, Bozkurt H, Tekİn AR. Effect of different cooking procedures on cholesterol and fat contents of selected meat products. Food Sci Technol. 2018;38:683–90. https://doi.org/10.1590/fst.13117.
58. Omojola AB, Ahmed SA, Attoh-Kotoku V, Wogar GI. Effect of cooking methods on cholesterol, mineral composition and formation of total heterocyclic aromatic amines in Muscovy drake meat. J Sci Food Agric. 2015;95(1):98–102. https://doi.org/10.1002/jsfa.6687.
59. Rather SA, Masoodi FA, Rather JA, Gani A, Wani SM, Ganaie TA, Akhter R. Impact of thermal processing and storage on fatty acid composition and cholesterol oxidation of canned traditional low-fat meat product of India. LWT. 2021;139:110503. https://doi.org/10.1016/j.lwt.2020.110503.
60. Attia YA, Bovera F, Abd-El-Hamid AE, Tag EL, Din AE, Al-Harthi MA, Nizza A, Elharidy RM. Effect of dietary protein concentrations, amino acids and conjugated linoleic acid supplementations on productive performance and lipid metabolism of broiler chicks. Ital J Anim Sci. 2017;16(4):563–72. https://doi.org/10.1080/1828051X.2017.1301228.

61. Sun L, Yuan JL, Chen QC, Xiao WK, Ma GP, Liang JH, Chen XK, Wang S, Zhou XX, Wu H, Hong CX. Red meat consumption and risk for dyslipidaemia and inflammation: a systematic review and meta-analysis. Front Cardiovasc Med. 2022;9:996467. https://doi.org/10.3389/fcvm.2022.996467.
62. Abrhaley A, Leta S. Medicinal value of camel milk and meat. J Appl Anim Res. 2018;46(1):552–8. https://doi.org/10.1080/09712119.2017.1357562.
63. Vahmani P, Ponnampalam EN, Kraft J, Mapiye C, Bermingham EN, Watkins PJ, Proctor SD, Dugan ME. Bioactivity and health effects of ruminant meat lipids. Invite Rev Meat Sci. 2020;165:108114. https://doi.org/10.1016/j.meatsci.2020.108114.
64. Tucker LA, LeCheminant JD, Bailey BW. Meat intake and insulin resistance in women without type 2 diabetes. J Diabetes Res. 2015;2015 https://doi.org/10.1155/2015/174742.
65. Li C, Yang L, Zhang D, Jiang W. Systematic review and meta-analysis suggest that dietary cholesterol intake increases risk of breast cancer. Nutr Res. 2016;36(7):627–35. https://doi.org/10.1016/j.nutres.2016.04.009.
66. Aveta A, Cacciapuoti C, Barone B, Di Zazzo E, Del Giudice F, Maggi M, Ferro M, Terracciano D, Busetto GM, Lucarelli G, Tataru OS. The impact of meat intake on bladder cancer incidence: is it really a relevant risk? Cancers. 2022;14(19):4775. https://doi.org/10.3390/cancers14194775.
67. Cullere M, Dalle ZA. Rabbit meat production and consumption: state of knowledge and future perspectives. Meat Sci. 2018;143:137–46. https://doi.org/10.1016/j.meatsci.2018.04.029.
68. Zamaratskaia G, Havrysh O, Korzeniowska M, Getya A. Potential and limitations of rabbit meat in maintaining food security in Ukraine. Meat Sci. 2023;26:109293. https://doi.org/10.1016/j.meatsci.2023.109293.
69. Tonial I, Aguia AC, Oliveira CC, Bonnafé EG, Visentainer JV, De Souza NE. Fatty acid and cholesterol content, chemical composition and sensory evaluation of horsemeat. South African. J Anim Sci. 2009;39(4) https://doi.org/10.4314/sajas.v39i4.51135.
70. Kudrnáčová E, Bartoň L, Bureš D, Hoffman LC. Carcass and meat characteristics from farm-raised and wild fallow deer (*Dama dama*) and red deer (*Cervus elaphus*): a review. Meat Sci. 2018;141:9–27. https://doi.org/10.1016/j.meatsci.2018.02.020.

Meat Products with Low Salt/Sodium

Ruchi Sharma

1 Introduction

Meat and meat products are important parts of a balanced diet because they are great sources of vitamins, proteins, and minerals (zinc and iron) [1]. The high sodium content of processed meat and meat products is well known; consuming too much sodium causes an increased risk of stroke, hypertension, *hypernatremia,* and heart disease [2, 3]. In the developed world, meat and meat products are thought to make up 20–30% of daily sodium intake, while processed food products contribute to about 75% of sodium intake [4]. Many developed nations have implemented programs and strategies aimed at reducing the consumption of dietary sodium, with a focus on meat products, because of the significant economic and health burden associated with the population's excessive sodium intake [5]. According to Kloss et al. (2015), the World Health Organization advises adults to consume no more than 5 g of salt per day, which is equal to up to 2 g of sodium [6]. Table salt provides the majority of the sodium in meat preparations and meat products, but other additives in the form of sodium salts, such as phosphates, glutamates, lactates, and nitrites, may contribute to the sodium burden during meat processing [5] A typical salt level for meat products is 2.0% NaCl, which translates to 8 g Na/kg of raw material. Most meat formulations must have a salt concentration of at least 2% in order to guarantee the ionic strength necessary for the solubilization and extraction of the salt-soluble proteins necessary for the stability of the emulsion and the texture of the final product because it affects the solubilization of myofibrillar proteins and the activity of muscle enzymes [7, 8]. For the flavor and texture of processed meat products like bacon and sausages, as well as for the control of spoilage microorganisms, a salt level of ≥2% has also been recommended [9]. Additionally, sodium

R. Sharma (✉)
School of Bioengineering & Food Technology, Shoolini University of Biotechnology and Management Sciences, Solan, India

A. Kumar Pathera et al. (eds.), *Healthier Meat Products*,
https://doi.org/10.1007/978-3-031-78215-2_16

chloride acts as a bacteriostatic (decreasing water activity and inhibiting pathogenic microorganisms) [10]. Because of these characteristics, it is challenging to directly reduce the amount of salt in meat products without seeing a negative impact on the products' quality. Meat products that have had their salt content directly reduced must have additional ingredients or treatments added to make up for the lost functional qualities. Using substitute ingredients, like non-sodium salts and organic acids, has proven to be an effective method for creating meat products with a lower sodium content [5]. Consumer awareness of the connection between nutrition and health is increasing these days. In light of this, international organizations recommend consuming fewer high-salt products in favour of healthier meat alternatives with lower salt content [11]. To reduce salt intake by 30% until 2025, a progressive reduction has been suggested in this context [12]. Non-thermal technologies, like high-pressure processing, pulse electric fields, and ultrasound, have also demonstrated their potential for sodium reduction applications. These technologies affect a number of factors, including the interaction between proteins and salt ions, the diffusion and release of sodium from the meat matrix, and sensory perceptions. As a result, they are thought to be a novel approach to lowering the sodium content of processed foods [5, 13].

2 Scenario of Meat Consumption

Consumption of meat has been rising since the 1960s, but it has increased most from the 1980s to the present [14]. While some researchers contend that the supply of meat products increased by 204% between 1960 and 2010, other studies show that meat consumption increased by as much as 500% between 1992 and 2016 [15, 16]. Between 1961 and 2019, the world's meat production and consumption almost increased by fourfold. The high-income western countries consumed the greatest amount of meat, two times as much per person as the global average, which is roughly 33 kg [17].

3 Sodium Intake and Recommendations

An essential nutrient, sodium is involved in blood pressure (BP), fluid and electrolyte balance, and the preservation of normal cellular homeostasis. Due to its significant osmotic action, it plays a critical role in maintaining ECF volume. It also plays a vital role in the excitability of muscle and nerve cells, as well as in the transfer of nutrients and substrates across plasma membranes [18]. Essential physiological processes like the movement of nutrients and the contraction of muscles are maintained by salt. Nonetheless, a high sodium diet is linked to a higher chance of hypertension, stroke, and cardiovascular disease [19]. The current low sodium recommendations have not been well experienced by humans because most individuals around the globe consume 3 to 6 g of sodium daily, with only 5–10% consuming less than 2.3 g [20]. This implies that in order for most people worldwide to follow the current recommendations, their diets will need to be significantly altered. The Global Burden of Disease collaboration meta-analysis of surveys from 187 countries estimated the average daily intake of sodium to be 3.95 g [21]. According to the INTERMAP study, Eastern Europe, Central Asia, and East Asia have intakes above 4.2 g/day. This study aimed to compare nutrient intakes between the Chinese, Japanese, UK, and US INTERMAP samples and evaluate potential associations between dietary patterns and variations in cardiovascular disease patterns between East Asian and Western nations. The Beijing sample, located in Northern China, had the highest mean intakes, with men consuming up to 6.9 g/day and women 5.8 g/day. The United States had sodium intake ranging from women (3.0–3.5 g/day) to men (4.1–4.4 g/day) [21, 22]. In order to lower blood pressure and the risk of cardiovascular diseases, WHO advises reducing sodium intake (strong recommendation). For adults, the WHO advises cutting back to <2 g/day of sodium (5 g/day of salt) (strong recommendation). In order to lower blood pressure in children, WHO advises reducing sodium intake (strong recommendation). The recommended daily maximum intake of 2 grams of sodium for adults should be lowered in accordance with the energy needs of children in comparison to those of adults. The term "adults" in this recommendation refers to people who are at least 16 years old. In the context of this recommendation, "children" refers to people ages 2–15, inclusive [3].

4 Role of Sodium Chloride in Meat Products

Since ancient times, salt, or sodium chloride (NaCl), has been used as a food preservative [23]. NaCl is a necessary ingredient in the processing of meat, and salt content has increased from 1.1 g/100 g in sausages to 4.6 g/100 g in salami. Myofibrillar proteins in meat are eliminated when meat products are chopped with the addition of NaCl [24, 25]. By enclosing the fat component and preventing the release of moisture, the extracted myofibrillar proteins help to stabilize the meat emulsion and give the finished product the desired properties [26]. In addition to this, the product's desired texture is achieved by the binding of meat, water, and fat by the myofibrillar proteins that are eluted by NaCl [27]. An improvement in the texture of processed meat products and a more stable meat emulsion have been observed by adding NaCl. It's also known that NaCl improves flavor and juiciness. The extraction of myofibrillar proteins soluble in salt causes an increase in juiciness by making up for the stable state of the proteins bound to water [28]. Furthermore, the growth of microorganisms can be inhibited by controlling electrolyte imbalance, osmotic shock, and water activity. So, when processed meat products are being manufactured, salt is frequently added [24].

Table 1 Daily value (DV %) and sodium content of meat products and common ingredients

Products	DV (%)	Sodium (g)
Processed meat		
Frankfurters	3	0.06
Sausages	4.5	0.09
Bacon	90	1.8
Chicken nuggets	2.5	0.05
Ham	90	1.8
Ingredients		
Hydrolyzed protein powder	1.5	0.03
Non-fat dry milk	2.5	0.05
Salt	195	3.9
Monosodium glutamate (MSG)	62.5	1.25
Whey protein powder	1	0.02

Adapted from [34]

5 Sources of Sodium from Meat Products

One of the main sources of sodium in the diet is common salt, or sodium chloride. The second-largest food category is meat and meat products, behind cereals and real foods, accounting for around 20.8% of intake, or 0.54 g of sodium or 1.38 g of salt per day [29]. Meat products account for the majority of this; fresh meat contributes only 0.05 g of sodium. Unprocessed meat makes up a very small portion of the diet because it has less than 0.1 g of sodium (0.25 g of salt) per 100 g. Meats from cattle, pigs, and poultry naturally contain 50–70 mg of sodium per 100 g [30]. Breidenstein (1982) studied that 2% salt is typically added to meat products and that salt accounts for roughly 79% of the total sodium content of the finished meat product [31]. Apart from the widely recognized source of sodium found in processed meats, commonly known as common salt, the products' sodium content is also influenced by other additives added during the processing process. Even though these ingredients contain far less sodium overall in the diet than regular salt does, their impact cannot be neglected. A 2019 study by Sych et al. found that 42.7 g of processed meat is typically consumed daily, adding to the daily sodium intake (Table 1) [32]. As a result, the meat industry is attempting to reduce the amount of sodium in meat products in various ways [7, 33].

6 Techniques for Lowering Sodium in Meat Products

The strategy of cutting back on salt consumption is crucial due to the general decline in dietary sodium, and it is inconvenient to produce low sodium meat products. The functional qualities that common salt provides are crucial for meat products, so it is important to take those into account. Reducing or replacing common salt would affect the quality of the final product [34]. Because salt-soluble proteins become less extractable during the manufacturing process, meat products with low-salt content have less capacity to hold moisture. Therefore, new ingredients must be added to replace the salt-soluble proteins that can cause water retention in meat products [25].

6.1 Use of Flavour Enhancers

Sodium ions may have a significant impact on taste perception during oral processing, which involves a number of sensory interactions [35]. According to McGough et al. (2012), these ingredients offer a special means of lowering the sodium content of food since the enhancers increase taste perception by activating umami flavor receptors [36]. Wallis and Chapman (2012) state that flavor enhancers can reduce the final product's sodium content by 40% [37]. Every year, more flavor enhancers become available as a result of the significant and expanding demand for food products to contain less sodium. According to research studies, various flavor enhancers can lessen the unpleasant sensory effects of substitute salts and/or the reduction of NaCl in meat products [38]. To increase the sensory acceptance of meat products with sodium reduction, it may be a good idea to employ flavor enhancers such as lysine, taurine, monosodium glutamate, yeast extract, lactates, disodium guanylate, and disodium inosinate [39]. In another study, a combination of K-lactate/glycine and KCl/glycine was used to partially replace NaCl in fermented sausage. According to Gallbert et al. (2003), substitution of salt was permitted until it reached 40–70% and caused distinguishable defects (texture and flavour) [40]. There was no discernible difference in the total microbial count when 45% NaCl, 25% KCl, 20% $CaCl_2$, and 10% $MgCl_2$ were substituted for NaCl and compared to 100% NaCl. In a manner similar to that of NaCl, the growth of aerobic mesophilic bacteria was inhibited by $CaCl_2$. However, a noticeably higher cell count was noted in these mixtures concerning the amount of halotolerant bacteria. An earlier study by Lorenzo et al. (2015) showed that NaCl prevents the evolution of lactic acid bacteria, pathogenic bacteria, and salt-tolerant bacteria [41]. Additionally, flavor enhancers hit the market and are commonly used in meat products to completely or partially replace salt. An unappealing flavor was produced when frankfurter was made by substituting 75% and 50% KCl for salt and adding flavor enhancers such as taurine, lysine, and monosodium glutamate (MSG). Moreover, yeast autolysate-derived Provesta® and Aromild® products can lessen NaCl and mask the bitter taste of KCl [42]. Certainly, it is important to take product defects and health concerns into account when using flavor enhancers. Furthermore, hyperactivity, illness, and migraines have all been linked to MSG, along with potential health consequences. Moreover, some products may not be suitable for the potent broth flavor of autolysates [37].

6.2 Use of Salt Mixtures

Salt mixtures have been used for sodium content reduction in meat products. Some of these products performed satisfactorily, demonstrating similar perceived saltiness [43]. A mineral salt mixture that can cut sodium by almost half is marketed under the brand name Pansalt® and contains KCl, the amino acid L-lysine hydrochloride, and magnesium sulfate. During the triangle test, it showed no signs of organoleptic problems. On the other hand, it did not perform well in terms of cooking loss value, but it looked like ground beef patties made with regular salt [44]. Furthermore, it is disclosed that Pansalt® tastes the same as regular table salt [37]. However, because of its high potassium content, it may prevent hypertension from developing, according to later epidemiological studies [45]. KcLeanTM and Sub4salt are the commercial salts that are commonly used to lower the sodium content of meat products. KcLeanTM salt blends KCl and NaCl with a naturally occurring proprietary ingredient. Consequently, NaCl, KCl, and sodium gluconate make up Sub4salt. Wallis and Chapman (2012) worked to reduce sodium by up to 50% and 35%, respectively [37]. There were no differences in the flavor, texture, or color between the salt and Sub4salt curing agents used in the manufacturing test. Moreover, both product stability and safety demonstrated an equal capacity to regulate microbial, pH, and water activity [5]. For this reason, it was noted that meat products maintained their protein hydration and retained their flavor, just like when salt was controlled [24]. In a different study, a combination of K-lactate/glycine and KCl/glycine was used to partially replace NaCl in fermented sausage. Gelabert et al. (2003) demonstrated that flavor and texture problems have been caused by partial replacement, which was allowed up until the salt substitution reached 40–70% [40].

6.3 Changing the Size and Shape of Salt

Aside from partially substituting salt with alternatives, altering the physical properties of salt is one of the primary methods for lowering salt consumption. Sodium is detected when the concentration of sodium in saliva surpasses the concentration of saliva when it exceeds the residual concentration from an earlier stimulus. Because they can't duplicate the specificity ion channels of salts, some alternative salts are less effective [46]. As a result, physical salt modification is a promising tactic for lowering the salt content of food. There is conjecture that the acceptance of sodium in the mouth is contingent upon either oral processing or efficient delivery. For delivery to be successful, free sodium needs to permeate through the tongue's static saliva barrier layer, and salt needs to dissolve from its crystal form. Since static diffusion cannot be affected, the remaining factor is the dissolution of the crystal into a liquid state. Reducing the crystal's size by increasing its surface area is one way to regulate this [47]. Salt crystals that are smaller in size may release sodium into the saliva more quickly, which could impact how salty a mouthful feels. According to earlier research, salt's physical form has a significant impact on how salty food tastes, so altering the size of the salt crystals may be a workable way to lower the sodium content. When 20-micron sea salt particles were used, there was more salt diffusion into the food matrix, which increased the perception of saltiness [34]. As an alternative to salt, Soda-Lo® salt microspheres are made with cystalline micospheres (free-flowing). By optimizing surface area in relation to volume, this physical formation of salt is said to lower the product's salt content by 20–50% [48]. An increase in saltiness through taste contrast and an accelerated sodium delivery measured in the mouth and in a model mastication simulator were observed upon the addition of coarse-grained salt with crystal sizes ranging from 0.4 to 1.4 mm. Because sodium was available more quickly when the samples were added to an aqueous salt solution, there was a larger disparity in sodium content and a stronger sense of saltiness. These approaches demonstrated a 25% reduction in sodium content without affecting flavor [49].

6.4 Use of Mineral Salts

The objective of partially substituting other mineral salts for traditional salt has been to lower the sodium level of meat products. Low-sodium meat products have employed calcium, magnesium, and potassium chlorides to stabilize meat emulsions [43]. In a study by Horita et al. (2011) the best emulsion stability was obtained with a mixture of 1% NaCl, 0.5% KCl, and 0.5% $MgCl_2$ in mortadella to partially reduce sodium [50]. However, the mixture had a poor flavor. Consequently, it is difficult to reformulate meat batters with low salt content as ionic compounds are needed to replace the sodium chloride that is removed from the recipe's ability to bind fat, retain water, and bind proteins [51]. Lithium is another chloride salt that was utilized in place of sodium chloride. The principal cations with a salty taste are lithium and sodium. To replace sodium in meat products, lithium is used, as it mimics the saltiness of sodium. However, it seems that the human body is highly toxic to lithium chloride [7]. An additional study was carried out to offer a possible quick way to find out how much lithium was present in meatballs when it was substituted for sodium. It was assumed that naturally occurring lithium occurs in food at levels that are safe for humans, while excessive concentrations are harmful to living things. Due to its significant effects on human health, it is crucial to quantify the amount of lithium chloride present in food products [52]. Diffusion of magnesium in muscle foods has been studied to lower the salt content of dry-cured hams. Higher water activity was seen inside the product when partial replacement of NaCl with $MgCl_2$ was observed, all for the same total amount of added salt. This might be explained by the fact that magnesium has a higher charge density than sodium Mg (0.082 units of charge/molecular weight) and Na (0.044 units), which contributed to the fact that magnesium found it more difficult to penetrate the interior of dry-cured ham. In addition, Mg^{2+} cations firmly attach to the polar groups of proteins, enhancing protein interaction and preventing salt penetration [53]. Research was done to find out if sausages made with K-lactate and Na-diacetate instead of salt would be accepted by consumers. The flavor of these samples was greatly enhanced by the addition of 1.5% K-lactate/Na-diacetate and 0.5% NaCl, although the texture became harder and less juicy as a result [54]. A similar textural alteration was also observed in earlier studies on fermented sausages that substituted lactate for salt. Because K-lactate has a lower cohesiveness than salt, it exhibits a different texture when salt is replaced by more than 30% of it [46]. Potassium, sodium, and calcium lactates have comparable effects to ordinary salt in terms of microbiological growth. Furthermore, by inhibiting the growth of bacteria in packaged meat products, the combination of potassium lactate and sodium diacetate decreased sodium levels by 40% and increased shelf life [55]. Finally, salt can be substituted with other mineral compounds to determine the low sodium content of processed foods, particularly meat. That being said, it's crucial to consider the steps involved in processing meat products to achieve the same qualities that salt promotes. In addition, their impact on human health could be negative.

6.5 Natural Replacers

It is more effective and efficient to use natural ingredients. These ingredients are not only reasonably priced but also encourage trust through the promotion of additional positive health effects. Plants known as halophytes have the capacity to decrease salinity due to their unique traits, morphology, and capacity to endure in extremely salinized environments. Growing in salt marshes, glasswort is a halophyte that has been used in many different types of food as a salt substitute. Glasswort powder was used to reduce the salt content of meat products by 50% and, in addition helps, to add texture, color, and other sensory qualities to meat products [56]. The ability of glasswort to reduce the salt level of meat products is connected to its ability to absorb salt from soil and its tolerance to salinity of up to 3% [57]. impact of mineral concentration in *Salicornia ramosissima* has been assessed in view of the recent significant rise in the use of halophytes as nutritious food. According to Lima et al. (2020), the findings showed that the sodium concentration rose with rising salinity [58]. Furthermore, it was noted how salinity affected *Kochia scoparia* germination in a greenhouse. This demonstrated that a higher salinity was associated with a higher sodium concentration in the plant's roots and shoots [59]. However, more work needs to be done to create novel halophyte-based, low-sodium meat product recipes. The effectiveness of these natural replacers in lowering the salt content of meat products and their potential benefits for improving human digestion remain undetermined. Studies that substituted seasoning salt with winter mushroom powder (WMP) in chicken nuggets found no adverse effects on the sensory aspects of the product. Thus, more research is needed on natural salt substitutes and their potential health benefits, including their antimicrobial and antioxidant properties

[60]. Other seasonings that have been suggested as salt substitutes are oregano and rosemary. They have a high concentration of substances like flavonoids, phenolic acids, and terpenes that combat free radicals, in addition to their capacity to reduce salt. Test subjects with normotension and hypertension were given oregano added to food. Treating hypertensive consumers with oregano supplements changed their preference for foods with less salt. Moreover, a study demonstrated that rosemary enhances taste acceptability and can replace about 48% of salt [61]. Garlic (*Allium sativum* L.) has been used as a substitute for salt in a wide range of culinary recipes because of its low cost. As a salt substitute to prevent cardiovascular diseases, garlic's bioactive components have gained popularity due to their potential role in maintaining health [62]. Fresh garlic is added to Brazilian frankfurters to lower the sodium content while also promoting microbiological stability, delaying lipid oxidation, and acting as an antimicrobial. Furthermore, it has been proposed that manufactured goods made from garlic derivatives can be used without losing their acceptability [63]. Even though the rabbit meat burgers had less sodium, the application of garlic powder did not prove to be sufficient to stop the oxidation-related loss of nutritional value caused by the addition of salt. However, another study Mancini et al. (2020) suggested raising the concentration of garlic [64]. As the seeds, leaves, and roots of coriander (*Coriandrum sativum* L.) are edible and have a distinct flavor, they are frequently used as seasonings. Coriander is an herb that has the potential to extend food shelf life, but its leaves and seeds also contain sodium. In addition to its anti-diabetic, anti-oxidant, antimicrobial, anti-anxiety, and anti-mutagenic properties, it helps in maintaining hormonal balance [65]. When coriander and garlic were added to smoked hilsa fish, the amount of fat and salt was reduced while the amount of moisture increased [66]. Asian cuisine frequently uses spices like chili, black pepper, and mustard, which offer tastes of sweetness, sourness, bitterness, salt, and umami [67]. Natural salt substitutes such as whey proteins, herbs, and spices were effective in lowering the sodium and salt content of ready-to-eat foods, especially chicken supreme, by nearly 50% without causing any negative effects [68].

7 Alternative Processing Technologies for Sodium Reduction in Meat Products

Alternative techniques, especially in the processing aspects, must be developed in addition to employing salt substitutes to lower the sodium content of meat products. Green or non-thermal technologies in the food industry, like pulsed electric field (PEF), ultrasounds (US), and high-pressure processing (HPP), can lower the amount of additives, particularly sodium chloride, in meat products. They function by changing the structure of proteins, increasing their functional quality, and lowering salt content. Below, we go into more detail about them.

7.1 High Pressure Processing

One technology that shows great promise for producing meat products is high-pressure processing (HPP) [68]. A hydrostatic pressure (100–800 MPa) at a specific temperature and for a specific amount of time is the fundamental characteristic of HPP. It functions by altering food structures to render undesirable microorganisms inactive, which helps to lower the salt content. Numerous investigations have been carried out to examine the relationship between high-pressure treatment and salt content in the functional characteristics of meat products [69]. When compared to control sausages, high-pressure treatment improved emulsion stability and decreased cooking loss [27]. However, reducing the salt to below 2.0% resulted in an increase in cooking loss. Additionally, with HPP 100 MPa, the cooked ham's salt content dropped to 45% (by substitution of 1.1% NaCl and KCl with 0.25% phosphate). In terms of consistency, appearance, and water holding, it is technologically acceptable. In another investigation, the sodium content of nitrite-free sausages was reduced at HPP conditions (600 MPa, 8 °C, 3 min), and further increases in shelf life were observed without adversely affecting the sensory qualities; additionally, the texture became less elastic and softer due to the high-water holding capacity [70]. Additional research demonstrated that HPP can both reduce the salt content and maintain cooking loss. Meat emulsion products treated with HPP showed reduced cooking loss due to a higher viscoelastic network and maintained color performance, particularly at 200 MPa. Analyses using scanning electron microscopy and assessments of texture, color, and senses all support the use of this approach. In conclusion, this technology can be used for salt reduction in meat products. In addition, by selecting processing variables like pressure, time, and temperature, the cooking losses can be maintained [71].

7.2 Ultrasound

One of the best ways to reduce the amount of salt in food is by using ultrasound. They use sound waves to create vibrational energy [72]. In this sense, applying ultrasound for salt reduction results in the modification of the cell membrane and the mass transfer process. During the processing of meat, this technique has aided in the marinating, drying, and tenderizing of tissues. Additionally, even though the overall NaCl concentration is decreased, it results in better salt allocation in the meat and simultaneously delivers a higher salt flavor [73]. The interaction between protein and water during the power ultrasound application of pork salting suggests that ultrasound can support protein outlay and accelerate mass transfer while enabling faster NaCl assimilation through microjetting; this could be thought of as a microscale inoculation system interacting with the interface [74]. The use of ultrasound to assess the microstructure, sensory qualities, and physicochemical characteristics of restructured cooked ham with varying salt concentrations was studied by

McDonnell et al. (2014). To lower the concentration of salt, 10 minutes of 600 W/cm^2 ultrasound treatments were given with four distinct salt concentrations. A 30% salt reduction could be obtained with the 0.75% NaCl treatment without sacrificing quality. Additionally, using ultrasound to reach a 2.25% level of NaCl at 72 W/cm^2 for two hours without affecting quality attributes was possible in the cured ham. The use of ultrasound to assess the microstructure, sensory qualities, and physicochemical characteristics of restructured cooked ham with varying salt concentrations. To lower the concentration of salt, 10 minutes of 600 W/cm^2 ultrasound treatment were given with four distinct salt concentrations. A 30% salt reduction could be obtained with the 0.75% NaCl treatment without sacrificing quality. Additionally, using ultrasound to reach a 2.25% level of NaCl at 72 W/cm^2 for two hours without affecting quality attributes was feasible in the cured ham [75].

7.3 Use of Pre-rigor Meat/Hot Processing/ Hot-Boning Technology

Following slaughter, skeletal muscles go through a number of metabolic processes that lower the pH of the muscles through anaerobic postmortem glycolysis. These processes are primarily brought on by the exhaustion of biological energy sources, such as ATP and creatine phosphate, and the disruption of the oxygen supply [76]. This leads to the development of rigor mortis, which lowers key processing properties like protein solubility and water-holding capacity [77]. In order to improve the quality attributes of the final products and minimize financial losses during storage and distribution, a hot-boning technique has been proposed that is utilized to obtain pre-rigor muscles with excellent processing characteristics, high pH, and relaxed microstructure [78]. However, because it needs special facilities that ensure a continuous process of slaughter, sectioning, deboning, and processing before rigor mortis begins, the production of pre-rigor meat is restricted in many countries. Furthermore, this process needs to be regulated due to the high temperature, microbial contamination, and faster rate of microbe growth involved in it than in conventional ones [79]. Pre-rigor salting is a crucial step when using hot-boning technology because it has a greater influence on the microbial growth of meat products than deboning time [80]. Desmond (2006) has described the idea of utilizing pre-rigor meat to enhance the qualities of low-salt meat products [24]. Prior research examining the effects of pre-rigor salting discovered that adding a minimum of 2% NaCl was required to ensure improvements in meat properties like protein solubility, emulsifying ability, water-holding capacity, and microbial growth regulation, which further aided in the inactivation of glycolytic enzymes [80]. In this context, the use of pre-rigor salted meat in low-salt meat products may be restricted by the addition of a minimum quantity of salt to inhibit postmortem anaerobic glycolysis. Choi et al. (2015) developed a meat emulsion with pre- and post-rigor salted meat at a lower salt concentration (1% NaCl) as a substitute tactic. Moreover, the amount of salt used in the pre-rigor salting procedure had a big impact on the finished meat product's quality [26]. The optimal amount of salt to add was found to be 3% when pre-rigor salt concentrations ranging from 1% to 5% were compared. Potassium chloride, or KCl, is a commonly substituted salt. Song et al. (2020) assessed the impact of KCl pre-rigor salting on meat's technical attributes and proposed that KCl was less harsh on meat than NaCl. When compared to post-rigor salting with NaCl, pre-rigor salting with KCl enhanced protein solubility, pH value, and water holding capacity [81].

7.4 Radiation

The food industry uses food irradiation technology to control microbiological growth and ripening. The ionizing sources (X-rays, gamma rays, and electron beams) used to irradiate food are effective in the shelf-life enhancement of meat products [82]. Yet, there has been evidence of an expedited decline in physicochemical characteristics, primarily due to oxidative stress. This includes biological molecule oxidation and discoloration, as well as the development of radiation off-odors and flavors in meat products exposed to radiation [83]. With its superior sterilizing effect, the food irradiation technique has been applied to a number of low-salt foods that could be susceptible to microbial growth-induced spoilage [84]. The gamma-ray treatment at a dose of 2 kGy has proven effective in significantly reducing the microbial counts in processed meat products. The aerobic mesophilic microbial counts decreased from an initial count of 7.6 log10 counts/g to a final count of 1.2 log10 counts/g. This reduction is particularly notable under 7 °C storage conditions for a duration of 1 week with a relatively low salt concentration of 2%. According to a recent study by Song et al. (2017), during four weeks of chilled storage, the ionizing sources at 5 kGy showed the absence of coliforms, aerobic microbes, Enterobacteriaceae and Pseudomonas spp. So, to ensure the microbiological safety of low-salt sausages, food irradiation techniques are practically feasible. Radiation could cause accelerated oxidation, which would hasten the deterioration of meat products, even though microbial growth could be effectively controlled. Furthermore, the unfavourable consumer reaction to foods contaminated with radiation and the substantial expenses associated with upkeep and operation remain crucial issues that must be taken into account [85].

7.5 Pulsed Electric Field Processing

Another non-thermal technology that can lower salt content and enhance food quality is PEF. According to the latest studies, PEF affects saltiness by promoting salt diffusion and the production of a salty flavor. PEF treatments, however, destroyed their salty flavor even though they were very promising in many ways when it came to reducing salt [86]. Specifically, the studies conducted indicated that PEF technology processes caused more salt to be absorbed because of the effects of mass transfer and interference with cellular tissues. PEF, in short, can shorten the process's duration and lower the concentration of salt because it improves uptake rates [87]. PEF treated beef meat showed a perception of released sodium during chewing and an improvement in salt diffusion. In fact, the salt content was removed simultaneously without compromising microbial stability, lipid oxidation, or sensory aspects. Furthermore, PEF markedly improved the diffusion of salt during the fresh pork curing process; otherwise, its application to meat safety is limited [65]. According to Toepfl et al. (2014), electric field strength of 1–10 kV/cm and an energy input of 0.5–10 kJ/kg could be used to achieve the proper level of meat tenderness [88]. Additionally, they revealed the pulsed electric field's mechanism of action, demonstrating that enhanced proteolysis was how the tenderization benefit of the pulsed electric field was attained. A study by Bhat et al. (2020) looked into the efficacy of pulsed electric field processing as a method for reducing the amount of salt in meat products. This study demonstrated that direct salt reduction was possible during the pulsed electric field processing of beef jerky without compromising the product's microbial stability, sensory appeal, or lipid oxidation. In addition to possibly releasing sodium, the pulsed electric field enhanced the perception of salt during chewing and its diffusion and distribution throughout the meat matrix [89].

8 Challenges and Future Perspectives

While it's true that meat and its products can provide essential nutrients, including high-quality protein, vitamins, and minerals, it's crucial to consider both the benefits and potential drawbacks of regular meat consumption. Considering this, it is crucial to lower sodium levels while maintaining the food's microbiological safety to make it healthier. To address existing health concerns, several strategies have been established to lower the consumption of sodium from these products. Currently, sodium decrement is a controversial area of international interest. The World Health Organization (WHO) indeed recognizes the impact of dietary factors, including excessive salt intake, on global health. A comprehensive initiative aimed at addressing the rising trends of non-communicable diseases (NCDs) worldwide by the global strategy on diet, physical activity, and health [90]. The reduction of sodium is still difficult, particularly in terms of customer acceptance. A reduction in quality parameters (i.e., flavour and texture) was observed when salt was replaced. On the other hand, substituting artificial substances for salt may have unfavourable health consequences. Food redevelopment, consumer education, and training are the primary strategies for salt reduction [91]. Governmental bodies and organizations must keep informing the public about the link between eating salt and health issues. Consumer preferences should also be taken into account. For example, the food sector has been investigating natural alternatives to synthetic preservatives due to concern about the safety of synthetic ones [92]. In order to find useful ingredients from natural derivatives that are acceptable in terms of quality and acceptability as salt substitutes for use in meat products, the food industry must cooperate with the pharmaceutical industry. To support these health-promoting qualities, a high standard of scientific evidence must also be available, which requires the cooperation of academic institutions and research organizations. Nonetheless, many meat products today have a comparatively healthy aspect and demonstrate exceptional potency in delivering nutritious ingredients. By taking those actions, meat manufacturers will be able to access more functional food markets [93].

References

1. Felisberto MH, Galvão MT, Picone CS, Cunha RL, Pollonio MA. Effect of prebiotic ingredients on the rheological properties and microstructure of reduced-sodium and low-fat meat emulsions. LWT-Food Sci Tech. 2015;60(1):148–55.
2. WHO. Reducing salt intake in populations. Paris: Paper Presented at the Report of a WHO Forum and Technical Meeting; 2006.
3. WHO. Guideline: sodium intake for adults and children. World Health Organization; 2012.
4. Aaslyng MD, Vestergaard C, Koch AG. The effect of salt reduction on sensory quality and microbial growth in hotdog sausages, bacon, ham and salami. Meat Sci. 2014;96(1):47–55.
5. Inguglia ES, Zhang Z, Tiwari BK, Kerry JP, Burgess CM. Salt reduction strategies in processed meat products—a review. Trends Food Sci Tech. 2017;59:70–8.
6. Kloss L, Meyer JD, Graeve L, Vetter W. Sodium intake and its reduction by food reformulation in the European Union—a review. NFS J. 2015;1(1):9–19.
7. Verma AK, Banerjee R. Low-sodium meat products: retaining salty taste for sweet health. Crit Rev Food Sci Nutr. 2012;52(1):72–84.
8. Bombrun L, Gatellier P, Carlier M, Kondjoyan A. The effects of low salt concentrations on the mechanism of adhesion between two pieces of pork semimembranosus muscle following tumbling and cooking. Meat Sci. 2014;96(1):5–13.
9. Bamforth CW, Ward RE, editors. The Oxford handbook of food fermentations. Oxford University Press; 2014.
10. Perez-Palacios T, Salas A, Munoz A, Ocana ER, Antequera T. Sodium chloride determination in meat products: comparison of the official titration-based method with atomic absorption spectrometry. J Food Compo Anal. 2022;108:104425.
11. Domínguez R, Pateiro M, Pérez-Santaescolástica C, Munekata PE, Lorenzo JM. In: Lorenzo JM, Carballo FJ, editors. Salt reduction strategies in meat products made from whole pieces. Strategies for Obtaining Healthier Foods; 2017. p. 267–89.
12. Pateiro M, Munekata PE, Cittadini A, Domínguez R, Lorenzo JM. Metallic-based salt substitutes to reduce sodium content in meat products. Curr Opinion Food Sci. 2021;38:21–31.
13. Bhat ZF, Morton JD, Mason SL, Bekhit AE. Current and future prospects for the use of pulsed electric field in the meat industry. Cri Rev Food Sci Nutr. 2019;59(10):1660–74.
14. González N, Marquès M, Nadal M, Domingo JL. Meat consumption: Which are the current global risks? A review of recent (2010–2020) evidences. Food Res Int. 2020;137:109341.
15. Basu S. The transitional dynamics of caloric ecosystems: changes in the food supply around the world. Crit Public Health. 2015;25(3):248–64.
16. Katare B, Wang HH, Lawing J, Hao N, Park T, Wetzstein M. Toward optimal meat consumption. Am J Agric Econ. 2020;102(2):662–80.
17. Jia J, Dawson TP, Wu F, Han Q, Cui X. Global meat demand projection: Quo Vadimus? J Cleaner Production. 2023;429:139460.
18. Strazzullo P, Leclercq C. Sodium. Adv Nutr. 2014;5(2):188–90.

19. Kaushik S, Kumar R, Kain P. Salt an essential nutrient: advances in understanding salt taste detection using drosophila as a model system. J Experiment Neurosci. 2018;12:1179069518806894.
20. Powles J, Fahimi S, Micha R, Khatibzadeh S, Shi P, Ezzati M, Engell RE, Lim SS, Danaei G, Mozaffarian D. Global Burden of Diseases Nutrition and Chronic Diseases Expert Group (NutriCoDE). Global, regional and national sodium intakes in 1990 and 2010: a systematic analysis of 24 h urinary sodium excretion and dietary surveys worldwide. BMJ Open. 2013;3(12):e003733.
21. Mente A, O'Donnell M, Yusuf S. Sodium intake and health: what should we recommend based on the current evidence? Nutrients. 2021;13(9):3232.
22. Zhou BF, Stamler J, Dennis B, Moag-Stahlberg A, Okuda N, Robertson C, Zhao L, Chan Q, Elliott P. Nutrient intakes of middle-aged men and women in China, Japan, United Kingdom, and United States in the late 1990s: the INTERMAP study. J Human Hypertension. 2003;17(9):623–30.
23. Yim DG, Shin DJ, Jo C, Nam KC. Effect of sodium-alternative curing salts on physicochemical properties during salami manufacture. Food Sci Ani Resources. 2020;40(6):946–56.
24. Desmond E. Reducing salt: a challenge for the meat industry. Meat Sci. 2006;74(1):188–96.
25. Kim TK, Yong HI, Jung S, Kim HW, Choi YS. Technologies for the production of meat products with a low sodium chloride content and improved quality characteristics—a Review. Food Secur. 2021;10(5):957.
26. Choi YS, Kim HW, Hwang KE, Song DH, Jeong TJ, Jeon KH, Kim YB, Kim CJ. Combined effects of presalted prerigor and postrigor batter mixtures on chicken breast gelation. Poultry Sci. 2015;94(4):758–65.
27. O'Flynn CC, Cruz-Romero MC, Troy D, Mullen AM, Kerry JP. The application of high-pressure treatment in the reduction of salt levels in reduced-phosphate breakfast sausages. Meat Sci. 2014;96(3):1266–74.
28. Tamm A, Bolumar T, Bajovic B, Toepfl S. Salt (NaCl) reduction in cooked ham by a combined approach of high pressure treatment and the salt replacer KCl. Innovative Food Sci Emerging Tech. 2016;36:294–302.
29. Phillips F. Don't pass the salt. Nutr Bulletin. 2003;28(4):339–40.
30. McCance RA, Widdowson EM. McCance and Widdowson's the composition of foods. Royal Society of Chemistry; 2014.
31. Breidenstein BC. Understanding and calculating the sodium content of your products. Meat Processing. 1982;21(5):62.
32. Sych J, Kaelin I, Gerlach F, Wróbel A, Le T, FitzGerald R, Pestoni G, Faeh D, Krieger JP, Rohrmann S. Intake of processed meat and association with sociodemographic and lifestyle factors in a representative sample of the Swiss population. Nutrients. 2019;11(11):2556.
33. Matthews K, Strong M. Salt–its role in meat products and the industry's action plan to reduce it, vol. 30; 2005. p. 55–61.
34. Aprilia GH, Kim HS. Development of strategies to manufacture low-salt meat products–a review. J Animal Sci Tech. 2022;64(2):218–34.
35. Delwiche J. The impact of perceptual interactions on perceived flavor. Food Qual Prefer. 2004;15(2):137–46.
36. McGough MM, Sato T, Rankin SA, Sindelar JJ. Reducing sodium levels in frankfurters using a natural flavor enhancer. Meat Sci. 2012;91(2):185–94.
37. Wallis K, Chapman S. Review of current salt replacing ingredients. Campden BRI; 2012.
38. Flores M, Corral S, Cano-García L, Salvador A, Belloch C. Yeast strains as potential aroma enhancers in dry fermented sausages. Int J Food Microbio. 2015;212:16–24.
39. Campagnol PC, dos Santos BA, Wagner R, Terra NN, Pollonio MA. The effect of yeast extract addition on quality of fermented sausages at low NaCl content. Meat Sci. 2011;87(3):290–8.
40. Gelabert J, Gou P, Guerrero L, Arnau J. Effect of sodium chloride replacement on some characteristics of fermented sausages. Meat Sci. 2003;65(2):833–9.
41. Lorenzo JM, Bermúdez R, Domínguez R, Guiotto A, Franco D, Purriños L. Physicochemical and microbial changes during the manufacturing process of dry-cured lacón salted with

potassium, calcium and magnesium chloride as a partial replacement for sodium chloride. Food Contr. 2015;50:763–9.
42. dos Santos BA, Campagnol PC, Morgano MA, Pollonio MA. Monosodium glutamate, disodium inosinate, disodium guanylate, lysine and taurine improve the sensory quality of fermented cooked sausages with 50% and 75% replacement of NaCl with KCl. Meat Sci. 2014;96(1):509–13.
43. Aliño M, Grau R, Toldrá F, Blesa E, Pagán MJ, Barat JM. Influence of sodium replacement on physicochemical properties of dry-cured loin. Meat Sci. 2009;83(3):423–30.
44. Ketenoğlu O, Candoğan K. Effect of low-sodium salt utilization on some characteristics of ground beef patties. Blood Press. 2011;4:5.
45. Katz A, Rosenthal T, Maoz C, Peleg E, Zeidenstein R, Levi Y. Effect of a mineral salt diet on 24-h blood pressure monitoring in elderly hypertensive patients. J Hum Hypertens. 1999;13(11):777–80.
46. Rama R, Chiu N, Carvalho Da Silva M, Hewson L, Hort J, Fisk ID. Impact of salt crystal size on in-mouth delivery of sodium and saltiness perception from snack foods. J Texture Studies. 2013;44(5):338–45.
47. Sun C, Zhou X, Hu Z, Lu W, Zhao Y, Fang Y. Food and salt structure design for salt reducing. Innov Food Sci EmergTech. 2021;67:102570.
48. Đorđević Đ, Buchtová H, Macharáčková B. Salt microspheres and potassium chloride usage for sodium reduction: case study with sushi. Food Sci Tech Int. 2018;24(1):3–14.
49. Mueller E, Koehler P, Scherf KA. Applicability of salt reduction strategies in pizza crust. Food Chem. 2016;192:1116–23.
50. Horita CN, Morgano MA, Celeghini RM, Pollonio MA. Physico-chemical and sensory properties of reduced-fat mortadella prepared with blends of calcium, magnesium and potassium chloride as partial substitutes for sodium chloride. Meat Sci. 2011;89(4):426–33.
51. Doyle ME, Glass KA. Sodium reduction and its effect on food safety, food quality, and human health. Comprehensive Rev Food Sci Food Safety. 2010;9(1):44–56.
52. Sezer B, Velioglu HM, Bilge G, Berkkan A, Ozdinc N, Tamer U, Boyaci IH. Detection and quantification of a toxic salt substitute (LiCl) by using laser induced breakdown spectroscopy (LIBS). Meat Sci. 2018;135:123–8.
53. Barat JM, Pérez-Esteve E, Aristoy MC, Toldrá F. Partial replacement of sodium in meat and fish products by using magnesium salts. a review. Plant Soil. 2013;368:179–88.
54. Paulsen MT, Nys A, Kvarberg R, Hersleth M. Effects of NaCl substitution on the sensory properties of sausages: temporal aspects. Meat Sci. 2014;98:164–70. https://doi.org/10.1016/j.meatsci.2014.05.020.
55. Devlieghere F, Vermeiren L, Bontenbal E, Lamers PP, Debevere J. Reducing salt intake from meat products by combined use of lactate and diacetate salts without affecting microbial stability. Int J Food Science Tech. 2009;44(2):337–41.
56. SeoNg PN, Seo HW, Cho SH, Kim YS, Kang SM, Kim JH, Kang GH, Park BY, Moon SS, Hoa VB. Potential use of glasswort powder as a salt replacer for production of healthier dry-cured ham products. Czech. J Food Sci. 2017;35(2):149–59.
57. Sïlybaeva BM, Mussabayeva BK, Zharykbasova KS, Kidyrmoldina AS, Kaygusuz O. Biologically active agents of Salicornia europaeal. Grown in East Kazakhstan. Research J Pharma Biolog Chem Sci; 2016.
58. Lima AR, Castaneda-Loaiza V, Salazar M, Nunes C, Quintas C, Gama F, Pestana M, Correia PJ, Santos T, Varela J, Barreira L. Influence of cultivation salinity in the nutritional composition, antioxidant capacity and microbial quality of Salicornia ramosissima commercially produced in soilless systems. Food Chem. 2020;333:127525.
59. Endo T, Kubo-Nakano Y, Lopez RA, Serrano RR, Larrinaga JA, Yamamoto S, Honna T. Growth characteristics of kochia (K ochia scoparia L.) and alfalfa (M edicago sativa L.) in saline environments. Grassl Sci. 2014;60(4):225–32.
60. Jo K, Lee J, Jung S. Quality characteristics of low-salt chicken sausage supplemented with a winter mushroom powder. K J Food Sci Animal Res. 2018;38(4):768.

61. Villela PT, de Oliveira EB, Villela PT, Bonardi JM, Bertani RF, Moriguti JC, Ferriolli E, Lima NK. Salt preferences of normotensive and hypertensive older individuals. J Clinical Hypertension. 2014;16(8):587–90.
62. Santhosha SG, Jamuna P, Prabhavathi SN. Bioactive components of garlic and their physiological role in health maintenance: a review. Food Biosci. 2013;3:59–74.
63. Horita CN, Farías-Campomanes AM, Barbosa TS, Esmerino EA, da Cruz AG, Bolini HM, Meireles MA, Pollonio MA. The antimicrobial, antioxidant and sensory properties of garlic and its derivatives in Brazilian low-sodium frankfurters along shelf-life. Food Res Int. 2016;84:1–8.
64. Mancini S, Mattioli S, Nuvoloni R, Pedonese F, Dal Bosco A, Paci G. Effects of garlic powder and salt on meat quality and microbial loads of rabbit burgers. Food Secur. 2020;9(8):1022.
65. Bhat S, Kaushal P, Kaur M, Sharma HK. Coriander (Coriandrum sativum L.): Processing, nutritional and functional aspects. Afr J Plant Sci. 2014 Jan;8(1):25–33.
66. Hossain M, Adhikary RK, Mahbub KR, Begum M, Ul Islam MR. Effect of 10% concentrations of salt, garlic and coriander on the quality of smoked hilsa fish (Tenualosa ilisha). A J Food Tech. 2012;7:501–5.
67. Taladrid D, Laguna L, Bartolomé B, Moreno-Arribas MV. Plant-derived seasonings as sodium salt replacers in food. Trends Food Sci Tech. 2020;99:194–202.
68. Mitchell M, Brutnon NP, Fitzgerald RJ, Wilkinson MG. The use of herbs, spices, and whey proteins as natural flavor enhancers and their effect on the sensory acceptability of reduced-salt chilled ready-meals. J Culinary Sci Tech. 2013;11(3):222–40.
69. Rodrigues FM, Rosenthal A, Tiburski JH, Cruz AG. Alternatives to reduce sodium in processed foods and the potential of high pressure technology. Food Sci Tech. 2015;36:01–8.
70. Pietrasik Z, Duda Z. Effect of fat content and soy protein/carrageenan mix on the quality characteristics of comminuted, scalded sausages. Meat Sci. 2000;56(2):181–8.
71. Yang H, Han M, Wang X, Han Y, Wu J, Xu X, Zhou G. Effect of high pressure on cooking losses and functional properties of reduced-fat and reduced-salt pork sausage emulsions. Innovative Food Sci Emerging Tech. 2015;29:125–33.
72. Morris C, Brody AL, Wicker L. Non-thermal food processing/preservation technologies: A review with packaging implications. Packaging Technology and Science: An Int J. 2007;20(4):275–86.
73. Alarcon-Rojo AD, Janacua H, Rodriguez JC, Paniwnyk L, Mason TJ. Power ultrasound in meat processing. Meat Sci. 2015;1(107):86–93.
74. McDonnell CK, Lyng JG, Arimi J, Allen P. The accelerated curing of pork using power ultrasound: a pilot-scale production trial. Int Congress Meat Sci Technol. 2013;26:191–8.
75. McDonnell CK, Allen P, Morin C, Lyng JG. The effect of ultrasonic salting on protein and water–protein interactions in meat. Food Chem. 2014;147:245–51.
76. Torres E, Pearson AM, Gray JI, Booren AM, Shimokomaki M. Effect of salt on oxidative changes in pre-and post-rigor ground beef. Meat Sci. 1988;23(3):151–63.
77. Kim HW, Hwang KE, Song DH, Kim YJ, Ham YK, Yeo EJ, Jeong TJ, Choi YS, Kim CJ. Effect of pre-rigor salting levels on physicochemical and textural properties of chicken breast muscles. Korean J Food Sci Ani Res. 2015;35(5):577–84.
78. Pisula A, Tyburcy A. Hot processing of meat. Meat Sci. 1996;43:125–34.
79. Seideman SC, Cross HR. The economics and palatability attributes of hot-boned beef: a review. J Food Quality. 1982;5(3):183–201.
80. Sukumaran AT, Holtcamp AJ, Englishbey AK, Campbell YL, Kim T, Schilling MW, Dinh TT. Effect of deboning time on the growth of Salmonella, E. coli, aerobic, and lactic acid bacteria during beef sausage processing and storage. Meat Sci. 2018;139:49–55.
81. Song DH, Ham YK, Ha JH, Kim YR, Chin KB, Kim HW. Impacts of pre-rigor salting with KCl on technological properties of ground chicken breast. Poult Sci. 2020;99(1):597–603.
82. O'bryan CA, Crandall PG, Ricke SC, Olson DG. Impact of irradiation on the safety and quality of poultry and meat products: a review. Cri Rev Food Sci Nutr. 2008;48(5):442–57.

83. Ham YK, Kim HW, Hwang KE, Song DH, Kim YJ, Choi YS, Song BS, Park JH, Kim CJ. Effects of irradiation source and dose level on quality characteristics of processed meat products. Phys Chem. 2017;130:259–64.
84. Song BS, Lee Y, Park JH, Kim JK, Park HY, Kim DH, Kim CJ, Kang IJ. Toxicological and radiological safety of chicken meat irradiated with 7.5 MeV X-rays. Rad Phy Chem. 2018;144:211–7.
85. Song DH, Kim HW, Hwang KE, Kim YJ, Ham YK, Choi YS, Shin DJ, Kim TK, Lee JH, Kim CJ, Paik HD. Impacts of irradiation sources on quality attributes of low-salt sausage during refrigerated storage. Korean J Food Sci Anim Resour. 2017;37:698–707.
86. Pinton MB, dos Santos BA, Lorenzo JM, Cichoski AJ, Boeira CP, Campagnol PC. Green technologies as a strategy to reduce NaCl and phosphate in meat products: An overview. Curr Opinion Food Sci. 2021;40:1–5.
87. Galindo FG, Dejmek P, Lundgren K, Rasmusson AG, Vicente A, Moritz T. Metabolomic evaluation of pulsed electric field-induced stress on potato tissue. Planta. 2009;230:469–79.
88. Toepfl S, Siemer C, Heinz V. Effect of high-intensity electric field pulses on solid foods. In Emerging technologies for food processing. Academic Press. 2014:147–54.
89. Bhat ZF, Morton JD, Mason SL, Bekhit AE. The application of pulsed electric field as a sodium reducing strategy for meat products. Food Chem. 2020 Feb;15(306):125622.
90. Annegowda HV, Bhat R, Yeong KJ, Liong MT, Karim AA, Mansor SM. Influence of drying treatments on polyphenolic contents and antioxidant properties of raw and ripe papaya (Carica papaya L.). Int J Food Prop. 2014;17(2):283–92.
91. Trieu K, Neal B, Hawkes C, Dunford E, Campbell N, Rodriguez-Fernandez R, Legetic B, McLaren L, Barberio A, Webster J. Salt reduction initiatives around the world—a systematic review of progress towards the global target. PLoS One. 2015;10(7):e0130247.
92. Nikmaram N, Budaraju S, Barba FJ, Lorenzo JM, Cox RB, Mallikarjunan K, Roohinejad S. Application of plant extracts to improve the shelf-life, nutritional and health-related properties of ready-to-eat meat products. Meat Sci. 2018;145:245–55.
93. Grasso S, Brunton NP, Lyng JG, Lalor F, Monahan FJ. Healthy processed meat products—regulatory, reformulation and consumer challenges. Trends Food Sci Tech. 2014;39(1):4–17.

Meat Products with Fat Replacers

Abu Bakar Asyrul-Izhar, Jamilah Bakar, Awis Qurni Sazili, Yong Meng Goh, and Mohammad Rashedi Ismail-Fitry

1 Introduction

Meat and meat products are crucial components of the human diet due to their diverse range of high-quality substances, including essential nutrients, that have significant biological value [1]. Consumers are drawn to these products due to their abundant content of protein, fat, carbon polymers, vitamins, minerals, and flavour. Additionally, they contain a substantial quantity of solid fats that contribute to the desired functionality, texture, flavour, and mouthfeel [2]. Nevertheless, their nutritional quality is called into doubt as they also include significant amounts of fat, saturated fatty acids (SFA) and cholesterol which are detrimental to human health [3]. Excessive and long-term consumption of these products contributes to illnesses such as heart disease, obesity and type 2 diabetes [4]. As these issues have tremendously become a worldwide issue, the World Health Organization (WHO) suggests

A. B. Asyrul-Izhar · J. Bakar
Department of Food Technology, Faculty of Food Science and Technology, Universiti Putra Malaysia, Serdang, Selangor, Malaysia

A. Q. Sazili
Department of Animal Science, Faculty of Agriculture, Universiti Putra Malaysia, Serdang, Selangor, Malaysia

Halal Products Research Institute, Universiti Putra Malaysia, Serdang, Selangor, Malaysia

Y. M. Goh
Department of Veterinary Preclinical Sciences, Faculty of Veterinary Medicine, Universiti Putra Malaysia, Serdang, Selangor, Malaysia

M. R. Ismail-Fitry (✉)
Department of Food Technology, Faculty of Food Science and Technology, Universiti Putra Malaysia, Serdang, Selangor, Malaysia

Halal Products Research Institute, Universiti Putra Malaysia, Serdang, Selangor, Malaysia
e-mail: ismailfitry@upm.edu.my

A. Kumar Pathera et al. (eds.), *Healthier Meat Products*,
https://doi.org/10.1007/978-3-031-78215-2_17

restricting total fat intake to less than 30% of total energy consumption and saturated fat (which consists of less than 10% of the total energy consumed) [5]. In conjunction with that, it is crucial to decrease the daily consumption of products that contain animal fat besides developing strategies to reduce fat to develop healthier meat products. The concept of "Let food be thy medicine and medicine be thy food" by Hippocrates is now being applied in the development of healthier and functional food products [6].

Animal fats which contain significant amounts of saturated fats function as a fat source in the production of meat products. The utilisation of these saturated fats in various roles within the products they are incorporated into poses a big challenge for the processed meat industry [7]. Saturated fats can enhance the mouthfeel, texture, juiciness, and sensory acceptability of foods like frankfurters [8]. Fat plays a crucial role in meat products by stabilising emulsions, minimising cooking loss, enhancing water retention, and contributing to juiciness and texture [9]. Nevertheless, consumers nowadays are becoming more aware of the consequences of consuming excessive amounts of meat products with animal fats. As a result, the pursuit of enhancing the healthiness and nutritional value of meat products has become a prominent objective and needs to be addressed [10]. On top of that, many governments and international institutions have proposed guidelines for replacing saturated fatty acids with unsaturated fatty acids to optimise total fat intake and decrease the risk of malnutrition and various diseases [3]. To meet the nutritional requirements set by nutritional standards established by international organizations, the meat sector needs to develop strategies for reformulating traditional meat products to enhance their health benefits [11].

A recent strategy to overcome this problem is the reformulation of meat products. Decreasing the meat product's fat content is one of the best ways to create a nutritious meat product, and adding fat replacers is one of the most important ways to replace animal fat in meat products [12]. This fat replacement strategy in meat products could be challenging as it alters the mouthfeel, flavour, texture, viscosity, and other sensory and processing aspects of the meat products [13]. Consequently, potential fat replacers ought to function effectively in terms of their nutritional content and their ability to provide desired structural and sensory qualities [14]. In general, fat replacers could be categorized into several classifications which include carbohydrate-, protein- or lipid-based fat replacers. Each of these classifications of fat replacers possesses unique qualities that distinguish them from one another.

Hence, this context mainly reviews the classifications and applications of fat replacers in meat products. To classify every fat replacement component, it is necessary to compare them according to technical quality factors. The chapter aims to highlight the different classifications and applications of various types of fat replacers that could be incorporated in meat products as well as their suitability for utilisation in meat products. As methods to enhance the promotion of a meat product that is both nutritious and healthy while also possessing the desired quality characteristics, it is anticipated that the results will provide important data for relevant industries and researchers.

2 Fat Replacer Definition and Classifications

Fat replacer is known as an ingredient that is used to replace animal fat as a source of fat in meat products. According to the American Dietetic Association, fat replacers are ingredients that can be utilized to perform the identical tasks as fat but with fewer calories [15]. Understanding the impact of various fat replacers on the sensory and physical attributes of meat products is crucial for professionals in the field of product development and food technology as this knowledge serves as a valuable guide in the creation of healthier alternative goods. It is also crucial to highlight that the word "fat replacer" refers to a substance that has some of the good physical or sensory qualities of fats but none of the bad ones [16]. There were multiple classifications of fat replacers, which frequently led to ambiguity in terms of their categorization. There are two categories of fat replacers, which are fat substitutes and fat mimetics. Fat substitutes are lipid-like substances developed to serve as alternatives to fats in a one-to-one ratio. On the other hand, fat mimetics are food components composed of proteins or carbohydrates that replicate the sensory attributes, such as texture, mouthfeel, and overall sensory experience, of actual fats.

2.1 Fat Substitutes

Fat substitutes are substances that imitate the functional and sensory characteristics of fat in food. Generally, these alternatives can provide approximately 9 kcal/g, potentially resulting in zero calorie intake if none is absorbed [16]. Consequently, they can be employed to substitute a portion or the entirety of the fat content typically found in a particular food item [17]. Fat substitutes have been found to have a dual effect, as they not only contribute to a reduction in dietary fat consumption and the prevalence of obesity [18], but they also offer comparable sensory characteristics to conventional food products. Recently, there has been an upward trend in interest in the development of lipid materials as a means to decrease or entirely replace fat content to produce meat products that promote health [19]. Due to the significant impact of the fatty acid profile and distribution on human health, incorporating fat substitutes into meat products is advantageous for consumers as it lowers fat content and enhances the distribution of fatty acids [20]. The utilisation of fat substitutes is increasingly widespread as a means to partially substitute the fat content in diverse animal products, with a special emphasis on plant-based alternatives [21].

2.1.1 Lipid-Based

Fat substitutes are macromolecules that resemble triglycerides in their physical and chemical properties, which are commonly found in fats and oils. These substitutes possess the capacity to replace fat in food products in an equivalent amount, measured in g. Fat substitutes, commonly known as lipid- or fat-based fat replacers, consist of two distinct categories: those that are chemically synthesised and those that are produced from ordinary fats and oils through enzymatic modification. Numerous fat substitutes exhibit stability when subjected to cooking and frying temperatures [22, 23]. In placing more emphasis, on their lipid composition, these substances exhibit stability at high boiling and frying temperatures and offer the same functionalities as fats. Moreover, they possess a caloric yield of less than 9 kcal/g, potentially resulting in zero caloric intake if no absorption occurs [16]. Replacing animal fats with lipid-based fat replacers leads to changes in the sizes of oil globules, which in turn affects the colour of the meat products [24]. Nevertheless, certain attributes of meat products remain unchanged. Incorporating lipid-based fat replacers like vegetable oils into meat products, not only compensates for the alterations caused by the lack of animal fat but also provides advantageous health effects to the consumer [25]. Gel formation is a novel approach to converting plant-based oils that are high in unsaturated fats into solid-like systems. These systems are promoted as healthier alternatives to saturated and trans-unsaturated fatty acids while still meeting consumers' expectations in terms of functionality and sensory experience [10]. This includes the process of creating oleogels and emulsion gels using liquid oils as the raw ingredients. In order to replicate the juiciness of meat products obtained from the change in the state of animal fats, an oleogel was utilised. This involved structuring vegetable oil with a network composed of whey protein isolates (WPI) and water [26]. Emulsion gels were also employed in the production of meat analogues to replicate the composition and technological properties of chicken products [27].

2.2 *Fat Mimetics*

Fat mimetics are considered as an ingredient that has different chemical structures from fat. Generally, these substances belong within other macronutrient groups that consist of carbohydrates and/or protein. They act as replacements for fats by providing comparable volume, texture and sensory sensations in the mouth [28]. They are also an ingredient which imitates the sensory and physical properties of dietary fat [16]. Fat mimetics can also be defined as substances that are used to partially substitute fat in food products by duplicating or mimicking some effects of fat. However, not all functions are replicated [29].

2.2.1 Carbohydrate-Based

Carbohydrate-based fat replacers tend to create a gel-like structure when mixed with water, resulting in lubricating and flow properties that are similar to those of fat [30]. Carbohydrate-based fat replacers imitate the properties of fat by stabilising the additional water inside a gel-like structure, enabling the release of water in a manner akin to the release of fat [31]. Carbohydrate-based fat replacers are produced from many sources such as cereals, grains, and plants. These ingredients encompass a range of complex carbohydrates, including both digestible and indigestible forms. The ingredients consist of starch and fat mimetics derived from fibre. The first category includes maltodextrins and modified starches, while the second category includes microcrystalline cellulose, methylcellulose, gums, pectins, hydrocolloid gums, and polydextrose [16]. The water-binding characteristics of carbohydrate-fat replacers have the ability to restrict the movement of water within food products, thereby replicating the sensory experience of the lipid-rich mouthfeel typically associated with animal fat [32]. A study by Aziz et al. [33] investigated carbohydrate-fat replacers and their water manipulation properties. The researchers discovered that these qualities are accomplished by the usage of a dynamic hydrocolloid randomly arranged coil polymer. This characteristic is particularly beneficial for low-fat diets since they typically have a higher water content compared to high-fat alternatives. Table 1 displays various carbohydrate-based fat replacers and their corresponding food applications.

2.2.2 Protein-Based

To enhance specific functional properties, decrease lipid content, increase yields, and reduce formulation costs, several plant- and animal-based proteins have been incorporated into processed meat products. Proteins derived from vegetables and animals are incorporated into a variety of meat products to enhance the emulsion stability of the substances by binding fat and water. For instance, proteins can undergo enzymatic denaturation, heat, or pH alteration, which modifies their three-dimensional structure and causes them to function more like fat [34]. This research examined the impact of fat level on the qualities of bologna sausage, specifically focusing on the effects of plasma protein and soy fibre concentrations. The presence of plasma protein and soy fibre in the composition facilitated the development of firmer and more resilient structures, exhibiting enhanced abilities to bind fat and water. In meat products, the reduction of animal fat resulted in a decrease in textural qualities. Thus, the binding and textural qualities were found to be more significantly influenced by the incorporation of protein-based fat replacers and as a result, it was suggested that limiting the impact of fat reduction would be best achieved by focusing on the usage of protein-based fat replacers. Research has shown that using protein-based fat replacers may lead to an increase in water retention [35]. The observed phenomenon is likely caused by the relationship that exists between the polar structures of matrix proteins and the formation of hydrogen chains with water

Table 1 Examples of carbohydrate-based, protein-based and lipid-based fat replacers

Types of fat replacer	Food application	General functions of different fat replacers	References
Guar-xanthan gum mixture	Meat emulsion	Enhanced emulsion stability and increased cooking yield	[37]
Prebiotic fibres (inulin, fructooligosaccharide, polydextrose, and resistant starch)	Meat emulsion	The presence of fibre in the experiment caused an interruption in the association of globular myosin heads, resulting in the formation of a gel at higher temperatures	[38]
Whey protein	Meatballs	Affected fat retention values	[39]
Soy protein concentrate	Frankfurter type sausage	The inclusion of soy protein concentrate resulted in an enhancement of the water retention ability and a reduction in cooking losses	[13]
Extra virgin olive oil	Salami	The products were minimally affected in terms of their chemical, physical, and sensory attributes, except for water activity and firmness	[40]
Pre-emulsified olive oil	Sucuk (fermented sausage)	Reduced the cholesterol level by approximately 41.3%	[41]

molecules that are not bound to other substances [36]. Table 1 displays various protein-based fat replacers and their corresponding food applications.

3 Strategies for Fat Reduction

Multiple research has investigated the utilization of different food ingredients to produce low-fat meat products. For instance, proteins have been investigated as potential sources of fat replacer [42]. Similarly, starches have been examined for their potential application in the development of low-fat goods [43]. Additionally, dietary fibres have been explored as alternative ingredients for the production of low-fat food items [44]. Using vegetable oil, such as sunflower, soy, flax, olive, rapeseed, or marine oil, as a substitute for fat is also a brilliant technique for creating more nutritious meat products [24]. The reformulation process strategies intend to replace animal fat with vegetable oil to reduce the fat content and improve the fatty acid composition. This means increasing the concentrations of monounsaturated and polyunsaturated fatty acids while decreasing saturated fatty acids (SFA) and cholesterol. Furthermore, the objective is to attain healthier ratios of n-6/n-3 PUFA/SFA [45].

An additional approach that may be adopted involves the application of fat replacers in the form of powders [35]. This has the potential to decrease the amount of animal fat, leading to a subsequent decrease in cholesterol, saturated fatty acid levels, and energy value. Additionally, the incorporation of powdered fat replacers can enhance the nutritional composition and physical characteristics of the meat products without negatively impacting the attributes of the quality of the products [46]. Carrot is an example of an ingredient that has been turned into powder to be incorporated as a fat replacement in meat products [47]. A potential approach for reducing fat content involves utilizing a fat replacer in the form of a paste [35]. This method incorporates plant-derived substances in intricate plant structures that consist of carbohydrates, proteins, lipids, or a combination of these components. Consequently, when employed as a replacement for animal fat, they can fulfil comparable functions while also influencing the characteristics of the final product, such as texture and sensory perception [48]. Nevertheless, fat plays an important function in a meat product by contributing to sensory perception, technological aspects, textural features, oxidative stability, and the generation of fragrance compounds [49]. Modifications in the quantity and/or makeup of fat typically have a detrimental impact on the fundamental characteristics of meat products, leading to decreased marketability and consumer approval [50]. Hence, it is crucial for fat replacers to mimic the structural composition and other qualitative characteristics of natural fat, while simultaneously providing more nutritional benefits than animal fat [51].

Recently, numerous studies have proposed novel methods to organise liquid oils, enabling the creation of reformulated products that closely resemble those manufactured with saturated fat [52, 53]. Hence, the utilisation of emulsion gels and oleogels exhibits significant potential for implementation in food products, particularly in the advancement of more nutritious meat products. Emulsion gels (EGs) are a class of soft-solid material that can be produced by aggregating scattered droplets within a particulate gel or by dispersing emulsified droplets within a continuous gel matrix [54]. Oleogel is created through the process of oleogelation, which is also known as organogelation. This method involves trapping liquid oil within a three-dimensional network generated by one or more oleogelator compounds, which are present in small amounts [55]. Glyceryl monostearate-based oleogels [56], rapeseed oil oleogels based on beeswax [57], chia and oat emulsion gels [58] and emulsion gels formulated with chestnut flour and chia oil [59] are examples of ingredients that were processed to form emulsion gel or oleogel to be incorporated as fat replacers in meat products. Table 2 indicates examples of strategies for fat replacement/reduction.

4 Applications in Selected Meat Products

The consumption of emulsified meat products is common primarily due to their convenient preparation, favourable sensory appeal, and their provision of proteins, essential amino acids, as well as minerals like iron, zinc, and vitamin B [68].

Table 2 Examples of fat reduction strategies for meat products

Fat reduction strategies	Ingredients	Types of meat products	Quality attributes	References
Vegetable oil	Linseed oil	Chicken patties	Better appearance and flavour of meat products	[25]
Vegetable oil	Sunflower oil	Chicken sausage	There were no noticeable differences observed in terms of cooking yield and water holding capacity	[60]
Powder	Chicory roots	Beef burger	The findings indicated that the inclusion of chicory root powder enhanced both the nutritional composition and culinary characteristics of the burger	[61]
Powder	Apple pomace	Goshtaba	The protein, moisture, and ash contents are significantly higher, while the fat content is lower	[62]
Paste	Poppy seed	Meat burgers	As the amount of ground poppy seed in the meat burgers increased, the addition of poppy seed significantly reduced the levels of saturated fatty acids	[63]
Paste	Walnut	Sucuk	The replacement of beef fat in sucuk with walnut did not have any impact on the sensory characteristics of the meat products	[64]
Oleogels	Peanut oil and ethylcellulose	Harbin red sausage	The texture, lipid oxidation, and sensory characteristics of reformulated sausages were maintained	[65]
Oleogels	Rapeseed oil and beeswax	Beef heart patties	Saturated fatty acids decreased from 54.08 to 8.34 and polyunsaturated fatty acids increased from 8.88 to 34.37 g/100 g in beef heart patties made with oleogel.	[57]
Emulsion gels	Sunflower oil, agar and κ-carrageenan	Sausages	The calorie content of the sausages decreased	[66]
Emulsion gels	Olive oil and chia mucilage	Patties	Reduced lipid oxidation and decreased fat content	[67]

Nevertheless, meat products contain a high amount of fat especially saturated fatty acids content. Excessive consumption of meat products could contribute to various kinds of non-communicable diseases such as hypertension, diabetes, and obesity [69]. In addition to that, the application of fat replacers in various types of meat products could result in various quality characteristics.

4.1 Burgers/Patties

Burgers, under different circumstances, are also commonly referred to as patties. The term "burgers" is derived from the word "hamburger," which is believed to have originated from Hamburg [70]. The meat patty or burger is a convenient meat product made from a comminuted meat emulsion [71]. The emulsion consists of homogenised muscle proteins, fat, water, salt, phosphates, and additional non-meat components that are combined in a mixer or bowl cutter [72]. The mixing process extracts the dispersed intracellular myofibrillar protein from the inside of the meat and spreads it onto the surface of the meat pieces. This protein acts as a natural binder and is distributed together with the fat, connective tissue, and other non-meat ingredients to enhance the properties of emulsion and gelation [73]. Burgers or patties can be made from various types of meat such as chicken [74], beef [75], lamb [76], pork [77] and duck [78]. The majority of European nations have implemented regulations stipulating that burgers must consist of around 80% meat and a fat content ranging from 20% to 30%. Presently, there is an increasing tendency among consumers towards low-fat products, which has generated concern among manufacturers of processed meat [79]. The practice of substituting certain ingredients with non-meat alternatives has been observed within the processed meat sector. The substitution is carried out for various reasons, including those related to quality, health, or economic considerations [70]. The substitution of animal-derived ingredients such as animal fat with plant-based alternatives has been implemented within the food industry, serving as an illustrative case. According to Kirkyol and Akköse [80], almond flour was used to replace animal fat in beef patties, and it resulted in the production of beef patties with a lower amount of saturated fats with no significant differences in the sensory properties of the product. A reformulated pork patty with emulsion gel showed that they have acceptable sensory attributes with an enhanced lipid oxidation quality [81]. The replacement of animal fat in beef patties with pomace olive oil and canola oil indicated that the product had a more desired flavour with a softer texture compared to animal fats [82].

In the production of buffalo patties with the incorporation of different types of squashes, all the fat-replaced patties could be categorized as low-fat patties as the fat content of the fat-replaced patties significantly decreased to below 3% [48]. Chicken patties replaced with 100% breadfruit resulted in the sensory panellists preferring the breadfruit-replaced chicken patties in terms of appearance, texture, flavour, juiciness, and aroma compared to chicken patties made with animal fat [83]. Different types of fat replacers were incorporated in chevon/goat patties, and it showed that the usage of fat replacers reduced the cholesterol content of the patties [84]. When compared to patties with animal fats, reformulated patties with emulsion gels as fat replacer had significantly improved lipid profiles (reduced levels of saturated fatty acids, a lower ratio of n-6 to n-3 fatty acids, and increased levels of long-chain polyunsaturated fatty acids) [81]. The usage of almond flour as a fat replacer in beef patties increased the amounts of oleic and linoleic acids besides decreasing the TBARS value of the patties [80].

4.2 Sausages

Sausages are commonly prepared by combining ground meat, such as beef, poultry, or pork, with additional ingredients such as fat, salt, spices, stabilisers, antioxidants, flavourings, and fillers. The specific additives used depend on the desired outcome. The mixture is then filled into a casing [85]. These products involve the alteration of fresh minced meats through different processing techniques in order to achieve specific characteristics [86]. Sausages exhibit a significant degree of diversity and exhibit subtle variances throughout different regions of the world. The sausages mentioned derived their names from the respective regions where they originated: the Frankfurter from Frankfurt, Germany; the bologna from Bologna, Italy; the genoa salami from Genoa, Italy; the Lyon sausage from Lyons, France; and the Berliner sausage from Berlin, Germany. In England, sausages became strongly associated with specific counties such as Berkshire, Wiltshire, Lincolnshire, and Cumberland [87]. On top of that, the classification of sausages is based on the methods of processing or manufacture. For instance, (1) fresh sausages consist of unprocessed meats that are typically not cured, smoked, fermented, or cooked. It is necessary to store fresh sausages in a refrigerator before consuming them. The consumer heats them prior to serving, such as fresh pork, breakfast sausage, Italian sausage and fresh chorizo. (2) Fermented sausages are produced using cured or uncured meats that have not undergone any form of heat processing. These meats are typically non-fermented and may also be smoked. The sausages possess a unique and sharp taste as a result of the lactic acid generated through fermentation. The meat is inserted into a casing and undergoes fermentation, a metabolic process in which bacteria convert sugars into acids and other compounds. The process of meat fermentation involves the utilisation of bacteria that produce lactic acid to create the tangy flavour found in dry sausages. These types of sausages, such as Genoa, Salami, Frizzes, Lebanon bologna, Cevelat, Thuringer, and Mortadella, are commonly known as "summer sausages" and are typically consumed cold. (3) Uncooked smoked sausages are primarily cured, non-fermented products. Their shelf-life is extended by heating, which partially reduces their moisture content. Typically, they are fully cooked before being consumed. Examples of sausages include kielbasas (Polish sausage), Chinese sausage, mettwurst, teawurst and smoked pork sausage. (4) Smoked sausages can be either cured or uncured, and they undergo heat processing and occasionally smoking. They frequently consist of a diverse range of meat or organ meat, such as the liver. Cooked sausages, being a ready-to-eat food, are typically consumed cold and can be refrigerated for a maximum of 7 days after opening. For example, Frankfurters, Cotto salami, and Berliner. (5) Cooked sausages are fully cooked products, typically made from pre-cooked fresh or occasionally cured raw ingredients, and undergo a final cooking process after being stuffed without any additional smoking. Examples include Liverwurst, Braunschweiger and blood sausage [87].

There has been a significant global rise in the consumption of processed meat products such as sausages. Fat influences the formation of aromatic compounds during fermentation through lipolysis, which helps meat products achieve their desired textural and organoleptic qualities [49]. According to Lee and Chin [88], animal fat can help stabilise the myofibrillar protein network, which improves the organoleptic sense of texture and juiciness in a variety of meat products. Furthermore, throughout the fermentation and dehydration process, animal fat particles help to regulate and continuously release moisture, enhancing the semi-dry sausages' firmness, juiciness, and mouthfeel and ultimately influencing customer acceptance of them [89]. As fat could also cause a lot of negative consequences to the consumer's health, fat replacers are being incorporated in sausages to enhance the properties of the meat product. For instance, research by Zhu et al. [90] indicated that the usage of Jerusalem artichoke powder and olive oil as fat replacer in Harbin dry sausage was much healthier than the sausage with animal fat as their fat content was lower and an enhanced fatty acid profile. Peanut and linseed oil were also used to make an emulsion gel to replace the animal fat in sausage which resulted in enhanced cooking characteristics and emulsion stability of the meat product [91]. The interactions between phytosterols and monoglycerides with sunflower oil as oleogels produced more robust oleogel networks with enhanced hardness and thus, produced sausage that does not compromise the physicochemical, textural and sensorial characteristics of the meat product [92].

Fermented sausage with the incorporation of emulsion gel showed a decrease in the cholesterol and lipid content of the meat product along with a reduction in the oxidation value of the fermented sausage [93]. Besides the reduction in the amount of saturated fats in Bologna sausage, the incorporation of emulsion gel as a fat replacer also produces a meat product with a more homogeneous and compact structure microscopically [68]. Incorporation of emulsion gels made from chia mucilage and olive oil as a replacement for pork back fat increased the amount of polyunsaturated fatty acids and the decrease in the fat and caloric content of the reformulated Bologna sausage enabled the reformulated products to make nutritional claims of "reduced fat" [94]. Replacing pork back fat with camellia oil gels had no impact on the cohesiveness, resilience, springiness, and protein content. However, it did lead to an increase in moisture content, lightness, and yellowness [95]. Fat replacer was used to replace animal fat in chorizo and it had a fat content ranging from 99 to 130 g/kg, while the normal fat sausages had a fat content of 316 g/kg. The incorporation of an oil-in-konjac matrix led to a notable reduction in saturated fatty acid levels and a notable increase in polyunsaturated fatty acids (PUFA), thereby improving the n-6/n-3 ratio in dry sausage [96]. Replacing fat with hydrophilic polysaccharides had a significant impact on the in vitro digestibility of proteins in emulsion-type sausage, particularly during the simulated gastric and initial intestinal stages [97]. Hazelnut oil was substituted for beef fat in the preparation of sucuk, a Turkish fermented sausage and it caused the cholesterol content to decrease significantly [98]. Another example, animal fat was replaced by ethylcellulose-based organogels to be incorporated in ground-cooked salami which resulted in salami having a similar nutritional profile to conventional ground products but with a lower amount of saturated fats [99].

4.3 Meatballs

Meatballs are ready-to-eat meat products made by combining minced beef, water, flour, spices, and fat and then moulded into a round shape, then processed through precooking, freezing, and warming to ensure they are ready for consumption [100]. Typically, meatballs are prepared by boiling, baking, or frying and are commonly served alongside pasta. However, they can also be enjoyed independently, with or without a dipping sauce [101]. Meatball is a widely consumed meat product that falls under various categories such as chicken balls, fish balls, beef balls, squid balls, and prawn balls. However, the most commonly available meatballs in the market are chicken balls, fish balls, and beef balls [102]. Meatballs are referred to by various names around the world. Examples of meatballs in other countries include bebola daging in Malaysia [103], bakso in Indonesia [104], kofta in India [105], cig kofte´ in Turkey [106] and kungwan in China [107], and numerous other names that reflect their country of origin and cultural context. The majority of consumers favour meatballs because of their better sensory qualities (such as cohesiveness, chewiness, and springiness), lower cost, easier processing, and distinct flavour. Due to the quantity of animal fat used in the production of meatballs, they have become known for their juiciness and flavorful taste [108]. Nevertheless, excessive consumption of this product is not advisable as it could contribute to various kinds of diseases.

The replacement of fat replacers for animal fat is an important step in enhancing the nutritional value of meat products. Different types of mushrooms were used in the production of meatballs to replace animal fat. This process not only enhances the meatballs' technological characteristics but also elevates their sensory attributes [109]. The replacement of animal fat with adzuki bean flour showed that it improved the cooking yield of the meatball [103]. The fatty acid profile of chicken meatballs with canola oil as a fat replacer showed an improvement besides having good acceptability in sensory properties [110]. The replacement of animal fat with an oleogel, which was formed by gelling carnauba wax with a mixture of sunflower oil and black seed oil, led to the development of a meatball that closely resembled the control in terms of both look and flavour [111].

Hydrocolloids were used as fat replacers in pork meatballs, and it indicated that the elevated intermolecular forces and organised secondary configuration of the system suggest that hydrocolloids have the potential to enhance the stability of the salt soluble protein and gel structure of the pork meatballs [112]. The soluble dietary fibre-gelatin composite gels can be incorporated into processed meat (meatballs) as a fat replacer, providing a favourable texture [113]. Pork and chicken meatballs with the presence of inulin as fat replacers showed acceptance among consumers, indicating that the replacement of animal fat with fat replacers was successful [114]. Perilla seeds, which are rich in unsaturated fat were used to replace animal fat in meatballs showed that incorporation of Perilla seeds at a concentration of 10% (w/w) resulted in a statistically significant enhancement of the texture, composition, and levels of polyunsaturated fatty acids (PUFAs), dietary fibre, and protein in meatballs [115].

4.4 Other Meat Products

The usage of fat replacers is also implemented in different types of meat products. For instance, pâté is a traditional cooked meat delicacy enjoyed in numerous countries, especially in Europe (Denmark, France, Germany, and Spain). The product is made by combining minced liver, fat, and meat with water and various additives. This mixture is then packaged in glass containers and subjected to thermal processing [116]. Replacing a portion of pork backfat with microencapsulated fish oil shows potential as a method to create more nutritious pâtés. This substitution decreases the amount of fat and improves the composition of fatty acids, namely in terms of monounsaturated fatty acids (MUFA) and polyunsaturated fatty acids (PUFA), as well as the ratio of *n*-6 to *n*-3 fatty acids [117]. On top of that, the hardness values of the pâtés made with canola oil organogels were comparable with the ones made with animal fats [118]. Replacement of pork fat with canola oil organogels showed almost a 40% reduction in saturated fat besides showing good maintenance in the textural properties of the pâtés [119]. Ethyl cellulose (EC-OG) and beeswax (W-OG) oleogels were created by blending a combination of olive, linseed, and fish oils with an optimal fatty acid content to enhance the nutritional value of pâtés. The purpose of these oleogels is to imitate the characteristics of solid animal fats and act as replacements for pork liver pâtés. The presence of α-linolenic acid and omega-3 fatty acids in pork liver pâtés enables health claims to be made following EU laws [120].

Another example of a meat product is luncheon meat, a type of compressed minced meat used as a filling which experienced a rise in popularity during World War II. Luncheon meat was showcased as a vital source of sustenance for the military. Currently, canned luncheon meat, which can be either pre-cooked or cured, continues to be widely favoured as a food item [121]. Typically, canned luncheon meat is primarily made from pork, beef, or chicken [122]. It is also known as sliced meat or canned meat and it is usually consumed cold. Luncheon meat is commonly consumed and typically enjoyed in sandwich form. In general, luncheon meat is an inexpensive type of cold-cut product that is typically packed into a waterproof casing with a diameter of approximately 80 mm. Typically, the all-in method is employed, either in a bowl cutter or by utilising a combination of a bowl cutter and emulsifier. Following the process of emulsification and filling, the product undergoes cooking in a hot-water bath or with steam at a temperature range of 76–80 °C until it reaches a core temperature of 70 °C. The refrigerated product is typically sliced and packaged in a vacuum-sealed manner, or sold to the final consumer as a single piece weighing between 300 and 500 g [123]. The findings of fat-replaced chicken luncheon meat demonstrated that incorporating fat replacers into cooked chicken luncheon meat resulted in an enhancement of the composition of saturated, monounsaturated, and polyunsaturated fatty acids, thereby improving its nutritional profile [124]. Luncheon meat was replaced with an oleogel derived from linseed oil, serving as a nutritious substitute for solid fats. The sensory evaluation indicated that the overall acceptance of luncheon meat was the best when 25% of the pork backfat was replaced with oleogel as the luncheon meat achieved the desired level of overall acceptance [125].

Conventionally, nuggets are manufactured by utilising mechanically separated chicken meat, chicken skin, and other components [126]. Recently, customers have been consuming breaded and battered foods, such as chicken nuggets, extensively due to their high nutritional value, appealing sensory characteristics, extended shelf-life, and low cost [127]. Nevertheless, the presence of a high amount of saturated fats in this product led to the development of a fat replacer to replace the animal fat in the mentioned meat product. The incorporation of apple pulp in chicken nuggets led to an increase in moisture percentage, dietary fibre, and product colour [128]. Replacement of chicken skin with chia flour in chicken nuggets increased polyunsaturated fatty acids, including the omega-3 fatty acid, α-linolenic acid, as well as dietary fibre. At the same time, it led to a decrease in moisture, saturated fatty acids along with monounsaturated fatty acids and reduced water activity [129]. The replacement of some of the meat and fat with pea fibre in chicken nuggets is feasible, as it does not significantly affect the physical and chemical properties or the sensory perception [130]. Chicken nuggets with animal fat were also replaced with palm fat and it resulted in a reduction of the cholesterol levels of the chicken nuggets [131]. In the formulation of a chicken nugget, the chicken skin was replaced with okara flour and rice bran which resulted in a chicken nugget with a higher amount of dietary fibre and lower fat content [126].

5 Advantages and Disadvantages of Fat Replacer

As low-calorie and reduced-fat foods, such as low-fat meat products, are gaining popularity in these markets, the meat industry benefits the most from fat replacers [29]. The usage of fat replacers in meat products has its advantages and disadvantages towards the characteristics of the meat products which are shown in Table 3.

6 Future Prospect

The strategy that is used to develop a healthier meat product, based on the incorporation of fat replacers could be a feasible option to improve the nutrient content, and physicochemical and sensorial properties of meat products. Nevertheless, the characteristics of the meat product could be enhanced by improving the properties of the fat replacer. For instance, the method of cross-linking was applied to oil-modified crosslinked starch which enhanced the properties of the fat replacer itself to be

Table 3 Advantages and disadvantages of fat replacers

Fat replacer	Meat product	Advantages	Disadvantages	Health benefits	References
Mango peel powder	Chicken patties	Higher cooking yield High moisture content	Lower emulsion stability Low sensory scores	Lower fat content Low cholesterol content	[132]
Safflower oil	Wieners	Do not alter the sensory characteristics of the meat product	Higher lipid oxidation	Lower cholesterol content Greater concentration of unsaturated fatty acids	[133]
Black jelly mushroom and olive oil	Chicken emulsion	Better water holding capacity High emulsion stability	Affected the textural properties of the meat product, especially the hardness value	Lower fat content Fat replacer contains a high amount of unsaturated fatty acids	[134]
Eggplants	Chicken sausage	No significant differences in water holding capacity and cooking loss	The colour of the sausage was affected especially the lightness (L*) value	Can be claimed as a reduced-fat chicken sausage	[135]
Organogel produced with ethylcellulose, sorbitan monostearate and canola oil	Frankfurters	This resulted in a similar hardness value compared to frankfurters produced with animal fat	Affected the color properties of the meat product especially the redness (a*) value	Reduce the amount of saturated fatty acids	[7]
Emulsion gels based on egg white protein curdlan	Meat batter	The thickening and lubricating effect of fat-replaced meat products indicated that they could mimic the properties of animal fat	Lower hardness Higher cooking loss	Have the ability to serve as a replacement for fat to create more nutritious meat products	[136]
Emulsion gels containing two different solid polyphenol extracts i.e. grape seed or grape seed and olive	Frankfurters	Good thermal stability	Higher lipid oxidation	Improved lipid content Contains high levels of phenolic compounds	[137]

(continued)

Table 3 (continued)

Fat replacer	Meat product	Advantages	Disadvantages	Health benefits	References
Oil-modified crosslinked starch	Pork meat batter	Enhances the gel properties of the meat product Improves the water holding capacity	No significant improvement in the amount of unsaturated fatty acids	Lower amount of total fat and saturated fat	[138]
Guar gum	Goshtaba	High in protein High in moisture	Lower lightness (L*) value Lower hardness value	Lower fat content	[139]
Rapeseed oleosomes	Meat model system	Moisture content increases The lightness (L*) value increases	Lower hardness value	Lower fat content	[140]

incorporated in meat products. The results showed that the fatty acid profile of the pork meat gels improved, leading to better rheological properties, texture, and water-holding capacity (WHC). This improvement was attributed to the establishment of a more organised and complex protein network structure, which also resulted in lower water mobility [138]. The usage of fat replacer especially the involvement of liquid oil in the formation of fat replacer was always associated with high lipid oxidation. Thus, it is suggested that the usage of physical technology such as the use of ultrasound technology be implemented in the formation of fat replacer. Ultrasound is widely used in emulsification (such as the formation of oleogels and emulsion gels), resulting in a variety of effects on the functional properties of the products (e.g., pH, colour, tenderness, WHC, and oxidation stability) [141].

7 Conclusion

With the growing popularity of the healthy diet concept among consumers, the demand for low-fat meat products has increased. The utilization of fat replacers in meat products can be an ultimate solution to reduce the total amount of fat besides enhancing the fatty acid profiles of the meat products, providing a nutritional alternative to prevent non-communicable diseases. On top of that, the newly formulated new products could also provide new technological characteristics and sensory properties for a low-fat meat product. Nevertheless, prior to implementing these fat-reduction techniques in low-fat meat products, it is necessary to address certain concerns, such as the physicochemical and sensory qualities, as well as the nutritional value. This is crucial to strike a balance between the fat replacement ratio and the overall quality of the meat product. This chapter indicated that formulations of fat replacers were very important because each type i.e. carbohydrate, protein and lipid has different effects on the final meat products. For instance, carbohydrate and protein-based fat replacers do not have a similar chemical structure with animal fat but on the other hand, lipid-based fat replacer has a similar chemical structure with animal fat. The fat reduction strategies also showed that the oil structuring method could give a better characteristic to the meat products compared to other strategies. When a fat replacer is used in meat products, lipid oxidation is unavoidable during fat replacement as a result of a decrease in SFA and an increase in PUFA. In conjunction with that, novel physical technology could be applied to assist in decelerating the lipid oxidation of meat products.

References

1. Lorenzo JM, Sarriés MV, Tateo A, Polidori P, Franco D, Lanza M. Carcass characteristics, meat quality and nutritional value of horsemeat: a review. Meat Sci. 2014;96:1478–88.
2. Bascuas S, Morell P, Hernando I, Quiles A. Recent trends in oil structuring using hydrocolloids. Food Hydrocoll. 2021;118:106612.
3. Vargas-Ramella M, Munekata PE, Pateiro M, Franco D, Campagnol PC, Tomasevic I, Domínguez R, Lorenzo JM. Physicochemical composition and nutritional properties of deer burger enhanced with healthier oils. Foods. 2020;9:571.
4. Zembyla M, Murray BS, Sarkar A. Water-in-oil emulsions stabilized by surfactants, biopolymers and/or particles: a review. Trends Food Sci Technol. 2020;104:49–59.
5. Domínguez R, Bohrer B, Munekata PE, Pateiro M, Lorenzo JM. Recent discoveries in the field of lipid bio-based ingredients for meat processing. Molecules. 2021;26:190.
6. Hygreeva D, Pandey MC, Radhakrishna K. Potential applications of plant based derivatives as fat replacers, antioxidants and antimicrobials in fresh and processed meat products. Meat Sci. 2014;98:47–57.
7. Barbut S, Wood J, Marangoni A. Potential use of organogels to replace animal fat in comminuted meat products. Meat Sci. 2016;122:155–62.
8. Herrero AM, Carmona P, Pintado T, Jiménez-Colmenero F, Ruiz-Capillas C. Lipid and protein structure analysis of frankfurters formulated with olive oil-in-water emulsion as animal fat replacer. Food Chem. 2012;135:133–9.
9. Yoo SS, Kook SH, Park SY, Shim JH, Chin KB. Physicochemical characteristics, textural properties and volatile compounds in comminuted sausages as affected by various fat levels and fat replacers. Int J Food Sci Technol. 2007;42:1114–22.
10. Guo J, Cui L, Meng Z. Oleogels/emulsion gels as novel saturated fat replacers in meat products: a review. Food Hydrocoll. 2023;137:108313.
11. Heck RT, Saldaña E, Lorenzo JM, Correa LP, Fagundes MB, Cichoski AJ, de Menezes CR, Wagner R, Campagnol PC. Hydrogelled emulsion from chia and linseed oils: a promising strategy to produce low-fat burgers with a healthier lipid profile. Meat Sci. 2019;156:174–82.
12. Kumar Y. Development of low-fat/reduced-fat processed meat products using fat replacers and analogues. Food Rev Int. 2021;37:296–312.
13. Cengiz E, Gokoglu N. Effects of fat reduction and fat replacer addition on some quality characteristics of frankfurter-type sausages. Int J Food Sci Technol. 2007;42:366–72.
14. López-Pedrouso M, Lorenzo JM, Gullón B, Campagnol PC, Franco D. Novel strategy for developing healthy meat products replacing saturated fat with oleogels. Curr Opin Food Sci. 2021;40:40–5.

15. Colla K, Costanzo A, Gamlath S. Fat replacers in baked food products. Foods. 2018;7:192.
16. Omayma ES, Youssef MM. Fat replacers and their applications in food products: a review. J Food Sci Technol. 2007;4:29–44.
17. Mattes RD. Position of the American Dietetic Association: fat replacers. J Am Diet Assoc. 1998;98:463–9.
18. Liu J, Han Y, Chen J, Zhang Z, Miao S, Zheng B, Zhang L. MCT/LCT mixed oil phase enhances the rheological property and freeze-thawing stability of emulsion. Foods. 2022;11:712.
19. Ruiz-Capillas C, Herrero AM. Development of meat products with healthier lipid content: vibrational spectroscopy. Foods. 2021;10:341.
20. Ren Y, Huang L, Zhang Y, Li H, Zhao D, Cao J, Liu X. Application of emulsion gels as fat substitutes in meat products. Foods. 2022;11:1950.
21. Zhao D, Liu S, Li H, Yoong JH, Miao Z, Cao J, Liu X. Effect of palm oil content and melting point on the freeze-thaw stability of fat substitutes. LWT Food Sci Technol. 2023;188:115452.
22. Cooper DA, Berry DA, Jones MB, Kiorpes AL, Peters JC. Olestra's effect on the status of vitamins A, D and E in the pig can be offset by increasing dietary levels of these vitamins. J Nutr. 1997;127:1589S–608S.
23. Ognean CF, Darie N, Ognean M. Fat replacers: review. J Agroaliment Process Technol. 2006;12:433–42.
24. Carvalho Barros J, Munekata PE, de Carvalho FA, Pateiro M, Barba FJ, Domínguez R, Trindade MA, Lorenzo JM. Use of tiger nut (*Cyperus esculentus L.*) oil emulsion as animal fat replacement in beef burgers. Foods. 2020;9:44.
25. Singh R, Chatli MK, Biswas AK, Sahoo J. Quality of ω-3 fatty acid enriched low-fat chicken meat patties incorporated with selected levels of linseed flour/oil and canola flour/oil. J Food Sci Technol. 2014;51:353–8.
26. Han C, Wang G, Guo J, Wang J, Yang X. Oral oil release improves lubrication and sensory properties of meat analogs with protein-stabilized oleogel. Food Hydrocoll. 2023;142:108788.
27. Tan Y, Zhang Z, McClements DJ. Preparation of plant-based meat analogs using emulsion gels: lipid-filled RuBisCo protein hydrogels. Food Res Int. 2023;167:112708.
28. Chandalia SH, Modi SV. Fat replacers and fat blockers. Int J Diabetes Dev Ctries. 1999;19:139–43.
29. Goswami M, Tanuja VP, Sharma H, Ojha S. Potential use of fat replacers for development of functional food of animal origin. J Anim Feed Sci Technol. 2019;7:81–6.
30. Chavan RS, Khedkar CD, Bhatt S. Fat replacer. In: Caballero B, Finglas P, Toldrá F, editors. The encyclopedia of food and health. London: Oxford; 2016. p. 589–95.
31. Brewer MS. Reducing the fat content in ground beef without sacrificing quality: a review. Meat Sci. 2012;91:385–95.
32. Daniel JR. Advances in development of fat replacers and low-fat products. In: Passos ML, Ribeiro CP, editors. Innovation in food engineering: new techniques and products. Boca Raton: CRC Press; 2010. p. 660–70.
33. Aziz NS, Sofian-Seng NS, Yusop SM, Kasim KF, Razali NS. Functionality of okra gum as a novel carbohydrate-based fat replacer in ice cream. Food Sci Technol Res. 2018;24:519–30.
34. Olcay HS. The methods reducing of fat content in meat and meat products. Int J Food Eng Res. 2020;6:1–6.
35. Asyrul-Izhar AB, Bakar J, Sazili AQ, Meng GY, Ismail-Fitry MR. Incorporation of different physical forms of fat replacers in the production of low-fat/reduced-fat meat products: which is more practical? Food Rev Int. 2023;39:6387–419.
36. Marchetti L, Andres SC, Califano AN. Low-fat meat sausages with fish oil: optimization of milk proteins and carrageenan contents using response surface methodology. Meat Sci. 2014;96:1297–303.
37. Rather SA, Masoodi FA, Akhter R, Rather JA, Gani A, Wani SM, Malik AH. Application of guar–xanthan gum mixture as a partial fat replacer in meat emulsions. J Food Sci Technol. 2016;53:2876–86.

38. Felisberto MH, Galvão MT, Picone CS, Cunha RL, Pollonio MA. Effect of prebiotic ingredients on the rheological properties and microstructure of reduced-sodium and low-fat meat emulsions. LWT Food Sci Technol. 2015;60:148–55.
39. Serdaroğlu M. Improving low fat meatball characteristics by adding whey powder. Meat Sci. 2006;72:155–63.
40. Severini C, De Pilli T, Baiano A. Partial substitution of pork backfat with extra-virgin olive oil in "salami" products: effects on chemical, physical and sensorial quality. Meat Sci. 2003;64:323–31.
41. Kayaardı S, Gök V. Effect of replacing beef fat with olive oil on quality characteristics of Turkish soudjouk (sucuk). Meat Sci. 2004;66:249–57.
42. Kamani MH, Meera MS, Bhaskar N, Modi VK. Partial and total replacement of meat by plant-based proteins in chicken sausage: evaluation of mechanical, physico-chemical and sensory characteristics. J Food Sci Technol. 2019;56:2660–9.
43. Pietrasik Z, Soladoye OP. Use of native pea starches as an alternative to modified corn starch in low-fat bologna. Meat Sci. 2021;171:108283.
44. Gibis M, Schuh V, Allard K, Weiss J. Influence of molecular weight and degree of substitution of various carboxymethyl celluloses on unheated and heated emulsion-type sausage models. Carbohydr Polym. 2017;159:76–85.
45. de Carvalho FA, Lorenzo JM, Pateiro M, Bermúdez R, Purriños L, Trindade MA. Effect of guarana (*Paullinia cupana*) seed and pitanga (*Eugenia uniflora L.*) leaf extracts on lamb burgers with fat replacement by chia oil emulsion during shelf life storage at 2 C. Food Res Int. 2019;125:108554.
46. Özer CO. Utilization of Jerusalem artichoke powder in production of low-fat and fat-free fermented sausage. Ital J Food Sci. 2019;31(2):301–10.
47. Öztürk-Kerimoğlu B, Kara A, Urgu-Öztürk M, Serdaroğlu M. A new inverse olive oil emulsion plus carrot powder to replace animal fat in model meat batters. LWT Food Sci Technol. 2021;135:110044.
48. Kahar SN, Ismail-Fitry MR, Yusoff MM, Rozzamri A, Bakar J, Ibadullah WZ. Substitution of fat with various types of squashes and gourds from the Cucurbitaceae family in the production of low-fat buffalo meat patties. Malays Appl Biol. 2021;50:169–79.
49. Wi G, Bae J, Kim H, Cho Y, Choi MJ. Evaluation of the physicochemical and structural properties and the sensory characteristics of meat analogues prepared with various non-animal based liquid additives. Foods. 2020;9:461.
50. Manzoor S, Masoodi FA, Rashid R, Naqash F, Ahmad M. Oleogels for the development of healthy meat products: a review. Appl Food Res. 2022;2:100212.
51. Domínguez R, Munekata PE, Pateiro M, López-Fernández O, Lorenzo JM. Immobilization of oils using hydrogels as strategy to replace animal fats and improve the healthiness of meat products. Curr Opin Food Sci. 2021;37:135–44.
52. Asyrul-Izhar AB, Bakar J, Sazili AQ, Goh YM, Ismail-Fitry MR. Emulsion gels formed by electrostatic interaction of gelatine and modified corn starch via pH adjustments: potential fat replacers in meat products. Gels. 2023;9:50.
53. de Souza Paglarini C, de Figueiredo Furtado G, Biachi JP, Vidal VA, Martini S, Forte MB, Cunha RL, Pollonio MA. Functional emulsion gels with potential application in meat products. J Food Eng. 2018;222:29–37.
54. Dickinson E. Emulsion gels: the structuring of soft solids with protein-stabilized oil droplets. Food Hydrocoll. 2012;28:224–41.
55. Lim J, Jeong S, Lee S. Evaluation of soybean oil-carnauba wax oleogels as an alternative to high saturated fat frying media for instant fried noodles. LWT Food Sci Technol. 2017;84:788–94.
56. Ferro AC, de Souza Paglarini C, Pollonio MA, Cunha RL. Glyceryl monostearate-based oleogels as a new fat substitute in meat emulsion. Meat Sci. 2021;174:108424.

57. Gao Y, Li M, Zhang L, Wang Z, Yu Q, Han L. Preparation of rapeseed oil oleogels based on beeswax and its application in beef heart patties to replace animal fat. LWT Food Sci Technol. 2021;149:111986.
58. Pintado T, Herrero AM, Jiménez-Colmenero F, Cavalheiro CP, Ruiz-Capillas C. Chia and oat emulsion gels as new animal fat replacers and healthy bioactive sources in fresh sausage formulation. Meat Sci. 2018;135:6–13.
59. Lucas-González R, Roldán-Verdu A, Sayas-Barberá E, Fernández-López J, Pérez-Álvarez JA, Viuda-Martos M. Assessment of emulsion gels formulated with chestnut (*Castanea sativa M.*) flour and chia (*Salvia hispanica L*) oil as partial fat replacers in pork burger formulation. J Sci Food Agric. 2020;100:1265–73.
60. Shin DJ, Lee HJ, Lee D, Jo C, Choe J. Fat replacement in chicken sausages manufactured with broiler and old laying hens by different vegetable oils. Poult Sci. 2020;99:2811–8.
61. El-Zeny T, Essa RY, Bisar BA, Metwalli SM. Effect of using chicory roots powder as a fat replacer on beef burger quality. Slov Vet Res. 2019;56:509–14.
62. Rather SA, Akhter R, Masoodi FA, Gani A, Wani SM. Utilization of apple pomace powder as a fat replacer in goshtaba: a traditional meat product of Jammu and Kashmir, India. J Food Meas Charact. 2015;9:389–99.
63. Gök V, Akkaya L, Obuz E, Bulut S. Effect of ground poppy seed as a fat replacer on meat burgers. Meat Sci. 2011;89:400–4.
64. Ercoşkun H, Demirci-ercoşkun TU. Walnut as fat replacer and functional component in sucuk. J Food Qual. 2010;33:646–59.
65. Shao L, Bi J, Li X, Dai R. Effects of vegetable oil and ethylcellulose on the oleogel properties and its application in Harbin red sausage. Int J Biol Macromol. 2023;239:124299.
66. Fontes-Candia C, Martínez-Sanz M, Gómez-Cortés P, Calvo MV, Verdú S, Grau R, López-Rubio A. Polysaccharide-based emulsion gels as fat replacers in frankfurter sausages: physicochemical, nutritional and sensorial evaluation. LWT Food Sci Technol. 2023;180:114705.
67. Liu S, Lu J, Zhang J, Su X, Peng X, Guan H, Shi C. Emulsion gels prepared with chia mucilage and olive oil as a new animal fat replacer in beef patties. J Food Process Preserv. 2022;46:e16972.
68. de Souza Paglarini C, de Figueiredo Furtado G, Honorio AR, Mokarzel L, da Silva Vidal VA, Ribeiro AP, Cunha RL, Pollonio MA. Functional emulsion gels as pork back fat replacers in Bologna sausage. Food Struct. 2019;20:100105.
69. McAfee AJ, McSorley EM, Cuskelly GJ, Moss BW, Wallace JM, Bonham MP, Fearon AM. Red meat consumption: an overview of the risks and benefits. Meat Sci. 2010;84:1–3.
70. Ramadhan K, Huda N, Ahmad R. Physicochemical characteristics and sensory properties of selected Malaysian commercial chicken burgers. Int Food Res J. 2011;18:1349–57.
71. Ismail MA, Chong GH, Ismail-Fitry MR. Comparison of the microstructural, physicochemical and sensorial properties of buffalo meat patties produced using bowl cutter, universal mixer and meat mixer. J Food Sci Technol. 2021;1:1–8.
72. Serdaroğlu M, Öztürk B, Urgu M. Emulsion characteristics, chemical and textural properties of meat systems produced with double emulsions as beef fat replacers. Meat Sci. 2016;117:187–95.
73. Sun XD, Holley RA. Factors influencing gel formation by myofibrillar proteins in muscle foods. Compr Rev Food Sci Food Saf. 2011;10:33–51.
74. Jeon DS, Moon YH, Park KS. Effects of gums on the quality of low fat chicken patty. J Korean Soc Food Sci Nutr. 2023;33(1):193–200.
75. Aoudeh E, Oz E, Oz F. Effect of beef patties fortification with black garlic on the polycyclic aromatic hydrocarbons (PAHs) content and toxic potency. Food Chem. 2023;428:136763.
76. Baker IA. Study on the effectiveness of adding some plant extracts with BHT on meat quality of lamb patties during chilling storage. Iraqi J Agric Sci. 2023;54:147–55.
77. Zhang H, Li X, Kang H, Peng X. Chitosan nanoparticles effectively improved quality stability of pork patties subjected to multiple freeze–thaw cycles. Meat Sci. 2023;196:109029.

78. Biswas S, Chakraborty A, Patra G, Dhargupta A. Quality and acceptability of duck patties stored at ambient and refrigeration temperature. Int J Livest Prod. 2011;1:1–6.
79. Weiss J, Gibis M, Schuh V, Salminen H. Advances in ingredient and processing systems for meat and meat products. Meat Sci. 2010;86:196–213.
80. Kirkyol M, Akköse A. Effects of animal fat replacement with almond flour on quality parameters of beef patties. Food Sci Nutr. 2023;11:7091–9.
81. Cîrstea N, Nour V, Boruzi AI. Effects of pork backfat replacement with emulsion gels formulated with a mixture of olive, chia and algae oils on the quality attributes of pork patties. Foods. 2023;12:519.
82. Los PR, Marson GV, Dutcosky SD, Nogueira A, Marinho MT, Simões DR. Optimization of beef patties produced with vegetable oils: a mixture design approach and sensory evaluation. Food Sci Technol. 2020;40:12–20.
83. Ismail-Fitry M, Abas N. Potential use of jackfruit (*Artocarpus heterophyllus*) and breadfruit (*Artocarpus altilis*) as fat replacer to produce low-fat chicken patties. Int J Eng Technol. 2018;7:292–6.
84. Nayak NK, Pathak V. Comparative study of different fat replacer in the development of low fat chevon patties. Asian J Dairy Food Res. 2023;42:373–9.
85. Nyaguthii KC, Omwamba M, Nduko JM. Gum arabic and soy protein concentrate as binding agents on quality and nutritional properties of mushroom-based sausage analogues. Food Humanit. 2023;1:1627–36.
86. Noor S, Radhakrishnan NS, Hussain K. Sausage formulations and effects of addition of different non-meat ingredients on sausages-a review. Int J Livest Res. 2016;6:1–9.
87. Akpan IP. Trends in sausage production. Afr J Food Sci Technol. 2017;8:081–4.
88. Lee CH, Chin KB. Evaluation of pork myofibrillar protein gel with pork skin gelatin on rheological properties at different salt concentrations. Food Sci Anim Resour. 2019;39:576.
89. Câmara AK, Geraldi MV, Okuro PK, Júnior MR, da Cunha RL, Pollonio MA. Satiety and in vitro digestibility of low saturated fat Bologna sausages added of chia mucilage powder and chia mucilage-based emulsion gel. J Funct Foods. 2020;65:103753.
90. Zhu Y, Guo L, Tang W, Yang Q. Beneficial effects of Jerusalem artichoke powder and olive oil as animal fat replacers and natural healthy compound sources in Harbin dry sausages. Poult Sci. 2020;99:7147–58.
91. Nacak B, Öztürk-Kerimoğlu B, Yıldız D, Çağındı Ö, Serdaroğlu M. Peanut and linseed oil emulsion gels as potential fat replacer in emulsified sausages. Meat Sci. 2021;176:108464.
92. Kouzounis D, Lazaridou A, Katsanidis E. Partial replacement of animal fat by oleogels structured with monoglycerides and phytosterols in frankfurter sausages. Meat Sci. 2017;130:38–46.
93. Öztürk-Kerimoğlu B, Kavuşan HS, Gürel DB, Çağındı Ö, Serdaroğlu M. Cold-set or hot-set emulsion gels consisted of a healthy oil blend to replace beef fat in heat-treated fermented sausages. Meat Sci. 2021;176:108461.
94. Câmara AK, Ozaki MM, Santos M, Vidal VA, Ribeiro WO, de Souza Paglarini C, Bernardinelli OD, Sabadini E, Pollonio MA. Olive oil-based emulsion gels containing chia (*Salvia hispanica L.*) mucilage delivering healthy claims to low-saturated fat Bologna sausages. Food Struct. 2021;28:100187.
95. Wang X, Xie Y, Li X, Liu Y, Yan W. Effects of partial replacement of pork back fat by a camellia oil gel on certain quality characteristics of a cooked style Harbin sausage. Meat Sci. 2018;146:154–9.
96. Jiménez-Colmenero F, Triki M, Herrero AM, Rodríguez-Salas L, Ruiz-Capillas C. Healthy oil combination stabilized in a konjac matrix as pork fat replacement in low-fat, PUFA-enriched, dry fermented sausages. LWT Food Sci Technol. 2013;51:158–63.
97. Zeng X, Lv B, Zhu Y, Li Q, Zhang K, Li C, Zhao D, Li C. Influence of hydrophilic polysaccharide fat replacers on the in vitro digestibility of protein in emulsion-type sausage. Food Res Int. 2023;170:113008.

98. Yıldız-Turp G, Serdaroğlu M. Effect of replacing beef fat with hazelnut oil on quality characteristics of sucuk–a Turkish fermented sausage. Meat Sci. 2008;78:447–54.
99. Woern C, Marangoni AG, Weiss J, Barbut S. Effects of partially replacing animal fat by ethylcellulose based organogels in ground cooked salami. Food Res Int. 2021;147:110431.
100. Meng X, Wu D, Zhang Z, Wang H, Wu P, Xu Z, Gao Z, Mintah BK, Dabbour M. An overview of factors affecting the quality of beef meatballs: processing and preservation. Food Sci Nutr. 2022;10:1961–74.
101. Huda N, Shen YH, Huey YL, Ahmad R, Mardiah A. Evaluation of physico-chemical properties of Malaysian commercial beef meatballs. Am J Food Technol. 2010;5:13–21.
102. Aripin N, Huda N. Quality characteristics of meatball prepared from different ratios of chicken and duck meat. Asia Pac J Sustain Agric Food Energy. 2018;6:6–9.
103. Aslinah LN, Mat Yusoff M, Ismail-Fitry MR. Simultaneous use of adzuki beans (*Vigna angularis*) flour as meat extender and fat replacer in reduced-fat beef meatballs (bebola daging). J Food Sci Technol. 2018;55:3241–8.
104. Purnomo H, Rahardiyan D. Indonesian traditional meatball. Int Food Res J. 2008;15:101–8.
105. Modi VK, Yashoda KP, Bhaskar N, Mahendrakar NS. Effect of carrageenan and oat flour on storage characteristics of fried mutton kofta. J Food Process Preserv. 2009;33:763–76.
106. Bingol EB, Colak H, Cetin O, Hampikyan H. Effects of sodium lactate on the shelf life and sensory characteristics of cig kofte–a Turkish traditional raw meatball. J Food Process Preserv. 2014;38:1024–36.
107. Kang ZL, Zou YF, Xu XL, Zhu CZ, Wang P, Zhou GH. Effect of a beating process, as a means of reducing salt content in Chinese-style meatballs (kung-wan): a physico-chemical and textural study. Meat Sci. 2014;96:147–52.
108. Asyrul-Izhar AB, Masran SH, Faridah MR, Sarbon NM, Ismail-Fitry MR. Incorporation of shiitake (Lentinula edodes) and brown button mushroom (Agaricus bisporus) as fat replacers in meatballs processed with different mixing times. Food Res. 2023;7:59–68.
109. Ramle NA, Zulkunain M, Ismail-Fitry MR. Replacing animal fat with edible mushrooms: a strategy to produce high-quality and low-fat buffalo meatballs. Int Food Res J. 2021;28:905–15.
110. Vieira AS, Souza XR, Rodrigues EC, Sousa DC. Replacement of animal fat by canola oil in chicken meatball. Braz J Poul Sci. 2019;21:eRBCA-2019.
111. Palamutoğlu R. Replacement of beef fat in meatball with Oleogels (black cumin seed oil/sunflower oil). J Hell Vet Med Soc. 2021;72:3031–40.
112. Li Y, Guo J, Wang Y, Zhang F, Chen S, Hu Y, Zhou M. Effects of hydrocolloids as fat-replacers on the physicochemical and structural properties of salt-soluble protein isolated from water-boiled pork meatballs. Meat Sci. 2023;204:109280.
113. Niu Y, Fang H, Huo T, Sun X, Gong Q, Yu L. A novel fat replacer composed by gelatin and soluble dietary fibers from black bean coats with its application in meatballs. LWT Food Sci Technol. 2020;122:109000.
114. Montoya L, Quintero N, Ortiz S, Lopera J, Millán P, Rodríguez-Stouvenel A. Inulin as a fat-reduction ingredient in pork and chicken meatballs: its effects on physicochemical characteristics and consumer perceptions. Foods. 2022;11:1066.
115. Ran M, Chen C, Li C, He L, Zeng X. Effects of replacing fat with Perilla seed on the characteristics of meatballs. Meat Sci. 2020;161:107995.
116. Pateiro M, Lorenzo JM, Amado IR, Franco D. Effect of addition of green tea, chestnut and grape extract on the shelf-life of pig liver pâté. Food Chem. 2014;147:386–94.
117. Vargas-Ramella M, Lorenzo JM, Zamuz S, Montes L, López EM, Moreira R, Franco D. Influence of pork backfat replacement by microencapsulated fish oil on physicochemical, rheological, nutritional, and sensory features of pork liver pâtés. LWT Food Sci Technol. 2022;163:113522.
118. Barbut S, Marangoni AG, Thode U, Tiensa BE. Using canola oil organogels as fat replacement in liver pâté. J Food Sci. 2019;84:2646–51.

119. Barbut S, Tiensa BE, Marangoni AG. Partial fat replacement in liver pâté using canola oil organogel. LWT Food Sci Technol. 2021;139:110428.
120. Gómez-Estaca J, Herrero AM, Herranz B, Álvarez MD, Jiménez-Colmenero F, Cofrades S. Characterization of ethyl cellulose and beeswax oleogels and their suitability as fat replacers in healthier lipid pâtés development. Food Hydrocoll. 2019;87:960–9.
121. Farag ZS. Improvement of red tuna luncheon quality attributes. J Food Dairy Sci. 2023;14:25–9.
122. Agarwal S, Fulgoni VL, Berg EP. Association of lunch meat consumption with nutrient intake, diet quality and health risk factors in US children and adults: NHANES 2007–2010. Nutr J. 2015;14:1–9.
123. Feiner G. Meat products handbook: practical science and technology. Cambridge: Elsevier; 2006.
124. Khalil AN, Ockerman HW. Fatty acids chemical quality of cooked chicken luncheon meat product produced by the use of fat replacers. Int J Eng Technol Manag Res. 2020;7:31–8.
125. Ma Y, Ye F, Chen J, Ming J, Zhou C, Zhao G, Lei L. The microstructure and gel properties of linseed oil and soy protein isolate based-oleogel constructed with highland barley β-glucan and its application in luncheon meat. Food Hydrocoll. 2023;140:108666.
126. Echeverria L, da Silva C, Danesi ED, Porciuncula BD, Barros BC. Characterization of okara and rice bran and their application as fat substitutes in chicken nugget formulations. LWT Food Sci Technol. 2022;161:113383.
127. Magdelaine P, Spiess MP, Valceschini E. Poultry meat consumption trends in Europe. Worlds Poult Sci J. 2008;64:53–64.
128. Verma AK, Sharma BD, Banerjee R. Effect of sodium chloride replacement and apple pulp inclusion on the physico-chemical, textural and sensory properties of low fat chicken nuggets. LWT Food Sci Technol. 2010;43:715–9.
129. Barros JC, Munekata PE, Pires MA, Rodrigues I, Andaloussi OS, da Costa Rodrigues CE, Trindade MA. Omega-3-and fibre-enriched chicken nuggets by replacement of chicken skin with chia (*Salvia hispanica L.*) flour. LWT Food Sci Technol. 2018;90:283–9.
130. Polizer YJ, Pompeu D, Hirano MH, Freire MT, Trindade MA. Development and evaluation of chicken nuggets with partial replacement of meat and fat by pea fibre. Braz J Food Technol. 2015;18:62–9.
131. Alina AR, Babji AS, Affandi S. Nutritional quality of palm fat substituted chicken nuggets. Nutr Food Sci. 2009;39:181–8.
132. Chappalwar AM, Pathak V, Goswami M, Verma AK. Development of functional chicken patties with incorporation of mango peel powder as fat replacer. Nutr Food Sci. 2020;50:1063–73.
133. Kaynakci E, Kilic B. Effect of replacing beef fat with safflower oil on physicochemical, nutritional and oxidative stability characteristics of wieners. Food Sci Technol. 2020;41:52–9.
134. Faridah MR, Salman NA, Asyrul-Izhar AB, Siti Aimi SZA, Ismail-Fitry MR. Effect of black jelly mushroom (*Auricularia polytricha*) and olive oil as fat replacers on the physicochemical and microstructural properties of chicken meat emulsion. Food Res. 2023;7:10–20.
135. Hamidon FN, Faridah MR, Asyrul-Izhar AB, Sani MSA, Ismail-Fitry MR. Effect of fat replacement with different types of eggplants on the physicochemical and sensorial properties of chicken sausages: a chemometric approach. Malays J Anal Sci. 2022;26:1313–31.
136. Wang B, Chen J, Wang Q, Zhu Y, Wang P, Xu X. Functional performance of a novel emulsion gel-based pork fat mimics in low-fat meat batter system: incorporation of physicochemical and oral processing. Food Struct. 2023;37:100335.
137. Pintado T, Muñoz-González I, Salvador M, Ruiz-Capillas C, Herrero AM. Phenolic compounds in emulsion gel-based delivery systems applied as animal fat replacers in frankfurters: physico-chemical, structural and microbiological approach. Food Chem. 2021;340:128095.
138. Zhao X, Guo R, Li X, Wang X, Zeng L, Wen X, Huang Q. Effect of oil-modified crosslinked starch as a new fat replacer on gel properties, water distribution, and microstructures of pork meat batter. Food Chem. 2023;409:135337.

139. Rather SA, Masoodi FA, Akhter R, Gani A, Wani SM, Malik AH. Effects of guar gum as fat replacer on some quality parameters of mutton goshtaba, a traditional Indian meat product. Small Rumin Res. 2016;137:169–76.
140. Bibat MA, Ang MJ, Eun JB. Impact of replacing pork backfat with rapeseed oleosomes–natural pre-emulsified oil–on technological properties of meat model systems. Meat Sci. 2022;186:108732.
141. Zhang H, Zhang W, Zeng X, Zhao X, Xu X. Recent progress of fat reduction strategies for emulsion type meat products. Food Mater Res. 2022;2:10.

Meat Products with Low Nitrites

Sadhna Ojha, Gauri Jairath, Suman Bishnoi, Gorakh Mal, Monika Verma, Mukesh Gangwar, Nilakshi Chauhan, and Ashok Kumar Pathera

1 Introduction

History has registered several preservative techniques before the invention of new-aged preservative techniques like refrigeration to prolong the shelf life of food products, including muscle foods. The initial objective of the preservation technique was to make the meat or food product available during times of scarcity; however, with time, a paradigm shift has occurred towards the food safety aspect. Amongst several traditional preservative techniques, curing, which originally emerged from the salt preservation technique, is a widely acceptable technique, practised since 3000 B.C. [1]. The curing process involves the application of several curing

S. Ojha · M. Gangwar
Department of Livestock Products Technology, College of Veterinary and Animal Sciences, Kishanganj, Bihar, India

G. Jairath (✉) · G. Mal
ICAR-Indian Veterinary Research Institute, Regional Station, Palampur, Himachal Pradesh, India
e-mail: gauri.jairath@icar.gov.in

S. Bishnoi · M. Verma
Department of Livestock Products Technology, College of Veterinary Sciences, LUVAS, Hisar, Haryana, India

N. Chauhan
Department of Food Science Nutrition and Technology, College of Community Science, CSKHPKV,
Palampur, Himachal Pradesh, India

A. K. Pathera
Amity Institute of Food Technology, Amity University Uttar Pradesh, Noida, Uttar Pradesh, India

A. Kumar Pathera et al. (eds.), *Healthier Meat Products*,
https://doi.org/10.1007/978-3-031-78215-2_18

ingredients, such as salt, nitrite, ascorbate, etc., while processing and comminution to prolong the shelf life and enhance the flavour profile and food safety [2].

Each curing ingredient plays a specific role in the marinade. Among the curing ingredients, nitrite salt plays a pivotal role as an essential additive in the process of meat curing. Though it can only be incorporated up to 150 parts per million (ppm) due to safety concerns, it not only serves as an effective preventive agent at this level against the proliferation of harmful microorganisms, especially, *Clostridium botulinum*, a food poisoning-causing pathogen in meat and meat products, but also acts as an antimicrobial agent besides imparting a blooming red hue (nitroso-hemoglobin) to a variety of meat products like sausages, hams and pickles. This additive further contributes to the overall quality of cured meat products by enhancing their texture, tenderness, and aroma [3–6]. Additionally, nitrite inhibits the occurrence of lipid oxidation in meat products. This function proves instrumental in effectively preventing the development of rancidity [7]. The enzymatic production of nitric oxide is also important within the human body, as it serves essential regulatory roles across a variety of physiological processes. Nitric oxide actively contributes to the regulation of blood pressure, facilitates wound healing, enhances immune responses, and plays a role in diverse neurological processes, as outlined in studies by Hunault and co-wokers [8]. However, it is imperative to note that elevated concentrations of nitrite pose a considerable risk, particularly to infants, as they can lead to a condition known as infant methemoglobinemia. Additionally, the interaction of nitrite with secondary or tertiary amines results in the formation of nitrosamines, recognized for their carcinogenic properties [9, 10].

In the year 2015, Richi et al. carried out an investigation uncovering that extended intake of growing amounts of processed meat is responsible for increased death rates and intensified vulnerabilities to diseases like Type-II diabetes, bowel (colorectal) cancer and cardiac ailments in human beings, irrespective of their gender [11]. The collaborative committee, comprising experts from FAO (Food and Agriculture Organization) and WHO (World Health Organization), has set a permissible daily allowance of 70 μg of nitrite per kg body weight, considered safe for infants, children, and adults in good health. However, it is important to emphasize the potential health risks associated with excessive nitrite intake and the importance of adhering to recommended doses [12]. Further, a joint publication by the World Cancer Research Fund and the American Institute of Cancer Research in 2007 also disclosed a moderate but significant association between heightened dietary intake of processed meat and an increased risk of colorectal cancer with excessive consumption further amplifying this risk [13].

Although reducing meat consumption may not be a complete solution for mitigating carcinogenic effects, it is challenging in terms of the potential loss of nutrients, especially iron. Consequently, actively investing in healthier meat production may be more feasible than relying solely on reducing processed meat consumption. In response to growing concerns about the long-term adverse effects of sodium nitrite, many developed countries have implemented stringent regulations governing the use of processed meat accordingly. Concepts such as "all-natural" and "clean labels" are gaining ground in the marketplace.

In today's market, consumers prefer natural additives over chemical additives in processed meat due to associated health risks. Thus, today's demands to explore natural alternatives to nitrite salt as a healthier option, as suggested by Sindelar et al. [14] are a need of the hour. Till now, no substitute for nitrite that may match the multi-faceted benefits of nitrite is available in the processed meat market. This challenge may be addressed to some extent by applying hurdle technology that offers the use of two or more hurdles at the same time, which would ultimately decrease or replace the nitrite content while processing [15]. Moreover, phytoextracts have shown the potential to substitute nitrite to some extent. Thus, an attempt has been made while compiling the chapter to discuss the safety concerns associated with nitrite and the possible substitutes, that may be used in the meat industry while curing to enhance their shelf-life with improved colour and flavour.

2 Origin and Safety Implication of N-Nitroso Compounds in Meat and Meat Products

The safety concerns are associated with the use of nitrites due to the formation of N-nitroso compounds, especially N-nitrosamines and N-nitrosamides. N-Nitrosodimethylamine and N-nitrosomethylethylamine are two N-nitrosamines, among others, that are frequently found in meat products and are hazardous due to their carcinogenic and genotoxic effects [16]. Recently, N-nitroso-diisopropylamine has also been detected in sausages and ham under GC/MS spectrometry, which are more carcinogenic than N-nitrosodimethylamine and N-nitrosomethylethylamine and thus have been classified under Group 1B carcinogen [17, 18]. This recent detection of N-nitroso-diisopropylamine has raised food safety concerns and speculations about the potential presence of other undetectable compounds in meat products. The formation of these nitroso compounds is further augmented by the incorporation of organic acids in the presence of naturally occurring thiocyanates. During the formation of these compounds, microbes with nitrate-reducing activity also play an important role and promote their formation process. These safety concerns can be well addressed by incorporating vitamins C and E during meat processing.

In 2007, the World Cancer Research Fund made a declaration stating that the consumption of red or processed meat likely increases the risk of developing colorectal cancer [19]. The International Agency for Research on Cancer classified all types of mammalian muscle meat, encompassing beef, veal, pork, lamb, mutton, horse, and goat, as probably carcinogenic (Group 2A). Interestingly, treated/processed meat, characterized as meat subjected to procedures such as salting, curing, fermenting, smoking, or other techniques to enrich taste or maintain quality, is recognized as carcinogenic to individuals (Group 1), recognizing the increased cancer hazard linked with the ingestion of red and processed meat [20]. As per Ferysiuk and Wojciak [18], roughly 80% of nitrite exposure and 85% of nitrate exposure stem from the dietary intake of vegetables, whereas intake of meat and meat products accounts for only about 5% of nitrite exposure.

3 Nitrite Sources

Nitrites contribute significantly to the intricate nitrogen biogeochemical cycle in natural aqueous contexts. They pervade various surroundings, encompassing earth, aquatic bodies, comestibles, flora, atmosphere (in the guise of nitrogen dioxide), and biological specimens [16]. Nitrites contribute significantly to the complex nitrogen cycle observed in the natural world, influencing various sources. In the biological nitrogen cycle, bacteria play a pivotal role in converting nitrogen to nitrate, which is subsequently assimilated by plants and integrated into their structures at tissue levels. Animals, in turn, utilize nitrate obtained from consuming these plants to synthesize proteins. The reintegration of nitrate into the environment occurs through processes such as animal excretion and microbial decomposition of deceased organisms, thereby sustaining the nitrogen cycle. Notably, microorganisms can convert nitrate or ammonium ions into nitrite, a phenomenon occurring both in environmental settings and inside the gastrointestinal systems of humans and non-humans.

Nitrite serves as the intermediary compound resulting from nitrification in the overall nitrogen cycle. The nitrogen cycle is completed as environmental bacteria convert nitrate into nitrites and then into nitrogen. This intricate microbiological process involves various bacterial species, including Nitrosomonas and Nitrobacter, orchestrating the transformation of ammonium into nitrates through the intermediate production of nitrite—a process commonly known as nitrification [21]. The detailed specifics of the sources of nitrite and nitrate, along with their status of consumers' exposure, are comprehensively outlined in Table 1.

4 Efficacy of Nitrite and Nitrate on Meat Excellence and Potential Substitutes

4.1 Colour Formation Effect

Researchers are actively investigating alternative substances containing natural nitrite or nitrate to impart a cured colour (pinkish-red colour) to food items. Prominent among these alternatives are vegetable extracts and microbial sources, which are undergoing thorough examination for their efficacy. A recent study conducted by Jarulertwattana and co-workers revealed that a combination of ginger paste and a nitrite dosage of 4×10^6 g/g was effective in marinating chicken meat. Similarly, Zhu et al. [22] explored the use of *Lactobacillus plantarum* as a substitute

Table 1 The sources of nitrite and nitrate and their exposure status to humans [17]

Sources	Approaches of invading the human body	Exposure status to human	References
Vegetable and its products	Vegetable products, cooked or fresh vegetables, and vegetable extracts	About 80% nitrite exposure, 85% nitrate exposure	[33, 72]
Oral reduction of nitrate	Nitrate-reducing bacteria can reduce about 5–7% of all ingested nitrate to nitrite	About 70–80% of the human total nitrite exposure	[74–76]
Water	Drinking water, beverage, aquatic products, liquid condiments, agriculture products fertilized with natural organic wastes or overused fertilizers	About 5–40% dietary nitrate intake, if the nitrate level in drinking-water is more than 50 mg, it will be the main source of exposure to nitrates	[33, 77]
Meat and meat products	Fresh meat, chilled meat, freezing meat, cured meat products, sauced and stewed meat, roasting meat burning smoke, dried meat, fried meat products, sausage meat products, prepared meat products, ham sausage, others	About 5% nitrite exposure, it depends the category of meat and meat products	[18, 44]
Dairy products	Liquid milk, milk powder, other dairy products	It depends on the amount of consumption	[78]
Fruit and its products	Fresh fruits, fruit juice beverage, canned fruit, dried fruit slices, jam, fruit powder, pickle fruit	-do-	[18, 79]
Herbs	Herbal medicine, plant seasoning, functional foods, beverage	-do-	[44]
Cereals	Fertilizer, irrigation water	It depends on the plant management	[33]
Medicine	Therapeutic drugs	It depends on the patient's conditions	[33]
Food additives	L-arginine, decarboxylase and amino acid	It depends on the addition amount and process.	[80]

for nitrite in pork sausages, demonstrating that the combination of 50 ppm sodium nitrite with 7 log CFU/g *Lactobacillus plantarum* yielded comparable results to 100 ppm sodium nitrite.

Recently, an attempt has been made by Ozaki et al. [23] to substitute nitrite with dried powder of radish and essential oil derived from oregano. Radish powder incorporated at 0.5% or 1.0% and oregano essential oil at 100 ppm in the fermented cooked sausages made up of pork, beef, and pork back fat, resulting in an enhanced red hue and an improved sensory and microbial profile. However, lipid oxidation and microbial growth were the concerns in the case of sausages treated with oregano essential oil.

An attempt was also made by Higuero et al. [24] to identify the minimal incorporation level of nitrite/nitrate in meat products by examining the change in colour dynamics, nitrosylmyoglobin content, and lipid oxidation with lower and higher levels of nitrite/nitrate incorporation and the duration of drying in dry-cured loins. The potassium nitrate and sodium nitrite studies were incorporated in ppm as 150 NO_3^- + 150 NO_2^-; 75 NO_3^- + 75 NO_2^-; 37.5 NO_3^- + 37.5 NO_2^- and NO_3^-/NO_2^- (0 ppm). The oxidative stability was significantly lower in products without nitrate/nitrite. The removal of these additives in the product affects the colour coordinates with higher nitrosylmyoglobin levels, which were sensorially perceived by consumers. The study concluded that nitrite can be reduced up to 37.5 mg/kg without any adverse effects.

Kim et al. attempted to replace sodium nitrite ($NaNO_2$) in emulsified pork sausages with a paprika oleoresin (PO) solution and studied the effect of replacement during refrigerated storage. Sodium nitrite was added at 150 ppm; 75 ppm $NaNO_2$; 75 ppm $NaNO_2$ plus 0.1% PO (1% paprika oleoresin and 99% sunflower seed oil); and 75 ppm $NaNO_2$ plus 0.1% POS (5% paprika oleoresin and 95% sunflower seed oil). PO increased the redness and yellowness and decreased the lightness significantly in comparison to sausages with 150 and 75 ppm of nitrite. The oxidative stability was minimal in nitrite-containing sausages compared to PO. The study concluded that PO can be effectively used as a nitrite replacer [25].

4.2 *Antimicrobial Effect*

Being bacteriostatic, the incorporation of nitrate and nitrite helps to extend the shelf life of meat and meat products by preventing the growth of spoilage-causing bacteria and food-borne pathogens like Achromobacter, Aerobacter, *Listeria*, *Salmonella*, *Clostridium botulinum*, *Clostridium perfringens*, *Flavobacterium*, *Micrococcus* spp, etc. The ability of nitrite to combat pathogens is due to (I) disruption of oxygen consumption and energy production processes, (II) blocking key enzymes such as aldolase, glyceraldehyde-3-phosphate dehydrogenase, and nitrogen-fixing enzymes, and (III) generating nitrite-based compounds with bactericidal properties [26]. The protective action of nitrite is twofold: it varies with concentration, effectively stopping *C. botulinum* spore growth, and it strengthens as the meat's pH decreases, with its antimicrobial power intensifying significantly with each drop in pH level. The lowest concentration of nitrite effective in preventing the growth of *Clostridium botulinum*, ranges from 4 to 8 ppm [26].

In a recent study conducted by Khatib et al. [27], nitrite was substituted with hop components (lupulon and xanthohumol) in cooked beef sausage using lupulon-xanthohumol loaded nanoliposomes (L–X-NL) created via sonication (10.8 min and 72.7 W) with a lecithin concentration of 140 mg/mL. The study concluded that a mixture of 30 ppm nitrite and 150 ppm L–X-NL remained safe over 30 days of storage at 4 °C in terms of microbial safety and lipid oxidation without any adverse effects on sensory qualities. The study augmented the reduction of nitrite by up to 50% and substituted it with L–X-NL for enhanced shelf life.

4.3 Flavour Improvement

Nitrite plays a great role in developing the cured flavour. Because of the safety concerns, a paradigm shift has been taken to explore the role of microbes in developing cured flavour.

In this context, the incorporation of *Debaryomyces hansenii*, a yeast, in dry fermented sausages has shown promising results in developing a cured aroma. The role of this yeast strain lies in its ability to synthesize esters and sulfur-containing compounds [28]. Perea-Sanz et al. revealed that *Debaryomyces hansenii* synthesizes ethyl ester and 3-methylbutanal in dry-fermented sausage that gives a cured flavour while improving the oxidative stability [28]. Tian et al. [29] also exhibited the role of *Lactobacillus fermentum* RC4 as a nitrite replacer in salted fermented meat. The study concluded that *Lactobacillus fermentum* enhances the flavour of meat during fermentation by producing 73 identified volatile compounds in comparison to the non-fermented control group, where only 67 volatile compounds were detected. Along similar lines, Zhao and co-workers utilized TR1-1-3 and ZF22 strains of *Lactobacillus helveticus* strains to enhance the flavour profile of mutton sausages. In this study, compounds like 1-pentene-3-ol, 1-octene-3-ol, and 3-hydroxy-2-butanone were detected that gave a distinctive aroma to the fermented sausages [30].

Furthermore, Huang et al. utilized *Lactobacillus* starter cultures like *Lactobacillus fermentum* RC4 and *L. plantarum* B6 to enhance the flavour of cured meat. The study found that these strains resulted in the production of volatile compounds like D-limonene, nonanal, (E, E)-2,4-decadienal, 1-octene-3-ol, and anethole in cured meat [31].

4.4 Antioxidant Effect

The antioxidant effects of nitrite are mediated by the release of nitric oxide from nitrite that, under oxidation, reduces the surrounding oxygen and attaches to heame iron, thereby inhibiting lipid oxidation [17]. The nitrite replacers like L–X-NL mentioned in Sect. 4.2. have been explored in beef sausage by Khatib et al. recently for their effectiveness in lipid oxidation prevention [27]. Other examples of evaluating phosphorylated nitrosohemoglobin efficacy as an antioxidant have also been reviewed by Zhang and co-workers [17], wherein the antioxidant potency of phosphorylated nitrosohemoglobin was compared with sodium nitrite in emulsified sausage. The study exhibited a non-statistical higher antioxidant potential of phosphorylated nitrosohemoglobin with thiobarbituric acid values of 0.62 and 0.67 ppm, respectively, for phosphorylated nitrosohemoglobin and sodium nitrite.

4.5 *Extended Shelf-Life*

Shelf-life of food items, including muscle foods, depends on multiple interlinked factors, like hygienic status while handling raw materials, processing, packaging, and storage conditions like temperature and humidity. The packaging material and type of packaging are also decisive factors [17]. The bacterial load majorly determines the shelf-life of meat as it serves as a growth medium for microbes, and the nitrite/nitrate enhances the shelf life by curbing the bacterial growth effectively [26]. Recently, in the year 2020 [32], nitrite was incorporated into packaging material with thermoplastic starch and low-density polyethylene and was utilized to package pork under vacuum conditions. The colour of vacuumed pork was found to be 30% improved throughout the retail display. The pork was found to be more oxidative and microbially stable. The shelf-life of cooked beef sausage can be enhanced by 30 days under retail display by incorporating L–X-NL as a nitrite replacer due to its antimicrobial and antioxidant properties [27].

5 Nitrites Incorporation in Meat: Legislation and Regulatory Framework

The irreplaceable multifunctionality of nitrite and nitrate in meat and animal products presents a challenge when considering alternative food additives. Completely prohibiting their use may not be a practical solution, given the diverse roles they play, and such a prohibition might encounter resistance from both producers and consumers [33]. The historical context of regulating nitrite in meat dates back to 1934 in Germany, where the Nitrite Curing Salt Law permitted its application, ranging from 0.5% to 0.6% when mixed with salt [34]. Danish legislation allows the addition of 60 mg/kg nitrite to most products [BEK nr 1247, 2020]. However, under Regulation (EC) No. 1333/2008 of the European Parliament and the Council, specific limits are set for nitrite in meat products: a maximum of 100 mg/kg for sterilized meat products and 150 mg/kg for other meat products [Regulation (EC) No. 1333/2008]. Navigating these regulations reflects the ongoing challenge of balancing the necessity of nitrite and nitrate in meat processing with the imperative of ensuring safety and adherence to standards.

On the contrary, in the United States, as stipulated by Federal Regulations [35], sodium nitrite levels in finished products should generally not surpass 200 ppm, and for sodium nitrate, the limit is set at 500 ppm, depending on the specific application and product type. In Canada, the total nitrite content, considering sodium nitrite alone or in combination with potassium nitrite, should not exceed 200 ppm for most products. However, for curing side bacon, the total nitrite amount should not exceed 120 ppm [36]. These regulations underscore the careful calibration required to balance the beneficial roles of nitrite and nitrate with the imperative of ensuring food safety standards.

The Australia New Zealand Food Standards Code meticulously outlines the regulatory landscape for nitrite levels in various meat products. For commercially sterile canned cured meat, the code imposes a stringent maximum permissible level of 50 mg/kg. Similarly, for dried meat, cured meat, slow-dried meat, and processed comminuted meat, poultry, and game products, the prescribed limit is 125 mg/kg, aligning with the specifications in Regulation (EC) No. 1333/2008. While these standards provide a crucial framework, Ferysiuk and Wojciak [18] propose a potential reduction in nitrite utilization in meat products. However, they acknowledge that achieving this reduction would necessitate the implementation of additional steps in the production process. Notably, India has set a distinct maximum limit for potassium nitrite in fermented, non-heated treated processed meat and poultry products at 80 mg/kg, as outlined by FSSAI [37]. Highlighting the global perspective, a study conducted by Abd Hamid et al. [38] in Brunei Darussalam uncovered that over 20% of school children surpassed the Acceptable Daily Intake of nitrite. This emphasizes the need for consistent adherence to regulations and directives, underscoring the importance of regular inspections and monitoring of market samples to ensure compliance. For a comprehensive understanding, Table 2 provides an overview of the prescribed limits for nitrite and nitrate addition in muscle products across the world.

6 Health Risks Associated with High Nitrite Meat Products

Elevated nitrite levels pose a significant health risk, particularly in inducing methemoglobinemia, a condition characterized by reduced oxygen-carrying capacity of blood. This concern is especially pronounced in infants below 6 months, who exhibit heightened vulnerability due to their immature methemoglobin reductase system and lower stomach acid production [39]. The intricate processes of nitrite exposure go beyond methemoglobinemia, extending to the production of

Table 2 Addition limit of nitrite and nitrate in meat and meat products in different countries [17]

Country or organization	Meat and meat products	Limit of nitrite	References
The United Nations	Cured ham and cooked cured pork shoulder	$\leq 8 \times 10^{-5}$	[81]
	Heat-treated processed meat, poultry, and game products in whole pieces or cuts	$\leq 8 \times 10^{-5}$	
	Processed comminuted meat, poultry, and game products	$\leq 8 \times 10^{-5}$	
	Luncheon meat, cooked cured chopped meat corned beef	$\leq 8 \times 10^{-5}$	

(continued)

Table 2 (continued)

Country or organization	Meat and meat products	Limit of nitrite	References
The United States	Pumped and/or massaged bacon	$\leq 1.2 \times 10^{-4}$	[82]
	Immersion-cured bacon	$\leq 1.2 \times 10^{-4}$	
	Comminuted meat and poultry products	$\leq 1.56 \times 10^{-4}$	
	Massaged or pumped meat and poultry products	$\leq 2 \times 10^{-4}$	
	Immersion cured meat and poultry products	$\leq 2 \times 10^{-4}$	
	Dry cured bacon The ingoing level of sodium (or potassium)	$\leq 2 \times 10^{-4}$	
	Dry cured meat and poultry products	$\leq 6.25 \times 10^{-4}$	
European Union	Canned meat products	$\leq 1.5 \times 10^{-4}$	[83]
	Meat products	$\leq 1.5 \times 10^{-4}$	[70]
	Other cured meat products	$\leq 1.5 \times 10^{-4}$ Residual amounts (Sodium nitrite) $\leq 1 \times 10^{-4}$	[55]
	Cured bacon	$\leq 1.5 \times 10^{-4}$ Residual amounts (Sodium nitrite) $\leq 1.75 \times 10^{-4}$	[55]
	Dry cured bacon	Residual amounts (nitrites and nitrates) 4.25×10^{-4}	[75]
	Non-heat-treated processed meat Maximum added amount (sodium) or	$\leq 1.5 \times 10^{-4}$	[55, 69]
	Sterilised meat products	$\leq 1.5 \times 10^{-4}$	[69]
	Sterilised meat products	$\leq 1 \times 10^{-4}$	[70]
	Other traditionally cured meat products	$\leq 1.8 \times 10^{-4}$ Maximum residual sodium nitrite level is $\leq 5 \times 10^{-5}$	[55, 70]
	Only sterilised meat products	$\leq 1 \times 10^{-4}$	[69]
	Traditional immersion cured meat products (number or products)	5×10^{-5}–1.75×10^{-4}	[70]
	Traditional dry cured meat products (number of products)	Maximum residual sodium nitrite level is 5×10^{-5}–1.75×10^{-4}	[70]

(continued)

Table 2 (continued)

Country or organization	Meat and meat products	Limit of nitrite	References
Asian countries			
China	Cured meat products [such as bacon, cured meat, salted duck, Chinese ham and sausage]	Maximum addition amount (sodium/potassium nitrite) is $\leq$0.5 g/kg Based on sodium nitrite, the residue is $\leq 3 \times 10^{-5}$	[84]
	Sauced and stewed meat products	Maximum addition amount (sodium/potassium nitrite) is $\leq$0.5 g/kg Based on sodium nitrite, the residue is $\leq 3 \times 10^{-5}$	[84]
	Smoked and roasted meat products	Maximum addition amount (sodium/potassium nitrite) is $\leq$0.5 g/kg Based on sodium nitrite, the residue is $\leq 3 \times 10^{-5}$	[84]
	Fried meat products	$\leq$0.5 g/kg Based on sodium nitrite, the residue is $\leq 3 \times 10^{-5}$	[84]
	Western style ham (smoked and roasted, smoked, stewed)	$\leq$0.5 g/kg Based on sodium nitrite, the residue is $\leq 7 \times 10^{-5}$	[84]
	Sausage meat products	$\leq$0.5 g/kg Based on sodium nitrite, the residue is $\leq 3 \times 10^{-5}$	[84]
	Fermented meat products	$\leq$0.5 g/kg Based on sodium nitrite, the residue is $\leq 3 \times 10^{-5}$	[84]
	Canned meat products	$\leq$0.5 g/kg Based on sodium nitrite, the residue is $\leq 5 \times 10^{-5}$	[84]
India	Fermented nonheated treated processed meat and poultry products in whole pieces or cuts	Maximum level $\leq 8 \times 10^{-5}$	[85]
	Heat-treated processed meat and Poultry products in whole pieces or cuts (canned chicken, canned mutton and goat meat)	Maximum level $\leq 8 \times 10^{-5}$	[85]
	Processed comminuted meat and poultry products	$\leq 8 \times 10^{-5}$	[85]

N-nitrosamines in meat. This phenomenon occurs not only during production and processing but also within the gastrointestinal tract post-consumption [40].

N-nitrosamines are broadly categorized into non-volatile (e.g., N-nitrosoproline, N-nitrososarcosine, and N-nitrosomethylaniline) and volatile types (e.g., N-nitrosodimethylamine, N-nitrosomethylethylamine, and N-nitrosodiethylamine)

[41, 42]. It's noteworthy that most volatile nitrosamines are acknowledged as carcinogenic to humans, as elucidated by EFSA [42]. Bryan et al. [39] studied the carcinogenic impact of N-nitrosamines on specific organs and their potential teratogenic effects. Intriguingly, the cooking process at high temperatures and the curing of meat can contribute to the formation of N-nitroso compounds. Furthermore, the presence of unabsorbed nitrosylated heme in the colon, originating from red meat, stands as an additional source of N-nitroso compounds. These multifaceted pathways underscore the intricate nature of nitrite-related health implications.

7 Potential Alternatives to Nitrite

In response to the involvement of nitrite in nitrosamine production, the meat industry is actively exploring innovative approaches to replace traditional sodium nitrite in cured meats, thereby mitigating nitrite exposure. There is a growing consumer interest in the pursuit of natural possible substitutes and potential preservative methods perceived as healthier. However, the versatile antimicrobial actions of nitrite pose a challenge in finding a singular substitute. A blend of nitrite and alternative antimicrobial agents emerges as a potential strategy to effectively replace nitrite in meat products [14]. While advancing consumption safety, it is imperative not to compromise the unique characteristics of organic and naturally processed meats, aligning with consumer preferences. The quest for a nitrite substitute and the development of novel products raise questions about the extent to which these changes can capture consumer interest. Preferences for meat products with lower nitrite levels underscore the significance of understanding nitrite's function, reasons for its application, and resultant outcomes in consumer purchasing decisions [43]. Consequently, successfully reducing nitrite levels in meat products, coupled with the incorporation of diverse alternatives, stands poised to yield various benefits for consumers, including a reduction in carcinogenic substances.

7.1 Extracts from Plants

Nitrate, a prevalent compound in green-leaf vegetables, is found abundantly in celery, lettuce, cress, spinach, and rucola, with concentrations exceeding 2500 mg nitrate/kg [44]. These vegetables present a viable option as either a partial or complete substitute for chemical nitrite in meat curing, given that various microorganisms can facilitate the reduction of nitrate to nitrite. Parsley, characterized by a substantial nitrate content ranging from 1000 to 2500 mg/kg on average, emerges as a potential replacer for nitrite in processed meat applications [45]. Notably, mortadella-type sausages produced with nitrite derived from parsley extract powder demonstrate equivalence to traditionally cured sausages regarding the inhibition of *Listeria monocytogenes* and the mitigation of microbes-originated spoilage during storage. The incorporation of parsley extract powder enables the achievement of reduced residual nitrite levels, presenting an avenue to lower customers' nitrite intake. The commercial viability of applying parsley extract powder is supported by consumer acceptability, as demonstrated in a recent study [16].

Celery stands out as a promising substitute for sodium nitrite in the context of cured or processed meats, as evidenced by sausages developed with celery powder yielding comparable results to the control group (sodium nitrite). Parameters such as thiobarbituric acid, pH, total microbial count, volatile basic nitrogen content, and sensory evaluations exhibited similar outcomes between sausages with celery powder and those treated with sodium nitrite [46]. Noteworthy is the additional benefit offered by celery powder in enhancing the meat's visual appeal while concurrently reducing the residual nitrite content. This dual effect stems from celery's natural composition, which encompasses pigments and nitrate. In a specific application, 75.6 g of its concentrate showcased the ability to impede the proliferation of *Listeria monocytogenes*, mirroring the efficacy of nitrite in sodium salt at a concentration of 100 ppm [46].

Additionally, the Swiss chard powder obtained after spray drying has surfaced as a natural source of nitrate akin to celery powder, exhibiting roughly 3–3.5 g of nitrate per 100 g. Swiss chard powder upon spray drying has emerged as a natural provider of nitrate comparable to celery powder, containing approximately 3.0–3.5% nitrate content, alongside phenolic acid and flavonoid constituents [47]. The incorporation of Swiss chard powder into the formulation not only contributes to colour consistency in cooked pork patties but also extends the product's shelf life due to its inhibitory effects on coliform bacteria and lipid oxidation [48]. These findings underscore the potential of natural sources like celery and Swiss chard to reformulate meat products to meet both aesthetic and safety criteria.

Barberry extract emerges as a noteworthy substitute for nitrite in processed meat applications. In the context of cooked sausages, the utilization of a barberry extract at 90 ppm, in conjunction with a nitrite level of either 30 ppm or 60 ppm, demonstrated a substantial extension of shelf life when subjected to cold storage conditions (4 °C). Beyond its preservative attributes, the incorporation of barberry extract exerted a positive influence on antioxidant properties, effectively curbing the formation of carcinogenic nitrosamines [49]. For the production of chouriço, red wine or a combination of red wine and garlic proves to be a viable strategy for controlling biological hazards. This addition not only enhances the destruction of Salmonella during meat processing but also imparts a robust cured flavour to the product. Furthermore, it contributes to the overall colour enhancement of chouriço, presenting itself as a potential alternative to conventional nitrite salt usage [2]. In Turkish fermented sausage manufacturing, beetroot powder has been explored as a substitute for nitrite. The introduction of beetroot powder brought about discernible alterations in various quality parameters, notably impacting the redness value of the product. Importantly, sensory analysis revealed no significant deviations when compared to the control group, attesting to the viability of beetroot powder as a nitrite alternative [50].

Aquilani et al. conducted a comprehensive study investigating the impact of grape seed and chestnut extract in combination with olive pomace hydroxytyrosol on the colour attributes of dry-fermented sausages (*Cinta senese*). The incorporation of grape seed extract led to elevated redness values, potentially attributed to the formation of Zn-protoporphyrin. Notably, these extracts exhibited inhibitory effects on lipid oxidation and demonstrated a reduction in microbial counts, encompassing *Listeria monocytogenes*, *E. coli*, *Clostridium spp.*, *Staphylococcus spp.*, and *Salmonella spp.*, throughout a 3-week ripening period [51]. In another context, the successful application of dried powder of tomato pulp to muscle foods was explored by Tan et al. [52]. This addition proved to be impactful without compromising the physicochemical characteristics of the meat. The incorporation of tomato pulp powder contributed to heightened colour values, enhanced microbiological stability, imparted a savoury taste, diminished the oxidation process, and overall enhanced the consumers' acceptability. Furthermore, the utilization of wastes (seeds and peels) from the tomato processing industries was highlighted by Eyiler et al. [53]. These by-products contain a myriad of bioactive compounds, notably carotenoids like β-carotene, lycopene, phytofluene, phytoene, and lutein, showcasing considerable bio-functional potential.

Furthermore, the incorporation of grape pomace also presents a viable option as a partial substitute for nitrite in meat processing. However, the introduction of substantial quantities may induce alterations in the colour and oxidation parameters of beef sausages. Despite these modifications, sensory panellists deemed the sausages acceptable. Furthermore, the addition of dry red grape pomace to sausages demonstrated its efficacy in preventing bacterial growth, reducing residual nitrite levels, lowering pH, and extending the overall shelf life. In a broader context, grape pomace powder emerges as a promising and health-conscious ingredient for dietary formulations [54]. On a similar note, annatto seed powder, recognized as a natural colourant prevalent in Asian cuisine, proves to be a potential substitute for nitrite in pork sausage production. Its inclusion enhances the colour and texture of sausages while concurrently exerting inhibitory effects on lipid oxidation and microbial proliferation.

7.2 Organic Acids and Salts

Organic acids, owing to their multifaceted benefits, include lowering pH and microbial load and accelerating the curing process during the development of processed cured products [55]. Varieties of organic acids, including lactate, sorbate, acetate, and benzoate, are extensively employed in the meat industry to enhance colour development and hinder microbial proliferation during the curing process [56]. Amongst organic acids, lactic acid is the most commonly used during processing and the most preferable one, as other acids serve as antimicrobial agents only. However lactic acid, in addition to its antimicrobial effect, introduces not only salty flavour to the meat but also preserves the colour profile and improves the water holding capacity [57].

Sodium lactate serves as a potent agent for preventing undesirable flavours in meat by effectively scavenging free radicals and impeding lipid oxidation. Its combination with sodium chloride demonstrates a remarkable reduction in oxidative reactions, thereby effectively mitigating lipid oxidation in stored ground pork [52]. In addition to its antioxidative properties, lactate plays a crucial role in enhancing the colour stability of fresh meat. This is achieved by elevating lactate dehydrogenase activity and fostering the generation of nitric oxide, contributing to the maintenance of a persistent pinkish-red colour and safeguarding myoglobin pigments from oxidation over time [56, 58, 59].

Furthermore, lactate exhibits noteworthy antimicrobial properties, leading to a reduction in bacterial populations within meat products. Elevated concentrations of lactate can modulate pyruvate conversion, and sodium lactate, functioning as a weak acid, permeates bacterial membranes, creating an acidic intracellular environment [60, 61]. The application of sodium lactate proves effective in diminishing bacterial growth and extending the shelf life of meat products. The inhibition of *Clostridium perfringens* growth is contingent upon various factors, and in comparison to sodium nitrite, sodium lactate demonstrates improved antimicrobial action and enhanced microbiological quality in sausages [61].

Sodium lactate, in combination with sodium chloride, exerts a decelerating effect on bacterial growth, extending the shelf life of uncooked ground beef while inhibiting the development of *Clostridium perfringens*. This antimicrobial action extends to processed beef goulash slices and poultry sausages, preventing the proliferation of various bacteria during cold storage. Additionally, the utilization of buffered sodium citrate or a combination of buffered sodium citrate and sodium diacetate proves effective in hindering the growth of *Clostridium perfringens* in injected pork and roast beef during chilling process [62].

The introduction of potassium lactate, at a concentration of 3%, to beef slices and frankfurters has demonstrated the ability to enhance or preserve their inherent colour, flavour, and sensory taste. Concurrently, this application ensures the maintenance of lipid oxidation levels, shear properties, and cooking characteristics. Even when applied to the outer surface of frankfurters, this concentration of potassium lactate causes only a slight reduction in water activity, with negligible or minimal alterations reported by sensory panellists. A study involving 3.3% sodium lactate exposure revealed the inhibition of *Lactobacillus alimentarius* and other lactic acid bacteria growth in both cooked and acidified chicken meat for 14 days. Furthermore, common additives like sorbate, propionate, and benzoate, known for their safety, exhibit mold-growth preventive properties in various foods and can also impede the development of Gram-positive bacteria, including *Staphylococcus aureus*, *Listeria monocytogenes*, and *Clostridium botulinum*, in meat systems [63, 64].

While sodium sorbate cannot entirely substitute for sodium nitrite in meat products, particularly in terms of colour development, a synergistic combination of sodium sorbate (2600 ppm) and sodium nitrite (80 ppm) emerges as a potential alternative to the sole use of nitrite (120 ppm). However, it's crucial to note that elevated concentrations of sorbate, specifically at 2600 mg/kg, may lead to increased product rancidity and have a detrimental impact on the sensory quality of mortadella when compared to lower concentrations (1000 mg/kg) or the use of nitrite alone. This finding emphasizes the importance of carefully considering the concentration levels of sodium sorbate in meat formulations to achieve the desired outcomes without compromising product quality [65].

7.3 Cooking Method

Nitrite is susceptible to oxidation in high-temperature aerobic environments, such as during cooking, leading to a decrease in its content in meat or meat products. In uncooked sausage models, nitrate reduction and nitrite accumulation were observed; however, Waga et al. [66] noted that nitrate inhibits nitrite auto-decomposition in pork ham when cooked in a water bath at 75 °C. In a study by Asioli et al. [67], investigating the effects of dry-frying temperatures (100, 150, 200, and 250 °C) on residual nitrite and N-nitrosamine levels in smoked bacon, it was found that the residual nitrite content initially increased (from unheated control to 150 °C) and then sharply decreased between 150 and 250 °C.

Furthermore, Sallan et al. [68] conducted a study on the impact of cooking on volatile nitrosamine formation in *sucuk*, a semi-dry fermented sausage. Their findings revealed that higher temperatures and heat intensity led to an increased conversion of nitrite into nitrosamines. The levels of nitrite and the cooking process also influenced the N-nitrosopiperidine content. The study suggested that cooking could be employed as a method to reduce residual nitrite in meat and meat products.

7.4 Irradiation

Food irradiation, utilizing gamma rays, X-rays, or electron beams, stands out as a highly promising method, although its progress and commercialization have been impeded by unfavourable public perceptions. Despite these challenges, the endorsement of food irradiation by various international food and health organizations has bolstered consumer confidence and sparked increased interest within the food industry. Gamma irradiation, in particular, has demonstrated efficacy in reducing residual nitrite and N-nitrosamines in meat products.

For instance, a study on a model sausage subjected to γ-irradiation revealed that high-dose irradiation (>10 kGy) significantly decreased residual nitrite levels [69]. Another investigation found that a dose of 5 kGy γ-irradiation could significantly reduce residual nitrite in Chinese Rugao ham before ripening [70]. However, it is noteworthy that there has been limited recent research into the impact of irradiation on the reduction of residual nitrite.

7.5 Other Technologies

The food industry is transforming to cater to the preferences of consumers who prioritize clean labels. In response, there is a current emphasis on the development of natural food additives, processing technologies with minimal harm, and non-destructive methodologies, all aimed at reducing or substituting the use of nitrite or nitrate in food products. With over 2000 edible vegetables available, the exploration of extracts from a majority of them remains underway for potential nitrite reduction.

Emerging research areas, such as ohmic heating and its integration with other non-destructive technologies, show promise in this regard. Additionally, bio-preservation methods, cultures producing bacteriocins, and genetically modified fermentation strains are being explored as minimally harmful processing technologies with substantial potential. The ongoing efforts in the food industry reflect a noteworthy commitment to providing consumers with more natural and healthier options. This evolving landscape is reflected in the works of researchers [18, 71–73].

In recent years, extensive research has been conducted on alternative methods to replace nitrite and nitrate in meat processing. Notably, the industrialization of substituting nitrite and nitrate with vegetable extracts has gained traction, and certain natural additives derived from these extracts have entered commercial markets. The exploration of alternative additives, bacterial fermentation, enhanced packaging techniques, and high hydrostatic pressure have been particularly prominent in scientific investigations. While numerous methods have been explored to enhance meat quality during production, it is crucial to acknowledge the potential cost implications and the challenge of achieving widespread consumer acceptance. Consequently, conventional food processing and storage techniques, such as cooking and optimizing storage conditions, continue to be widely employed. Table 3 indicates the recent advancements in various aspects of nitrites, including research and regulatory frameworks.

8 Conclusion

Nitrite is the key component of curing in the meat industry, known for its triple effect, including antimicrobial and antioxidant effects and improvement of sensory attributes (colour and flavour). Its bacteriostatic effect against botulinum toxins from *Clostridium botulinum*, inhibitory effect against oxidation of heam iron, and beneficial effect of imparting a pinkish-red colour and unique flavour to cured meat products make it a preferable key ingredient. However, despite its several benefits, sodium nitrite has been a topic of discussion due to its potential carcinogenic effects on humans, as indicated by several studies. It is important to be aware that excessive consumption of nitrite can lead to methemoglobinemia in children and increase the risk of colon cancer in adults. However, there is a growing demand for meat that has reduced levels of nitrite or is organic. The meat industry is currently working on

Table 3 Recent updates on nitrite/nitrates

S.no.	Recent research	Reference
1.	Nitrite consumption leads to type 2 diabetes	[86]
2.	Incorporation of L-arginine amino acid in meat to activate the intrinsic nitric oxide synthase enzyme of meat to generate NO for curing effects with an objective to replace nitrite.	[87]
3.	EFSA examined 32 nitrosamines out of which 10 nitrosamines were identified as both carcinogenic and genotoxic and emphasized on maximum level of incorporation of sodium nitrite (50 mg/kg) and of sodium nitrate (250 mg/kg)	[88]
4.	The processing techniques like water bath boiling or ohmic heating of sausages before consumption or storing the raw or partially cooked sausages under frozen conditions for some period can lower the nitrite exposure in consumers.	[89]
5.	Regulations on nitrite and nitrate in meat and fish products have been revised [Commission Regulation (EU) 2023/2108] on October, 2023 and will be complied from October, 2025 to minimize risks associated with nitrosamine exposure. For nitrites, meat preparations and non-heat-treated meat products will have an ingoing limit reduced from 150 to 80 mg/kg (equivalent to 120 mg/kg as $NaNO_2$), with a residual limit of 45 mg/kg expressed as NO_2 ion. Heat-treated meat products, specifically sterilised ones, will see an ingoing limit decrease from 100 to 55 mg/kg (equivalent to 82.5 mg/kg as $NaNO_2$), and other specified products from 150 to 80 mg/kg, with a new residual limit of 25 mg/kg for sterilised products. Traditional immersion and dry-cured products will have revised residual limits to 105 and 65 mg/kg for ham, aiming for safer levels in finished products. For nitrates, the changes focus on non-heat-treated meat products, with the ingoing limit going down from 150 to 90 mg/kg (equivalent to 126 mg/kg as $NaNO_3$) and introducing a limit of 110 mg/kg for large bacon primals and dry sausages without added nitrites. Traditional immersion cured products like Wiltshire bacon and similar items will have their maximum residual amount reduced from 250 to 150 mg/kg (equivalent to 210 mg/kg as $NaNO_3$).	[90]
6.	S-nitrosothiols may replicate nitrite's antioxidant effects in cooked hams at levels above 50 mg NEq/kg, produce equivalent nitrite residue at over 150 mg NEq/kg, and achieve similar curing colour at 300 mg NEq/kg, though with slightly lower colour stability.	[91]

developing effective methods for lowering the amount of residual nitrite in meat products and finding safer alternatives to nitrite for the production of organic meat products. In processed meats, plant extracts and organic acids such as lactate and sorbate can be used as substitutes for nitrite. However, no single substitute has been discovered yet that can perform all of nitrite's functions at once. The hurdle technology approach, which involves combining reduced levels of nitrite with other additives or processing techniques, may hold promise in achieving antimicrobial effects against the most common microbial pathogens while also enhancing sensory characteristics. However, further research is needed to identify a single alternative to nitrite that can provide the broad-spectrum activities of nitrite cost-effectively, and new-aged rapid detection methods are needed to be required to detect undetectable compounds and monitor public safety closely.

References

1. Pegg RB, Honikel KO. Principles of curing. In: Fidel T, editor. Handbook of fermented meat and poultry. Hoboken: Wiley; 2014. p. 19–30.
2. Patarata L, Martins S, Silva JA, Fraqueza MJ. Red wine and garlic as a possible alternative to minimize the use of nitrite for controlling *Clostridium sporogenes* and Salmonella in a cured sausage: safety and sensory implications. Foods. 2020;9(2):206.
3. Candan T, Bağdatlı A. Natural applications for nitrite/nitrate reduction in meat products. Pamukkale Univ Muh Bilim Derg. 2018;24(7):1382–7. https://doi.org/10.5505/pajes.2017.80269.
4. Patarata L, Carvalho F, Fraqueza MJ. Nitrite-free implications on consumer acceptance and the behavior of pathogens in cured pork loins. Foods. 2022;11(6):796.
5. Altunay N, Elik A. A green and efficient vortex-assisted liquid-phase microextraction based on supramolecular solvent for UV-VIS determination of nitrite in processed meat and chicken products. Food Chem. 2020;332:127395.
6. Lopez CM, Dallolio G, Bonilauri P, Rebecchi A. Strategies for nitrite replacement in fermented sausages and effect of high pressure processing against Salmonella spp. and Listeria innocua. Foods. 2021;10(11):2617.
7. Aminzare M, Hashemi M, Ansarian E, Bimakr M, Hassanzad Azar H, Mehrasbi MR, Daneshamooz S, Raeisi M, Jannat B, Afshari A. Using natural antioxidants in meat and meat products as preservatives: a review. Adv Anim Vet Sci. 2019;7(5):417–26.
8. Hunault CC, van Velzen AG, Sips AJ, Schothorst RC, Meulenbelt J. Bioavailability of sodium nitrite from an aqueous solution in healthy adults. Toxicol Lett. 2009;190(1):48–53.
9. Brender JD. Human health effects of exposure to nitrate, nitrite, and nitrogen dioxide. In: Just enough nitrogen: perspectives on how to get there for regions with too much and too little nitrogen. Cham: Springer; 2020. p. 283–94.
10. Oliveira SM, Lopes TI, Rangel AO. Spectrophotometric determination of nitrite and nitrate in cured meat by sequential injection analysis. J Food Sci. 2004;69(9):C690–5.
11. Richi EB, Baumer B, Conrad B, Darioli R, Schmid A, Keller U. Health risks associated with meat consumption: a review of epidemiological studies. Int J Vitam Nutr Res. 2015;85(1–2):70–8.
12. Casoni D, Badiu RR, Frentiu T. Spectrophotometric determination and assessment of potential health risk of nitrite from meat and processed meat products. Stud Univ Babes Bolyai Chem. 2019;2:265–77.
13. Hung Y, de Kok TM, Verbeke W. Consumer attitude and purchase intention towards processed meat products with natural compounds and a reduced level of nitrite. Meat Sci. 2016;121:119–26.
14. Sindelar JJ, Milkowski AL. Sodium nitrite in processed meat and poultry meats: a review of curing and examining the risk/benefit of its use. Am Meat Sci Assoc White Paper Ser. 2011;3:1–4.
15. Alahakoon AU, Jayasena DD, Ramachandra S, Jo C. Alternatives to nitrite in processed meat: up to date. Trends Food Sci Technol. 2015;45(1):37–49.
16. Shakil MH, Trisha AT, Rahman M, Talukdar S, Kobun R, Huda N, Zzaman W. Nitrites in cured meats, health risk issues, alternatives to nitrites: a review. Foods. 2022;11(21):3355.
17. Zhang Y, Zhang Y, Jia J, Peng H, Qian Q, Pan Z, Liu D. Nitrite and nitrate in meat processing: functions and alternatives. Curr Res Food Sci. 2023;6:100470.
18. Ferysiuk K, Wójciak KM. Reduction of nitrite in meat products through the application of various plant-based ingredients. Antioxidants. 2020;9(8):711.
19. Rajkovic A, Tomasevic I, De Meulenaer B, Devlieghere F. The effect of pulsed UV light on Escherichia coli O157: H7, *Listeria monocytogenes*, *Salmonella typhimurium*, *Staphylococcus aureus* and staphylococcal enterotoxin A on sliced fermented salami and its chemical quality. Food Control. 2017;73:829–37.

20. Crowe W, Elliott CT, Green BD. A review of the in vivo evidence investigating the role of nitrite exposure from processed meat consumption in the development of colorectal cancer. Nutrients. 2019;11(11):2673.
21. Li Y, Chapman SJ, Nicol GW, Yao H. Nitrification and nitrifiers in acidic soils. Soil Biol Biochem. 2018;116:290–301.
22. Zhu Y, Guo L, Yang Q. Partial replacement of nitrite with a novel probiotic *Lactobacillus plantarum* on nitrate, color, biogenic amines and gel properties of Chinese fermented sausages. Food Res Int. 2020;137:109351.
23. Ozaki MM, Dos Santos M, Ribeiro WO, de Azambuja Ferreira NC, Picone CS, Domínguez R, Lorenzo JM, Pollonio MA. Radish powder and oregano essential oil as nitrite substitutes in fermented cooked sausages. Food Res Int. 2021;140:109855.
24. Higuero N, Moreno I, Lavado G, Vidal-Aragón MC, Cava R. Reduction of nitrate and nitrite in Iberian dry cured loins and its effects during drying process. Meat Sci. 2020;163:108062.
25. Kim GH, Chin KB. Evaluation of quality characteristics of low-nitrite pork sausages with paprika oleoresin solution during refrigerated storage. Food Sci Anim Resour. 2021;41(3):428–39.
26. Rivera N, Bunning M, Martin J. Uncured-labeled meat products produced using plant-derived nitrates and nitrites: chemistry, safety, and regulatory considerations. J Agric Food Chem. 2019;67(29):8074–84.
27. Khatib N, Varidi MJ, Mohebbi M, Varidi M, Hosseini SM. Replacement of nitrite with lupulon-xanthohumol loaded nanoliposome in cooked beef-sausage: experimental and model based study. J Food Sci Technol. 2020;57:2629–39.
28. Perea-Sanz L, López-Díez JJ, Belloch C, Flores M. Counteracting the effect of reducing nitrate/nitrite levels on dry fermented sausage aroma by *Debaryomyces hansenii* inoculation. Meat Sci. 2020;164:108103.
29. Tian Q, Wang X, Ye C, Pan Q, Zeng X, Pan D, Shen J. Application of *Lactobacillus fermentum* RC4 on salted meat for nitrite degradation and quality improvement. J Ningbo Univ Sci Technol Ed. 2020;33(1):38–44.
30. Zhao Y. Study on the antioxidative function of lactic acid bacteria and its application in mutton fermented sausage. Master's thesis, Inner Mongolia Agricultural University; 2020.
31. Huang L, Sun Z, Zeng XQ, Pan DD, He J, Shen JL. Effects of multi-ingredients for nitrite on the volatile flavor compounds in cured meat. J Chin Inst Food Sci Technol. 2021;21(3):324–33.
32. Chatkitanan T, Harnkarnsujarit N. Development of nitrite compounded starch-based films to improve color and quality of vacuum-packaged pork. Food Packag Shelf Life. 2020;25:100521.
33. Karwowska M, Kononiuk A. Nitrates/nitrites in food-risk for nitrosative stress and benefits. Antioxidants. 2020;9(3):241.
34. Merino L, Darnerud PO, Toldrá F, Ilbäck NG. Time-dependent depletion of nitrite in pork/beef and chicken meat products and its effect on nitrite intake estimation. Food Addit Contam Part A. 2016;33(2):186–92.
35. Federal Regulations. Title 21–Food and Drugs, Chapter I–Food and Drug Administration, Department of Health and Human Services, Subchapter B—Food for Human Consumption (Continued), Part 172—Food Additives Permitted for Direct Addition to Food for Human Consumption, Subpart B—Food Preservatives. Sec 172.175 Sodium nitrite. [Internet]. [cited 2024 Mar 6]. Available from: https://www.accessdata.fda.gov/scripts/cdrh/cfdocs/cfcfr/CFRSearch.cfm?CFRPart=172.
36. Canadian Food Inspection Agency. Preventive control recommendations on the use of nitrites in the curing of meat products. [Internet]. 2021 [cited 2024 Mar 6]. Available from: https://inspection.canada.ca/preventive-controls/meat/nitrites/eng/1522949763138/1522949763434.
37. Food Safety and Standards Authority of India. Food Safety and Standards (Food Products Standards and Food Additives) Regulations, 2011, Version-XII (08.09.2020). [Internet]. 2020 [cited 2024 Mar 6]. Available from: https://www.fssai.gov.in/upload/uploadfiles/files/Compendium_Food_Additives_Regulations_08_09_2020-compressed.pdf.

38. Abd Hamid NF, Khan MM, Lim LH. Assessment of nitrate, nitrite and chloride in selected cured meat products and their exposure to school children in Brunei Darussalam. J Food Compos Anal. 2020;91:103520.
39. Bryan NS, van Grinsven H. The role of nitrate in human health. Adv Agron. 2013;119:153–82.
40. European Commission. Final Report on a desk study to monitor the implementation of Directive 2006/52/EC in the EU Member States as regards the use of nitrites by the industry in the different categories of meat products and the organization of national controls. [Internet]. 2013 [cited 2024 Mar 6]. Available from: https://ec.europa.eu/food/food-feed-portal/backend/api/document?filename=5924-final-report-on-nitrites.pdf&id=649801.
41. Herrmann SS, Granby K, Duedahl-Olesen L. Formation and mitigation of N-nitrosamines in nitrite preserved cooked sausages. Food Chem. 2015;174:516–26.
42. Scientific Panel on Food Additives and Nutrient Sources added to Food (ANS). Minutes of the 81st Plenary meeting Held on 15–17 May 2018, Parma (Italy) Meeting open for Observers (Open session: 16 May 2018, 09:00–18:00h; 17 May 2018, 09:00–15:30h) (Agreed on 07 June 2018). [Internet]. European Food Safety Authority; 2018 [cited 2024 Mar 6]. Available from: https://ec.europa.eu/food/food-feed-portal/backend/api/document?filename=5924-final-report-on-nitrites.pdf&id=649801.
43. Hung Y, Verbeke W, de Kok TM. Stakeholder and consumer reactions towards innovative processed meat products: insights from a qualitative study about nitrite reduction and phytochemical addition. Food Control. 2016;60:690–8.
44. Colla G, Kim HJ, Kyriacou MC, Rouphael Y. Nitrate in fruits and vegetables. Sci Hortic. 2018;237:221–38.
45. Riel G, Boulaaba A, Popp J, Klein G. Effects of parsley extract powder as an alternative for the direct addition of sodium nitrite in the production of mortadella-type sausages-impact on microbiological, physicochemical and sensory aspects. Meat Sci. 2017;131:166–75.
46. Horsch AM, Sebranek JG, Dickson JS, Niebuhr SE, Larson EM, Lavieri NA, Ruther BL, Wilson LA. The effect of pH and nitrite concentration on the antimicrobial impact of celery juice concentrate compared with conventional sodium nitrite on *Listeria monocytogenes*. Meat Sci. 2014;96(1):400–7.
47. Pyo YH, Lee TC, Logendra L, Rosen RT. Antioxidant activity and phenolic compounds of Swiss chard (Beta vulgaris subspecies cycla) extracts. Food Chem. 2004;85(1):19–26.
48. Sebranek JG, Jackson-Davis AL, Myers KL, Lavieri NA. Beyond celery and starter culture: advances in natural/organic curing processes in the United States. Meat Sci. 2012;92(3):267–73.
49. Khalehgi A, Kasaai R, Khosravi-Darani K, Rezaei K. Combined use of black barberry (*Berberis crataegina* L.) extract and nitrite in cooked beef sausages during the refrigerated storage. J Agric Sci Technol. 2016;18(3):601–14.
50. Sucu C, Turp GY. The investigation of the use of beetroot powder in Turkish fermented beef sausage (sucuk) as nitrite alternative. Meat Sci. 2018;140:158–66.
51. Aquilani C, Sirtori F, Flores M, Bozzi R, Lebret B, Pugliese C. Effect of natural antioxidants from grape seed and chestnut in combination with hydroxytyrosol, as sodium nitrite substitutes in *Cinta Senese* dry-fermented sausages. Meat Sci. 2018;145:389–98.
52. Tan S, Ke Z, Chai D, Miao Y, Luo K, Li W. Lycopene, polyphenols and antioxidant activities of three characteristic tomato cultivars subjected to two drying methods. Food Chem. 2021;338:128062.
53. Eyiler E, Oztan A. Production of frankfurters with tomato powder as a natural additive. LWT Food Sci Technol. 2011;44(1):307–11.
54. Riazi F, Zeynali F, Hoseini E, Behmadi H, Savadkoohi S. Oxidation phenomena and color properties of grape pomace on nitrite-reduced meat emulsion systems. Meat Sci. 2016;121:350–8.
55. Honikel KO. The use and control of nitrate and nitrite for the processing of meat products. Meat Sci. 2008;78(1–2):68–76.
56. McClure BN, Sebranek JG, Kim YH, Sullivan GA. The effects of lactate on nitrosylmyoglobin formation from nitrite and metmyoglobin in a cured meat system. Food Chem. 2011;129(3):1072–9.

57. Vijayakumar J, Aravindan R, Viruthagiri T. Recent trends in the production, purification, and application of lactic acid. Chem Biochem Eng Q. 2008;22(2):245–64.
58. Kim YH, Keeton JT, Smith SB, Berghman LR, Savell JW. Role of lactate dehydrogenase in metmyoglobin reduction and color stability of different bovine muscles. Meat Sci. 2009;83(3):376–82.
59. Sedghi H, Mohammadi T, Shariati MA, Atarod S, Branch Q. The effect of substituting sodium nitrite with sodium acetate on microbial, physiochemical and sensory characteristics of meat products. Agric Food Sci. 2014. Available from: https://www.semanticscholar.org/paper/The-effect-of-substituting-sodium-nitrite%2C-with-on-Sedghi-Mohammadi/a689e9e16ff3660a1b90adc57a3d1cf1fc865db5.
60. Carpenter CE, Broadbent JR. External concentration of organic acid anions and pH: key independent variables for studying how organic acids inhibit growth of bacteria in mildly acidic foods. J Food Sci. 2009;74(1):R12–5.
61. Bradley EM, Williams JB, Schilling MW, Coggins PC, Crist C, Yoder S, Campano SG. Effects of sodium lactate and acetic acid derivatives on the quality and sensory characteristics of hot-boned pork sausage patties. Meat Sci. 2011;88(1):145–50.
62. Cegielska-Radziejewska R, Pikul J. Sodium lactate addition on the quality and shelf life of refrigerated sliced poultry sausage packaged in air or nitrogen atmosphere. J Food Prot. 2004;67(3):601–6.
63. Quilo SA, Pohlman FW, Brown AH, Crandall PG, Dias-Morse PN, Baublits RT, Aparicio JL. Effects of potassium lactate, sodium metasilicate, peroxyacetic acid, and acidified sodium chlorite on physical, chemical, and sensory properties of ground beef patties. Meat Sci. 2009;82(1):44–52.
64. Glass K, Preston D, Veesenmeyer J. Inhibition of Listeria monocytogenes in Turkey and pork-beef bologna by combinations of sorbate, benzoate, and propionate. J Food Prot. 2007;70(1):214–7.
65. Al-Shuibi AM, Al-Abdullah BM. Substitution of nitrite by sorbate and the effect on properties of mortadella. Meat Sci. 2002;62(4):473–8.
66. Waga M, Takeda S, Sakata R. Effect of nitrate on residual nitrite decomposition rate in cooked cured pork. Meat Sci. 2017;129:135–9.
67. Asioli D, Aschemann-Witzel J, Caputo V, Vecchio R, Annunziata A, Næs T, Varela P. Making sense of the "clean label" trends: a review of consumer food choice behavior and discussion of industry implications. Food Res Int. 2017;99:58–71.
68. Sallan S, Kaban G, Oğraş SS, Çáelik M, Kaya M. Nitrosamine formation in a semi-dry fermented sausage: effects of nitrite, ascorbate and starter culture and role of cooking. Meat Sci. 2020;159:107917.
69. Ahn HJ, Kim JH, Jo C, Lee CH, Byun MW. Reduction of carcinogenic N-nitrosamines and residual nitrite in model system sausage by irradiation. J Food Sci. 2002;67(4):1370–3.
70. Wei F, Xu X, Zhou G, Zhao G, Li C, Zhang Y, Chen L, Qi J. Irradiated Chinese Rugao ham: changes in volatile N-nitrosamine, biogenic amine and residual nitrite during ripening and post-ripening. Meat Sci. 2009;81(3):451–5.
71. Fraqueza MJ, Borges A, Patarata L. Strategies to reduce the formation of carcinogenic chemicals in dry cured meat products. In: Food control and biosecurity. London: Academic Press; 2018. p. 295–342.
72. Ranasinghe RA, Marapana RA. Nitrate and nitrite content of vegetables: a review. J Pharmacogn Phytochem. 2018;7(4):322–8.
73. Zhang Y, Zhang L, Venkitasamy C, Pan Z, Ke H, Guo S, Wu D, Wu W, Zhao L. Potential effects of umami ingredients on human health: pros and cons. Crit Rev Food Sci Nutr. 2020;60(13):2294–302.
74. Chan TYK. Vegetable-borne nitrate and nitrite and the risk of methemoglobinaemia. Toxicol Lett. 2011;200(1–2):107–8.
75. Gassara F, Kouassi AP, Brar SK, Belkacemi K. Green alternatives to nitrates and nitrites in meat-based products – a review. Crit Rev Food Sci Nutr. 2016;56(13):2133–48.

76. Leth T, Fa Gt S, Nielsen S, Andersen R. Nitrite and nitrate content in meat products and estimated intake in Denmark from 1998 to 2006. Food Addit Contam Part A Chem Anal Control Expo Risk Assess. 2008;25(10):1237–45.
77. Elias A, Jalakas S, Roasto M, Reinik M, Nurk E, Kaart T, Elias T. Nitrite and nitrate content in meat products and estimated nitrite intake by the Estonian children. Food Addit Contam Part A Chem Anal Control Expo Risk Assess. 2020;37(8):1229–37.
78. Gangolii S, Brandt P, Feron V, Janz-Wsky C, Speijers G. Assessment of nitrate, nitrite, and N-nitroso-compounds. Eur J Pharmacol. 1994;292(11):1–38.
79. Ding ZS, Johanningsmeier SD, Price R, Reynolds R, Truong VD, Payton SC, Breidt F. Evaluation of nitrate and nitrite contents in pickled fruit and vegetable products. Food Control. 2018;90:304–11.
80. Flores M, Toldra F. Chemistry, safety, and regulatory considerations in the use of nitrite and nitrate from natural origin in meat products – invited review. Meat Sci. 2021;171(1):1–12.
81. Additives Permitted for Direct Addition to Food for Human Consumption. Subpart B—Food Preservatives. Sec 172.175 Sodium nitrite. [Internet]. [cited 2024 Mar 6]. Available from: https://www.ecfr.gov/current/title-21/chapter-I/subchapter-B/part-172.
82. USDA. Processing inspectors' calculations handbook. Washington, DC: Food Safety and Inspection Service; 1995.
83. Liu N, Zhu Q, Zeng X, Yang B, Liang M, Hu P, Zhang R. Influences of pulsed light-UV treatment on the storage period of dry-cured meat and shelf life prediction by ASLT method. J Food Sci Technol. 2019;56(4):1744–56.
84. National Health and Family Planning Commission of the People's Republic of China. National Food Safety Standard – Standards for Uses of Food Additives. GB 2760-2014. Issued December 24, 2014; Implemented May 24, 2015. Available from: https://www.mpi.govt.nz/dmsdocument/42706-China-GB-Standard-A.10-Uses-of-food-additives.
85. Ministry of Health and Family Welfare, Food Safety and Standards Authority of India. Food Safety and Standards (Food Products Standards and Food Additives) Regulations, 2011. Notification No. F.No. 2-15015/30/2010. New Delhi, dated 1st August 2011. [Internet]. Available from: https://www.fssai.gov.in/upload/uploadfiles/files/Food_Additives_Regulations.pdf.
86. Srour B, Chazelas E, Druesne-Pecollo N, Esseddik Y, De Edelenyi FS, Agaësse C, et al. Dietary exposure to nitrites and nitrates in association with type 2 diabetes risk: results from the NutriNet-Santé population-based cohort study. Imamura F, editor. PLoS Med. 2023;20(1):e1004149. Available from: https://dx.plos.org/10.1371/journal.pmed.1004149.
87. Food Safety Magazine. Researchers aim to make cured meat safer with "no nitrite-added" curing process. Food Safety [Internet]. 2023 [cited 2024 Mar 5]. Available from: https://www.food-safety.com/articles/8373-researchers-aim-to-make-cured-meat-safer-with-no-nitrite-added-curing-process.
88. SGS Digicomply. Food safety snapshot: nitrosamines in meat products [Internet]. 2023 [cited 2024 Mar 6]. Available from: https://www.digicomply.com/blog/food-safety-snapshot-nitrosamines-in-meat-products.
89. Jantapirak S, Vangnai K, Tumpanuvatr T, Jittanit W. Effects of heating method, temperature, initial nitrite level, and storage time on residual nitrite, pigments, and curing efficiency of chicken sausages. Int J Food Prop. 2023;26(1):2186–200. Available from: https://www.tandfonline.com/doi/full/10.1080/10942912.2023.2244687.
90. European Commission. Regulation-EU-2023/2108-EN-EUR-Lex [Internet]. Official Journal of the European Union; 2023 [cited 2024 Mar 6]. Available from: https://eur-lex.europa.eu/eli/reg/2023/2108/oj.
91. Andrade BF, Guimarães AS, Do Carmo LR, Tanaka MS, Fontes PR, Ramos ADLS, et al. S-nitrosothiols as nitrite alternatives: effects on residual nitrite, lipid oxidation, volatile profile, and cured color of restructured cooked ham. Meat Sci. 2024;209:109397.

Marinated Meat Products

Irfan Khan, Akanksha Soni, Sweta Joshi, Tariq Khan, Tohfa Siddiqui, and Mohammad Umar Khan

1 Historical Perspective

Marinating Meat is a long-standing culinary custom that has evolved over many cultures and historical periods. Although its precise origins are difficult to determine, the art of marination has roots that extend back thousands of years [1]. Since decades, people have been marinating meat with ingredients, and these days, the food business as well as homes are using them more and more. Historically, salt, herbs, and spices were frequently employed in marinades in antiquity (Fig. 1) [2].

Meat was made more palatable and tender by soaking it in liquids, like water or acidic fruit juices, during ancient civilizations. In order to preserve and enhance the flavor of meat and meat products, the Egyptians and Middle Ages Europeans also

I. Khan
Department of Bioengineering, Integral University, Lucknow, India

GVICAA-A Foodtech Research & Development Consultancy, Bareilly, Uttar Pradesh, India

A. Soni
GVICAA-A Foodtech Research & Development Consultancy, Bareilly, Uttar Pradesh, India

S. Joshi (✉)
Department of Food Technology, School of Interdisciplinary Sciences and Technology, Jamia Hamdard, New Delhi, India
e-mail: swetajoshi@jamiahamdard.ac.in

T. Khan
Department of Post Harvest Engineering and Technology, Aligarh Muslim University, Aligarh, Uttar Pradesh, India

T. Siddiqui
Galgotias College of Pharmacy, Greater Noida, India

M. U. Khan
Dehlvi Natural, New Delhi, India

A. Kumar Pathera et al. (eds.), *Healthier Meat Products*,
https://doi.org/10.1007/978-3-031-78215-2_19

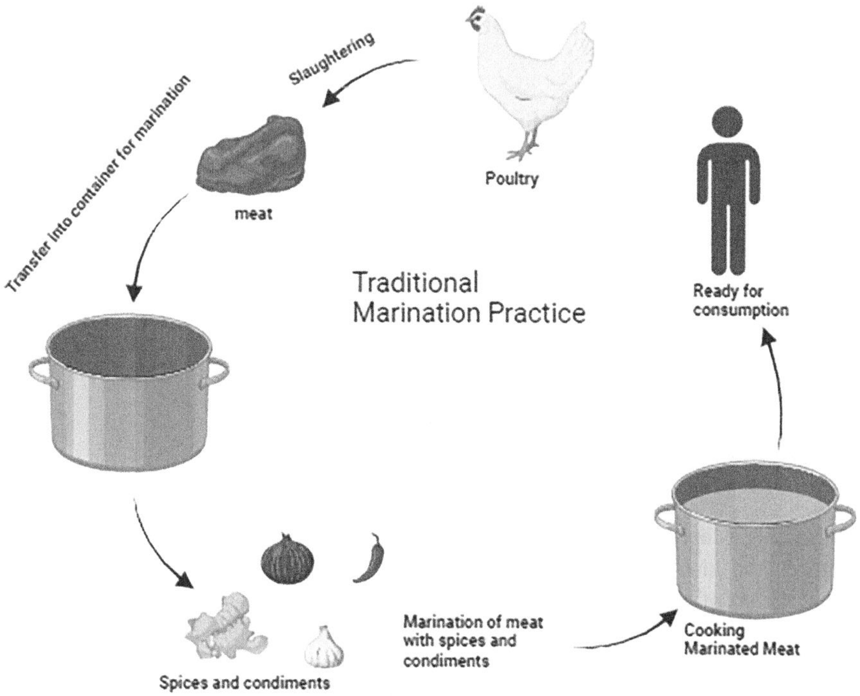

Fig. 1 Traditional meat marination practices

used vinegar, wine, and other alcoholic beverages; these methods were also seen in ancient India and Persia [3]. Wine, herbs, and spices have become staples in marinades, used not just as culinary tools but also to enhance the sensory attributes of meat products [4]. The discovery of new spices and ingredients by European explorers in the Americas, Asia, and Africa resulted in a global fusion of culinary traditions that influenced marinade recipes. These countries used soy sauce, vinegar, and a variety of herbs and spices, as evidenced by meals like Chinese char siu and Korean bulgogi.

The options for enhancing meat quality have increased due to a variety of ingredients like traditional acids and spices and cutting-edge techniques like sous-vide cooking and vacuum marination. Marinating meat stands as a testament to the ever-evolving and diverse world of gastronomy, connecting cultures across time and borders [5]. Marinated meats and meat products have gained widespread popularity and hold immense cultural significance across the globe, playing a central role in various cuisines and traditions. Beyond flavor, marinationculinary serves as a transformative process, tenderizing meats through the inclusion of acids, spices, enzymes and other elements. The culinary ideas and ingredients continue to exchange globally, the fusion of marination traditions from different cultures results in innovative and exciting flavor combinations, reflecting the dynamic nature of the culinary landscape [5]. The components used in marinade formulations are essential to obtaining the targeted sensory qualities, like as flavor and texture, which in the end define the caliber of the finished product [6]. Numerous techniques for marinating have been employed, including vacuum tumbling, injection, messaging, and immersion [2].

2 Introduction

Marination is a meat preparation technique that involves soaking or infusing meat with seasoned liquid marinades. The Latin/Italian word "marinara," which meaning "of the sea," is where the word marinates originates. Centuries ago, briny or salty liquids were called as marinades. These marinades help in preserve the foods for longer period and also provide tenderization and flavour to the foods. These marinades contain a variety of elements, including acids, enzymes, and spices to enhance the flavour and texture of meat [1]. The acid component of the marinades break down the meat fibres internally viz., M-line which improves the tenderization and water holding capacity of meat. The meat pH is the important aspect that control many sensory attributes viz., tenderness, texture, appearance/colour and flavour of the meat ad meat products [7]. All these organoleptic properties are improved by meat marination. Out of several meat attributes, meat tenderization is one of the most essential aspects of marination to make the meat more palatable for consumption. The acid treatment denatured the microstructure of meat muscle and also inhibits the lipid oxidation particularly with citric acid.

Another important aspect of marination is to provide the flavours to the meat for making it more tasty and appealing. It is one of the methods to improve the meat tenderness in making the product more palatable while chewing. Marinades are the mixture of ingredients that improves taste, texture, flavor, palatability and overall acceptability of the meat products. This method is applied on variety of muscle food such as poultry products, seafoods, red meat products etc. [2, 8–10]. The focus of marinades has shifted from preservation and pickling to primarily improving tenderness and flavour of meat and meat products [2]. Marination can be done with help of biological agents such as kefir, yogurt, or butter [10, 11], while other methods include application of spices and condiments such as turmeric, torch ginger, lemongrass, curry leaves, chili, pepper and garlic [12, 13], or exotic choices like karuk juice [14]. In addition to natural ingredients, a variety of different chemicals are also used viz., citric acid, acetic acid, tartaric acid, apple cider vinegar, agraz-verjus wine, and common staples like sodium chloride (NaCl), sodium nitrite ($NaNO_2$), and sodium bicarbonate ($NaHCO_3$) [15]. Recent innovations in meat marination have also been introduced the physical methods such as ultrasound, tumbling, and gas chromatography etc. [16, 17]. The choice of ingredients in marinades plays a pivotal role in achieving the desired sensory characteristics, including flavor and texture, which is crucial in determining the overall quality of the end product [6]. These ingredients influence the pH of the meat which ultimately stimulates enzymatic and proteolytic activity during meat maturation process. The change in meat pH also affects meat ageing by reducing the time of tenderness [5]. The acids can be used in marination process to breakdown the muscle connective tissue that ultimately improved the tenderness of meat. Apart from its impact on organoleptic properties, the spices also retard the growth of pathogenic and food spoilage microorganisms by reducing the water activity of meat and with antimicrobial components of spices such as polyphenols, organic acids, ethanol, and antimicrobial agents [2, 10].

3 Types of Marination

There are many commercially available types of marinades that use a variety of ingredients, including onion, garlic, and chili powder, to marinate meat. Hence it is divided into three main categories that based on type of ingredient used in the process.

3.1 *Biological Marination*

In biological marination, biological or bio-based products are used to get the desired attributes such as taste, texture, tenderness, flavour etc. The typical most biological marinades are kefir, curd, yoghurt and other dairy based products including probiotics. Onion, turmeric, lemongrass, garlic, coriander, cinnamon, oregano, parsley, mustard, pepper, rosemary, and lemon juice have been staples in marination for ages, as evidenced by studies (Table 1) [18–21]. The Biological marinates are healthier and due to presence of probiotics it pose extra health benefits on human health. The biological marination is more common at domestic level where people do not have access of chemical marinates and physical methods for the marination. These ingredients not only enhance taste and tenderness but also contribute to

Table 1 List of marinades used in biological marination

Biological marinades		
Category	Ingredients	References
Seed/oil	Olive oil, sesame seed oil, soybean oil, mustard seed oil, flax seed oil, Plantago seeds etc.	
Spice/herb	Chili, onion, turmeric, ginger, lemongrass, star anise, garlic, coriander, cinnamon, cumin, oregano, parsley, mustard, pepper, rosemary, lemon juice, curry leaves, basil leaves, karuk juice, red beetroot juice	[12–14]
Dairy products	Curd, yogurt, kefir, buttermilk, fermented milk product, probiotic	[10, 11]

Source: [27]

reduce the formation of hazardous elements like polycyclic aromatic hydrocarbons (PAHs) and heterocyclic amines (HCAs) in marinated meat. PAHs and HCAs, known for their carcinogenic and mutagenic activities, arise from conventional cooking methods like smoking, frying, roasting, grilling, or barbecuing, especially when organic matter is involved.

Although the exact mechanism for generating PAHs and HCAs is not known, while the experiments suggested that the synthesis of these compounds at very high temperature i.e. around 300 °C due to pyrolysis of amino acids and proteins in meat(Alvarado and McKee, 2007).Various techniques have been developed to reduce the synthesis of HCAs and PAHs through meat marination especially by using indigenous spices like torch ginger, curry leaves, lemon grassat a 10% concentration, and turmeric at 4% [18, 22]. In the investigation it was found that on cooking the mutton with 10% torch ginger, 86.6% reduction was found in the formation of HCA & PAH. The elevated production of HCAs during frying is attributed to reduced moisture content and simultaneously release of fat molecules, proteins, free amino acids, creatine and sugar from the food matrix which on high temperature produces the heterocyclic amines [23]. In the investigation, it was found that the optimum level of torch ginger and turmeric such as 10 g/100 g and 4 g/100 g along with lemon grass and curry leaves significantly reduces the synthesis of HCA and PAH [18].

Additionally, incorporating 1.2% lemon juice in the marination process was found to reduce PAHs content in grilled beef to a negligible extent although the exact mechanism is yet to be found [24]. Presently marinated meat with fermented dairy products and probiotics has gained popularity due to its extra health benefits on digestive system and boosting the innate immunity of human body. Among biological marinades, yogurt, kefir, and buttermilk are the most commonly employed marinades to improve the meat quality now days. Buttermilk and yogurt are highly recommended by the food scientists and researchers as suitable marinades even at commercial scale. Despite their potential, the exact mechanism of fermented dairy products as a biological marinade is still unknown i.e. mechanism of improving tenderness, textural quality, flavour is yet to be found [25]. The most commonly used probiotic strains are *L. rhamnosus, L. casei, L. acidophilus, L. plantarum, Bifidobacteria* and their combination significantly were found to enhance the sensory appeal and safety standards of marination (Pradhan et al. 2020; Zhang et al. 2015). Biological marination significantly reduce the food borne pathogens *Salmonella typhimurium, Listeria monocytogenes, and Escherichia coli O157: H7*, from 0.8, 2.1, and 0.7 to 2.0, 3.3 and 2.7 log CFU/g, respectively [26]. Furthermore, meat samples marinated with *L. casei* displayed high satisfaction in terms of color, appearance, flavor, and overall acceptability [26].

3.2 Chemical Marination

The most widely used method in the commercial meat and meat product marinating industries for adding value is the application of various chemical marinades. That's why, chemical marination is commonly employed at industrial level where value addition of meat is done to improve meat tenderness, textural quality etc. The most commonly used chemical marinades at commercial scale are organic acids, sodium chloride (NaCl), sodium bicarbonate ($NaHCO_3$), sodium triphosphate, calcium chloride ($CaCl_2$), ammonium hydroxide (NH_4OH), and alcoholic beverages and soy sauce. The alcoholic beverages including wine is not permissible as per Islamic jurisprudence and that's why these are not used as chemical marinades in Islamic countries such as Saudi Arabia, Dubai, Qatar etc. Marinades are traditionally made with salt, water, and acidic ingredients including wine, vinegar, and fruit juice. Acidic marinades are popular for use in tough meats to promote tenderness [28]. Consequently, many studies have been conducted on chicken and rabbit meat marination and it was found that the acidic marinades such as lactic acid, acetic acid, soy sauce, sodium lactate and red wine were found to improve the tenderness and increased the palatability of the meat [29–33]. In order to improve the water-holding capacity (WHC), acceptable elasticity, and rigidity of meat, proteins must subsequently aggregate and denaturate. Key players in this process include calcium chloride, sodium bicarbonate, sodium triphosphate, ammonium hydroxide, and different phosphates [34–38].

The quality of meat reflects through its tenderness which depends upon the water holding capacity of meat proteins especially myofibrillar proteins. At commercial scale to reduce the drip loss and improve the water holding capacity various compounds are used and the most commonly used is sodium phosphate. For improving the tenderness of firm meat cut, different acids are used at commercial scale [39].

3.3 Physical Marination

To get meat with a particular quality and set of qualities, it takes a long time to marinate meat and meat products. To put it another way, physical marinades are those kinds of marinades that are used in conjunction with biological, chemical, or both biological and chemical marinades to expedite the marination process and improve the quality and safety of meat. In the meat business, physical marinades combined with biological or chemical marinades have grown in favor recently as a way to outperform the conventional approach and improve performance. Physical marinades, or emerging technologies, include CO_2 micro-perforation, ultrasound treatment, high-pressure treatment, and pulsed vacuum impregnation [3, 30, 40–42]. Due to expansion, compression, and cavitation, physical marinades shorten processing times, increase diffusion rates, speed up mass transfer, lessen microbiological spoilage, lower lipid oxidation, alter the composition of fatty acids, and improve product quality [3, 30, 41–43]. Scanning electron microscopy (SEM) analysis was not used to evaluate the micro structural impacts during the physical marination procedure. Furthermore, no longer-lasting products have been reported in prior research using these methods.

4 Methods of Marination

Two types of marination methods are there wet and dry marination.

4.1 Wet Marination

Wet marination is a culinary technique that involves immersion of meat in liquid-based marinade. This method is used to enhance flavor, tenderize tougher cuts of meat, and preserve the meat [44]. The process of wet marination is that where liquid marinades such as organic acids like vinegar, citrus juice, yogurt, buttermilk and wine are added. These acidic components serve a dual purpose: they tenderize the meat by breaking down its tough muscle fibers and also impart flavor. Complementary to the acidic base, oils such as olive oil, vegetable oil and sesame oil which contribute richness and help retain moisture in the food during the cooking process to further enhance the flavor profile, a variety of seasonings are included in the marinade such as herb like rosemary, thyme, basil, and spices such as garlic, ginger, paprika, cumin, salt, and pepper.The marination process involves coating the meat thoroughly with marinade, either by submerging it in a container or placing it in a resealable bag. Based on the variety of food item, the marination time may vary from 30 min to 12 h under refrigerated storage condition. During marination process, flavour molecules would get enough time to permeate the food system which results in a flavourful and tender meat product.

Wet marination has advantage to produce the flavourful meat product due to liquid nature of marinade along with improved tenderness due to acid components in the marinade. Moreover, certain marinades with oils or acidic components aid in retaining moisture and preventing the food from drying during the cooking process [45].

4.2 Dry Marination

Dry marination is a distinctive culinary method which does not utilize liquid based marinades rather dry seasoning material or spices. Dry seasonings comprise a blend of dry components like spices, herbs, salts, and sugars occasionally. This mixture, including spices such as paprika, cumin, chilli powder, and herbs like thyme and rosemary, used to develop a flavourful crust or bark on the meat surface. Dry brining, on the other hand, involves the application of a mixture of salt (and sometimes sugar) directly onto the food's surface, drawing out initial moisture before allowing the meat to reabsorb the seasoning. This process improves flavor and texture by enhancing natural juices within the meat.

Dry marination does not utilize liquid-based marinades; rather, it works on the surface of the food while wet marination deeply permeates the food, affecting both its internal structure and moisture content, dry marination primarily impacts the food's exterior, enhancing its taste and texture without any liquid immersion [46].

5 Injection Marination

Injection marination is widely used technique for meat marination due to less time-consuming process, precision and effectivity. This method involves inserting probes or needles into meat, allowing for the controlled administration of marinade. After injecting the marinade the injecting material disperse uniformly throughout the bulk of meat [1]. In comparison to tumbling process, injection marination was found better in respect of higher marinade intake from 6% to 8% respectively [39]. This improved marinade intake was attributed to the mechanical process of injection before immersing the meat in the marinade, possibly linked to increased breakdown of muscle protein, enhancing marinate absorption. However, despite the greater marinate absorption, this method could lead to a higher cooking loss. The injection process might create pathways within the meat, causing leaks during cooking, reducing water-holding capacity and resulting in more cooking and purge loss [47]. Furthermore it is also advocated in some investigations that by using phosphate or salt marinade at a pH value 11, no significant difference was found between tumbling and injecting technique regarding marinade intake [48]. Existing research articles have presented findings primarily on retention time, marinade uptake, and cooking loss concerning this method. However, there's room for extensive research to explore potential alterations in sensory attributes, nutritional composition, biochemical properties, and safety aspects of meat and meat products.

6 Functions of Marinade

Variety of ingredients or marinades are used to improve the functional properties of meat and meat products with marination technique with varying effect on meat matrix. Some ingredients control the colour, taste, flavour and some tenderness and textural properties of meat to make it more appealing and palatable. Because they can improve the activity of myofibrillar proteins, sodium chloride and phosphates are the most commonly employed functional components [36, 49]. The effects of different marinades are discussed below in order to understand its effect on different sensory properties individually.

6.1 *Tenderizing Effect of Acids*

Tenderness has a significant influence on the organoleptic and qualitative properties of meat and meat products. Meat and meat products' juiciness and tenderness are greatly influenced by marinades. Consumers' decisions to repurchase are influenced by variations in tenderness, and they are willing to pay more for meat that is juicy and tender. Organic acids like those found in vinegars or citrus juices help break down the tough muscle fibers in meat by denaturing proteins, resulting in a more tender texture. Acids increase the porosity of meat cells, allowing the marinade's flavors to penetrate deeper into the meat and enhancing tenderness [50]. The application of acidic marinating to enhance the texture of meat products is being investigated by researchers. In another investigation, it was found that application of 60% red dates improved tenderness of beef effectively [51]. Tough buffalo meat could be marinated in a 0.5% ammonium hydroxide solution to make it more tender [35]. In a study, it was found that seasoning of buffalo longissimus thoracis steaks with 200 mmol/L $CaCl_2$ improved the tenderness when applied after rigor mortis [52].

6.2 *Flavor Enhancement*

A variety of herbs and spices used in the marination process contains characteristic flavour compounds that penetrate inside the meat system during marination to produce flavourful meat product as per the consumer demand. Essential meat sensory characteristics, flavors can significantly influence customer acceptance and meat choice [53]. These aromatics and flavourings play a crucial role in enhancing the taste profile of marinated meats. Garlic and onions offer robust and distinct flavors, while herbs such as rosemary, thyme, basil, and cilantro infuse a characteristic fragrances and tastes, imparting depth and complexity to the meat [54]. One approach to satisfy consumer need for variety, new flavors, spiciness, and outstanding presentation while also assisting in the production of value addition in products is to incorporate seasonings and flavorings into the marinating process. By improving basic taste, regaining flavor lost during processing, creating a unique flavor profile, and stifling and neutralizing overheated, excessive flavor, flavorings and spices applied while marinating improve the flavor of meat products [39]. The flavor of pre-rigor boneless broiler breast meat marinated with 1% NaCl salt was disliked by fewer than 21% of customers.

It should be stated that biological marinating improves the flavor of meat and meat products.

Though very few studies have been conducted on the enhancement of flavor using chemical marinades and other marination procedures, this conclusion was reached from the majority of the investigations. Lots of investigations are carried out to find out newer options for efficient and effective marination with chemical and physical marination techniques. Exploring these aromatics further reveals their unique contributions in marination process such as garlic is renowned for its pungent and savoury essence adds a rich umami flavour [55]. Onions, whether caramelized or raw, bring a sweet and sometimes sharp flavor. Herbs, with their fresh and aromatic qualities, lend a vibrant and nuanced taste that complements the meat's natural flavors. Meanwhile, spices like cumin, paprika, coriander, and chilli powder offer distinct quality, adding warmth, heat, or earthiness to the marinade. Conclusively, the research is going in pursue of finding sustainable solution that could be more safe for human.

6.3 *Water Holding Capacity*

One of the most important properties of marination is that it improves water holding capacity of the meat. In meat, proteins play important role in improving water holding capacity. The two most significant myofibrillar proteins linked to water binding and meat quality attributes are actin (thin filament) and myosin (thick filament). Brine or marinade components cannot restore all of the lost protein activity in meat containing highly denatured proteins, such as PSE. Water holding capacity has a strong correlation with pH value. Meat has minimum water holding capacity (about 5.2–5.3 in poultry meat) at its PI value while it has increased value below ad above the PI value. Since phosphates like sodium tripolyphosphate remove proteins and change the pH from the isoelectric point of muscle proteins, they improve the capacity of the material to store water [56]. In order to optimize water retention and the meat gels' acceptable stiffness and elasticity, sodium chloride is essential for the solubilization of myofibrillar proteins, which then lead to denaturation and aggregation.

Meat (beef and pork) generally undergoes irreversible alterations that result in lower protein functionality and decreased WHC at pH values below 4.5 and above 10 [57]. In poultry products, sodium pyrophosphates and STP have been shown to enhance the WHC and salty taste in frankfurters formulated with reduced salt levels by 20% and 40% [58]. A number of marinades such as salts are used to improve water holding capacity and simultaneously it also affect tenderness of the meat. While salt does make meat more tender, how well it improves meat qualities will depend on the condition of the myofibrillar proteins [59].

6.4 *Moisture Retention and Spice Absorption*

Oils and fats create a protective layer on the meat's surface, preventing moisture loss during cooking and keeping the meat juicy and tender. When the meat is tenderized by acid (such as vinegar or lemon juice) or enzymes (found in fruits like pineapple or papaya), the oil helps keep the meat wet and transfers flavors from herbs and spices. A variety of vegetable oils are used in marination. Oils and fats help spices adhere to the meat's surface, ensuring that the flavors are absorbed and distributed evenly throughout the bulk of meat cuts. Aromatics, encompassing elements like garlic, onions, herbs, and a spectrum of spices, serve as fundamental components in elevating the taste, aroma, and overall appeal of marinated meats.

7 Microbial Ecology of Marinated Meat

The marinated meat has variety of microflora predominantly psychrotrophic lactic acid bacteria (LAB) under standard refrigerated storage condition that can control the growth of *Enterobacteriaceae* [60, 61]. Lactic acid bacteria (LAB) are the most prevalent microorganisms in the marinated poultry meat that is packed in MAP [62]. The LABs were found to be the source of spoiling in fresh meat, chicken marinated in tomatoes, and marinated meat packed in MAP [63].

Leuconostoc, Lactobacillus sakei, Lactobacillus curvatus, and a few other isolated bacteria with phenotypic similarity to the heterofermentative *Lactobacillus* types were among the dominant *Lactobacillus species.* Microorganisms that cause spoiling in marinated poultry include *Pseudomonas* spp., *Enterobacteriaceae, Staphilococcus aureus, Salnonella* spp., and others [64]. Thyme, oregano, and horseradish added to the marinade solution [63] effectively prevented the growth of LAB in chickens. Gram-negative psychrotrophic bacteria, specifically *Pseudomonas* spp., are the primary cause of deterioration in marinated poultry that is stored aerobically [65].

In many countries, it was found that *Brochothrix thermosphecta* was associated with spoilage of marinated meat [60]. This species has not been shown to make up a significant portion of the population responsible for spoiling Finnish MAP marinated broiler meat strips [66]. This is the reason why Finnish meat products that have been marinated are typically packaged in modified atmospheres to inhibit the growth of aerobic spoiling germs. The product spoiled due to a mixed LAB population that was dominated by a *Leuconostoc species* that resembled *Leuconostoc gelidum.* The other major organisms found in this rotting population included *Lactobacillus sakei, Lactobacillus curvatus,* and a Gram-positive rod that phenotypically resembled heterofermentative *Lactobacillus species.* Marinating alters the product's initial population of spoilage LABs. At the other side, when we see the non-marinated meat, the development of bacteria in non-marinated meat constantly increases during refrigerated storage [67]. Pseudomonas and the psychrotrophic bacteria count (PTC) are considerably reduced when 1% lactic acid is added to the marinating solution [64].

Lactic acid, a byproduct of the LAB, significantly inhibits the growth of bacteria that produce H2S, such as *Pseudomonas species* [68]. The psychrotrophic bacterium count (PTC) in chickens is reduced by 0.2 and 0.4 logarithmic units (logs) when 1% or 3% sodium lactate is added [64]. Pseudomonas is susceptible to the antibacterial effects of several spices and their extracts. To suppress the growth of *Pseudomonas*, 0.5% pimento leaf oil or 0.5% clove oleoresin can be added to the marinade solution [65]. The gram-negative bacteria were thought to be more susceptible to acidic environments than non-gram-negative bacteria [69]. As a result, *L. monocytogenes* was more robust than *Salmonella strains* at low pH [70], and survived longer in acidic circumstances than *E. coli O157: H7* or *S. enterica* [71]. Nisin was combined with EDTA, sodium lactate, or potassium sorbate to prevent Salmonella from growing [72]. Further there is a lot of space for such investigations to find out the effective marinade combinations so as to minimize the spoilage microflora in marinated meat and could provide the food safety assurance to the meat lovers.

8 Health and Sustainable Trends Globally

Meat marination is the oldest traditional practice that is followed by most of the countries. Marination not only provides the tenderness and texture but also the flavour and aroma to the consumers. Inspite of all these characteristics of marination, it should also provide the food safety assurance to the consumers. A good sensory profile of the marinated meat may influence the consumer to eat it again and again but its microbial safety and chemical safety may deteriorate the health of the consumers. Ethnic and global flavors heavily influence marinating techniques and ingredient choices. Each culture brings unique elements to marinades, such as the use of soy sauce and torch ginger in Asian marinades or the robustness of cumin and paprika in Middle Eastern blends. These varieties of ingredients offer numerous health that make marinade a functional food entity [54]. Blending various aromatics and spices in the right proportions creates a symphony of taste, ensuring that no single element overwhelms the palate, but rather harmonizes to create a well-rounded and flavourful marinade [73]. Incorporating the superfoods, probiotics, and low-sodium in marination process not only improves the nutritional profile of the meat but also assures the good human health. Moreover, using organic and sustainable ingredients supports ethical and environmentally friendly practices, aligning with the growing interest in conscious consumption and reducing the carbon footprint in the culinary world [74]. Innovative trends in marinades also include a focus on healthier, sustainable, and locally sourced ingredients. Incorporating superfoods, probiotics, and low-sodium options not only adds nutritional value but also assures the consumer health. Moreover, using locally sourced and sustainable ingredients supports ethical and environmentally friendly practices, aligning with the growing interest in conscious consumption and reducing the carbon footprint in the culinary world [74].

9 Techniques of Meat Marination

9.1 Marination Period

For effective marination process and desirable sensory attributes of meat, the marination period plays a crucial role that depends upon variety of internal and external factors. Thinner cuts like fish or shrimp might only need 15–30 min, whereas chicken breasts or pork chops can benefit from 30 min to 2 h. For thicker cuts like steaks or roasts, it is suggested to marinate for at least 2 h or even overnight to ensure the flavors penetrate the meat completely [39]. However, acidic marinades, such as citrus juices or vinegar can digest the protein part of meat and may result a mossy texture which is not desirable. Therefore, for optimum marination, the synergistic effect of acidity and the marinating time must be optimized for different types of meat cuts [75].

9.2 Marinating Equipment

A number of tools and equipment's are used in the meat marination process from simple to complex depending on type of requirement. The containers or resealable plastic bags are the suitable material for immersing the meat in marinades. Vacuum sealers remove air from the package, allowing the marinade to penetrate more efficiently. Injection devices, such as marinade injectors or meat needles, are used to directly inject marinade into the meat, ensuring deeper penetration of flavours, particularly beneficial for larger cuts. The vacuum tumbler, sometimes referred to as the vacuum marinating machine, is primarily used to increase the flavor of meat by processing meats including beef, lamb, chicken, duck, goose, hog, and fish. The meat will thoroughly and consistently absorb the pickling liquid under the impact of collision, friction, and roll kneading, which will enhance its flavor and flexibility. The Vacuum Marinator machine drastically cuts down on the amount of time needed to marinate proteins by using a tumbling movement in conjunction with vacuum pressure. As the meat tumbles, its flat edge tenderizes it, and the vacuum draws air out of the chamber, allowing the marinade to seep deeply into the meat. In as little as 20–25 min, this suction and tumble action can mimic overnight marinating. For restaurants with limited refrigeration space but still needing to marinate meats, or as an addition to an overnight marination program, the commercial vacuum food marinator is perfect.

10 Future Prospects

The future of marinated meat products is set to undergo intriguing shifts driven by technological advancements, sustainability goals, and evolving consumer tastes. Smart packaging, equipped with sensors to monitor freshness and quality, could become the norm that ensures food safety and may reduce the waste for global sustainability. Moreover, the potential integration of 3D printing technology might revolutionize marination processes, allowing for more effective penetration of marinades or even personalized marinade blends tailored to individual preferences. Sustainability will likely play a pivotal role in shaping the future of marinated meats. As concerns about environmental impact persist, the market might witness an expansion of alternative protein sources such as algal protein, plant protein and blue protein sources. This could include the incorporation of plant-based or lab-grown meats into marinated products, providing consumers with a broader range of options while addressing sustainability issues associated with traditional meat production. Additionally, the corporate world may adopt circular economy initiatives, finding innovative ways to repurpose unused marinades or utilize by-products from other food processes, thus reducing waste. Consumer preferences are evolving, particularly with a growing focus on health-conscious choices and diverse culinary experiences. There's a discernible trend toward cleaner labels, prompting the development of healthier marinade recipes with reduced salt, sugar, and additives to meet the demands of health-conscious consumers. Simultaneously, the exploration of global flavours and fusion cuisine is on the rise, leading to increased demand for marinated meats infused with unique tastes from different cultures. This shift could drive companies to innovate and offer a wider variety of ethnic marinades. In essence, the future of marinated meat products appears promisingly dynamic, with technology, sustainability, and evolving consumer preferences driving innovation and transformation within the market.

11 Conclusion

Since ancient times, marination was initially employed as a preservation technique, utilizing herbs, spices, and acidic elements to extend the shelf life of meats and meat products. This practice evolved beyond preservation, transforming into an innovative method of infusing flavor, tenderizing, and enhancing the overall sensory profile of meats. Marinated meat products stand as a cornerstone in modern culinary practices, seamlessly blending rich historical roots with cutting-edge trends. In contemporary times, marinated meat products have taken on new dimensions, adapting to the preferences and lifestyles of consumers in current scenario. Technological advancements and innovation have propelled marination techniques into the realm of precision and convenience. The cutting-edge methods leverage technology to ensure even distribution of flavors, precise marination times, and packaging solutions that maintain freshness and quality of meat and meat products. It offers an array of choices, including organic, gluten-free, and plant-based alternatives, meeting the demands of health-conscious consumers without any compromise with the quality of meat products. It willled to the exploration of alternative proteins and the development of innovative, eco-conscious marinades that reduce waste and environmental impact with ethical value in the interest of consumers. With increasing the consumer demand for such kind of products with personalised nutrition, the products will be prepared or developed with the help of 3D food printing technology in the near future along with impregnation of other emerging food processing technologies such as ultra-sound processing, high pressure processing, pulse-electric technology, NMR technology etc.

References

1. Ehsanur Rahman SM, Islam S, Pan J, Kong D, Xi Q, Du Q, et al. Marination ingredients on meat quality and safety – a review. Food Qual Saf. 2023;7
2. Lopes SM, da Silva DC, Tondo EC. Bactericidal effect of marinades on meats against different pathogens: a review. Crit Rev Food Sci Nutr. 2022;62(27):7650–8.
3. Figueroa C, Ramírez C, Núñez H, Jaques A, Simpson R. Application of vacuum impregnation and CO2-laser microperforations in the potential acceleration of the pork marinating process. Innovative Food Sci Emerg Technol. 2020;66:102500.
4. Brian A, Nummer PD. Historical origins of food preservation. National Center for Home Food Preservation. 2002; CQ Press Food Safety and Inspection Service In: Federal Regulatory Guide 2020. p. 612–6.
5. Gómez I, Janardhanan R, Ibañez FC, Beriain MJ. The effects of processing and preservation technologies on meat quality: sensory and nutritional aspects. Food Secur. 2020;9:1416.
6. Birk T, Grønlund AC, Christensen BB, Knøchel S, Lohse K, Rosenquist H. Effect of organic acids and marination ingredients on the survival of Campylobacter jejuni on meat. J Food Prot. 2010;73(2):258–65.
7. Min B, Ahn DU. Mechanism of lipid oxidation in meat and meat products – a review. Food Sci Biotechnol. 2005;14:152–63.
8. Baltić T, Mitrović R, Todorović E, Ivanović J, Lončina J, Baltić MŽ. The influence of vacuum-packaging and marination on the behaviour of total Enterobacteriaceae on chicken breast fillets. International conference biological food safety and quality, Belgrade, 4–5. October 2012 (pp. 95–7). Novi Sad: Faculty of Agriculture.
9. Gómez I, Ibañez FC, Beriain MJ. Physicochemical and sensory properties of sous vide meat and meat analog products marinated and cooked at different temperature-time combinations. Int J Food Prop. 2019;22(1):1693–708.
10. Latoch A. Effect of meat marinating in kefir, yoghurt and buttermilk on the texture and color of pork steaks cooked sous-vide. Ann Agric Sci. 2020;65(2):129–36.
11. Latoch A, Libera J, Stasiak DM. Physicochemical properties of pork loin marinated in kefir, yoghurt or buttermilk and cooked sous vide. Acta ScientiarumPolonorumTechnologia Alimentaria. 2019;18(2):163–71.
12. Gamage HG, Mutucumarana RK, Andrew MS. Effect of Marination method and holding time on physicochemical and sensory characteristics of broiler Meat. J Agric Sci. 2017;12(3):172–84. https://doi.org/10.4038/jas.v12i3.8264.
13. Iqbal SZ, Talib NH, Hasnol ND. Heterocyclic aromatic amines in deep fried lamb meat: the influence of spices marination and sensory quality. J Food Sci Technol. 2016;53(3):1411–7.

14. Ozturk B, Sengun IY. Inactivation effect of marination liquids prepared with koruk juice and dried koruk pomace on Salmonella typhimurium, Escherichia coli O157: H7 and Listeria monocytogenes inoculated on meat. Int J Food Microbiol. 2019;304:32–8.
15. Petracci M, Laghi L, Rimini S, Rocculi P, Capozzi F, Cavani C. Chicken breast meat marinated with increasing levels of sodium bicarbonate. J Poult Sci. 2014;51(2):206–12.
16. Gao T, Li J, Zhang L, Jiang Y, Ma R, Song L, Gao F, Zhou G. Effect of different tumbling marination treatments on the quality characteristics of prepared pork chops. Asian Australas J Anim Sci. 2015;28(2):260.
17. Inguglia ES, Burgess CM, Kerry JP, Tiwari BK. Ultrasound-assisted marination: role of frequencies and treatment time on the quality of sodium-reduced poultry meat. Food Secur. 2019;8(10):473.
18. Farhadian A, Jinap S, Faridah A, Zaidul IS. Effects of marinating on the formation of polycyclic aromatic hydrocarbons (benzo [a] pyrene, benzo [b] fluoranthene and fluoranthene) in grilled beef meat. Food Control. 2012;28(2):420–5.
19. Kumar Y, Singh P, Tanwar VK, Ponnusamy P, Singh PK, Shukla P. Augmentation of quality attributes of chicken tikka prepared from spent hen meat with lemon juice and ginger extract marination. Nutr Food Sci. 2015;45(4):606–15.
20. Manful CF, Pham TH, Nadeem M, Wheeler E, Warren KJ, Vidal NP, Thomas RH. Assessing unfiltered beer-based marinades effects on ether and ester linked phosphatidylcholines and phosphatidylethanolamines in grilled beef and moose meat. Meat Sci. 2021;171:108271.
21. Siroli L, Baldi G, Soglia F, Bukvicki D, Patrignani F, Petracci M, Lanciotti R. Use of essential oils to increase the safety and the quality of marinated pork loin. Food Secur. 2020;9(8):987.
22. Jinap S, Iqbal SZ, Talib NH, Hasnol NDS. Heterocyclic aromatic amines in deep fried lamb meat: the influence of spices marination and sensory quality. J Food Sci Technol. 2016;53(3):1411–7.
23. Adeyeye SAO, Ashaolu TJ. Heterocyclic amine formation and mitigation in processed meat and meat products: a mini-review. J Food Prot. 2021;84:1868–77.
24. Smith JS, Ameri F, Gadgil P. Effect of marinades on the formation of heterocyclic amines in grilled beef steaks. J Food Sci. 2008;73(6)
25. Fernández M, Hudson JA, Korpela R, De Los Reyes-Gavilán CG. Impact on human health of microorganisms present in fermented dairy products: an overview. Biomed Res Int. 2015;2015
26. Gargi A, Sengun IY. Marination liquids enriched with probiotics and their inactivation effects against food-borne pathogens inoculated on meat. Meat Sci. 2021;182
27. Rahman SME, Islam S, Pan J, Kong D, Xi Q, Du Q, Yang Y, Wang J, Oh D, Han R. Marination ingredients on meat quality and safety. A review. Food Qual Saf. 2023;7:1–17. https://doi.org/10.1093/fqsafe/fyad027.
28. Kim HW, Hwang KE, Song DH, Kim YJ, Lim YB, Choi JH, Choi YS, Kim HY, Kim CJ. Effects of soy sauce on physicochemical and textural properties of tumbled chicken breast. Poult Sci. 2014;93(3):680–6.
29. Chou RGR, Tseng TF, Chen MT. Acceleration of post-mortem changes in Tsaiya duck (Anas platyrhynchos) breast muscle by lactic acid marination. Br Poult Sci. 1997;38(1):78–83.
30. Gómez-Salazar JA, Ochoa-Montes DA, Cerón-García A, Ozuna C, Sosa-Morales ME. Effect of acid marination assisted by power ultrasound on the quality of rabbit meat. J Food Qual. 2018;2018
31. Lin YC, Chen WT, Chou RG. Postmortem changes in mule duck muscle marinated in red wine. J Food Sci. 2000;65(5):906–8.
32. Kijowski J, Mast MG. Tenderization of spent fowl drumsticks by marination in weak organic solutions. Int J Food Sci Technol. 1993;28(4):337–42.
33. Smaoui S, Ben Hlima H, Ghorbel R. The effect of sodium lactate and lactic acid combinations on the microbial, sensory, and chemical attributes of marinated chicken thigh. Poult Sci. 2012;91(6):1473–81.
34. Li X, Sun Y, Pan D, Wang Y, Cao J. The effect of CaCl2 marination on the tenderizing pathway of goose meat during conditioning. Food Res Int. 2017;102:487–92.

35. Naveena BM, Kiran M, Reddy KS, Ramakrishna C, Vaithiyanathan S, Devatkal SK. Effect of ammonium hydroxide on ultrastructure and tenderness of buffalo meat. Meat Sci. 2011;88(4):727–32.
36. Petracci M, Rimini S, Mulder RWAW, Cavani C. Quality characteristics of frozen broiler breast meat pretreated with increasing concentrations of sodium chloride. J Poult Sci. 2013;50:282–6.
37. Sharedeh D, Gatellier P, Astruc T, Jean-Dominique D. Effects of pH and NaCl levels in a beef marinade on physicochemical states of lipids and proteins and on tissue microstructure. Meat Sci. 2015;110:24–31.
38. Smith DP, Young LL. Marination pressure and phosphate effects on broiler breast fillet yield, tenderness, and color. Poult Sci. 2007;86(12):2666–70.
39. Yusop SM, O'Sullivan MG, Kerry JP. Marinating and enhancement of the nutritional content of processed meat products. In: Processed meats: improving safety, nutrition and quality; 2011. p. 421–49.
40. Alvarado PE, Barrios RMM, Xóchihua JAM, Harnandez J. Fast and reliable DNA extraction protocol for identification of species in raw and processed meat products sold on the commercial market. Open Agric. 2017;2(1):469–72.
41. Ramírez N, Vega-Castro O, Simpson R, et al. Effect of pulsed vacuum and laser microperforations on the potential acceleration of chicken meat marination. J Food Process Eng. 2021;44(3):e13627.
42. Yang Y, Sun Y, Pan D, Wang Y, Cao J. Effects of high pressure treatment on lipolysis-oxidation and volatiles of marinated pork meat in soy sauce. Meat Sci. 2018;145:186–94.
43. González-González L, Luna-Rodríguez L, Carrillo-López LM, Alarcon-Rojo AD, Garcia-Galicia I, Reyes-Villagrana R. Ultrasound as an alternative to conventional marination: acceptability and mass transfer. J Food Qual. 2017:1–8.
44. Fencioglu H, Oz E, Turhan S, Proestos C, Oz F. The effects of the marination process with different vinegar varieties on various quality criteria and heterocyclic aromatic amine formation in beef steak. Food Secur. 2022;11:3251.
45. Rahman U, ur, Sahar A, Khan MI, Nadeem M. Production of heterocyclic aromatic amines in meat: chemistry, health risks and inhibition. A review. LWT. 2014;59(1):229–33.
46. Heinz G, Hautzinger P. Meat processing technology for small- to medium-scale producers. 2007. 36 p.
47. Yusop SM, O'Sullivan MG, Kerry JF, Kerry JP. Influence of processing method and holding time on the physical and sensory qualities of cooked marinated chicken breast fillets. LWT. 2012;46(1):363–70.
48. Alvarado CZ, Sams AR. Injection marination strategies for remediation of pale, exudative broiler breast meat. Poult Sci. 2003;82(8):1332–6.
49. Barbut S. Poultry products—formulation and gelation. In: Barbut S, editor. Poultry products processing. An industry guide. New York: CRC Press; 2002. p. 249–88.
50. Hinkle J, Calkins CR, de Mello AS Jr, Senaratne LS, Pokharel S. Acid marination for tenderness enhancement of beef bottom round. Nebraska Beef Cattle Report. 2010;
51. Ismail MA, Ibrahim MA, Ismail-Fitry MR. Application of Ziziphus Jujube (red date), Camellia Sinensis (black tea) and Aleurites Moluccana (candle nut) marinades as beef meat tenderizer. Int J Eng Technol. 2018;7(4.14):307–11.
52. Rajagopal K, Oommen GT, Kuttinarayanan P, George S. Effect of post rigor marination with calcium chloride on the tenderness, colour and palatability traits of buffalo meat. Nutr Food Sci. 2015;45(1):20–38.
53. Saha A, Perumalla AVS, Lee Y, Meullenet JF, Owens CM. Tenderness, moistness, and flavor of pre- and postrigor marinated broiler breast fillets evaluated by consumer sensory panel. Poult Sci. 2009;88(6):1250–6.
54. Raghavan S. Handbook of spices, seasonings, and flavorings. 2006.
55. Ansary J, Forbes-Hernández TY, Gil E, Cianciosi D, Zhang J, Elexpuru-Zabaleta, Simal Gandara J, Giamieri F, Battino M. Potential health benefit of garlic based on human intervention studies: A brief overview. Vol. 9, Antioxidants. 2020. p. 1–35.

56. Barbut S. Measuring sensory and functional properties. In: Poultry products processing. An industry guide. New York: CRC Press; 2002. p. 467–511.
57. Pearson AM, Young RB. Some condition occurring in muscle/meat. In: Muscle and meat biochemistry. New York: Academic; 1989. p. 420–39.
58. Barbut S, Mauer AJ, Lindsay RC. Effects of reduced sodium chloride and added phosphates on physical and sensory properties of Turkey frankfurters. J Food Sci. 1988;53:62–6.
59. Camou JP, Sebranek JG. Gelation characteristics of muscle proteins from pale, soft, exudative (PSE) pork. Meat Sci. 1991;30:207–20.
60. Borch E, Kant-Muermans ML, Blixt Y. Bacterial spoilage of meat and cured meat products. A review. Int J Food Microbiol. 1996;33:103–20.
61. Smolander M, Alakomi HL, Ritvanen T, Vainionpää J, Ahvenainen R. Monitoring of the quality of modified atmosphere packaged broiler chicken cuts stored in different temperature conditions. A. Time–temperature indicators as quality-indicating tools. Food Control. 2004;15:217–29.
62. Schirmer BC, Langsrud S. Evaluation of natural antimicrobials on typical meat spoilage bacteria in vitro and in vacuum-packed pork meat. J Food Sci. 2010;75:98–102.
63. Björkroth KJ. Microbial ecology of marinated meat products. Meat Sci. 2005;70:477–80.
64. Smaoui S, Hlima HB, Salah RB, Ghorbe R. Effects of sodium lactate and lactic acid on chemical, microbiological and sensory characteristics of marinated chicken. Afr J Biotechnol. 2011;10:11317–26.
65. Carlos AMA, Harrison MA. Inhibition of selected microorganisms in marinated chicken by pimento leaf oil and clove oleoresin. J Appl Poult Res. 1999;8:100–9.
66. Susiluoto T, Korkeala H, Bjo¨rkroth J. Leuconostoc gasicomitatum is the dominating lactic acid bacterium in retail modified-atmosphere-packaged marinated broiler meat strip products on sell by day. Int J Food Microbiol. 2002;80:89–97.
67. Knöchel S, Mach N, and, Karlsson A. Microbial and physical changes in marinated beef of high and normal pH during storage in different atmospheres. In: 53 rd International Congress of Meat Science and Technology, 2007, 1: 37–38.
68. Gonzalez-Fandos E, Herrera B, Maya N. Efficacy of citric acid against Listeria monocytogenes attached to poultry skin during refrigerated storage. Int J Food Sci Technol. 2009;44:262–8.
69. Choi YM, Bae YY, Kim KH, Kim BC, and Rhee MS. Effects of supercritical carbon dioxide treatment against generic Escherichia coli, Listeria monocytogenes, Salmonella typhimurium, and E. coli O157:H7 in marinades and marinated pork. Meat Sci, 2009, 82: 419–424.
70. Tan FJ, Ockerman HW. Microbiological changes of marinated broiler drum sticks treated with the lactoperoxidase system and with or without thermal treatment. Asian-Australian J Anim Sci. 2006;19:109–12.
71. Lin CM, Kim K, Du WX, Wei CI. Bactericidal activity of isothiocyanate against pathogens on fresh produce. J Food Prot. 2000;63:25–30.
72. Ukuku DO, Fett WF. Effect of nisin in combination with EDTA, sodium lactate, and potassium sorbate for reducing Salmonella on whole and fresh-cut cantaloupet. J Food Protect. 67:2143–50.
73. Gidwani B, Bhattacharya R, Shukla SS, Pandey RK. Indian spices: past, present and future challenges as the engine for bio-enhancement of drugs: impact of COVID-19. J Sci Food Agric. 2022;102:3065–77.
74. Meneses R, Teixeira P. Marination as a hurdle to microbial pathogens and spoilers in poultry meat products: a brief review. Appl Sci (Switzerland). 2022;12
75. Fiore AG, Palmieri L, Orlando I, De Giorgi A, Derossi A, Severini C. Comparison of different combined processes of preservation on the nutritional and sensory changes of "ready to eat" mackerel fillets. J Food Process Preserv. 2019;43(3)

Ethnic/Cultural Meat Products

Mamta Devi Sharma and Anup Kumar Sinha

1 Introduction

Meat is the most highly nutritious component in the list of foods. It is the flesh of the animals that can be consumed as food. Being animal flesh, it contains all the nutrition that is consumed by animals and stored in their flesh. It contains 70–75% water in it; a high content of protein is present in meat, around 20–22% and remaining 5% includes all other components like fats and carbohydrates [1]. One of the major factors in meat consumption is its perishable nature. These meats are actually animals' tissue of cells, which has a tendency to start degrading soon after it stops getting sufficient oxygen and nutrients through blood. So, after the death of animal, its meat is needed to be harvested properly, avoiding any microbial degradation.

These can be obtained from many animals, including rabbits, sheep, chickens, goats, pigs, and cattle. People living in different areas of the world don't have access to all kinds of meat from all these animals, so they look for easily accessible options available around them, and they start consuming that accessible animal's meat [2].

Meat is one of the most important food requirements on the market due to its beneficial effects on the human body after consumption. It is one of the major protein foods for humans which makes it suitable to build a highly attractive market in the food industry. This meat food industry now has a big face, including many ethnic meat products from across the world. Every community that started consuming meat as a food developed their own way to cook meat, which later developed into their own ethnic meat product. Every cuisine around the world, like Indian, Mexican, African, and Chinese includes ethnic or traditional meat products [3]. Ethnic meat products, carried forward from ancestors to next generations, represent the unique

M. D. Sharma (✉) · A. K. Sinha
Shoolini Institute of Life Sciences and Business Management,
Solan, Himachal Pradesh, India
e-mail: mamtasharma@silb.org

A. Kumar Pathera et al. (eds.), *Healthier Meat Products*,
https://doi.org/10.1007/978-3-031-78215-2_20

flavours, preservation techniques, benefits of food, and delicious heritage of various ethnic groups. Global cuisine is present all around, from the smoky barbecue of South America to the aromatic curries of India, which represent the diversity of ethnic meat products. Each product from across the world reflects local ingredients, recipes, cooking technique, type of meat used, and local cultural influence. All these factors shape and showcase the culinary identity of a particular region [4].

In the Middle East, tender kebabs, seasoned with aromatic spices, mouth-watering replicas and grilled to perfection are a culinary centre masterpiece [5]. North America put forward flavourful merguez sausages, filled with a mixture of herbs and spices, and tagines, slow-cooked stews that are full of senses with rich flavours. Asia has richest cuisine in the world, including Indian and Chinese. Indian subcontinents offer a symphony of flavours with tandoori chicken, marinated in yogurt and spices, and spicy vindaloo curries. Chinese cuisine showcases a variety of preserved meats, such as savoury sausages and smoked duck, while Korean barbecue tantalizes with its sizzling marinated meats.

Across the globe, ethnic meat products serve as cultural cornerstones, uniting communities around shared culinary traditions passed down through generations as cherished family recipes. These products not only provide sustenance but also embody the cultural identity and heritage of diverse ethnic groups. These delicacies not only tantalize the taste buds but also serve as powerful vessels of heritage and tradition, connecting individuals to their roots and fostering a sense of belonging within their communities. Whether it's the savory aroma of a beloved family recipe wafting through the kitchen or the communal gathering around a table laden with these cherished dishes, the impact of ethnic meat products transcends mere sustenance, serving as a tangible expression of cultural pride and unity. As we continue to celebrate and preserve these culinary traditions, we ensure that the rich tapestry of our global cultural heritage remains vibrant and enduring for generations to come.

2 History and Development

Meat is derived from the world "mete," which means food in general. In the studies of palaeontology, it was found that meat was a staple food; earlier hunters and gatherers used to hunt down local wild and grazing animals to consume the flesh as meat from them. Hunters mostly look for deer and bison for meat. From 10,000 BC, evidence of animal domestication was found in studies, where people used to keep the animals and allow them to breed to improve meat production [6]. Sheep, cattle, domesticated pigs, goats, and chickens are the most common meat providers. Sheep was domesticated at first in around 8000–7001 BC. It originated in Western Asia; later, in 3500–3000 BC, in Mesopotamia and Egypt, several breeds of sheep were established. Around 5000 BC, Mesopotamians established settled agriculture, which led to the domestication of cattle. Domesticated pigs came into existence in 2500 BC in Hungary. Mountain goats were also common in those times and were available in large numbers, so people started domesticating them as well. Chickens got domesticated in 6000 BC [6].

Hunters and gatherers started consuming raw meat from animals; soon they realized the taste enhancement by adding ingredients like herbs present around them, which led to the development of local meat products. Developments of several civilizations were already got information about the meat as staple food. Kashmir in 516 BC—1398 played a crucial role in the development of Kashmiri cuisine [7]. From Indian mythology, Mahabharata to the Kashmir invasion by durian, and then Mauryans who established Sri nagara, developed these cuisines. Every part of the world showed the remains of hunters and gatherers who used to catch the animals and consume their meat [8].

Adding a local punch of flavour to all these common meats led to development of local meats products. Like Africans, Asians and American, all countries among these nations used their local ingredients and techniques to enhance the flavour and taste of the meat products, which are introduced as their ethnic meat products.

3 Regions and Their Techniques of Making Cultural Meat Product

In the world, almost every country has a specific ethnic meat product. Indians have most common biryani with specific meat; Chinese people have four regions depending upon their meat products; Americans have sausages; and Africans have a list of products.

4 Asian Countries Region

The largest continent in Asia includes four major cuisines; Indian, Chinese, Korean, and Japanese. From the historical studies, it has been found that many civilizations were founded in the Indian region. This implicates great information about the start of ethnic meat products in Indian regions [9].

4.1 India

From the earliest civilization, the Indus Valley Civilization (2500–1900 BC), India has a great, long, and fascinating history of ethnic meat products. People introduced the culture of domesticated animals like sheep, cattle, and goats, which led to the introduction of meat later became a staple food. The Indus Valley Civilization introduced trade and also cultural exchange due to it, through which they learned many new cooking techniques and used different ingredients. That was the first imprint of such an introduction to India. This led to the development of an exotic variety of ethnic meat products. For example, the introduction of chili peppers from the Americas in the sixteenth century gave rise to spicy curries like vindaloo, while the arrival of the Mughals in the sixteenth century brought with it the tandoor oven, which is used to cook tandoori chicken and other dishes [10].

Today, Indian ethnic meat products are enjoyed all over the world. They are a testament to the country's rich culinary heritage and the diversity of its people.

4.1.1 Indian Regions with Products

Majorly 2 regions were present in India in which traditional meat products were developed.

4.1.1.1 Himalayan Regions Products

The Himalayan region is rich in a diverse display of ethnic meat products, each with a unique taste, flavours, texture, and cultural significance. All these products display the region's rich culinary heritage and the ingenuity of its people in adapting to the unbearable mountain environment [11].

(a) **Yak Meat Products:**
In Himalayas people feed on local animals available in such harsh localities, one such animal is yak. Yak meat is one in the staple foods for people of Himalayas. There are many types of Yak meat, like Yak jerky, Yak sausage, & Yak dumplings, are some of the main meat products [12].

(b) **Sheep and Goat Meat Products:**
Sheep and goat meat are also important sources of protein for people in the Himalayas, and they are used to make a variety of products, including mutton curry, goat kebab, and sheep cheese [13].

(c) **Pork Products:**
Pork is a popular meat in many parts of the Himalayas, and it is used to make a variety of products, including smoked pork, pork sausage, and pork dumplings [14].

These are just a few examples of the many ethnic meat products that are enjoyed in the Himalayan region.

Common and most famous ethnic meat dishes

I. Kargyong is a type of sausage-like product that is made from pork, beef, and yak meat. This is mostly consumed in Indian subcontinents such as Bhutan, Tibet, Drupka, Lepcha, and Sherpa in the Himalayan terrain [14]. There are total 3 types of Kargyong made from different meats, i.e., yak kargyong (from yak meat), lang kargyong (from beef), and faak kargyong (from pork).

II. Satchu is prepared from yak and beef meat by the technique of air drying and smoked meat products. It is prepared in two varieties, i.e. lang satchu and yak satchu.

III. Suka ko masu is an air-dried or smoked chevon or buffalo meat ethnic product. People living in the Western Himalayas consume foods goat meat and its scraps to prepare and consume such as chartayshya, arjia, and jamma.
IV. Chartayshya, is a meat product prepared in the festival called 'Kolatch' which is celebrated by Bhutia of Dharchula and Munsiyari in the district of Pithoregarh in Uttarakhand [14].

4.1.1.2 Southern Region Products

South India is a region with diverse ethnic meat products; these products play a significant role in many regional cuisines.

Some of the well-known and popular ethnic products from South India are:

I. Chettinad Chicken: This product got its name from its origin place, the Chettinad region of Tamil Nadu. This is made from marinated chicken with a blend of spices, which includes a list of herbs like black pepper, cumin seeds, and chillis, and then cooked in a rich and flavourful gravy.
II. Kerala Lamb Stew: This stew is prepared from Lamb meat perfectly favoured with spices and vegetables. It is a popular dish during the cold season.
III. Hyderabadi Biryani: Most iconic dish famous all over the world, maintains its legacy and Hyderabad's identity. This dish is the product of a influence of the Mughals on Hydarabadi cuisine. It is prepared by 2 main indentify ingredients of India, i.e., Basmati rice and tender chicken meat, which are cooked in a flavourful gravy infused with saffron and other spices. This dish is sealed in and cooked in dum to get perfectly cooked aromatic biryani.
IV. Hoan Pork Vindaloo: This is a classic Goan dish with a spicy and tangy curry. Pork is marinated and blended in vinegar, chilli, and ginger, then cooked in rich and flavourful gravy.
V. Mangalorean Beef Roast: This is a speciality dish of Mangalore, Karnataka. This is a product of beef meat marinated, blended with spices, and slowly roasted until tender and juicy.
VI. Keralan Fish Molee: Flavourful and spicy fish curry is a speciality of Kerala.
VII. Andhra Chicken Fry: This is a chicken dish from Andhra Pradesh.
VIII. Konkani Pork Sorpotel: This is a pork dish from Goa.
IX. Tamilian Mutton Kulambu: This is a speciality mutton curry from Tamil Nadu [15].

4.1.2 Benefits

Indian-ethnic meat products offer a variety of benefits, including:

Nutritional benefits: Meat is already a good source of proteins, iron, and other nutrients, and when we add more herbs and spices, it boosts the overall nutritional benefits by adding many antioxidants to it. It also acts as a source of probiotic in some cases due to the presence of beneficial bacteria.

Cultural significance: Indian ethnic meat products reflect the cultural diversity of different parts of India by making different products on special occasions, festivals, holidays, and celebrations. The preparation and consumption of meat products are important parts of cultural identity.

Economic benefits: Production and saleof these ethnic meat products are crucial many people for as their source of income. It also contributes to a country's economy.

Environmental benefits: Many part of Indian started making these ethnic meat products from sustainable meat options like goats and buffalos, which do not harm the environmental balance.

Here are some specific examples of the benefits of Indian-ethnic meat products:

Tandoori chicken, Rogas josh, Viindaloo, Haleem, Kebabs, Biryani, Momos, Kolhapuri Mutton, Chettinad, Chicken, and Kerala Lamb Stew.

Overall, Indian ethnic meat products are rich in nutrition and flavours. All these products build the cultural heritage of Indian food [8, 15].

4.1.3 Preservation

Traditional Preserving Methods in India
Indian represents multiple preservation techniques that were passed down from ancestors depending upon the meat type, local climate, accessible ingredients, and culinary traditions.

In Himalayas people mostly consume yak meat which is often preserved by the process of drying or smoking. The pickling technique is used to preserve the pork meat. Similarly, in South India, people use the methods of sun drying for fish meat, poultry, and goat meat [16].

4.2 China

China has a rich variety of cultural/ethnic meat products with more than 3000 years of history. Chinese products slowly enhanced in flavour, taste, aroma, preservation, and overall quality. Distribution of these meat products presents all [17].

4.2.1 Regions

4.2.1.1 North

The region is located around Beijing, and they are known for their "Fresh & crisp" meat. During the development of its identity, it gets integrated with many other flavours such as Muslims and Gong Ting foods, which gives and its uniqueness. With this highly acknowledged and proud history, North region meat is famous internationally and has gained huge popularity in China.

Some of the typical North flavour ethnic meat products include

Peking Roasted Duck,
Meat in sauce
Plain Boiled Pork

Peking duck is being made with a unique cooking technique in which it requires special local force-fed ducks, which are later cooked in one of the traditional closed stove of China. In this stoves ducks are hugged over an open fire beneath it. For this cooking process, they use special fruit trees as fuel to roast, which gives a golden, crispy skin with a product unique sweet and fragrant flavour. This product is frequently served at state banquets [18].

4.2.1.2 South

The origins of South flavor arose from the Tnag and Song dynasties, which usually contain the main elements, i.e., Jinling, Huaiyang, Shuxi, and Xuhai flavors. It was developed around the Nanjing and Suzhou areas and slowly gained popularity. It includes Jinhua ham, Rugao ham, and dried salted duck with sweet flavors. The shape of these ethnic products also has its uniqueness, like Jinhua ham, which is shaped into the traditional Chinese instrument, the lute. Due to its unique taste and shape, it has been supplied as a tribute to kings of different dynasties [17]. The process of this product is also unique, involving 3 steps of curing in which the ingredients are rubbed on a green ham at 4 °C. In the next step, the hams are hung in a smokehouse over a cold smoke between 21 and 32 °C. Due to this process, the internal temperature of the hams is maintained 5 °C lower than the smokehouse. This process is carried out until the meat obtains a desirable amber and mahogany color. Usually, this process takes 2–3 days. In the last step, aging was done for 6–9 months, which allowed the development of the flavor profile of the hams [19]. In sub-Yangtze areas of China, ham is a must-have. Traditional ethnic products or dishes are Yan Du Xian (where Yan means salted pork) and Da Ban Xiang (cured pink pork cubes that emit the alluring fragrance of camphor trees used for smoking) [17].

4.2.1.3 Guang

Tone of the most famous flavour which has international popularity, originated and developed around the Lingnan area of China which mainly consists of 3 sub-regions under it, i.e., Guangzhou, Kejia, and Chaozhou. This region's products are regarded as the representatives of Chinese ethnic meat products across the world, more than in China. The process of making such refreshing, tender, and fresh meat products is often more complex, flexible, and diversified than in all different regions [20]. A signature dish of this region includes sausage of guang flavour, chinese bacon, roast goose, roast suckling pig, and salt baked chicken.

4.2.1.4 Chuan

Another important flavour of Chinese ethnic meat products originates and is developed in the region around Chengdu and Chongqing which consists 3 sub flavours, i.e., Shanghe, Xiaohe, and Xiahe. Som spicy flavor with chilli pepper and Chinese prickly ash is this signature of seasoning added to its flavour [20]. Due to its flavour it started growing across China. Some special dishes are sausage in Chuan flavour, Chinese bacon, Butterfly Snout, Lamp-Shadow Beef Jerky, and Tangled Wire Rabbit. Making of every product is unique and use different flavour contents [21].

4.2.1.5 Others

There is a large list of regions available in China with their signature ethnic meat products. Above are the main regions, followed by many minor flavours which represent a wide area of China.

4.2.2 Benefits

Chinese ethnic meat products have major benefits for Chinese culture and world ethnic meat products community. The process of making these products requires specific dedication, which is needed to develop both world and country's economy.

4.2.3 Preservation

Preservation techniques in every country have their own specificity; among all the processes, chemical preservation with sodium lactate has been used in Chinese traditional or ethnic meat products because this technique improves the shelf life by improving the microbiological safety of the food [22, 23]. Naturally derived food-preserving chemical components are more preferred and are being used in the Chinese food-preserving techniques, which include three categories: plant-derived, animal-derived, and microbial-derived. Plants are mostly extracted from fruits, vegetables, herbs, and spices. Animals derived have a most common example of chitosan this is an animal derived antimicrobial. In many microorganisms, like LAB, which has health benefits for humans [24].

Japan

Japan has a rich history of food cuisine, especially meat consumption; this development in food cuisine in Japan started in the prehistoric Jōmon period. They were also limited to meat consumption due to Buddhist and some other political influences [25].

During the Edo period (1603–1868), people started consuming meat, but it was still a still a comparatively rare treat. Japan has a very short history of meat consumption. Japan mostly consumes fish as their staple.

Early History

The Jōmon period (14,000–300 BCE) provided some evidence of hunters and gatherers who consumed meat from animals like deer, bears, and other wild animals. A study says that meat was believed to be the source of protein and fats for all people at that time. Evidence came from the Kofun period, which says that evidence of Chinese culture in Japan and the development of ethnic meat products. Buddhism plays an important role in Japanese food culture. Under the Tang dynasty, Japan started imitating the taboos of meat products [25].

Buddhist Influence

Buddhism's belief in vegetarianism, which was introduced in Japan in the sixth century, plays an important role in the decline in meat consumption in Japan [26].

Edo Period

The Edo period from 1603–1868 again increased the production of meat consumption in Japan. Soon after, the loosening of the restrictions in Buddhism influenced Japan to go for developing new feedstock farming techniques, which led to meat consumption in Japan [27].

Modern Era

A major boost in meat consumption in Japan took place in the modern era, due to the influence of westernization of Japanese cuisine [27]. Due to this, it also increases the availability and affordability of meat in Japan.

Today

Meat has become more common in Japanese cuisine, and it is being used in a variety of dishes in Japan now, including yakiniku (grilled meat), sushi, and ramen.

4.2.4 Major Products

Some of the Japanese ethnic meat products are:

- Yakitori: This meat dish is grilled over charcoal. It is most common as street food and is served with a dipping sauce.
- Shabu-shabu: This meat product is cooked by making thin slices, and it is eaten with vegetables and sauces.
- Tonkatsu: It is a pork cutlet deep-fried with bread. It is popular in Japanese restaurants.

Table 1 Major characteristics of sushi and sashimi

Feature	Sushi	Sashimi
Main ingredient	Vinegared rice	Raw fish
Cooked	No	Yes
Typical toppings	Raw fish, vegetables, other ingredients	Raw fish
Sauce	Wasabi, pickled ginger	Dipping sauce

- Gyu-don: This is a rice dish with beef and onions.
- Ramen is a type of noodles made with pork [27].

These are just a few examples of the many different types of Japanese ethnic meat products. Meat is a versatile ingredient that is used in a variety of dishes in Japanese cuisine.

4.2.4.1 Sushi and Sashimi

Sushi and sashimi are two identical but unique features that feature raw seafood, with some acknowledging differences between both.

4.2.4.2 Sushi and Sashimi

Sushi and Sashimi are two identical but unique features which features raw seafood with some acknowledging differences between both (Table 1).

Here are some of the most popular types of sushi and sashimi:

Types

- Nigiri sushi: a piece of raw fish placed on top of a small mound of vinegared rice.
- Maki sushi: a roll of vinegared rice, nori (seaweed), and other ingredients, such as raw fish, vegetables, or avocado.
- Uramaki sushi: inside-out maki sushi, in which the rice is on the outside and the nori is on the inside.
- Temaki sushi: a cone-shaped piece of seaweed filled with vinegared rice, raw fish, and other ingredients.
- Sashimi: thinly sliced raw fish, typically served on its own or with a dipping sauce [28].

4.2.5 Benefits

Japanese cuisine has a rich cuisine of seafood and nutritional substances grown on sea surfaces, like seaweed. These ethnic meat products from Japan provide a wide range of nutrients, like iodine, which is much needed for human overall growth, and adding this nutrient to ethnic meat products increases the overall taste, flavor, and nutrients of the food [27].

4.2.6 Preservation

Among all types of food preservation techniques in Japan, they mostly use the air-drying technique. Before drying, they salt and cut the meat into thin strips.

5 African Subcontinent Region: North-African and Mediterranean Countries

- African Ethnic Meat Products: A Culinary Tapestry
  North African countries, including Egypt, Morocco, Algeria, Libya, and Tunisia, have a great influence on the ethnic meat product culture of this region. From the historic studies, it was found that early people were greatly dependent on meat and its products, like animal meat, intestines, and blood, as a good source of food. Various civilizations that invaded North African nations, i.e., the Phoenicians (332 B.C.), Romans (30 B.C.), Arabs (642 A.D.), Italians, and finally the French and British (eighteenth and nineteenth centuries), influenced the food practices in all 137 countries of this region. This all diversified the culture of ethnic food products [29].
- Preservation Techniques
  The preservation of meat was essential for survival in Africa's diverse environments. Traditional preservation techniques included sun-drying, salting, smoking, and fermentation. These methods allowed communities to store meat for extended periods, ensuring a steady supply of protein throughout the year [5].
- Cultural Significance of Ethnic Meat Products
  Meat products hold deep cultural significance in many African societies. They are often served at special occasions, such as weddings, festivals, and ceremonies. Additionally, meat products are often associated with hospitality and generosity.

5.1 Major Products with Making Techniques

Among them, some of the major ethnic meat products are:

- Biltong: This product is made from beef meat seasoned with spices like salt, pepper, coriander, and nutmeg. This is made in South Africa and Namibia.
- Suya: This product is made from beef, lamb, or goat meat, which is marinated in a blend of spices.
- Kilishi is a famous West African product made by cutting the meat into thin slices and then seasoning it with a spicy blend of spices.
- Chakalaka: This is a stew made from beef and chicken meat, full of vegetables and herbs.

- Kebab: One of the of the most famous dishes all around the world with local significance is kebab, which has its own high value. In this region, it is made from beef, lamb, and chicken meat marinated in a blend of spices, including turmeric, which acts as an antibiotic as well.
- Msimu: Famous Tanzanian and Zanzibar met stew with falafel-ful gravy [30].
- Tagine: lamb or beef dish made by slow cooking technique in a tagine pot.
- Isaw is a Nigerian dish made with chicken meat. It is seasoned with spices and grilled over charcoal till it gets crispy and flavorful [5].

These examples represent a small portion of the vast array of ethnic meat products found across Africa. Each region boasts its own unique culinary traditions and flavors, making African cuisine a diverse and vibrant tapestry of flavors and cultural influences.

5.2 Benefits

African society has a high influence on ethnic meat products. They made different styles and types of meat products for different occasions, like celebrations, weeding, etc. They believe that meat products are associated with their hospitality and generosity [5].

5.3 Preservation

Traditional preservation techniques like sun drying, salting, smoking, and fermentation are crucial methods of meat preservation in African nations. They maintain their taste through a combination of these techniques.

6 European Countries with Their Products

6.1 Germany-Bratwurst

The most beloved sausage dish in Germany is bratwurst. Met products have a long history of origin dating back to ancient times, and after that, they evolved from that to present as people's flavor choices.

This product got its name from the German words "braten" (to fry) and "wurst" (sausage). Its traces can be found back in the Bronze Age, through the exploration of archaeologists. It is mainly made from pork meat, and it is full of flavors. It is served as a celebration dish in Germany as well. With time, this product evolved as per different regions throughout the country. Each region put its own punch of flavor into the original product [31].

Modern bratwurst can be made from various meats, like pork, beef, and veal.

Cultural Significance of Bratwurst

This product is more than just a normal food in Germany; rather, it is also a national identity and culinary heritage. This dish is a souvenir for tourists.

6.2 France: Charcuterie

During the sixteenth century, butchers in Paris found that they needed to do something that worked as a preservation technique for their raw meat. Soon they came up with the word charcuterie, which means "cured or cooked d meat." They cooked different meat products like pork, fish, beef, and poultry. Then, during the Middle Ages, this product gained popularity in France. Later, they developed a variety of techniques like salting, curing, and making sausage filled with meat, herbs, and spices. Over the course of time, each region of France developed its own blend of this product by blending different spices. A total of 400 different types of products have been developed, from traditional to modern [32].

Types of products and making techniques:

- Hams
  Taking note of French charcuterie practices, hams should be made from the hind limbs of the ping only, which are further kept in two types: cooked or dry [32].
- Cooked hams
- In the cooked hams, many ingredients are added, like herbs and spices, wines, alcohols, liqueurs, sugars, and condiments, as per the requirements of the recipe and its category, as well as some particularly additives in strictly limited numbers and quantities. These cooked hams are further divided into 3 types based on the basic gelling agents and phosphate.

  1. Superior-cooked ham is present all over the French market, occupying almost 80% of the total market, which signifies its popularity. This product does not contain any gelling agents or phosphate.
  2. Choice cooked ham: this is the second most famous cooked ham, which contains phosphate not exceeding 0.2% in total, but there is no gelling agent added to it.
  3. Standard cooked hams: this product contains both gelling agents and phosphate and other additives as per prescribed standards [32].
  4. Dry-cured hams

This type of ham is prepared from the hind limb; in making the hind limb from the whole, it is salted. There are two techniques used for slating, i.e., the first technique is when the ham is immersed in the salt for a limited time, and in the other process, salt is rubbed on the hams for salting. The salting process follows the working of salt, which is absorbed by the ham and slowly gets inside it everywhere. This prevents the ham from deteriorating and promotes the maturation phase, which takes 130–210 days (for superior ham). All these processes enhance the taste, flavor, aroma, and overall product quality. Then it is cooked on low heat for a long time [32].

The Sausages
Sausages are one of the specialties of the charcuterie products; they are made by casings filled with different types of meats like pork, beef, veal, and poultry.

Dried Sausages
Dry sausages are prepared from a mixture of lean and pork fat chopped finely, salted, and then put in casings after being seasoned. This keeps the product good for a long time.

Pâtés and Terrines
It is a diverse variety of product that is made by the process of marinating meat with herbs and spices and then cooking it [32].

Black Pudding
Black pudding is made from pork blood and fat, onions, and husks for cooking. It is a flavorful product.

6.3 Spain: Paella

Paella is a national dish of Valencia with international popularity due to its great taste, flavors, and fresh ingredients. It was traditionally made with chicken and rabbit meat, complemented with fresh vegetables, including all kinds of vegetables. But today's most common paella is seafood paella, which consists of a myriad of seafood, including calamari, mussels, fish, clams, etc. [33].

6.4 UK Sausage Rolls

As a broad definition, sausages are chopped meat, mainly beef and pork meat, with salt and apices, and stuffed inside casings. The UK consumes this sausage every day, and they have many varieties of sausages depending upon the region, making, and ingredients [34].

Gloucestershire sausage: It is made from pork meat with a full range of flavors.
Oxford sausage: one of the savory sausages with sage, this sausage is made from pork and nicely mixed with spices and lemon [34].

7 Present World Trend

In the present time, people are moving more toward a vegetarian diet and a vegan diet to avoid the meat industry's cruelty to animals. Looking at the data made by my US Meat Department in 2023 shows that more than $122 billion of total sales of meat have been done so far. This is 3.2% up from last year [35]. There are many reasons on which this meat industry depends, causing effects on ethnic meat products as well. Some key factors or trends in the meat industry are:

- Plant-based meat demand
  Among the whole population of America, almost 15.5 million people follow a vegetarian diet, including two million who follow a who follow a completely vegan diet. These changes in meat consumption lead to changes in old grocery purchases. People are generous in buying plant-based meat products, taking concern for animals's welfare, environmental issues, and personal health. Now the market has also started storing possible plant-based meat, which comes in a wide range from beef to chicken nuggets, but on the other side, these meat products don't have all the characteristics of different animals' meat products. Thus, this leads to an increased demand for ethnic meat products.
- Sustainable market demand
  Focusing on the trend of meat availability in the market and animals' meat availability, people became more conscious of food sourcing. In the current time, people are trying to know the production of the meat that is coming into the market and prefer to buy such meat and meat products, in which animals are grown healthy and certified as completely organic. Organic meat products are making their way into the market with the name of different brands.
- Increased demand for ethnic meat products
  In the current trend, people are buying local ethnic meat products from different areas and cultures with local flavors. These products include those made from lamb from the Middle East or pork from Asia.
  Markets and retailers are increasing the demand for local ethnic meat products throughout the world. Interests are building among the people about ethnic meat products, and they are looking to have the products from different sources, like online markets.
- Convenience
  Today's people are trying to get pre-prepared meat products that are already pre-prepared and marinated. This market is developing among all markets. Three types of products are gaining value for many events, celebrations, etc. About 41.4% increase was made in these products from last year [36].

8 Conclusion

Ethnic meat products are one of the cultural culinary products that are being brought forward by different societies across the world and are still attracting a large number of populations. Due to raising concerns about meat cooking and hygiene, people are adapting to vegan and vegetarian diets, but still, a large number of populations are attached to ethnic meat products. The current trend of traveling from one place to another makes people more attracted to specific and ethnic products in any area, including meat products. So it's hard for people to forget the world's famous ethnic meat products from different parts of the world.

References

1. Ahmad RS, Imran A, Hussain MB. Nutritional composition of meat. Meat Sci Nutr. 2018;61(10.5772):61–75.
2. Steinfeld H, Gerber P, Wassenaar TD, Castel V, De Haan C. Livestock's long shadow: environmental issues and options. Food & Agriculture Org; 2006.
3. Leroy F, Smith NW, Adesogan AT, Beal T, Iannotti L, Moughan PJ, Mann N. The role of meat in the human diet: evolutionary aspects and nutritional value. Anim Front. 2023;13(2):11–8.
4. Prakash V. Introduction: the importance of traditional and ethnic food in the context of food safety, harmonization, and regulations. In: Regulating safety of traditional and ethnic foods. Academic Press; 2016. p. 1–6.
5. Gagaoua M, Boudechicha HR. Ethnic meat products of the North African and Mediterranean countries: an overview. J Ethn Foods. 2018;5(2):83–98.
6. Lawrie RA, Ledward DA. The spoilage of meat by infecting organisms. In: Lawrie's meat science. 7th ed. Sawston, Cambridgeshire: Woodhead Publishing; 2006. p. 157–88.
7. KI News. Greeks in Kashmir. Kashmir Images Newspaper; 30 Mar 2021. https://thekashmirimages.com/2021/03/30/greeks-in-kashmir/.
8. Rather SA, Masoodi FA, Akhter R. Ethnic meat products of Kashmiri wazwan: a review. J Ethn Foods. 2016;3(4):246–50.
9. Asian Spices and Seasonings. https://spicesinc.com/blogs/asian-spices-and-seasonings.
10. Indus Valley Civilisation. https://en.wikipedia.org/wiki/Indus_Valley_Civilisation.
11. Tamang JP, Okumiya K, Kosaka Y. Cultural adaptation of the Himalayan ethnic foods with special reference to Sikkim, Arunachal Pradesh and Ladakh. Himal Stud Monogr. 2010;11:177–85.
12. Wilson A. Essential eating in the Himalayas. 15 Sept 2013. https://live-less-ordinary.com/himalayan-food-eating-himalayas/.
13. Teixeira A, Silva S, Guedes C, Rodrigues S. Sheep and goat meat processed products quality: a review. Food Secur. 2020;9(7):960.
14. Rai AK, Palni U, Tamang JP. Traditional knowledge of the ethic Himalayan people on production of indigenous meat products. Indian J Tradit Knowl. 2009;8:586–91.
15. Bhandare S, Gadekar Y. Ethnic meat products/Indian subcontinent. In: Reference module in food science. Elsevier; 2022.
16. Seervi P, Singhal H, Ingale S, Bhati A, Prazapati P. Study of traditional methods of food preservation. 2014.
17. Zeng W, Wen W, Deng Y, Tian Y, Sun H, Sun Q. Chinese ethnic meat products: continuity and development. Meat Sci. 2016;120:37–46.

18. Ch'en MT, Ockerman HW. Chinese-style meat products processing science and technology. Fulin; 2012.
19. Zhou GH, Zhao GM. History and heritage of Jinhua ham. Anim Front. 2012;2(4):62–7.
20. Zeng Wei Cai ZW, Wen Wen Ting WW, Deng Yue DY, Tian Yuan Yuan TY, Sun Hong Hu SH, Sun Qun SQ. Chinese ethnic meat products: continuity and development. Meat Sci. 2016;120:37–46.
21. Zhou M. Development and processing of low-salt Sichuan style bacon. Meat Res. 2012;26:17–22. www.rlyj.net.cn, https://doi.org/10.7506/rlyj1001-8123-201202005.
22. Lin KW, Lin SN. Effects of sodium lactate and trisodium phosphate on the physicochemical properties and shelf life of low-fat Chinese-style sausage. Meat Sci. 2002;60(2):147–54.
23. Wang FS. Effects of three preservative agents on the shelf life of vacuum packaged Chinese-style sausage stored at 20 C. Meat Sci. 2000;56(1):67–71.
24. Hugo CJ, Hugo A. Current trends in natural preservatives for fresh sausage products. Trends Food Sci Technol. 2015;45(1):12–23.
25. Kim MK, Sasaki S, Otani T, Tsugane S. Japan public health center-based prospective study group. Dietary patterns and subsequent colorectal cancer risk by subsite: a prospective cohort study. Int J Cancer. 2005;115(5):790–8.
26. Yang HJ. The world food culture—Asia. Seoul: Clover; 2006. p. 23–30.
27. Nam KC, Jo C, Lee M. Meat products and consumption culture in the East. Meat Sci. 2010;86(1):95–102.
28. Rath EC. Oishii: the history of sushi. Reaktion Books; 2021.
29. Lien ME. Changing food habits: case studies from Africa, South America and Europe. Am. Ethnol. 2002;29(4):1037–8.
30. Weiss J, Gibis M, Schuh V, Salminen H. Advances in ingredient and processing systems for meat and meat products. Meat Sci. 2010;86(1):196–213.
31. Heinzelmann U. Beyond bratwurst: a history of food in Germany. Reaktion Books; 2014.
32. Louis-Sylvestre J, Krempf M, Lecerf JM. Les charcuteries. Cahiers de nutrition et de diététique. 2010;45(6):327–37.
33. The Editors of Encyclopaedia Britannica. Paella | definition, origin, & preparation. Encyclopedia Britannica; 1998. https://www.britannica.com/topic/paella.
34. The Editors of Encyclopaedia Britannica. Sausage | description, types, & ingredients. Encyclopedia Britannica; 2023. https://www.britannica.com/topic/sausage.
35. Nielsen IQ. Meat Department trends: what's new for 2023 – NIQ. NIQ. https://nielseniq.com/global/en/insights/education/2023/meat-department-trends-2023/.
36. Luong RH. Plant-based meat alternatives consumption and acculturation among international college students. Doctoral dissertation, Iowa State University. 2022.

Index

A. Kumar Pathera et al. (eds.), *Healthier Meat Products*,
https://doi.org/10.1007/978-3-031-78215-2

The manufacturer's authorised representative in the EU is Springer Nature Customer Service Centre GmbH, Europaplatz 3, 69115 Heidelberg, Germany. If you have any concerns regarding our products, please contact ProductSafety@springernature.com

Printed and bound by CPI Group (UK) Ltd, Croydon, CR0 4YY
30/09/2026
02225589-0018